Computational Electromagnetics and Model-Based Inversion

Scientific Computation

For further volumes:
http://www.springer.com/series/718

Harold A. Sabbagh • R. Kim Murphy
Elias H. Sabbagh • John C. Aldrin
Jeremy S. Knopp

Computational Electromagnetics and Model-Based Inversion

A Modern Paradigm for Eddy-Current Nondestructive Evaluation

Springer

Harold A. Sabbagh
Victor Technologies, LLC
Bloomington, IN, USA

R. Kim Murphy
Victor Technologies, LLC
Bloomington, IN, USA

Elias H. Sabbagh
Victor Technologies, LLC
Bloomington, IN, USA

John C. Aldrin
Computational Tools
Gurnee, IL, USA

Jeremy S. Knopp
Air Force Research Laboratory
(AFRL/RXLP)
Wright-Patterson AFB, OH, USA

ISSN 1434-8322
ISBN 978-1-4419-8428-9 ISBN 978-1-4419-8429-6 (eBook)
DOI 10.1007/978-1-4419-8429-6
Springer New York Heidelberg Dordrecht London

Library of Congress Control Number: 2013937723

Printed on acid-free paper

Springer is part of Springer Science+Business Media (www.springer.com)

Preface

A few years ago, a colleague wrote a note in which he reviewed the state of the art in eddy-current probes that were to address new problems in the nation's steam generators. With respect to calibration standards, he stated

> The performance of these probes will not be realized unless calibration standards that accurately simulate the range of expected defects are used. This means a series of axial and/or circumferential notches, on both the tube od and id must be used to accurately calibrate these probes. The ASME Section XI standard with flat bottomed holes is a very poor representation of cracks, particularly for directional probes. In addition, the cable between the instrument and the probe should be as short as reasonable and should be low capacitance, low noise and low loss.

The purpose of this book is to address the entirety of issues raised in this quote, and beyond, and to effectively resolve them favorably through the use of computational electromagnetics and model-based inversion methods as a replacement for expensive and unreliable standards. Indeed, we hope to demonstrate that our assertion, that the computer will soon be the most important instrument in eddy-current nondestructive evaluation (NDE), is only a mild stretch.

The book, like Gaul, is divided into three parts whose intention is to show that computational electromagnetics is more than simply solving Maxwell's equations for various configurations of "anomalies" in hosts. Rather, it becomes a part of system science of engineering, which is where we hope to elevate the notion of 'nondestructive evaluation'. Maxwell, himself, might not have envisioned the ramifications of his theory, but surely he would have admired the results. (We are reminded that a well-known aerospace NDE engineer once told us that had he realized that you could make money solving Maxwell's equations, he would have paid more attention in his undergraduate E&M course.)

In the first part, we compute the Green's dyad by working with the field equations directly, without the intervening use of potential functions (Heaviside would be proud!). Indeed, it isn't until the derivation of the "vector form" of the volume-integral equations in Chap. 3 that we first see potential functions explicitly stated. We applied this approach originally to model electromagnetic responses in plane-layered anisotropic bodies, such as advanced composites made of carbon-fiber reinforced polymers, which will be an important part of the second volume in this series. This approach works well in cylindrical coordinates, as is demonstrated in Chap. 9.

The discretization of the volume-integral equation via Galerkin's method on a regular grid is certainly well known in the method of moments, and much of the remainder of the book attempts to show the advantage of this approach in solving large problems with reasonable computer resources.

Starting in Part I, and continuing throughout the remainder of the book, we have sought to impress upon the reader the value of using equivalent electric circuits or networks to interpret the physical response of the volume-integral equation. This makes eddy-current NDE a subset of electrical engineering and should allow those familiar with the basic concepts of electrical engineering lead into the further development of eddy-current NDE.

The development of advanced probe models in Chap. 6 is original, especially in treating coils, of whatever shape or orientation, as generalized magnetic dipoles comprising solenoidal currents flowing in closed loops. This has allowed us to model rectangular coils, or D-shaped coils, in a consistent numerical manner. We have validated several of these models against benchmark data, as shown in Chap. 6. What was especially pleasant, however, was the realization that we could take the dual of the magnetic-dipole approach and model planar spiral-coil probes assuming that the source is an electric dipole (Chap. 7). This allows us to compute capacitive effects, as well as the usual inductive effects, thereby enabling a more efficient design process for high-frequency applications. Problems of this type—for example, spiral antennas—are usually treated by boundary-integral equations, assuming perhaps that the metallic traces are perfect conductors. We have applied the volume-integral approach, treating the spiral traces as an "anomaly" in free space that is excited by an electric dipole, rather than an anomaly in a host that is excited by a magnetic dipole, and are able to compute the capacitive and inductive reactances and resistance of the probe as a function of frequency. Clearly, this approach allows a single code to solve a greater variety of problems than originally assumed. We can now treat resonance phenomena rigorously on the computer, without relying on trial-and-error laboratory mockups.

Because of the increasing use of transmit–receive (T/R) arrays in eddy-current NDE, we have included a discussion of N-port analysis of T/R arrays in Chap. 8. We tie N-port theory of microwave networks with chain matrices to derive equivalent networks for these arrays. This should help in designing and understanding the behavior of these networks, especially at higher frequencies, where resonance effects become important.

Part II is really where the fun begins, and computational electromagnetics becomes a part of NDE system theory. We believe that the application of sophisticated signal-processing and inversion algorithms lies at the heart of the future development of NDE into a solid engineering discipline that is capable of handling the challenging problems that new structures and materials introduce. Indeed, it will lead NDE into the digital age, in which the computer replaces the oscilloscope as the instrument of choice.

The mathematical algorithms of Part II, of course, were developed independently of computational EM, but it is here that the advantages of modern developments in computational EM are manifest. For example, the use of volume-integral equations leads quickly to the efficient calculation of "surrogate structures," such as the interpolation tables that are used in NLSE, the nonlinear least-squares estimator that is introduced in Chap. 12. NLSE is the workhorse for the rest of the book and has proven to be quite flexible and robust in solving a large class of problems, namely, those in which the anomalous region can be defined through the use of models containing a few parameters. (In the second volume in this series, we will introduce certain voxel-based inversion algorithms which are more powerful than model-based algorithms. The use of volume-integral methods will be even more crucial in the application of these algorithms. The reader will have to take our word for it, and be patient.)

The iterative algorithm described in Fig. 14.3 of Chap. 14 is reminiscent of the projection onto convex sets (POCS) that is described in Fig. 14.1, so we refer to the former algorithm as "POCS" occasionally in the remainder of the book, even though we have not yet proved that the relevant sets are in fact convex. At other times, bowing to the purists among us, we simply refer to it as "the iterative algorithm." We have plans to resolve the matter in the future, but in any case, the algorithms cited in Chap. 14 are not there simply for their looks; we will use all of them in this book or its sequel, especially in connection with voxel-based inversion algorithms.

The high point of the book, from our perspective, is Part III, where we tie the material of the first two parts into a demonstration of the power of computer-aided modeling and design in solving realistic problems in eddy-current NDE. The examples described in this part are taken from real-life problems that the authors have explored (and continue to explore), principally in the areas of aerospace and nuclear power. It is here that we hope to demonstrate the future of eddy-current NDE and make a case for the aforementioned assertion about the supremacy of the computer in that future. But the computer is useless without algorithms that are based on fundamental physics, and this is nowhere illustrated better than in Chap. 19, where we demonstrate that a straightforward application of electromagnetic theory via Maxwell's equations solves the problem of ferritic heat-exchanger tubes in an elegant manner. Prior to this application, the orthodox view of "eddy currents" was an inspection technique that relied on analog instruments and could only be applied to isotropic, nonpermeable conductors. This chapter, therefore, shows that the phrase "eddy-current NDE" really implies "electromagnetic NDE," in which "electric" and "magnetic" are unified, as Maxwell envisioned a century-and-a-half ago.

At times, the book may appear to be a user's manual, theoretical manual, or simply an advertisement for Victor Technologies' proprietary volume-integral code, **VIC-3D®** [40]. There are two reasons for this: first, **VIC-3D®** was written to solve precisely the problems that are described in the book and is the code best known to us for doing that, and second, we believe that it is important for the reader and industry to understand that computational electromagnetics is not reserved for graduate theses and academic papers but is a commercially viable tool for solving those problems that the industry needs to solve. **VIC-3D®** is our contribution to the list of codes that solve Maxwell's equations for profit.

This book is not an introductory text; it will require a good background in electric circuit theory, especially in understanding the concepts of phasors, impedance and admittance, magnetic fields, magnetic induction and inductances, and magnetically coupled circuits, as well as electromagnetic field theory, including Maxwell's equations. Material on electric circuits is usually covered in the sophomore year in electrical engineering courses, whereas the required background in electromagnetic fields is covered in upper-level undergraduate courses.

In teaching courses on electric circuits, we have found *Circuit Analysis*, by Elias M. Sabbagh, Ronald Press, 1961, to be excellent preparation. It covers all aspects listed above. Although long out of print, it is available on the internet at *very* low prices. Of more recent vintage is *Linear Circuit Analysis*, by Raymond A. DeCarlo and Pen-Min Lin, Oxford University Press, 2001.

When it comes to senior-level undergraduate texts on electromagnetic theory, we can do no better than to recommend the classic *Fields and Waves in Communication Electronics*, by Simon Ramo, John R. Whinnery, and Theodore Van Duzer, John Wiley & Sons, New York, 1965. It not only gives a precise development of electromagnetic theory, including Maxwell's equations and their applications, but also derives circuit concepts that are consistent with Maxwell's equations.

If the reader wishes to develop his/her own code for Green's functions, he/she will need to become familiar quite quickly with the notion of "special functions." The classic reference on this subject, *Handbook of Mathematical Functions*, edited by Milton Abramowitz and Irene A. Stegun, National Bureau of Standards, 1970, has been superseded by *NIST Handbook of Mathematical Functions*, edited by F.W.J. Olver, D.W. Lozier, R.F. Boisvert, and C.W. Clark, Cambridge University Press, 2010. We have also found *Table of Integrals, Series, and Products*, by I.S. Gradshteyn and I.M. Ryzhik, Academic Press, 1980, to be useful. Finally, to tie all of this together in meaningful computer codes, we recommend *Numerical Recipes: The Art of Scientific Computing*, by W.H. Press, B.P. Flannery, S.A. Teukolsky, and W.T. Vetterling, Cambridge University Press, 1986.

We hope that you, the reader, will find this book useful and that you will agree that it brings us closer to the goal of making eddy-current NDE a systematic branch of engineering science, resulting in a more reliable system for making products and materials safer.

Bloomington, IN, USA Harold A. Sabbagh
R. Kim Murphy
Elias H. Sabbagh
Gurnee, IL, USA John C. Aldrin
Wright-Patterson AFB, OH, USA Jeremy S. Knopp

Acknowledgments

The development of **VIC-3D®** has been supported by the Departments of Defense and Energy, and the National Science Foundation through the Small Business Innovation Research (SBIR) program, and by commercial sources, including General Electric, Hercules Aerospace, Rolls-Royce, United Technologies Research Center, and the Electric Power Research Institute (EPRI). More information can be found at http://www.sabbagh.com.

Our colleagues at Sabbagh Associates Inc., namely David Sabbagh, Jeff Treece, and Anthony Chan, contributed significantly to the development of algorithms that made volume-integral methods so effective in eddy-current NDE problems. The current version of **VIC-3D®** contains code based on these algorithms, as well as a number of other features that they developed. It is a testament to their dedicated efforts.

Our good friend, Kenji Krzywosz of the Electric Power Research Institute (EPRI), suggested and supported the problem areas described in Sect. 18.3, Sect. 19.4, and Chap. 20, which allowed us to get a much better handle on the problems faced by the nuclear power industry. The material described in Sect. 15.1 was supported by Jim Benson, also with EPRI. The benchmark validation experiment described there was the first for the "modern" version of **VIC-3D®**, and the clutter rejection algorithm that evolved from this validation effort continues to be an important application in our current research.

Mark Blodgett, Eric Lindgren, and Charlie Buynak of the Air Force Research Lab have supported our research into the application of inverse methods and stochastic processes to complex aerospace structures. This support will allow us to bring our ideas into practice in a much shorter period of time than would otherwise have been possible. Our colleagues at the University of Dayton Research Institute (UDRI), Shamachary Sathish, Ryan Mooers, and Matt Cherry, contributed to this book in a number of ways, from theoretical discussions to carrying out validation exercises using benchmark data.

This is a good point at which to look back some 30 years and recall the contributions of two of our late colleagues, Walt Bantz of General Electric Aircraft

Engines and Sue Vernon of the Naval Surface Warfare Center, White Oak Labs. Both were early supporters of model-based eddy-current NDE, as well as its manifestation in **VIC-3D®**. Walt introduced us to the Split-D probe and informed us of its importance in inspecting aircraft engines. He encouraged its inclusion into **VIC-3D®**, and we followed his advice (see Sect. 6.6). It continues to be an important part of our current research. One of our first papers on modeling eddy-current probes with ferrite cores using volume-integral equations was coauthored with Sue. She supplied the data, and we supplied the model calculations. The results were encouraging, so we thought that we were on to something. Sue introduced us to carbon-fiber advanced composites and asked if we could do anything with eddy currents to characterize them. This led us to some very interesting problems in the electromagnetic modeling of anisotropic media and validating the models with experimental data. Again, the results were positive and will be a significant part of our second book.

Our men at Springer, Chris Coughlin and Ho Ying Fan, were a joy to work with. Erik Wallace, a doctoral student in mathematics at Indiana University, is going to investigate the "POCS-Iterative" issue that we described above (look for an answer in our next book), but of more immediate import is the fact that he is a LaTeX guru and assembled all the document preparation files for Ho Ying to pass to his people at Springer.

Contents

Part I
Computational Electromagnetics Background

Chapter 1
Overview of Methods of Computational Electromagnetics

1.1 Introduction

Nondestructive evaluation (NDE) is to materials and structures what CAT scanning is to the human body—an attempt to look inside without opening up the body. As in CAT scanning, modern NDE requires sophisticated mathematical software to perform its function. This is especially true with regard to quantitative NDE, wherein we attempt to quantify defects, that is, determine their size, location, even shape, rather than just to detect their presence. Low-frequency electromagnetic methods using eddy-currents are a traditional mode of doing NDE (approximately 35 % of NDE uses eddy-currents, depending upon the specific application), but the technology still suffers from a lack of algorithms and software to allow its full potential to be realized. The electromagnetics code, **VIC-3D®**, was developed to alleviate that problem, and in this book we apply **VIC-3D®** to solve forward and inverse problems in eddy-current NDE.

To put eddy-current NDE into historical perspective, we quote Burrows [1]

> Eddy-current testing first received systematic study during World War II in Germany. Its wider recognition, however, did not come until 1952. In that year Förster and his associates published the results of their extensive theoretical, experimental and industrial investigations in the first of an important series of papers [2, 3].
>
> These papers did not, however, include a quantitative theory for flaw detection, and apart from some useful English language accounts of the Förster papers. . . .no further theoretical work on flaw detection has appeared.

Burrows thesis, then, filled the gap in quantitative theoretical works on eddy-current NDE, but even then, his theory of flaw detection was restricted to flaws that were small compared to the skin depth of the material. This depth (or, equivalently, diffusion length) is the quantity which in conducting matter corresponds to wavelength in free space. By assuming that the flaw is very small, Burrows was able to represent its electrical behavior by an equivalent electrical dipole. His theory, though important in establishing the preeminence of Maxwell's

H.A. Sabbagh et al., *Computational Electromagnetics and Model-Based Inversion*, Scientific Computation, DOI 10.1007/978-1-4419-8429-6_1,

equations in eddy-current NDE, is not sufficient to solve contemporary problems, wherein we must compute the response of eddy-currents to flaws of arbitrary size and shape.

That's where modern computational electromagnetics comes in. Though researchers had been solving problems in electromagnetics since the time of Maxwell, it has only been since the mid-1960s that computational electromagnetics has evolved to be a distinct discipline. Research in this area was spurred largely by the US Department of Defense to solve practical problems in the radiation, scattering, and propagation of electromagnetic fields. Other areas were not neglected, however; considerable effort has been expended since roughly the mid-1970s in the application of microwaves to treat certain cancers, and this requires accurate models to compute the interaction of electromagnetic fields with biological media. The techniques for solving these problems range from formulations involving integral equations to those involving differential equations.

It is not surprising, therefore, that eddy-current NDE should benefit from research into computational electromagnetics, and **VIC-3D®** is the natural outgrowth of that research. It is based on the volume-integral mathematical approach to solving electromagnetics problems. This approach enables **VIC-3D®** to solve problems more efficiently on smaller computers, and with less preprocessing time, than other approaches, and, more importantly, it allows one to compute the response of flaws of arbitrary size and shape.

1.2 Oak Ridge National Laboratory

Following Burrows, work at the Oak Ridge National Laboratory included efforts in applying rigorous electromagnetic theory and developing computer programs to solve problems in eddy-current NDE, thereby bringing the discipline into the computer age. The primary authors of many of these papers were C.V. Dodd and W.E. Deeds. See [4–8].

1.3 Method of Moments

As a prelude to the work on computational electromagnetics that evolved in the 1970s, we should mention the pioneering and seminal work of R.F. Harrington on the method of moments (MOM) [12]. This book spurred the development and application of the modern theory of integral equations to electromagnetic fields, and was extremely influential in the development of **VIC-3D®**, as will be described in this book.

1.4 Model-Based Parameter Estimation in Electromagnetics

In the 1990s, E.K. Miller developed an analysis technique called model-based parameter estimation (MBPE) (see [9] and the references therein), whose purpose is to substitute a system approach for solving electromagnetics problems instead of the actual solution from first principle models. Thus, MBPE presents a sort of "surrogate" for Maxwell's equations. We will see a simple form of this when we apply interpolation tables to speed up the calculations for inverse problems in Parts II and III of this book.

1.5 Other Approaches to Computational Electromagnetics in NDE

Closely related to the volume-integral approach is the boundary-integral method, in which the boundary of the region of interest is discretized, rather than the entire volume. The theory of boundary-integrals, and their relation to the volume-integral approach in eddy-current NDE, is given in [10]. Recent developments in the theory and application of the boundary-integral approach to NDE are given in [11].

The method of finite-elements is also used for solving eddy-current NDE problems [104, 117]. We give an example of it in Chap. 17 ["Applications to Aerospace Structures"] in connection with the Cessna Sandwich (see Fig. 17.4). The finite-element method is a "field-solver," in which the unknowns are the electromagnetic field components, whereas integral equations, whether volume or boundary, have sources as their unknowns. Fields extend to infinity, so a field-solver must have a method for truncating the problem region by "approximating the field at infinity." This is unnecessary, of course, in integral equations in which the sources lie within a compact domain of space.

The key to developing a computationally efficient theory of integral equations is computing a Green's function, which is the starting point of this book.

Chapter 2
Green's Dyad for Plane-Layered Media

The Green's dyad, which is the electric-field response to a delta-function vector current source, plays a principal role in volume-integral equations, as we shall see later. In this chapter we develop the theory of the Green's dyad for plane-parallel layered media. In Chap. 9 we extend the development to multilayered media with cylindrical geometry.

2.1 Eigenmodes of Anisotropic Media

We will consider plane-parallel bodies of infinite extent in the (x,y) plane, which are made up of layers of homogeneous, anisotropic material. To be specific, we consider magnetic host materials that are characterized by the following biaxial generalized electrical permittivity matrix:

$$\varepsilon_h = \begin{bmatrix} \varepsilon_x & \varepsilon_{xy} & 0 \\ \varepsilon_{yx} & \varepsilon_y & 0 \\ 0 & 0 & \varepsilon_z \end{bmatrix}, \tag{2.1}$$

where the entries are generalized permittivities $\varepsilon + \sigma/j\omega$.

Maxwell's equations for an electrically anisotropic body are

$$\begin{aligned} \nabla\times\mathbf{E} &= -j\omega\mu_h\mathbf{H} - j\omega(\mu(\mathbf{r})-\mu_h)\mathbf{H} \\ &= -j\omega\mu_h\mathbf{H} + \mathbf{J}_m \\ \nabla\times\mathbf{H} &= j\omega\varepsilon_h\cdot\mathbf{E} + j\omega(\varepsilon(\mathbf{r})-\varepsilon_h)\cdot\mathbf{E} \\ &= j\omega\varepsilon_h\cdot\mathbf{E} + \mathbf{J}_e, \end{aligned} \tag{2.2}$$

H.A. Sabbagh et al., *Computational Electromagnetics and Model-Based Inversion*, Scientific Computation, DOI 10.1007/978-1-4419-8429-6_2,

where $\mathbf{J}_m$ and $\mathbf{J}_e$ are anomalous magnetic and electric currents that account for the presence of flaws, or anomalies, in the otherwise-uniform host material. From here on we drop the subscript h on the generalized host permittivity and permeability.

Because of the material anisotropy, it is convenient to work with a matrix formulation of these equations that has been useful in crystal optics, plasmas, and microwave devices [13–21]. If the body is homogeneous with respect to (x,y), then Maxwell's equations can be Fourier transformed with respect to (x,y) and written as the following four-vector matrix differential equation in the spectral domain:

$$\partial_z \tilde{\mathbf{e}} = \mathbf{S}\cdot\tilde{\mathbf{e}} + \mathbf{U}\cdot\tilde{\mathbf{J}}, \tag{2.3}$$

$$\tilde{E}_z = \frac{k_y}{\varepsilon_z\omega}\tilde{H}_x - \frac{k_x}{\varepsilon_z\omega}\tilde{H}_y + \frac{j}{\varepsilon_z\omega}\tilde{J}_{ez}, \tag{2.4}$$

$$\tilde{H}_z = \frac{-k_y}{\mu\omega}\tilde{E}_x + \frac{k_x}{\mu\omega}\tilde{E}_y - \frac{j}{\mu\omega}\tilde{J}_{mz}, \tag{2.5}$$

where the tilde denotes a function defined in the transform domain (k_x,k_y), and

$$\tilde{\mathbf{e}} = \begin{bmatrix} \tilde{E}_x \\ \tilde{E}_y \\ \tilde{H}_x \\ \tilde{H}_y \end{bmatrix}; \tilde{\mathbf{J}} = \begin{bmatrix} \tilde{J}_{ex} \\ \tilde{J}_{ey} \\ \tilde{J}_{ez} \\ \tilde{J}_{mx} \\ \tilde{J}_{my} \\ \tilde{J}_{mz} \end{bmatrix}. \tag{2.6}$$

The subscript e denotes an electric current and m denotes a magnetic current. The matrices in (2.3) are given by

$$\mathbf{S} = -\begin{bmatrix} 0 & 0 & a & b \\ 0 & 0 & c & d \\ \alpha & \beta & 0 & 0 \\ \gamma & \delta & 0 & 0 \end{bmatrix}; \mathbf{U} = \begin{bmatrix} 0 & 0 & k_x/\omega\varepsilon_z & 0 & 1 & 0 \\ 0 & 0 & k_y/\omega\varepsilon_z & -1 & 0 & 0 \\ 0 & 1 & 0 & 0 & 0 & -k_x/\omega\mu \\ -1 & 0 & 0 & 0 & 0 & -k_y/\omega\mu \end{bmatrix}. \tag{2.7}$$

The entries of $\mathbf{S}$ are given in terms of the entries of (2.1) by

$$\begin{aligned} a &= \frac{j}{\omega\varepsilon_z}k_xk_y; \alpha = \frac{j}{\omega\mu}(-\mu\varepsilon_{yx}\omega^2 - k_xk_y) \\ b &= \frac{j}{\omega\varepsilon_z}(\mu\varepsilon_z\omega^2 - k_x^2); \beta = \frac{j}{\omega\mu}(-\mu\varepsilon_y\omega^2 + k_x^2) \\ c &= \frac{j}{\omega\varepsilon_z}(-\mu\varepsilon_z\omega^2 + k_y^2); \gamma = \frac{j}{\omega\mu}(\mu\varepsilon_x\omega^2 - k_y^2) \\ d &= -\frac{j}{\omega\varepsilon_z}k_xk_y; \delta = \frac{j}{\omega\mu}(\mu\varepsilon_{xy}\omega^2 + k_xk_y). \end{aligned} \tag{2.8}$$

When $\tilde{\mathbf{J}}$ is a surface current confined to $z = z'$, i.e., $\tilde{\mathbf{J}} = \tilde{\mathbf{J}}_s \delta(z - z')$, then integration of (2.3) produces

$$\tilde{\mathbf{e}}^{(+)} - \tilde{\mathbf{e}}^{(-)} = \mathbf{U} \cdot \tilde{\mathbf{J}}_s, \tag{2.9}$$

which is called the equation of discontinuity. The superscript $(+)$ denotes the limit z approaches z' from above and the superscript $(-)$ denotes the limit from below. Equation (2.9) will be used in the next section to compute the Green's dyad for a layered workpiece.

Starting with these equations, Roberts [24] has developed a fairly complete theory of normal modes of biaxial anisotropic media. This work is based on, and extends, earlier work performed at Sabbagh Associates [22, 23, 25–29]. From here on we specialize the theory developed in [24] to the case to be considered here, in which the media involved are transversely isotropic to the z-coordinate. The dielectric permittivity tensor, in its principal-axis coordinate system, then takes the form

$$\varepsilon = \begin{bmatrix} \varepsilon_t & 0 & 0 \\ 0 & \varepsilon_t & 0 \\ 0 & 0 & \varepsilon_z \end{bmatrix} . \tag{2.10}$$

The entries in $\mathbf{S}$ now become

$$\begin{aligned} a &= \frac{j}{\omega\varepsilon_z} k_x k_y; \alpha = \frac{j}{\omega\mu}(-k_x k_y) \\ b &= \frac{j}{\omega\varepsilon_z}(\mu\varepsilon_z\omega^2 - k_x^2); \beta = \frac{j}{\omega\mu}(-\mu\varepsilon_t\omega^2 + k_x^2) \\ c &= \frac{j}{\omega\varepsilon_z}(-\mu\varepsilon_z\omega^2 + k_y^2); \gamma = \frac{j}{\omega\mu}(\mu\varepsilon_t\omega^2 - k_y^2) \\ d &= -\frac{j}{\omega\varepsilon_z} k_x k_y; \delta = \frac{j}{\omega\mu}(k_x k_y). \end{aligned} \tag{2.11}$$

Let's introduce some notation: $k_x^2 + k_y^2 = k_t^2$, $\omega^2\mu\varepsilon_t = \Omega_t^2$, $\omega^2\mu\varepsilon_z = \Omega_z^2$, $\varepsilon = \varepsilon_t/\varepsilon_z$. Then the eigenvalues of $\mathbf{S}$ are

$$\lambda_1 = \sqrt{k_t^2 - \Omega_t^2} \; \lambda_2 = -\lambda_1 \; \lambda_3 = \sqrt{\varepsilon}\sqrt{k_t^2 - \Omega_z^2} \; \lambda_4 = -\lambda_3 \; . \tag{2.12}$$

The linearly-independent eigenvectors that correspond to these eigenvalues are:

$$\mathbf{v}_1 = \begin{bmatrix} -j\omega\mu_0 k_y \\ j\omega\mu_0 k_x \\ \lambda_1 k_x \\ \lambda_1 k_y \end{bmatrix} \mathbf{v}_2 = \begin{bmatrix} -j\omega\mu_0 k_y \\ j\omega\mu_0 k_x \\ -\lambda_1 k_x \\ -\lambda_1 k_y \end{bmatrix} \mathbf{v}_3 = \begin{bmatrix} \lambda_3 k_x \\ \lambda_3 k_y \\ j\omega\varepsilon_t k_y \\ -j\omega\varepsilon_t k_x \end{bmatrix} \mathbf{v}_4 = \begin{bmatrix} \lambda_3 k_x \\ \lambda_3 k_y \\ -j\omega\varepsilon_t k_y \\ j\omega\varepsilon_t k_x \end{bmatrix} . \tag{2.13}$$

When $k_x = k_y = 0$, the following are linearly-independent eigenvectors:

$$\mathbf{v}_1 = \begin{bmatrix} 1 \\ 0 \\ 0 \\ -\sqrt{\dfrac{\varepsilon_t}{\mu_0}} \end{bmatrix} \quad \mathbf{v}_2 = \begin{bmatrix} 1 \\ 0 \\ 0 \\ \sqrt{\dfrac{\varepsilon_t}{\mu_0}} \end{bmatrix} \quad \mathbf{v}_3 = \begin{bmatrix} 0 \\ \sqrt{\dfrac{\mu_0}{\varepsilon_t}} \\ 1 \\ 0 \end{bmatrix} \quad \mathbf{v}_4 = \begin{bmatrix} 0 \\ -\sqrt{\dfrac{\mu_0}{\varepsilon_t}} \\ 1 \\ 0 \end{bmatrix}. \tag{2.14}$$

When we substitute $\mathbf{v}_1$, $\mathbf{v}_2$ of (2.13) into (2.4), with the source currents set to zero, we find that $\tilde{E}_z = 0$; hence, $\mathbf{v}_1$, $\mathbf{v}_2$ are transverse electric (TE) modes, with respect to z. Similarly, $\mathbf{v}_3$, $\mathbf{v}_4$ are transverse magnetic (TM) modes. Note that the TE modes are orthogonal to the TM modes. This will facilitate the computation of the Green's dyadic. $\mathbf{v}_1$ and $\mathbf{v}_3$ are downward-traveling waves in the z-direction; i.e., they represent waves that travel in the negative z-direction. $\mathbf{v}_2$, $\mathbf{v}_4$ are upward-traveling waves (in the positive z-direction). We see from (2.4) and (2.5) that all modes are TEM (transverse electric and magnetic) with respect to z for $k_x = k_y = 0$.

2.2 Green Dyad for Plane-Parallel Layered Media

Though the theory of eigenmodes that was developed in the preceding section is applicable to the class of transverse-isotropic media, we will apply it only to isotropic media. Hence, $\varepsilon_t = \varepsilon_z = \varepsilon$ in (2.10), which means that $\lambda_1 = \lambda_3$ in (2.13).

The flaw, or other anomaly, in a workpiece produces an anomalous current which is to be determined as the solution of a volume-integral equation. The kernel of this equation is a Green dyad for a plane-parallel layered workpiece, and we turn our attention to determining this dyad.

Consider the system shown in Fig. 2.1. The point-source of electric or magnetic current is in region 0, and we want to compute the fields in this region, or in any other region. This will give us the Green dyad. If the source is an electric current, and the field electric, then the resulting dyad is called electric–electric. If the source is a magnetic current, and the field electric, then the dyad is called electric–magnetic, and so-on.

We write for the system of four-vectors in the ith region

$$\mathbf{v}_{1i} = \begin{bmatrix} -j\omega\mu_i k_y \\ j\omega\mu_i k_x \\ \lambda_i k_x \\ \lambda_i k_y \end{bmatrix} \quad \mathbf{v}_{2i} = \begin{bmatrix} -j\omega\mu_i k_y \\ j\omega\mu_i k_x \\ -\lambda_i k_x \\ -\lambda_i k_y \end{bmatrix} \quad \mathbf{v}_{3i} = \begin{bmatrix} \lambda_i k_x \\ \lambda_i k_y \\ j\omega\hat{\varepsilon}_i k_y \\ -j\omega\hat{\varepsilon}_i k_x \end{bmatrix} \quad \mathbf{v}_{4i} = \begin{bmatrix} \lambda_i k_x \\ \lambda_i k_y \\ -j\omega\hat{\varepsilon}_i k_y \\ j\omega\hat{\varepsilon}_i k_x \end{bmatrix}. \tag{2.15}$$

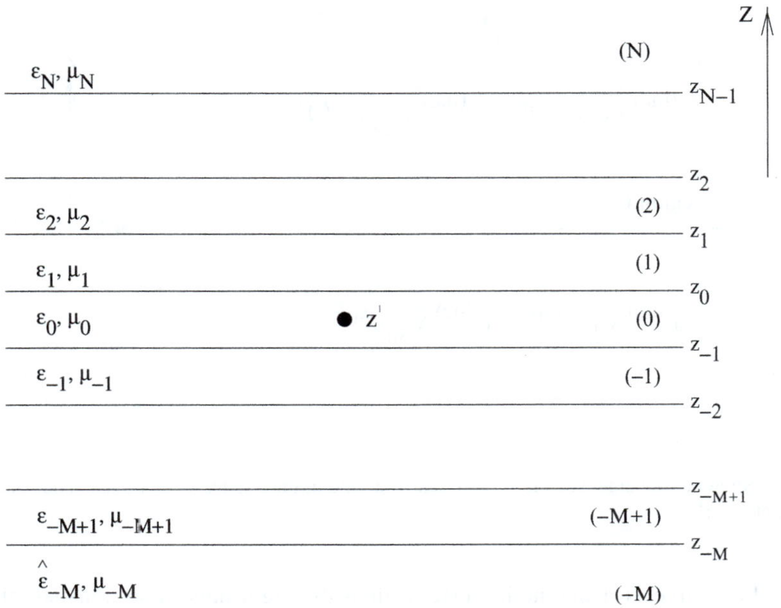

Fig. 2.1 Plane-parallel layered workpiece. The source is in region 0

2.2.1 *Infinite-Space Green Dyad*

First of all, we will compute the infinite-space Green dyad, which is the dyad associated with an infinite, uniform, unlayered medium. This corresponds to the situation wherein region 0 of Fig. 2.1 extends to $z = \pm\infty$; i.e., $z_0 = \infty$, and $z_{-1} = -\infty$.

Figure 2.2 shows a source located at $z = z'$ in region 0, together with the appropriate field eigenvectors on each side of z'. This choice of the eigenvectors is consistent with the fact that $\mathbf{v}_1$ and $\mathbf{v}_3$ travel in the negative z-direction, and $\mathbf{v}_2$ and $\mathbf{v}_4$ are positively traveling waves. The superscript, (inc), denotes "incident" fields due to the source.

If the source is an electric current vector, then the right-hand side of (2.9) consists of the three electric excitation vectors:

$$\mathbf{f}_x^{(e)} = \begin{bmatrix} 0 \\ 0 \\ 0 \\ -\tilde{J}_{ex} \end{bmatrix} \quad \mathbf{f}_y^{(e)} = \begin{bmatrix} 0 \\ 0 \\ \tilde{J}_{ey} \\ 0 \end{bmatrix} \quad \mathbf{f}_z^{(e)} = \begin{bmatrix} k_x\tilde{J}_{ez}/\omega\hat{\varepsilon}_0 \\ k_y\tilde{J}_{ez}/\omega\hat{\varepsilon}_0 \\ 0 \\ 0 \end{bmatrix}, \tag{2.16}$$

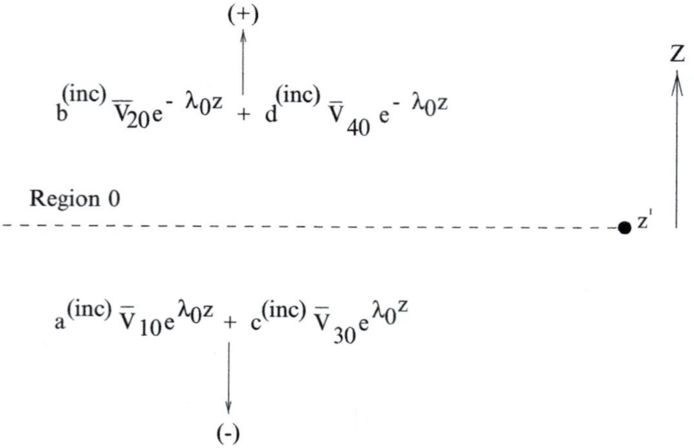

Fig. 2.2 Showing the eigenvectors used to calculate the infinite-space Green dyad. The source is located at $z = z'$

and if the source is a magnetic current, then the right-hand side consists of the magnetic excitation vectors:

$$\mathbf{f}_x^{(m)} = \begin{bmatrix} 0 \\ -\tilde{J}_{mx} \\ 0 \\ 0 \end{bmatrix} \quad \mathbf{f}_y^{(m)} = \begin{bmatrix} \tilde{J}_{my} \\ 0 \\ 0 \\ 0 \end{bmatrix} \quad \mathbf{f}_z^{(m)} = \begin{bmatrix} 0 \\ 0 \\ -k_x\tilde{J}_{mz}/\omega\mu_0 \\ -k_y\tilde{J}_{mz}/\omega\mu_0 \end{bmatrix} . \tag{2.17}$$

Equation (2.9) becomes therefore

$$-a^{(\mathrm{inc})}\mathbf{v}_{10}e^{\lambda_0 z'} + b^{(\mathrm{inc})}\mathbf{v}_{20}e^{-\lambda_0 z'} - c^{(\mathrm{inc})}\mathbf{v}_{30}e^{\lambda_0 z'} + d^{(\mathrm{inc})}\mathbf{v}_{40}e^{-\lambda_0 z'} = \mathbf{f}\,, \tag{2.18}$$

where $\mathbf{f}$ denotes one of the six excitation vectors in (2.16) or (2.17). We will return to this shortly.

Take the dot product of (2.18) with respect to the TE-mode vectors, $\mathbf{v}_{10}$, $\mathbf{v}_{20}$, and get

$$\begin{aligned} -A^{(i)}\alpha_{11} + B^{(i)}\alpha_{12} &= F_1 \\ -A^{(i)}\alpha_{12} + B^{(i)}\alpha_{22} &= F_2\,, \end{aligned} \tag{2.19}$$

where we have used the orthogonality of the TE and TM modal vectors to eliminate $c^{(\mathrm{inc})}$ and $d^{(\mathrm{inc})}$. Here

$$\begin{aligned} A^{(i)} &= a^{(\mathrm{inc})}e^{\lambda_0 z'} \\ B^{(i)} &= b^{(\mathrm{inc})}e^{-\lambda_0 z'} \\ F_1 &= \mathbf{v}_{10}\cdot\mathbf{f} \end{aligned}$$

$$
\begin{aligned}
F_2 &= \mathbf{v}_{20} \cdot \mathbf{f} \\
\alpha_{11} &= \mathbf{v}_{10} \cdot \mathbf{v}_{10} = k_t^2(\lambda_0^2 - \omega^2 \mu_0^2) \\
\alpha_{12} &= \mathbf{v}_{10} \cdot \mathbf{v}_{20} = -k_t^2(\lambda_0^2 + \omega^2 \mu_0^2) \\
\alpha_{22} &= \mathbf{v}_{20} \cdot \mathbf{v}_{20} = k_t^2(\lambda_0^2 - \omega^2 \mu_0^2).
\end{aligned} \tag{2.20}
$$

Upon solving (2.19) for $A^{(i)}$, $B^{(i)}$, and then using (2.20) we get

$$
\begin{aligned}
a^{(\text{inc})} &= \frac{F_1\alpha_{22} - F_2\alpha_{12}}{-\alpha_{11}\alpha_{22} + \alpha_{12}^2} e^{-\lambda_0 z'} \\
b^{(\text{inc})} &= \frac{-F_2\alpha_{11} + F_1\alpha_{12}}{-\alpha_{11}\alpha_{22} + \alpha_{12}^2} e^{\lambda_0 z'} \; .
\end{aligned} \tag{2.21}
$$

Hence, the contribution of the TE-modes to the infinite-space Green dyad is

$$
\begin{aligned}
&\frac{-F_2\alpha_{11} + F_1\alpha_{12}}{-\alpha_{11}\alpha_{22} + \alpha_{12}^2} \mathbf{v}_{20} e^{-\lambda_0(z-z')} \;, \; z \geq z' \\
&\frac{F_1\alpha_{22} - F_2\alpha_{12}}{-\alpha_{11}\alpha_{22} + \alpha_{12}^2} \mathbf{v}_{10} e^{\lambda_0(z-z')} \;, \; z \leq z' \; .
\end{aligned} \tag{2.22}
$$

To compute the TM-mode contribution, take the dot product of (2.18) with respect to $\mathbf{v}_{30}$, $\mathbf{v}_{40}$. Proceeding as before, we get for the TM-mode contribution

$$
\begin{aligned}
&\frac{F_4\beta_{33} - F_3\beta_{34}}{\beta_{33}\beta_{44} - \beta_{34}^2} \mathbf{v}_{40} e^{-\lambda_0(z-z')} \;, \; z \geq z' \\
-&\frac{F_3\beta_{44} - F_4\beta_{34}}{\beta_{33}\beta_{44} - \beta_{34}^2} \mathbf{v}_{30} e^{\lambda_0(z-z')} \;, \; z \leq z' \; ,
\end{aligned} \tag{2.23}
$$

where

$$
\begin{aligned}
F_3 &= \mathbf{v}_{30} \cdot \mathbf{f} \\
F_4 &= \mathbf{v}_{40} \cdot \mathbf{f} \\
\beta_{33} &= \mathbf{v}_{30} \cdot \mathbf{v}_{30} = k_t^2(\lambda_0^2 - \omega^2 \hat{\varepsilon}_0^2) \\
\beta_{34} &= \mathbf{v}_{30} \cdot \mathbf{v}_{40} = k_t^2(\lambda_0^2 + \omega^2 \hat{\varepsilon}_0^2) \\
\beta_{44} &= \mathbf{v}_{40} \cdot \mathbf{v}_{40} = k_t^2(\lambda_0^2 - \omega^2 \hat{\varepsilon}_0^2) \; .
\end{aligned} \tag{2.24}
$$

Therefore, upon combining (2.22) and (2.23) we get for the composite infinite-field Green dyad:

$$\left[\frac{-F_2\alpha_{11}+F_1\alpha_{12}}{-\alpha_{11}\alpha_{22}+\alpha_{12}^2}\mathbf{v}_{20}+\frac{F_4\beta_{33}-F_3\beta_{34}}{\beta_{33}\beta_{44}-\beta_{34}^2}\mathbf{v}_{40}\right]e^{-\lambda_0(z-z')} \quad , z\geq z'$$

$$\left[\frac{F_1\alpha_{22}-F_2\alpha_{12}}{-\alpha_{11}\alpha_{22}+\alpha_{12}^2}\mathbf{v}_{10}-\frac{F_3\beta_{44}-F_4\beta_{34}}{\beta_{33}\beta_{44}-\beta_{34}^2}\mathbf{v}_{30}\right]e^{\lambda_0(z-z')} \quad , z\leq z' . \tag{2.25}$$

The Fs that appear in (2.25) are defined in (2.20) and (2.24). They are computed by taking the dot product of the eigenvectors of (2.15) (with $i=0$) with the excitation vectors in (2.16) and (2.17). For example, the value of F_1 that is associated with electric excitation in the x-direction is given by $F_1^{(ex)}=\mathbf{v}_{10}\cdot\mathbf{f}_x^{(e)}=-\lambda_0 k_y\tilde{J}_{ex}$. Continuing in this manner, we find:

$$\begin{aligned}
F_1^{(ex)} &= -\lambda_0 k_y\tilde{J}_{ex}, F_1^{(mx)} = -j\omega\mu_0 k_x\tilde{J}_{mx}\\
F_2^{(ex)} &= \lambda_0 k_y\tilde{J}_{ex}, F_2^{(mx)} = -j\omega\mu_0 k_x\tilde{J}_{mx}\\
F_3^{(ex)} &= j\omega\hat{\varepsilon}_0 k_x\tilde{J}_{ex}, F_3^{(mx)} = -\lambda_0 k_y\tilde{J}_{mx}\\
F_4^{(ex)} &= -j\omega\hat{\varepsilon}_0 k_x\tilde{J}_{ex}, F_4^{(mx)} = -\lambda_0 k_y\tilde{J}_{mx}\\
F_1^{(ey)} &= \lambda_0 k_x\tilde{J}_{ey}, F_1^{(my)} = -j\omega\mu_0 k_y\tilde{J}_{my}\\
F_2^{(ey)} &= -\lambda_0 k_x\tilde{J}_{ey}, F_2^{(my)} = -j\omega\mu_0 k_y\tilde{J}_{my}\\
F_3^{(ey)} &= j\omega\hat{\varepsilon}_0 k_y\tilde{J}_{ey}, F_3^{(my)} = \lambda_0 k_x\tilde{J}_{my}\\
F_4^{(ey)} &= -j\omega\hat{\varepsilon}_0 k_y\tilde{J}_{ey}, F_4^{(my)} = \lambda_0 k_x\tilde{J}_{my}\\
F_1^{(ez)} &= 0, F_1^{(mz)} = -\frac{\lambda_0 k_t^2}{\omega\mu_0}\tilde{J}_{mz}\\
F_2^{(ez)} &= 0, F_2^{(mz)} = \frac{\lambda_0 k_t^2}{\omega\mu_0}\tilde{J}_{mz}\\
F_3^{(ez)} &= \frac{\lambda_0 k_t^2}{\omega\hat{\varepsilon}_0}\tilde{J}_{ez}, F_3^{(mz)} = 0\\
F_4^{(ez)} &= \frac{\lambda_0 k_t^2}{\omega\hat{\varepsilon}_0}\tilde{J}_{ez}, F_4^{(mz)} = 0 .
\end{aligned} \tag{2.26}$$

When these are combined with the α's and β's of (2.20) and (2.24), we get for the infinite-space Green dyad

$$[J_2\mathbf{v}_{20} + J_4\mathbf{v}_{40}]\, e^{-\lambda_0(z-z')}, z \geq z'$$
$$[J_1\mathbf{v}_{10} + J_3\mathbf{v}_{30}]\, e^{\lambda_0(z-z')}, z \leq z' \,, \tag{2.27}$$

where

$$\begin{array}{lcccc}
 & J_1 & J_2 & J_3 & J_4 \\
(ex): & \dfrac{k_y\tilde{J}_{ex}}{2\lambda_0 k_t^2} & \dfrac{k_y\tilde{J}_{ex}}{2\lambda_0 k_t^2} & \dfrac{jk_x\tilde{J}_{ex}}{2\omega\hat{\varepsilon}_0 k_t^2} & \dfrac{jk_x\tilde{J}_{ex}}{2\omega\hat{\varepsilon}_0 k_t^2} \\
(ey): & -\dfrac{k_x\tilde{J}_{ey}}{2\lambda_0 k_t^2} & -\dfrac{k_x\tilde{J}_{ey}}{2\lambda_0 k_t^2} & \dfrac{jk_y\tilde{J}_{ey}}{2\omega\hat{\varepsilon}_0 k_t^2} & \dfrac{jk_y\tilde{J}_{ey}}{2\omega\hat{\varepsilon}_0 k_t^2} \\
(ez): & 0 & 0 & -\dfrac{\tilde{J}_{ez}}{2\lambda_0\omega\hat{\varepsilon}_0} & \dfrac{\tilde{J}_{ez}}{2\lambda_0\omega\hat{\varepsilon}_0} \\
(mx): & -\dfrac{jk_x\tilde{J}_{mx}}{2k_t^2\omega\mu_0} & \dfrac{jk_x\tilde{J}_{mx}}{2k_t^2\omega\mu_0} & \dfrac{k_y\tilde{J}_{mx}}{2\lambda_0 k_t^2} & -\dfrac{k_y\tilde{J}_{mx}}{2\lambda_0 k_t^2} \\
(my): & -\dfrac{jk_y\tilde{J}_{my}}{2k_t^2\omega\mu_0} & \dfrac{jk_y\tilde{J}_{my}}{2k_t^2\omega\mu_0} & -\dfrac{k_x\tilde{J}_{my}}{2\lambda_0 k_t^2} & \dfrac{k_x\tilde{J}_{my}}{2\lambda_0 k_t^2} \\
(mz): & \dfrac{\tilde{J}_{mz}}{2\lambda_0\omega\mu_0} & \dfrac{\tilde{J}_{mz}}{2\lambda_0\omega\mu_0} & 0 & 0.
\end{array} \tag{2.28}$$

When we introduce the vectors of (2.15) into (2.27), and then use (2.28), we obtain an expression for the x- and y-components of the field in terms of the source components. The z-component of the fields are obtained from (2.4) and (2.5). For instance, we will compute the x-component of the electric field, due to the x-component of electric current:

$$\begin{cases}
\left[\dfrac{-j\omega\mu_0 k_y^2\tilde{J}_{ex}}{2\lambda_0 k_t^2} + \dfrac{jk_x^2\lambda_0\tilde{J}_{ex}}{2\omega\hat{\varepsilon}_0 k_t^2}\right] e^{-\lambda_0(z-z')}, \; z \geq z' \\[2ex]
\left[\dfrac{-j\omega\mu_0 k_y^2\tilde{J}_{ex}}{2\lambda_0 k_t^2} + \dfrac{jk_x^2\lambda_0\tilde{J}_{ex}}{2\omega\hat{\varepsilon}_0 k_t^2}\right] e^{\lambda_0(z-z')}, \quad z \leq z'
\end{cases}$$

$$= -j\omega\mu_0\left(1 - \frac{k_x^2}{k_0^2}\right)\frac{e^{-\lambda_0|z-z'|}}{2\lambda_0}\tilde{J}_{ex}\,. \tag{2.29}$$

The coefficient of $\tilde{J}_{ex}$ in (2.29) is the xx-component of the Fourier-domain, infinite-space, electric–electric Green dyad $\tilde{G}^{(ee)}_{(0)}$. The complete expression for the dyad is

$$\tilde{\mathbf{G}}^{(ee)}_{(0)}(k_x,k_y;z,z') = -j\omega\mu_0 \begin{bmatrix} 1-k_x^2/k_0^2 & -k_xk_y/k_0^2 & \pm jk_x\lambda_0/k_0^2 \\ -k_xk_y/k_0^2 & 1-k_y^2/k_0^2 & \pm jk_y\lambda_0/k_0^2 \\ \pm jk_x\lambda_0/k_0^2 & \pm jk_y\lambda_0/k_0^2 & 1+\lambda_0^2/k_0^2 \end{bmatrix} \frac{e^{-\lambda_0|z-z'|}}{2\lambda_0}$$

$$+j\frac{\delta(z-z')}{\omega\hat{\varepsilon}_0}\mathbf{a}_z\mathbf{a}_z\,, \tag{2.30}$$

where $(+)$ sign goes with $z > z'$ and $(-)$ with $z < z'$. The term $j\frac{\delta(z-z')}{\omega\hat{\varepsilon}_0}\mathbf{a}_z\mathbf{a}_z$ is called the "depolarizing" term [34] and follows from the last term in (2.4). A similar analysis holds for the magnetic–magnetic Green dyad and produces the same expression as (2.30), except that μ_0 is everywhere replaced by $-\hat{\varepsilon}_0$.

The spectral-domain, infinite-space, magnetic–electric Green dyad is given by

$$\tilde{\mathbf{G}}^{(em)}_{(0)}(k_x,k_y;z,z') = \begin{bmatrix} 0 & \pm\lambda_0 & -jk_y \\ \mp\lambda_0 & 0 & jk_x \\ jk_y & -jk_x & 0 \end{bmatrix} \frac{e^{-\lambda_0|z-z'|}}{2\lambda_0}\,. \tag{2.31}$$

It is not difficult to show that

$$\tilde{\mathbf{G}}^{(me)}_{(0)} = \tilde{\mathbf{G}}^{(em)}_{(0)}\,. \tag{2.32}$$

2.2.2 *Layered-Space Green Dyad*

The infinite-space dyad that we computed in the previous section serves as the "incident" field in the layered medium of Fig. 2.1. We are interested in computing the fields produced when the incident dyad is scattered from the layers.

As before, we focus our attention on region 0, but this time assume that it has a finite upper-boundary, z_0, and lower-boundary, z_{-1}, as shown in Fig. 2.3. Region 1, which lies above region 0, is a half-space, as is region -1, which lies below region 0.

The infinite-field Green dyad is shown in Fig. 2.3. Because of the singularity of this dyad, we split it into that part, $[J_2\mathbf{v}_{20}+J_4\mathbf{v}_{40}]e^{-\lambda_0(z-z')}$, which is valid for $z \geq z'$, and $[J_1\mathbf{v}_{10}+J_3\mathbf{v}_{30}]e^{\lambda_0(z-z')}$, which is valid for $z \leq z'$. The scattered field, $(a_0\mathbf{v}_{10}+c_0\mathbf{v}_{30})e^{\lambda_0 z}+(b_0\mathbf{v}_{20}+d_0\mathbf{v}_{40})e^{-\lambda_0 z}$, is continuous throughout region 0.

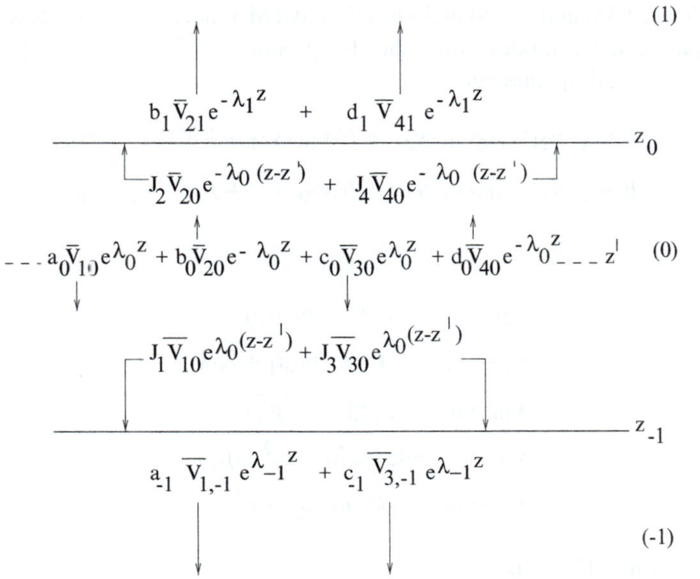

Fig. 2.3 Showing the source, at z', in region 0, together with the incident and scattered fields. Regions 1 and -1 are half-spaces

Continuity of the fields at $z = z_0$ requires that

$$b'_1\mathbf{v}_{21} + d'_1\mathbf{v}_{41} - a'_0\mathbf{v}_{10} - c'_0\mathbf{v}_{30} = b'_0\mathbf{v}_{20} + d'_0\mathbf{v}_{40} + J'_2 e^{\lambda_0 z'}\mathbf{v}_{20} + J'_4 e^{\lambda_0 z'}\mathbf{v}_{40}\ , \tag{2.33}$$

where

$$\begin{array}{c} b'_1 = b_1 e^{-\lambda_1 z_0},\ d'_1 = d_1 e^{-\lambda_1 z_0},\ a'_0 = a_0 e^{\lambda_0 z_0},\ b'_0 = b_0 e^{-\lambda_0 z_0} \\ c'_0 = c_0 e^{\lambda_0 z_0},\ \ d'_0 = d_0 e^{-\lambda_0 z_0},\ J'_2 = J_2 e^{-\lambda_0 z_0},\ J'_4 = J_4 e^{-\lambda_0 z_0} \end{array} . \tag{2.34}$$

b'_1, d'_1, a'_0, c'_0 are scattered fields at z_0 and b'_0, d'_0, J'_2, and J'_4 are incident upon z_0.

Continuity of the fields at the lower boundary, $z = z_{-1}$, requires that

$$a'_{-1}\mathbf{v}_{1,-1} + c'_{-1}\mathbf{v}_{3,-1} - b'_0\mathbf{v}_{20} - d'_0\mathbf{v}_{40} = a'_0\mathbf{v}_{10} + c'_0\mathbf{v}_{30} + J'_1 e^{-\lambda_0 z'}\mathbf{v}_{10} + J'_3 e^{-\lambda_0 z'}\mathbf{v}_{30}, \tag{2.35}$$

where now

$$\begin{array}{c} a'_{-1} = a_{-1}e^{\lambda_{-1}z_{-1}},\ c'_{-1} = c_{-1}e^{\lambda_{-1}z_{-1}},\ a'_0 = a_0 e^{\lambda_0 z_{-1}},\ b'_0 = b_0 e^{-\lambda_0 z_{-1}} \\ c'_0 = c_0 e^{\lambda_0 z_{-1}},\ \ \ d'_0 = d_0 e^{-\lambda_0 z_{-1}},\ \ J'_1 = J_1 e^{\lambda_0 z_{-1}},\ \ J'_3 = J_3 e^{\lambda_0 z_{-1}} \end{array} . \tag{2.36}$$

In this case, a'_{-1}, c'_{-1}, b'_0, d'_0 are scattered fields at z_{-1} and c'_0, a'_0, J'_1, and J'_3 are incident upon z_{-1}.

Equations (2.33) and (2.35) include TE- and TM-modes. They are easier to solve if we separate these modes. Take the dot product of (2.33) with respect to the TE-modes, $\mathbf{v}_{21}$ and $\mathbf{v}_{10}$ and get

$$\begin{aligned} b_1'\mathbf{v}_{21}\cdot\mathbf{v}_{21} - a_0'\mathbf{v}_{10}\cdot\mathbf{v}_{21} &= b_0'\mathbf{v}_{20}\cdot\mathbf{v}_{21} + J_2'e^{\lambda_0 z'}\mathbf{v}_{20}\cdot\mathbf{v}_{21} \\ b_1'\mathbf{v}_{21}\cdot\mathbf{v}_{10} - a_0'\mathbf{v}_{10}\cdot\mathbf{v}_{10} &= b_0'\mathbf{v}_{20}\cdot\mathbf{v}_{10} + J_2'e^{\lambda_0 z'}\mathbf{v}_{20}\cdot\mathbf{v}_{10}, \end{aligned} \tag{2.37}$$

where

$$\begin{aligned} \mathbf{v}_{21}\cdot\mathbf{v}_{21} &= k_t^2(\lambda_1^2 - \omega^2\mu_1^2) \\ \mathbf{v}_{10}\cdot\mathbf{v}_{21} &= -k_t^2(\omega^2\mu_0\mu_1 + \lambda_0\lambda_1) \\ \mathbf{v}_{10}\cdot\mathbf{v}_{10} &= k_t^2(\lambda_0^2 - \omega^2\mu_0^2) \\ \mathbf{v}_{20}\cdot\mathbf{v}_{21} &= k_t^2(\lambda_0\lambda_1 - \omega^2\mu_0\mu_1) \\ \mathbf{v}_{20}\cdot\mathbf{v}_{10} &= -k_t^2(\omega^2\mu_0^2 + \lambda_0^2)\,. \end{aligned} \tag{2.38}$$

The solution of (2.37) is

$$\begin{aligned} b_1 &= \frac{2\mu_0\lambda_0 e^{(\lambda_1-\lambda_0)z_0}}{\mu_0\lambda_1 + \mu_1\lambda_0}(b_0 + e^{\lambda_0 z'}J_2) \\ a_0 &= \frac{(\mu_1\lambda_0 - \mu_0\lambda_1)e^{-2\lambda_0 z_0}}{\mu_1\lambda_0 + \mu_0\lambda_1}(b_0 + e^{\lambda_0 z'}J_2)\,. \end{aligned} \tag{2.39}$$

Next, compute the TM-modes by taking the dot product of (2.33) with $\mathbf{v}_{41}$ and $\mathbf{v}_{30}$:

$$\begin{aligned} d_1'\mathbf{v}_{41}\cdot\mathbf{v}_{41} - c_0'\mathbf{v}_{30}\cdot\mathbf{v}_{41} &= d_0'\mathbf{v}_{40}\cdot\mathbf{v}_{41} + J_4'e^{\lambda_0 z'}\mathbf{v}_{40}\cdot\mathbf{v}_{41} \\ d_1'\mathbf{v}_{41}\cdot\mathbf{v}_{30} - c_0'\mathbf{v}_{30}\cdot\mathbf{v}_{30} &= d_0'\mathbf{v}_{40}\cdot\mathbf{v}_{30} + J_4'e^{\lambda_0 z'}\mathbf{v}_{40}\cdot\mathbf{v}_{30}\,, \end{aligned} \tag{2.40}$$

where

$$\begin{aligned} \mathbf{v}_{41}\cdot\mathbf{v}_{41} &= k_t^2(\lambda_1^2 - \omega^2\hat{\varepsilon}_1^2) \\ \mathbf{v}_{30}\cdot\mathbf{v}_{41} &= k_t^2(\omega^2\hat{\varepsilon}_0\hat{\varepsilon}_1 + \lambda_0\lambda_1) \\ \mathbf{v}_{30}\cdot\mathbf{v}_{30} &= k_t^2(\lambda_0^2 - \omega^2\hat{\varepsilon}_0^2) \\ \mathbf{v}_{40}\cdot\mathbf{v}_{41} &= k_t^2(\lambda_0\lambda_1 - \omega^2\hat{\varepsilon}_0\hat{\varepsilon}_1) \\ \mathbf{v}_{40}\cdot\mathbf{v}_{30} &= k_t^2(\omega^2\hat{\varepsilon}_0^2 + \lambda_0^2)\,. \end{aligned} \tag{2.41}$$

The solution of (2.40) is

$$
\begin{aligned}
d_1 &= \frac{2\hat{\varepsilon}_0\lambda_0 e^{(\lambda_1-\lambda_0)z_0}}{\hat{\varepsilon}_0\lambda_1+\hat{\varepsilon}_1\lambda_0}(d_0+J_4e^{\lambda_0 z'}) \\
c_0 &= \frac{(\lambda_1\hat{\varepsilon}_0-\lambda_0\hat{\varepsilon}_1)e^{-2\lambda_0 z_0}}{\lambda_1\hat{\varepsilon}_0+\lambda_0\hat{\varepsilon}_1}(d_0+J_4e^{\lambda_0 z'})\,. \qquad (2.42)
\end{aligned}
$$

We repeat the analysis for (2.35). First, for the TE-modes:

$$
\begin{aligned}
a'_{-1}\mathbf{v}_{1,-1}\cdot\mathbf{v}_{1,-1} - b'_0\mathbf{v}_{1,-1}\cdot\mathbf{v}_{20} &= (a'_0+J'_1e^{-\lambda_0 z'})\mathbf{v}_{10}\cdot\mathbf{v}_{1,-1} \\
a'_{-1}\mathbf{v}_{1,-1}\cdot\mathbf{v}_{20} - b'_0\mathbf{v}_{20}\cdot\mathbf{v}_{20} &= (a'_0+J'_1e^{-\lambda_0 z'})\mathbf{v}_{10}\cdot\mathbf{v}_{20}\,, \qquad (2.43)
\end{aligned}
$$

where

$$
\begin{aligned}
\mathbf{v}_{1,-1}\cdot\mathbf{v}_{1,-1} &= k_t^2(\lambda_{-1}^2-\omega^2\mu_{-1}^2) \\
\mathbf{v}_{20}\cdot\mathbf{v}_{1,-1} &= -k_t^2(\omega^2\mu_0\mu_{-1}+\lambda_0\lambda_{-1}) \\
\mathbf{v}_{20}\cdot\mathbf{v}_{20} &= k_t^2(\lambda_0^2-\omega^2\mu_0^2) \\
\mathbf{v}_{10}\cdot\mathbf{v}_{1,-1} &= k_t^2(\lambda_0\lambda_{-1}-\omega^2\mu_0\mu_{-1}) \\
\mathbf{v}_{20}\cdot\mathbf{v}_{10} &= -k_t^2(\omega^2\mu_0^2+\lambda_0^2)\,. \qquad (2.44)
\end{aligned}
$$

The solution of (2.43) is

$$
\begin{aligned}
a_{-1} &= \frac{2\mu_0\lambda_0 e^{(\lambda_0-\lambda_{-1})z_{-1}}}{\mu_0\lambda_{-1}+\mu_{-1}\lambda_0}(a_0+e^{-\lambda_0 z'}J_1) \\
b_0 &= \frac{(\mu_{-1}\lambda_0-\mu_0\lambda_{-1})e^{2\lambda_0 z_{-1}}}{\mu_{-1}\lambda_0+\mu_0\lambda_{-1}}(a_0+e^{-\lambda_0 z'}J_1)\,. \qquad (2.45)
\end{aligned}
$$

The TM-mode equations

$$
\begin{aligned}
c'_{-1}\mathbf{v}_{3,-1}\cdot\mathbf{v}_{3,-1} - d'_0\mathbf{v}_{40}\cdot\mathbf{v}_{3,-1} &= (c'_0+J'_3e^{-\lambda_0 z'})\mathbf{v}_{30}\cdot\mathbf{v}_{3,-1} \\
c'_{-1}\mathbf{v}_{40}\cdot\mathbf{v}_{3,-1} - d'_0\mathbf{v}_{40}\cdot\mathbf{v}_{40} &= (c'_0+J'_3e^{-\lambda_0 z'})\mathbf{v}_{30}\cdot\mathbf{v}_{40}\,, \qquad (2.46)
\end{aligned}
$$

where

$$
\begin{aligned}
\mathbf{v}_{3,-1}\cdot\mathbf{v}_{3,-1} &= k_t^2(\lambda_{-1}^2-\omega^2\hat{\varepsilon}_{-1}^2) \\
\mathbf{v}_{40}\cdot\mathbf{v}_{3,-1} &= k_t^2(\omega^2\hat{\varepsilon}_0\hat{\varepsilon}_{-1}+\lambda_0\lambda_{-1}) \\
\mathbf{v}_{40}\cdot\mathbf{v}_{40} &= k_t^2(\lambda_0^2-\omega^2\hat{\varepsilon}_0^2) \\
\mathbf{v}_{30}\cdot\mathbf{v}_{3,-1} &= k_t^2(\lambda_0\lambda_{-1}-\omega^2\hat{\varepsilon}_0\hat{\varepsilon}_{-1}) \\
\mathbf{v}_{40}\cdot\mathbf{v}_{30} &= k_t^2(\omega^2\hat{\varepsilon}_0^2+\lambda_0^2)\,, \qquad (2.47)
\end{aligned}
$$

have for their solutions

$$c_{-1} = \frac{2\hat{\varepsilon}_0 \lambda_0 e^{(\lambda_0 - \lambda_{-1}) z_{-1}}}{\hat{\varepsilon}_0 \lambda_{-1} + \hat{\varepsilon}_{-1} \lambda_0} (c_0 + J_3 e^{-\lambda_0 z'})$$

$$d_0 = \frac{(\hat{\varepsilon}_0 \lambda_{-1} - \hat{\varepsilon}_{-1} \lambda_0) e^{2\lambda_0 z_{-1}}}{\hat{\varepsilon}_0 \lambda_{-1} + \hat{\varepsilon}_{-1} \lambda_0} (c_0 + J_3 e^{-\lambda_0 z'}) \, . \tag{2.48}$$

We introduce the following transmission and reflection coefficients:

$$T_1^{(E)} = \frac{2\mu_0 \lambda_0}{\mu_0 \lambda_1 + \mu_1 \lambda_0}; T_1^{(M)} = \frac{2\hat{\varepsilon}_0 \lambda_0}{\hat{\varepsilon}_0 \lambda_1 + \hat{\varepsilon}_1 \lambda_0}$$

$$R_1^{(E)} = \frac{\mu_1 \lambda_0 - \mu_0 \lambda_1}{\mu_1 \lambda_0 + \mu_0 \lambda_1}; R_1^{(M)} = \frac{\hat{\varepsilon}_0 \lambda_1 - \hat{\varepsilon}_1 \lambda_0}{\hat{\varepsilon}_1 \lambda_0 + \hat{\varepsilon}_0 \lambda_1}$$

$$T_{-1}^{(E)} = \frac{2\mu_0 \lambda_0}{\mu_0 \lambda_{-1} + \mu_{-1} \lambda_0}; T_{-1}^{(M)} = \frac{2\hat{\varepsilon}_0 \lambda_0}{\hat{\varepsilon}_0 \lambda_{-1} + \hat{\varepsilon}_{-1} \lambda_0}$$

$$R_{-1}^{(E)} = \frac{\mu_{-1} \lambda_0 - \mu_0 \lambda_{-1}}{\mu_{-1} \lambda_0 + \mu_0 \lambda_{-1}}; R_{-1}^{(M)} = \frac{\hat{\varepsilon}_0 \lambda_{-1} - \hat{\varepsilon}_{-1} \lambda_0}{\hat{\varepsilon}_{-1} \lambda_0 + \hat{\varepsilon}_0 \lambda_{-1}} \tag{2.49}$$

The final expressions for the mode coefficients are given in terms of these scattering parameters:

$$a_0 = \frac{R_1^{(E)} R_{-1}^{(E)} e^{-\lambda_0 (2T + z')} J_1 + R_1^{(E)} e^{-\lambda_0 (2z_0 - z')} J_2}{1 - R_1^{(E)} R_{-1}^{(E)} e^{-2\lambda_0 T}}$$

$$b_0 = \frac{R_{-1}^{(E)} e^{\lambda_0 (2z_{-1} - z')} J_1 + R_1^{(E)} R_{-1}^{(E)} e^{-\lambda_0 (2T - z')} J_2}{1 - R_1^{(E)} R_{-1}^{(E)} e^{-2\lambda_0 T}}$$

$$c_0 = \frac{R_1^{(M)} R_{-1}^{(M)} e^{-\lambda_0 (2T + z')} J_3 + R_1^{(M)} e^{-\lambda_0 (2z_0 - z')} J_4}{1 - R_1^{(M)} R_{-1}^{(M)} e^{-2\lambda_0 T}}$$

$$d_0 = \frac{R_{-1}^{(M)} e^{\lambda_0 (2z_{-1} - z')} J_3 + R_1^{(M)} R_{-1}^{(M)} e^{-\lambda_0 (2T - z')} J_4}{1 - R_1^{(M)} R_{-1}^{(M)} e^{-2\lambda_0 T}}$$

$$b_1 = \frac{T_1^{(E)} R_{-1}^{(E)} e^{-\lambda_0 (T - z_{-1} + z')} J_1 + T_1^{(E)} e^{-\lambda_0 (z_0 - z')} J_2}{1 - R_1^{(E)} R_{-1}^{(E)} e^{-2\lambda_0 T}} e^{\lambda_1 z_0}$$

$$d_1 = \frac{T_1^{(M)} R_{-1}^{(M)} e^{-\lambda_0(T-z_{-1}+z')} J_3 + T_1^{(M)} e^{-\lambda_0(z_0-z')} J_4}{1 - R_1^{(M)} R_{-1}^{(M)} e^{-2\lambda_0 T}} e^{\lambda_1 z_0}$$

$$a_{-1} = \frac{T_{-1}^{(E)} e^{-\lambda_0(z'-z_{-1})} J_1 + T_{-1}^{(E)} R_1^{(E)} e^{-\lambda_0(T+z_0-z')} J_2}{1 - R_1^{(E)} R_{-1}^{(E)} e^{-2\lambda_0 T}} e^{-\lambda_{-1} z_{-1}}$$

$$c_{-1} = \frac{T_{-1}^{(M)} e^{-\lambda_0(z'-z_{-1})} J_3 + T_{-1}^{(M)} R_1^{(M)} e^{-\lambda_0(T+z_0-z')} J_4}{1 - R_1^{(M)} R_{-1}^{(M)} e^{-2\lambda_0 T}} e^{-\lambda_{-1} z_{-1}}, \tag{2.50}$$

where $T = z_0 - z_{-1}$ is the height of the source region.

In order to determine the scattered fields within the source region, we start with the expression $a_0 \mathbf{v}_{10} e^{\lambda_0 z} + b_0 \mathbf{v}_{20} e^{-\lambda_0 z} + c_0 \mathbf{v}_{30} e^{\lambda_0 z} + d_0 \mathbf{v}_{40} e^{-\lambda_0 z}$ and use (2.15):

$$\tilde{E}_x = e^{\lambda_0 z}(-j\omega\mu_0 k_y a_0 + \lambda_0 k_x c_0) + e^{-\lambda_0 z}(-j\omega\mu_0 k_y b_0 + \lambda_0 k_x d_0)$$

$$\tilde{E}_y = e^{\lambda_0 z}(j\omega\mu_0 k_x a_0 + \lambda_0 k_y c_0) + e^{-\lambda_0 z}(j\omega\mu_0 k_x b_0 + \lambda_0 k_y d_0)$$

$$\tilde{E}_z = jk_t^2(c_0 e^{\lambda_0 z} - d_0 e^{-\lambda_0 z}) - \frac{\delta(z-z')}{j\omega\hat{\varepsilon}_0} \tilde{J}_{ez}$$

$$\tilde{H}_x = e^{\lambda_0 z}(\lambda_0 k_x a_0 + j\omega\hat{\varepsilon}_0 k_y c_0) + e^{-\lambda_0 z}(-\lambda_0 k_x b_0 - j\omega\hat{\varepsilon}_0 k_y d_0)$$

$$\tilde{H}_y = e^{\lambda_0 z}(\lambda_0 k_y a_0 - j\omega\hat{\varepsilon}_0 k_x c_0) + e^{-\lambda_0 z}(-\lambda_0 k_y b_0 + j\omega\hat{\varepsilon}_0 k_x d_0)$$

$$\tilde{H}_z = jk_t^2(a_0 e^{\lambda_0 z} + b_0 e^{-\lambda_0 z}) + \frac{\delta(z-z')}{j\omega\mu_0} \tilde{J}_{mz}. \tag{2.51}$$

From here on we will drop the delta-function terms in $\tilde{E}_z$ and $\tilde{H}_z$, because they will be associated with the infinite-field Green dyad.

The coefficients a_0, b_0, c_0, d_0 that appear in (2.51) have been previously expressed in terms of J_1, J_2, J_3, J_4 in (2.50). When these results are substituted into (2.51) we get:

$$\tilde{E}_x = e^{-\lambda_0(2T+(z'-z))} \left[\frac{-R_1^{(E)} R_{-1}^{(E)} j\omega\mu_0 k_y J_1}{1 - R_1^{(E)} R_{-1}^{(E)} e^{-2\lambda_0 T}} + \frac{R_1^{(M)} R_{-1}^{(M)} \lambda_0 k_x J_3}{1 - R_1^{(M)} R_{-1}^{(M)} e^{-2\lambda_0 T}} \right]$$

$$+ e^{-\lambda_0(2T-(z'-z))} \left[\frac{-R_1^{(E)} R_{-1}^{(E)} j\omega\mu_0 k_y J_2}{1 - R_1^{(E)} R_{-1}^{(E)} e^{-2\lambda_0 T}} + \frac{R_1^{(M)} R_{-1}^{(M)} \lambda_0 k_x J_4}{1 - R_1^{(M)} R_{-1}^{(M)} e^{-2\lambda_0 T}} \right]$$

$$+ e^{-\lambda_0(2z_0-(z'+z))}\left[\frac{-R_1^{(E)}j\omega\mu_0 k_y J_2}{1-R_1^{(E)}R_{-1}^{(E)}e^{-2\lambda_0 T}}+\frac{R_1^{(M)}\lambda_0 k_x J_4}{1-R_1^{(M)}R_{-1}^{(M)}e^{-2\lambda_0 T}}\right]$$

$$+ e^{\lambda_0(2z_{-1}-(z'+z))}\left[\frac{-R_{-1}^{(E)}j\omega\mu_0 k_y J_1}{1-R_1^{(E)}R_{-1}^{(E)}e^{-2\lambda_0 T}}+\frac{R_{-1}^{(M)}\lambda_0 k_x J_3}{1-R_1^{(M)}R_{-1}^{(M)}e^{-2\lambda_0 T}}\right]$$

$$\tilde{E}_y = e^{-\lambda_0(2T+(z'-z))}\left[\frac{R_1^{(E)}R_{-1}^{(E)}j\omega\mu_0 k_x J_1}{1-R_1^{(E)}R_{-1}^{(E)}e^{-2\lambda_0 T}}+\frac{R_1^{(M)}R_{-1}^{(M)}\lambda_0 k_y J_3}{1-R_1^{(M)}R_{-1}^{(M)}e^{-2\lambda_0 T}}\right]$$

$$+ e^{-\lambda_0(2T-(z'-z))}\left[\frac{R_1^{(E)}R_{-1}^{(E)}j\omega\mu_0 k_x J_2}{1-R_1^{(E)}R_{-1}^{(E)}e^{-2\lambda_0 T}}+\frac{R_1^{(M)}R_{-1}^{(M)}\lambda_0 k_y J_4}{1-R_1^{(M)}R_{-1}^{(M)}e^{-2\lambda_0 T}}\right]$$

$$+e^{-\lambda_0(2z_0-(z'+z))}\left[\frac{R_1^{(E)}j\omega\mu_0 k_x J_2}{1-R_1^{(E)}R_{-1}^{(E)}e^{-2\lambda_0 T}}+\frac{R_1^{(M)}\lambda_0 k_y J_4}{1-R_1^{(M)}R_{-1}^{(M)}e^{-2\lambda_0 T}}\right]$$

$$+ e^{\lambda_0(2z_{-1}-(z'+z))}\left[\frac{R_{-1}^{(E)}j\omega\mu_0 k_x J_1}{1-R_1^{(E)}R_{-1}^{(E)}e^{-2\lambda_0 T}}+\frac{R_{-1}^{(M)}\lambda_0 k_y J_3}{1-R_1^{(M)}R_{-1}^{(M)}e^{-2\lambda_0 T}}\right]$$

$$\tilde{E}_z = e^{-\lambda_0(2T+(z'-z))}\frac{R_1^{(M)}R_{-1}^{(M)}jk_t^2 J_3}{1-R_1^{(M)}R_{-1}^{(M)}e^{-2\lambda_0 T}}$$

$$+ e^{-\lambda_0(2T-(z'-z))}\frac{-R_1^{(M)}R_{-1}^{(M)}jk_t^2 J_4}{1-R_1^{(M)}R_{-1}^{(M)}e^{-2\lambda_0 T}}$$

$$+ e^{-\lambda_0(2z_0-(z'+z))}\frac{R_1^{(M)}jk_t^2 J_4}{1-R_1^{(M)}R_{-1}^{(M)}e^{-2\lambda_0 T}}$$

$$+ e^{\lambda_0(2z_{-1}-(z'+z))}\frac{-R_{-1}^{(M)}jk_t^2 J_3}{1-R_1^{(M)}R_{-1}^{(M)}e^{-2\lambda_0 T}}$$

$$\tilde{H}_x = e^{-\lambda_0(2T+(z'-z))}\left[\frac{R_1^{(E)}R_{-1}^{(E)}\lambda_0 k_x J_1}{1-R_1^{(E)}R_{-1}^{(E)}e^{-2\lambda_0 T}}+\frac{R_1^{(M)}R_{-1}^{(M)}j\omega\hat{\varepsilon}_0 k_y J_3}{1-R_1^{(M)}R_{-1}^{(M)}e^{-2\lambda_0 T}}\right]$$

$$+ e^{-\lambda_0(2T-(z'-z))}\left[\frac{-R_1^{(E)}R_{-1}^{(E)}\lambda_0 k_x J_2}{1-R_1^{(E)}R_{-1}^{(E)}e^{-2\lambda_0 T}}+\frac{-R_1^{(M)}R_{-1}^{(M)}j\omega\hat{\varepsilon}_0 k_y J_4}{1-R_1^{(M)}R_{-1}^{(M)}e^{-2\lambda_0 T}}\right]$$

$$+ e^{-\lambda_0(2z_0-(z'+z))}\left[\frac{R_1^{(E)}\lambda_0 k_x J_2}{1-R_1^{(E)}R_{-1}^{(E)}e^{-2\lambda_0 T}}+\frac{R_1^{(M)}j\omega\hat{\varepsilon}_0 k_y J_4}{1-R_1^{(M)}R_{-1}^{(M)}e^{-2\lambda_0 T}}\right]$$

$$+ e^{\lambda_0(2z_{-1}-(z'+z))}\left[\frac{-R_{-1}^{(E)}\lambda_0 k_x J_1}{1-R_1^{(E)}R_{-1}^{(E)}e^{-2\lambda_0 T}}+\frac{-R_{-1}^{(M)}j\omega\hat{\varepsilon}_0 k_y J_3}{1-R_1^{(M)}R_{-1}^{(M)}e^{-2\lambda_0 T}}\right]$$

$$\tilde{H}_y = e^{-\lambda_0(2T+(z'-z))}\left[\frac{R_1^{(E)}R_{-1}^{(E)}\lambda_0 k_y J_1}{1-R_1^{(E)}R_{-1}^{(E)}e^{-2\lambda_0 T}}+\frac{-R_1^{(M)}R_{-1}^{(M)}j\omega\hat{\varepsilon}_0 k_x J_3}{1-R_1^{(M)}R_{-1}^{(M)}e^{-2\lambda_0 T}}\right]$$

$$+ e^{-\lambda_0(2T-(z'-z))}\left[\frac{-R_1^{(E)}R_{-1}^{(E)}\lambda_0 k_y J_2}{1-R_1^{(E)}R_{-1}^{(E)}e^{-2\lambda_0 T}}+\frac{R_1^{(M)}R_{-1}^{(M)}j\omega\hat{\varepsilon}_0 k_x J_4}{1-R_1^{(M)}R_{-1}^{(M)}e^{-2\lambda_0 T}}\right]$$

$$+ e^{-\lambda_0(2z_0-(z'+z))}\left[\frac{R_1^{(E)}\lambda_0 k_y J_2}{1-R_1^{(E)}R_{-1}^{(E)}e^{-2\lambda_0 T}}+\frac{-R_1^{(M)}j\omega\hat{\varepsilon}_0 k_x J_4}{1-R_1^{(M)}R_{-1}^{(M)}e^{-2\lambda_0 T}}\right]$$

$$+ e^{\lambda_0(2z_{-1}-(z'+z))}\left[\frac{-R_{-1}^{(E)}\lambda_0 k_y J_1}{1-R_1^{(E)}R_{-1}^{(E)}e^{-2\lambda_0 T}}+\frac{R_{-1}^{(M)}j\omega\hat{\varepsilon}_0 k_x J_3}{1-R_1^{(M)}R_{-1}^{(M)}e^{-2\lambda_0 T}}\right]$$

$$\tilde{H}_z = e^{-\lambda_0(2T+(z'-z))}\frac{R_1^{(E)}R_{-1}^{(E)}jk_t^2 J_1}{1-R_1^{(E)}R_{-1}^{(E)}e^{-2\lambda_0 T}}$$

$$+ e^{-\lambda_0(2T-(z'-z))}\frac{R_1^{(E)}R_{-1}^{(E)}jk_t^2 J_2}{1-R_1^{(E)}R_{-1}^{(E)}e^{-2\lambda_0 T}}$$

$$+ e^{-\lambda_0(2z_0-(z'+z))}\frac{R_1^{(E)}jk_t^2 J_2}{1-R_1^{(E)}R_{-1}^{(E)}e^{-2\lambda_0 T}}$$

$$+ e^{\lambda_0(2z_{-1}-(z'+z))}\frac{R_{-1}^{(E)}jk_t^2 J_1}{1-R_1^{(E)}R_{-1}^{(E)}e^{-2\lambda_0 T}}\,. \tag{2.52}$$

When the sources, J_1, J_2, J_3, J_4, are expressed in terms of the electric and magnetic currents, as in (2.28), then we can express the fields in (2.52) in terms of layered-space Green dyads:

$$\tilde{\mathbf{E}}(k_x,k_y;z,z') = \tilde{\mathbf{G}}_{(s)}^{(ee)}(k_x,k_y;z,z')\cdot\tilde{\mathbf{J}}_e(k_x,k_y)$$

$$\tilde{\mathbf{E}}(k_x,k_y;z,z') = \tilde{\mathbf{G}}_{(s)}^{(em)}(k_x,k_y;z,z')\cdot\tilde{\mathbf{J}}_m(k_x,k_y)$$

$$\tilde{\mathbf{H}}(k_x,k_y;z,z') = \tilde{\mathbf{G}}_{(s)}^{(me)}(k_x,k_y;z,z')\cdot\tilde{\mathbf{J}}_e(k_x,k_y)$$
$$\tilde{\mathbf{H}}(k_x,k_y;z,z') = \tilde{\mathbf{G}}_{(s)}^{(mm)}(k_x,k_y;z,z')\cdot\tilde{\mathbf{J}}_m(k_x,k_y)\,. \tag{2.53}$$

The result for the spectral-domain, electric–electric, dyadic Green function is shown here:

$$\begin{aligned}
\tilde{G}_{xx}^{(ee)(s)} &= -j\omega\mu_0\Bigg\{\frac{c(z'-z)}{2\lambda_0}e^{-2\lambda_0 T}\left[G_{1,-1}^{(E)} - k_x^2\left(G_{1,-1}^{(E)}\frac{1}{k_t^2} + G_{1,-1}^{(M)}\frac{\lambda_0^2}{k_0^2k_t^2}\right)\right]\\
&\quad + \frac{e^{\lambda_0(z+z')}}{2\lambda_0}e^{-2\lambda_0 z_0}\left[G_1^{(E)} - k_x^2\left(G_1^{(E)}\frac{1}{k_t^2} + G_1^{(M)}\frac{\lambda_0^2}{k_0^2k_t^2}\right)\right]\\
&\quad + \frac{e^{-\lambda_0(z+z')}}{2\lambda_0}e^{2\lambda_0 z_{-1}}\left[G_{-1}^{(E)} - k_x^2\left(G_{-1}^{(E)}\frac{1}{k_t^2} + G_{-1}^{(M)}\frac{\lambda_0^2}{k_0^2k_t^2}\right)\right]\Bigg\}
\end{aligned}$$

$$\begin{aligned}
\tilde{G}_{xy}^{(ee)(s)} &= -j\omega\mu_0(-k_xk_y)\Bigg\{\frac{c(z'-z)}{2\lambda_0}e^{-2\lambda_0 T}\left[G_{1,-1}^{(E)}\frac{1}{k_t^2} + G_{1,-1}^{(M)}\frac{\lambda_0^2}{k_0^2k_t^2}\right]\\
&\quad + \frac{e^{\lambda_0(z+z')}}{2\lambda_0}e^{-2\lambda_0 z_0}\left[G_1^{(E)}\frac{1}{k_t^2} + G_1^{(M)}\frac{\lambda_0^2}{k_0^2k_t^2}\right]\\
&\quad + \frac{e^{-\lambda_0(z+z')}}{2\lambda_0}e^{2\lambda_0 z_{-1}}\left[G_{-1}^{(E)}\frac{1}{k_t^2} + G_{-1}^{(M)}\frac{\lambda_0^2}{k_0^2k_t^2}\right]\Bigg\}
\end{aligned}$$

$$\begin{aligned}
\tilde{G}_{xz}^{(ee)(s)} &= -j\omega\mu_0(-jk_x\lambda_0)\Bigg\{-\frac{s(z'-z)}{2\lambda_0}e^{-2\lambda_0 T}G_{1,-1}^{(M)}\frac{1}{k_0^2} - \frac{e^{\lambda_0(z+z')}}{2\lambda_0}e^{-2\lambda_0 z_0}G_1^{(M)}\frac{1}{k_0^2}\\
&\quad + \frac{e^{-\lambda_0(z+z')}}{2\lambda_0}e^{2\lambda_0 z_{-1}}G_{-1}^{(M)}\frac{1}{k_0^2}\Bigg\}
\end{aligned}$$

$$\tilde{G}_{yx}^{(ee)(s)} = \tilde{G}_{xy}^{(ee)(s)}$$

$$\begin{aligned}
\tilde{G}_{yy}^{(ee)(s)} &= -j\omega\mu_0\Bigg\{\frac{c(z'-z)}{2\lambda_0}e^{-2\lambda_0 T}\left[G_{1,-1}^{(E)} - k_y^2\left(G_{1,-1}^{(E)}\frac{1}{k_t^2} + G_{1,-1}^{(M)}\frac{\lambda_0^2}{k_0^2k_t^2}\right)\right]\\
&\quad + \frac{e^{\lambda_0(z+z')}}{2\lambda_0}e^{-2\lambda_0 z_0}\left[G_1^{(E)} - k_y^2\left(G_1^{(E)}\frac{1}{k_t^2} + G_1^{(M)}\frac{\lambda_0^2}{k_0^2k_t^2}\right)\right]\\
&\quad + \frac{e^{-\lambda_0(z+z')}}{2\lambda_0}e^{2\lambda_0 z_{-1}}\left[G_{-1}^{(E)} - k_y^2\left(G_{-1}^{(E)}\frac{1}{k_t^2} + G_{-1}^{(M)}\frac{\lambda_0^2}{k_0^2k_t^2}\right)\right]\Bigg\}
\end{aligned}$$

$$\tilde{G}_{yz}^{(ee)(s)} = -j\omega\mu_0(-jk_y\lambda_0)\left\{-\frac{s(z'-z)}{2\lambda_0}e^{-2\lambda_0 T}G_{1,-1}^{(M)}\frac{1}{k_0^2}-\frac{e^{\lambda_0(z+z')}}{2\lambda_0}e^{-2\lambda_0 z_0}G_1^{(M)}\frac{1}{k_0^2}\right.$$
$$\left.+\frac{e^{-\lambda_0(z+z')}}{2\lambda_0}e^{2\lambda_0 z_{-1}}G_{-1}^{(M)}\frac{1}{k_0^2}\right\}$$

$$\tilde{G}_{zx}^{(a)(ee)(s)} = \tilde{G}_{xz}^{(a)(ee)(s)}, \quad \tilde{G}_{zx}^{(b)(ee)(s)} = -\tilde{G}_{xz}^{(b)(ee)(s)}$$
$$\tilde{G}_{zy}^{(a)(ee)(s)} = \tilde{G}_{yz}^{(a)(ee)(s)}, \quad \tilde{G}_{zy}^{(b)(ee)(s)} = -\tilde{G}_{yz}^{(b)(ee)(s)}$$
$$\tilde{G}_{zz}^{(ee)(s)} = -j\omega\mu_0\left\{\frac{c(z'-z)}{2\lambda_0}e^{-2\lambda_0 T}G_{1,-1}^{(M)}\frac{k_t^2}{k_0^2}-\frac{e^{\lambda_0(z+z')}}{2\lambda_0}e^{-2\lambda_0 z_0}G_1^{(M)}\frac{k_t^2}{k_0^2}\right.$$
$$\left.-\frac{e^{-\lambda_0(z+z')}}{2\lambda_0}e^{2\lambda_0 z_{-1}}G_{-1}^{(M)}\frac{k_t^2}{k_0^2}\right\}, \tag{2.54}$$

where $c(z'-z) = e^{\lambda_0(z'-z)} + e^{-\lambda_0(z'-z)}$ and $s(z'-z) = e^{\lambda_0(z'-z)} - e^{-\lambda_0(z'-z)}$. The superscript (a) denotes terms that are convolutional ("Töplitz") in z and z', i.e., depend upon $z-z'$, whereas (b) denotes terms that are correlational ("Hankel") in z and z', i.e., depend upon $z+z'$. The Gs are defined in terms of the TE and TM-mode reflection coefficients:

$$G_{1,-1}^{(E)} = \frac{R_1^{(E)}R_{-1}^{(E)}}{1-R_1^{(E)}R_{-1}^{(E)}e^{-2\lambda_0 T}}, \quad G_1^{(E)} = \frac{R_1^{(E)}}{1-R_1^{(E)}R_{-1}^{(E)}e^{-2\lambda_0 T}}, \quad G_{-1}^{(E)} = \frac{R_{-1}^{(E)}}{1-R_1^{(E)}R_{-1}^{(E)}e^{-2\lambda_0 T}}$$
$$G_{1,-1}^{(M)} = \frac{R_1^{(M)}R_{-1}^{(M)}}{1-R_1^{(M)}R_{-1}^{(M)}e^{-2\lambda_0 T}}, \quad G_1^{(M)} = \frac{R_1^{(M)}}{1-R_1^{(M)}R_{-1}^{(M)}e^{-2\lambda_0 T}}, \quad G_{-1}^{(M)} = \frac{R_{-1}^{(M)}}{1-R_1^{(M)}R_{-1}^{(M)}e^{-2\lambda_0 T}}. \tag{2.55}$$

The physical origin of the Töplitz and Hankel terms follows from the definitions in (2.55), and a graphical illustration is shown in Fig. 2.4.

The Töplitz structure, as shown in the top of Fig. 2.4, arises when the path between the source point, z', and field point, z, includes reflections from both boundaries. The total $z-$directed path length between z' and z is $z-z'+2T$ for path A and $z'-z+2T$ for path B. In each case, the length includes the difference between the z-coordinate of the source and field points.

The Hankel structure, as shown in the bottom of Fig. 2.4, arises when the path between the source and field points includes reflections from only one of the boundaries. The total path length between z' and z is $2Z_0-(z+z')$ for path A and $z+z'-2Z_{-1}$ for path B. In each case, the length includes the sum of the source and field z-coordinates.

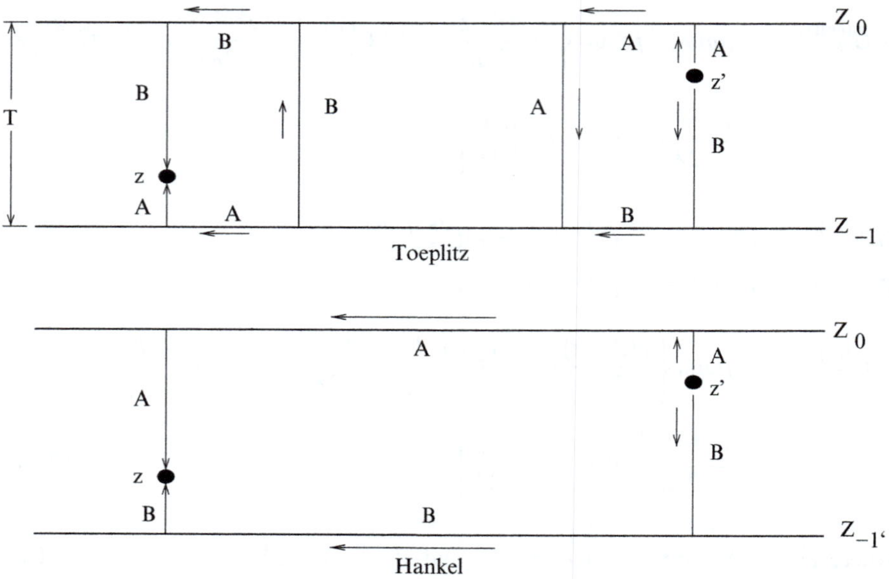

Fig. 2.4 Illustrating the difference between Töplitz (*top*) and Hankel (*bottom*) Green's functions

The spectral-domain, magnetic–magnetic, dyadic Green function is given by the dual of (2.54):

$$\tilde{G}_{xx}^{(mm)(s)} = j\omega\hat{\varepsilon}_0 \left\{ \frac{c(z'-z)}{2\lambda_0} e^{-2\lambda_0 T} \left[G_{1,-1}^{(M)} - k_x^2 \left(G_{1,-1}^{(M)} \frac{1}{k_t^2} + G_{1,-1}^{(E)} \frac{\lambda_0^2}{k_0^2 k_t^2} \right) \right] \right.$$
$$- \frac{e^{\lambda_0(z+z')}}{2\lambda_0} e^{-2\lambda_0 z_0} \left[G_1^{(M)} - k_x^2 \left(G_1^{(M)} \frac{1}{k_t^2} + G_1^{(E)} \frac{\lambda_0^2}{k_0^2 k_t^2} \right) \right]$$
$$\left. - \frac{e^{-\lambda_0(z+z')}}{2\lambda_0} e^{2\lambda_0 z_{-1}} \left[G_{-1}^{(M)} - k_x^2 \left(G_{-1}^{(M)} \frac{1}{k_t^2} + G_{-1}^{(E)} \frac{\lambda_0^2}{k_0^2 k_t^2} \right) \right] \right\}$$

$$\tilde{G}_{xy}^{(mm)(s)} = j\omega\hat{\varepsilon}_0 (-k_x k_y) \left\{ \frac{c(z'-z)}{2\lambda_0} e^{-2\lambda_0 T} \left[G_{1,-1}^{(M)} \frac{1}{k_t^2} + G_{1,-1}^{(E)} \frac{\lambda_0^2}{k_0^2 k_t^2} \right] \right.$$
$$- \frac{e^{\lambda_0(z+z')}}{2\lambda_0} e^{-2\lambda_0 z_0} \left[G_1^{(M)} \frac{1}{k_t^2} + G_1^{(E)} \frac{\lambda_0^2}{k_0^2 k_t^2} \right]$$
$$\left. - \frac{e^{-\lambda_0(z+z')}}{2\lambda_0} e^{2\lambda_0 z_{-1}} \left[G_{-1}^{(M)} \frac{1}{k_t^2} + G_{-1}^{(E)} \frac{\lambda_0^2}{k_0^2 k_t^2} \right] \right\}$$

$$
\begin{aligned}
\tilde{G}_{xz}^{(mm)(s)} &= j\omega\hat{\varepsilon}_0(-jk_x\lambda_0)\left\{-\frac{s(z'-z)}{2\lambda_0}e^{-2\lambda_0 T}G_{1,-1}^{(E)}\frac{1}{k_0^2}+\frac{e^{\lambda_0(z+z')}}{2\lambda_0}e^{-2\lambda_0 z_0}G_1^{(E)}\frac{1}{k_0^2}\right.\\
&\quad\left.-\frac{e^{-\lambda_0(z+z')}}{2\lambda_0}e^{2\lambda_0 z_{-1}}G_{-1}^{(E)}\frac{1}{k_0^2}\right\}\\
\tilde{G}_{yx}^{(mm)(s)} &= \tilde{G}_{xy}^{(mm)(s)}\\
\tilde{G}_{yy}^{(mm)(s)} &= j\omega\hat{\varepsilon}_0\left\{\frac{c(z'-z)}{2\lambda_0}e^{-2\lambda_0 T}\left[G_{1,-1}^{(M)}-k_y^2\left(G_{1,-1}^{(M)}\frac{1}{k_t^2}+G_{1,-1}^{(E)}\frac{\lambda_0^2}{k_0^2k_t^2}\right)\right]\right.\\
&\quad-\frac{e^{\lambda_0(z+z')}}{2\lambda_0}e^{-2\lambda_0 z_0}\left[G_1^{(M)}-k_y^2\left(G_1^{(M)}\frac{1}{k_t^2}+G_1^{(E)}\frac{\lambda_0^2}{k_0^2k_t^2}\right)\right]\\
&\quad\left.-\frac{e^{-\lambda_0(z+z')}}{2\lambda_0}e^{2\lambda_0 z_{-1}}\left[G_{-1}^{(M)}-k_y^2\left(G_{-1}^{(M)}\frac{1}{k_t^2}+G_{-1}^{(E)}\frac{\lambda_0^2}{k_0^2k_t^2}\right)\right]\right\}\\
\tilde{G}_{yz}^{(mm)(s)} &= j\omega\hat{\varepsilon}_0(-jk_y\lambda_0)\left\{-\frac{s(z'-z)}{2\lambda_0}e^{-2\lambda_0 T}G_{1,-1}^{(E)}\frac{1}{k_0^2}+\frac{e^{\lambda_0(z+z')}}{2\lambda_0}e^{-2\lambda_0 z_0}G_1^{(E)}\frac{1}{k_0^2}\right.\\
&\quad\left.-\frac{e^{-\lambda_0(z+z')}}{2\lambda_0}e^{2\lambda_0 z_{-1}}G_{-1}^{(E)}\frac{1}{k_0^2}\right\}\\
\tilde{G}_{zx}^{(a)(mm)(s)} &= \tilde{G}_{xz}^{(a)(mm)(s)},\quad \tilde{G}_{zx}^{(b)(mm)(s)} = -\tilde{G}_{xz}^{(b)(mm)(s)}\\
\tilde{G}_{zy}^{(a)(mm)(s)} &= \tilde{G}_{yz}^{(a)(mm)(s)},\quad \tilde{G}_{zy}^{(b)(mm)(s)} = -\tilde{G}_{yz}^{(b)(mm)(s)}\\
\tilde{G}_{zz}^{(mm)(s)} &= j\omega\hat{\varepsilon}_0\left\{\frac{c(z'-z)}{2\lambda_0}e^{-2\lambda_0 T}G_{1,-1}^{(E)}\frac{k_t^2}{k_0^2}+\frac{e^{\lambda_0(z+z')}}{2\lambda_0}e^{-2\lambda_0 z_0}G_1^{(E)}\frac{k_t^2}{k_0^2}\right.\\
&\quad\left.+\frac{e^{-\lambda_0(z+z')}}{2\lambda_0}e^{2\lambda_0 z_{-1}}G_{-1}^{(E)}\frac{k_t^2}{k_0^2}\right\}.
\end{aligned}
\tag{2.56}
$$

Finally, we list the mixed dyadic functions, starting with the electric–magnetic:

$$
\begin{aligned}
\tilde{G}_{xx}^{(em)(s)} &= -k_xk_y\left\{-s(z'-z)\frac{e^{-2\lambda_0 T}}{2}\left[\frac{G_{1,-1}^{(E)}-G_{1,-1}^{(M)}}{k_t{}^2}\right]-\frac{e^{\lambda_0(z'+z)}}{2}e^{-2\lambda_0 z_0}\left[\frac{G_1^{(E)}-G_1^{(M)}}{k_t{}^2}\right]\right.\\
&\quad\left.+\frac{e^{-\lambda_0(z'+z)}}{2}e^{2\lambda_0 z_{-1}}\left[\frac{G_{-1}^{(E)}-G_{-1}^{(M)}}{k_t{}^2}\right]\right\}
\end{aligned}
$$

$$\tilde{G}_{xy}^{(em)(s)} = e^{-2\lambda_0 T}\frac{s(z'-z)}{2}G_{1,-1}^{(M)} + e^{-2\lambda_0 z_0}\frac{e^{\lambda_0(z'+z)}}{2}G_{1}^{(M)} - e^{2\lambda_0 z_{-1}}\frac{e^{-\lambda_0(z'+z)}}{2}G_{-1}^{(M)}$$
$$-k_y{}^2\left\{-\frac{s(z'-z)}{2}e^{-2\lambda_0 T}\left[\frac{G_{1,-1}^{(E)}-G_{1,-1}^{(M)}}{k_t{}^2}\right] - \frac{e^{\lambda_0(z'+z)}}{2}e^{-2\lambda_0 z_0}\left[\frac{G_{1}^{(E)}-G_{1}^{(M)}}{k_t{}^2}\right]\right.$$
$$\left.+\frac{e^{-\lambda_0(z'+z)}}{2}e^{2\lambda_0 z_{-1}}\left[\frac{G_{-1}^{(E)}-G_{-1}^{(M)}}{k_t{}^2}\right]\right\}$$

$$\tilde{G}_{xz}^{(em)(s)} = -jk_y\left\{e^{-2\lambda_0 T}\frac{c(z'-z)}{2\lambda_0}G_{1,-1}^{(E)} + e^{-2\lambda_0 z_0}\frac{e^{\lambda_0(z'+z)}}{2\lambda_0}G_{1}^{(E)} + e^{2\lambda_0 z_{-1}}\frac{e^{-\lambda_0(z'+z)}}{2\lambda_0}G_{-1}^{(E)}\right\}$$

$$\tilde{G}_{yx}^{(em)(s)} = -e^{-2\lambda_0 T}\frac{s(z'-z)}{2}G_{1,-1}^{(M)} - e^{-2\lambda_0 z_0}\frac{e^{\lambda_0(z'+z)}}{2}G_{1}^{(M)} + e^{2\lambda_0 z_{-1}}\frac{e^{-\lambda_0(z'+z)}}{2}G_{-1}^{(M)}$$
$$-k_x{}^2\left\{\frac{s(z'-z)}{2}e^{-2\lambda_0 T}\left[\frac{G_{1,-1}^{(E)}-G_{1,-1}^{(M)}}{k_t{}^2}\right] + \frac{e^{\lambda_0(z'+z)}}{2}e^{-2\lambda_0 z_0}\left[\frac{G_{1}^{(E)}-G_{1}^{(M)}}{k_t{}^2}\right]\right.$$
$$\left.-\frac{e^{-\lambda_0(z'+z)}}{2}e^{2\lambda_0 z_{-1}}\left[\frac{G_{-1}^{(E)}-G_{-1}^{(M)}}{k_t{}^2}\right]\right\}$$

$$\tilde{G}_{yy}^{(em)(s)} = -\tilde{G}_{xx}^{(em)(s)}$$

$$\tilde{G}_{yz}^{(em)(s)} = -jk_x\left\{-e^{-2\lambda_0 T}\frac{c(z'-z)}{2\lambda_0}G_{1,-1}^{(E)} - e^{-2\lambda_0 z_0}\frac{e^{\lambda_0(z'+z)}}{2\lambda_0}G_{1}^{(E)} - e^{2\lambda_0 z_{-1}}\frac{e^{-\lambda_0(z'+z)}}{2\lambda_0}G_{-1}^{(E)}\right\}$$

$$\tilde{G}_{zx}^{(em)(s)} = -jk_y\left\{-e^{-2\lambda_0 T}\frac{c(z'-z)}{2\lambda_0}G_{1,-1}^{(M)} + e^{-2\lambda_0 z_0}\frac{e^{\lambda_0(z'+z)}}{2\lambda_0}G_{1}^{(M)} + e^{2\lambda_0 z_{-1}}\frac{e^{-\lambda_0(z'+z)}}{2\lambda_0}G_{-1}^{(M)}\right\}$$

$$\tilde{G}_{zy}^{(em)(s)} = -jk_x\left\{e^{-2\lambda_0 T}\frac{c(z'-z)}{2\lambda_0}G_{1,-1}^{(M)} - e^{-2\lambda_0 z_0}\frac{e^{\lambda_0(z'+z)}}{2\lambda_0}G_{1}^{(M)} - e^{2\lambda_0 z_{-1}}\frac{e^{-\lambda_0(z'+z)}}{2\lambda_0}G_{-1}^{(M)}\right\}$$

$$\tilde{G}_{zz}^{(em)(s)} = 0\,, \tag{2.57}$$

and ending with the magnetic–electric:

$$\tilde{G}_{xx}^{(me)(s)} = -k_x k_y \left\{ s(z'-z)\frac{e^{-2\lambda_0 T}}{2}\left[\frac{G_{1,-1}^{(E)}-G_{1,-1}^{(M)}}{k_t{}^2}\right] - \frac{e^{\lambda_0(z'+z)}}{2}e^{-2\lambda_0 z_0}\left[\frac{G_1^{(E)}-G_1^{(M)}}{k_t{}^2}\right] \right.$$
$$\left. +\frac{e^{-\lambda_0(z'+z)}}{2}e^{2\lambda_0 z_{-1}}\left[\frac{G_{-1}^{(E)}-G_{-1}^{(M)}}{k_t{}^2}\right]\right\}$$

$$\tilde{G}_{xy}^{(me)(s)} = e^{-2\lambda_0 T}\frac{s(z'-z)}{2}G_{1,-1}^{(E)} - e^{-2\lambda_0 z_0}\frac{e^{\lambda_0(z'+z)}}{2}G_1^{(E)} + e^{2\lambda_0 z_{-1}}\frac{e^{-\lambda_0(z'+z)}}{2}G_{-1}^{(E)}$$
$$-k_y{}^2\left\{\frac{s(z'-z)}{2}e^{-2\lambda_0 T}\left[\frac{G_{1,-1}^{(E)}-G_{1,-1}^{(M)}}{k_t{}^2}\right] - \frac{e^{\lambda_0(z'+z)}}{2}e^{-2\lambda_0 z_0}\left[\frac{G_1^{(E)}-G_1^{(M)}}{k_t{}^2}\right]\right.$$
$$\left. +\frac{e^{-\lambda_0(z'+z)}}{2}e^{2\lambda_0 z_{-1}}\left[\frac{G_{-1}^{(E)}-G_{-1}^{(M)}}{k_t{}^2}\right]\right\}$$

$$\tilde{G}_{xz}^{(me)(s)} = -jk_y\left\{e^{-2\lambda_0 T}\frac{c(z'-z)}{2\lambda_0}G_{1,-1}^{(M)} - e^{-2\lambda_0 z_0}\frac{e^{\lambda_0(z'+z)}}{2\lambda_0}G_1^{(M)} - e^{2\lambda_0 z_{-1}}\frac{e^{-\lambda_0(z'+z)}}{2\lambda_0}G_{-1}^{(M)}\right\}$$

$$\tilde{G}_{yx}^{(me)(s)} = -e^{-2\lambda_0 T}\frac{s(z'-z)}{2}G_{1,-1}^{(E)} + e^{-2\lambda_0 z_0}\frac{e^{\lambda_0(z'+z)}}{2}G_1^{(E)} - e^{2\lambda_0 z_{-1}}\frac{e^{-\lambda_0(z'+z)}}{2}G_{-1}^{(E)}$$
$$-k_x{}^2\left\{-\frac{s(z'-z)}{2}e^{-2\lambda_0 T}\left[\frac{G_{1,-1}^{(E)}-G_{1,-1}^{(M)}}{k_t{}^2}\right] + \frac{e^{\lambda_0(z'+z)}}{2}e^{-2\lambda_0 z_0}\left[\frac{G_1^{(E)}-G_1^{(M)}}{k_t{}^2}\right]\right.$$
$$\left. -\frac{e^{-\lambda_0(z'+z)}}{2}e^{2\lambda_0 z_{-1}}\left[\frac{G_{-1}^{(E)}-G_{-1}^{(M)}}{k_t{}^2}\right]\right\}$$

$$\tilde{G}_{yy}^{(me)(s)} = -\tilde{G}_{xx}^{(me)(s)}$$

$$\tilde{G}_{yz}^{(me)(s)} = -jk_x\left\{-e^{-2\lambda_0 T}\frac{c(z'-z)}{2\lambda_0}G_{1,-1}^{(M)} + e^{-2\lambda_0 z_0}\frac{e^{\lambda_0(z'+z)}}{2\lambda_0}G_1^{(M)} + e^{2\lambda_0 z_{-1}}\frac{e^{-\lambda_0(z'+z)}}{2\lambda_0}G_{-1}^{(M)}\right\}$$

$$\tilde{G}_{zx}^{(me)(s)} = -jk_y\left\{-e^{-2\lambda_0 T}\frac{c(z'-z)}{2\lambda_0}G_{1,-1}^{(E)} - e^{-2\lambda_0 z_0}\frac{e^{\lambda_0(z'+z)}}{2\lambda_0}G_1^{(E)} - e^{2\lambda_0 z_{-1}}\frac{e^{-\lambda_0(z'+z)}}{2\lambda_0}G_{-1}^{(E)}\right\}$$

$$\tilde{G}_{zy}^{(me)(s)} = -jk_x\left\{e^{-2\lambda_0 T}\frac{c(z'-z)}{2\lambda_0}G_{1,-1}^{(E)} + e^{-2\lambda_0 z_0}\frac{e^{\lambda_0(z'+z)}}{2\lambda_0}G_1^{(E)} + e^{2\lambda_0 z_{-1}}\frac{e^{-\lambda_0(z'+z)}}{2\lambda_0}G_{-1}^{(E)}\right\}$$

$$\tilde{G}_{zz}^{(me)(s)} = 0\,. \tag{2.58}$$

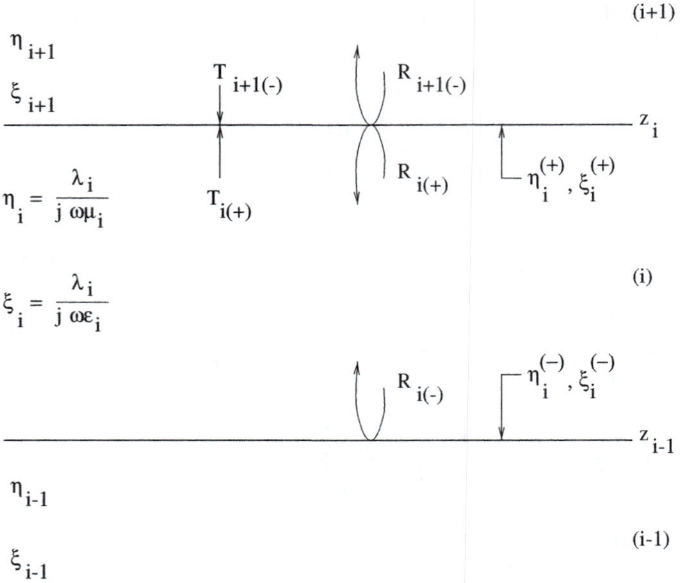

Fig. 2.5 Definition of scattering parameters and intrinsic wave-immittances in layered structures

2.2.3 *A Recursion Relation for Stratified Media*

The reflection and transmission coefficients of (2.49) are those of a slab surrounded by a half-space above and below it. We will derive a recursion relation that will allow the computation of the reflection coefficients, $R_{\pm 1}^{(E,M)}$, that are used in (2.55) when there are an arbitrary number of layers above and below the source slab [35]. Of course, there will ultimately be half-spaces that terminate the system at $\pm\infty$.

Let region i be bounded above by z_i and below by z_{i-1}. Region $i+1$ lies immediately above region i and region $i-1$ lies immediately below. The intrinsic wave-admittance, η_i, and wave-impedance, ξ_i, for the TE and TM modes in region i are

$$\text{(TE)}\ \eta_i = \frac{\lambda_i}{j\omega\mu_i}\ ,\ \xi_i = \frac{\lambda_i}{j\omega\hat{\varepsilon}_i}\ \text{(TM)}. \tag{2.59}$$

We define $R_{i\pm}^{(E,M)}$ to be the reflection coefficient in layer i at the interface with layer $i\pm 1$, and $T_{i\pm}^{(E,M)}$ to be the transmission coefficient, as shown in Fig. 2.5. Then, as shown in (2.49), for a slab sandwiched between two half-spaces

$$R_{i\pm}^{(E)} = \frac{\mu_{i\pm1}\lambda_i - \mu_i\lambda_{i\pm1}}{\mu_i\lambda_{i\pm1} + \mu_{i\pm1}\lambda_i} = \frac{\eta_i - \eta_{i\pm1}}{\eta_i + \eta_{i\pm1}}$$

$$R_{i\pm}^{(M)} = \frac{\hat{\varepsilon}_i\lambda_{i\pm1} - \hat{\varepsilon}_{i\pm1}\lambda_i}{\hat{\varepsilon}_i\lambda_{i\pm1} + \hat{\varepsilon}_{i\pm1}\lambda_i} = \frac{\xi_{i\pm1} - \xi_i}{\xi_i + \xi_{i\pm1}}$$

$$T_{i\pm}^{(E)} = \frac{\mu_i}{\mu_{i\pm1}} \frac{2\eta_i}{\eta_i + \eta_{i\pm1}}$$

$$T_{i\pm}^{(M)} = \frac{\hat{\varepsilon}_i}{\hat{\varepsilon}_{i\pm1}} \frac{2\xi_i}{\xi_i + \xi_{i\pm1}} \ . \tag{2.60}$$

When we have layers above and below region i, we must replace the intrinsic wave-parameters, $\xi_{i\pm1}$, $\eta_{i\pm1}$, by equivalent "load parameters," $\eta_i^{\pm}$, $\xi_i^{\pm}$, which are defined to be the surface admittance and impedance in layer i at the interface with layer $i \pm 1$. The surface admittance is defined to be $\tilde{H}_y/\tilde{E}_x$ at the appropriate interface, and the surface impedance is defined to be $\tilde{E}_x/\tilde{H}_y$. These ratios are the same regardless of which side of the interface they are evaluated at, because $\tilde{E}_x$ and $\tilde{H}_y$ are each continuous at the interface. Hence, we replace (2.60) by

$$R_{i\pm}^{(E)} = \frac{\eta_i - \eta_i^{\pm}}{\eta_i + \eta_i^{\pm}} \ , \ R_{i\pm}^{(M)} = \frac{\xi_i^{\pm} - \xi_i}{\xi_i^{\pm} + \xi_i} . \tag{2.61}$$

Let the TE-field in region i be

$$\begin{aligned} &a_i \mathbf{v}_{1i} e^{\lambda_i(z-z_i)} + b_i \mathbf{v}_{2i} e^{-\lambda_i(z-z_i)} \\ &= b_i \left[R_{i+}^{(E)} \mathbf{v}_{1i} e^{\lambda_i(z-z_i)} + \mathbf{v}_{2i} e^{-\lambda_i(z-z_i)} \right], \end{aligned} \tag{2.62}$$

then the field components at the lower-boundary, $z = z_{i-1}$, are

$$\begin{aligned} \tilde{E}_x &= b_i \left[R_{i+}^{(E)} (-j\omega\mu_i k_y) e^{-\lambda_i T_i} - j\omega\mu_i k_y e^{\lambda_i T_i} \right] \\ \tilde{H}_y &= b_i \left[R_{i+}^{(E)} (\lambda_i k_y) e^{-\lambda_i T_i} - \lambda_i k_y e^{\lambda_i T_i} \right]. \end{aligned} \tag{2.63}$$

The load-admittance at $z = z_{i-1}$, therefore, is given by

$$\begin{aligned} \eta_i^{(-)} = \eta_{i-1}^{(+)} &= \frac{\tilde{H}_y}{\tilde{E}_x} \\ &= \frac{\lambda_i R_{i+}^{(E)} e^{-\lambda_i T_i} - \lambda_i e^{\lambda_i T_i}}{-j\omega\mu_i \left(R_{i+}^{(E)} e^{-\lambda_i T_i} + e^{\lambda_i T_i} \right)} \end{aligned}$$

$$= \eta_i \frac{e^{\lambda_i T_i} - R_{i+}^{(E)} e^{-\lambda_i T_i}}{e^{\lambda_i T_i} + R_{i+}^{(E)} e^{-\lambda_i T_i}}$$

$$= \eta_i \frac{e^{\lambda_i T_i} - \dfrac{\eta_i - \eta_i^{(+)}}{\eta_i + \eta_i^{(+)}} e^{-\lambda_i T_i}}{e^{\lambda_i T_i} + \dfrac{\eta_i - \eta_i^{(+)}}{\eta_i + \eta_i^{(+)}} e^{-\lambda_i T_i}}$$

$$= \eta_i \frac{\eta_i^{(+)} + \eta_i \tanh(\lambda_i T_i)}{\eta_i + \eta_i^{(+)} \tanh(\lambda_i T_i)} . \tag{2.64}$$

When we use $\mathbf{v}_{3i}$ and $\mathbf{v}_{4i}$, we find that the components of the TM-field at $z = z_{i-1}$ are given by

$$\begin{aligned} \tilde{E}_x &= d_i \left[R_{i+}^{(M)} \lambda_i k_x e^{-\lambda_i T_i} + \lambda_i k_x e^{\lambda_i T_i} \right] \\ \tilde{H}_y &= d_i \left[R_{i+}^{(M)} (-j\omega\hat{\varepsilon}_i k_x) e^{-\lambda_i T_i} + j\omega\hat{\varepsilon}_i k_x e^{\lambda_i T_i} \right]. \end{aligned} \tag{2.65}$$

Hence, by an analysis similar to (2.64), we get

$$\begin{aligned} \xi_i^{(-)} = \xi_{i-1}^{(+)} &= \frac{\tilde{E}_x}{\tilde{H}_y} \\ &= \xi_i \frac{\xi_i^{(+)} + \xi_i \tanh(\lambda_i T_i)}{\xi_i + \xi_i^{(+)} \tanh(\lambda_i T_i)} . \end{aligned} \tag{2.66}$$

Let the TE-field in region i below the source region be given by

$$\begin{aligned} & a_i \mathbf{v}_{1i} e^{\lambda_i (z - z_{i-1})} + b_i \mathbf{v}_{2i} e^{-\lambda_i (z - z_{i-1})} \\ &= a_i \left[\mathbf{v}_{1i} e^{\lambda_i (z - z_{i-1})} + R_{i-}^{(E)} \mathbf{v}_{2i} e^{-\lambda_i (z - z_{i-1})} \right], \end{aligned} \tag{2.67}$$

where we treat the negatively-traveling wave, $a_i \mathbf{v}_{1i} e^{\lambda_i (z - z_{i-1})}$, as being incident on the surface $z = z_{i-1}$. Therefore, we have

$$\begin{aligned} \tilde{E}_x &= a_i \left[-j\omega\mu_i k_y e^{\lambda_i T_i} + R_{i-}^{(E)} (-j\omega\mu_i k_y) e^{-\lambda_i T_i} \right] \\ \tilde{H}_y &= a_i \left[\lambda_i k_y e^{\lambda_i T_i} + R_{i-}^{(E)} (-\lambda_i k_y) e^{-\lambda_i T_i} \right], \end{aligned} \tag{2.68}$$

for the fields at the upper-boundary, z_i, of the ith region. Hence, the load admittance at $z = z_i$ is

$$\begin{aligned}
\eta_i^{(+)} = \eta_{i+1}^{(-)} &= -\frac{\tilde{H}_y}{\tilde{E}_x} \\
&= \frac{\lambda_i R_{i-}^{(E)} e^{-\lambda_i T_i} - \lambda_i e^{\lambda_i T_i}}{-j\omega\mu_i \left(R_{i-}^{(E)} e^{-\lambda_i T_i} + e^{\lambda_i T_i} \right)} \\
&= \eta_i \frac{e^{\lambda_i T_i} - R_{i-}^{(E)} e^{-\lambda_i T_i}}{e^{\lambda_i T_i} + R_{i-}^{(E)} e^{-\lambda_i T_i}} \\
&= \eta_i \frac{e^{\lambda_i T_i} - \dfrac{\eta_i - \eta_i^{(-)}}{\eta_i + \eta_i^{(-)}} e^{-\lambda_i T_i}}{e^{\lambda_i T_i} + \dfrac{\eta_i - \eta_i^{(-)}}{\eta_i + \eta_i^{(-)}} e^{-\lambda_i T_i}} \\
&= \eta_i \frac{\eta_i^{(-)} + \eta_i \tanh(\lambda_i T_i)}{\eta_i + \eta_i^{(-)} \tanh(\lambda_i T_i)} \, .
\end{aligned} \tag{2.69}$$

Similarly, for regions below the source region

$$\xi_i^{(+)} = \xi_{i+1}^{(-)} = \xi_i \frac{\xi_i^{(-)} + \xi_i \tanh(\lambda_i T_i)}{\xi_i + \xi_i^{(-)} \tanh(\lambda_i T_i)} \, . \tag{2.70}$$

In order to avoid large numbers in the computation of the hyperbolic tangent, we write it as

$$\tanh(\lambda_i T_i) = \frac{e^{\lambda_i T_i} - e^{-\lambda_i T_i}}{e^{\lambda_i T_i} + e^{-\lambda_i T_i}} = \frac{1 - e^{-2\lambda_i T_i}}{1 + e^{-2\lambda_i T_i}} \, .$$

Equations (2.64), (2.69), (2.66), and (2.70) are the iterations that produce the immittances that go into the expressions for the reflection coefficients, (2.61). The iterations are started at the interface of the last slab with an infinite half-space, for which

$$\begin{aligned}
&\eta_{N-1}^{+} = \eta_N^{+} = \eta_N, \xi_{N-1}^{+} = \xi_N^{+} = \xi_N \\
&\eta_{-(M-1)}^{-} = \eta_{-M}^{-} = \eta_{-M}, \xi_{-(M-1)}^{-} = \xi_{-M}^{-} = \xi_{-M} \, .
\end{aligned} \tag{2.71}$$

Chapter 3
The Volume-Integral Equations for Plane-Layered Media

3.1 Transformation into the Spatial-Domain

Although the development of the infinite-space and layered-space dyadic Green functions is done in the Fourier-domain (or, as it is often referred to, the spectral-domain), the further development of the integral equation, together with its discretization, is best done in the spatial domain. We will now give an example of the transformation from the Fourier-domain to the spatial domain.

Consider the zz-term of (2.30); the kernel of the integral operator, $\frac{\lambda_0^2}{k_0^2}\frac{e^{-\lambda_0|z-z'|}}{2\lambda_0}$, is equivalent to the kernel $\frac{1}{k_0^2}\frac{d^2}{dz'^2}\frac{e^{-\lambda_0|z-z'|}}{2\lambda_0}+\frac{\delta(z-z')}{k_0^2}$, as simple differentiation proves. That is, both kernels behave identically as integral operators. Hence,

$$
\begin{aligned}
\tilde{E}_z(k_x,k_y;z) &= -j\omega\mu_0\int dz'\left[\left(1+\frac{\lambda_0^2}{k_0^2}\right)\frac{e^{-\lambda_0|z-z'|}}{2\lambda_0}-\frac{\delta(z-z')}{k_0^2}\right]\tilde{J}_{ez}(k_x,k_y;z') \\
&= -j\omega\mu_0\int dz'\left[\left(1+\frac{1}{k_0^2}\frac{d^2}{dz'^2}\right)\frac{e^{-\lambda_0|z-z'|}}{2\lambda_0}\right]\tilde{J}_{ez}(k_x,k_y;z') \\
&= -j\omega\mu_0\left(1+\frac{1}{k_0^2}\frac{d^2}{dz^2}\right)\int dz'\frac{e^{-\lambda_0|z-z'|}}{2\lambda_0}\tilde{J}_{ez}(k_x,k_y;z')\,. \qquad (3.1)
\end{aligned}
$$

Hence, the spatial-domain field is obtained by taking the inverse Fourier transform of (3.1)

$$
E_z(\mathbf{r}) = -j\omega\mu_0\left(1+\frac{1}{k_0^2}\frac{\partial^2}{\partial z^2}\right)\int dz'\frac{1}{4\pi^2}\iint_{-\infty}^{\infty}\frac{e^{-\lambda_0|z-z'|}}{2\lambda_0}\tilde{J}_{ez}(k_x,k_y;z')e^{-j[k_xx+k_yy]}dk_xdk_y
$$

H.A. Sabbagh et al., *Computational Electromagnetics and Model-Based Inversion*, Scientific Computation, DOI 10.1007/978-1-4419-8429-6_3,

$$= \left(1 + \frac{1}{k_0^2}\frac{\partial^2}{\partial z^2}\right) \int d\mathbf{r}' \Phi^{(e)}(\mathbf{r}-\mathbf{r}') J_{ez}(\mathbf{r}') , \tag{3.2}$$

where

$$\Phi^{(e)}(\mathbf{r}-\mathbf{r}') = \frac{-j\omega\mu_0}{4\pi^2} \int\int_{-\infty}^{\infty} \frac{e^{-\lambda_0|z-z'|}}{2\lambda_0} e^{-j[k_x(x-x')+k_y(y-y')]} dk_x dk_y . \tag{3.3}$$

The transform function in (3.3) is spherically symmetric, because it is a function of $\lambda_0 = \sqrt{k_t^2 - k_0^2}$. Hence, we can transform the two-dimensional integrals into one-dimensional integrals in the following way: transform to cylindrical coordinates in both physical and **k**-space

$$\begin{aligned} x - x' &= r\cos\phi \ ; \ k_x = l\cos\alpha \\ y - y' &= r\sin\phi \ ; \ k_y = l\sin\alpha . \end{aligned} \tag{3.4}$$

(We are replacing the variable, k_t, by l.) Then the integral in (3.3) becomes

$$\Phi^{(e)}(\mathbf{r}-\mathbf{r}') = \frac{-j\omega\mu_0}{4\pi^2} \int_0^{2\pi} d\alpha \int_0^{\infty} \frac{e^{-\lambda_0|z-z'|}}{2\lambda_0} e^{-jlr\cos(\alpha-\phi)} l dl . \tag{3.5}$$

According to a well-known identity involving Bessel functions, we have

$$e^{-jlr\cos(\alpha-\phi)} = J_0(lr) + 2\sum_{n=1}^{\infty}(-j)^n J_k(lr)\cos n(\alpha-\phi) . \tag{3.6}$$

When (3.6) is substituted into (3.5), only the J_0 term survives the integration over 2π radians, so that

$$\begin{aligned} \Phi^{(e)}(\mathbf{r}-\mathbf{r}') &= \frac{-j\omega\mu_0}{2\pi} \int_0^{\infty} \frac{e^{-\lambda_0|z-z'|}}{2\lambda_0} J_0(rl) l dl \\ &= -j\omega\mu_0 \frac{e^{-jk_0|\mathbf{r}-\mathbf{r}'|}}{4\pi|\mathbf{r}-\mathbf{r}'|} . \end{aligned} \tag{3.7}$$

3.2 The Electric Differential Volume-Integral Equation

We decompose the total Green dyad into the "infinite" part, which is the field produced by the source in infinite space, and the "layered" part, which is due to the presence of the various layers of the workpiece. If we consider only nonmagnetic problems ($\mu = \mu_0$) right now, and let $\mathbf{J}(\mathbf{r}')$ be the unknown anomalous electric current produced by the flaw, then the infinite part of the dyad produces the "infinite-space" contribution to the total electric field:

$$
\begin{aligned}
E_x^{(0)}(\mathbf{r}) &= A_x^{(e)} + \frac{1}{k_0^2}\frac{\partial^2}{\partial x^2}A_x^{(e)} + \frac{1}{k_0^2}\frac{\partial^2}{\partial x \partial y}A_y^{(e)} + \frac{1}{k_0^2}\frac{\partial^2}{\partial x \partial z}A_z^{(e)} \\
E_y^{(0)}(\mathbf{r}) &= \frac{1}{k_0^2}\frac{\partial^2}{\partial y \partial x}A_x^{(e)} + A_y^{(e)} + \frac{1}{k_0^2}\frac{\partial^2}{\partial y^2}A_y^{(e)} + \frac{1}{k_0^2}\frac{\partial^2}{\partial y \partial z}A_z^{(e)} \\
E_z^{(0)}(\mathbf{r}) &= \frac{1}{k_0^2}\frac{\partial^2}{\partial z \partial x}A_x^{(e)} + \frac{1}{k_0^2}\frac{\partial^2}{\partial z \partial y}A_y^{(e)} + A_z^{(e)} + \frac{1}{k_0^2}\frac{\partial^2}{\partial z^2}A_z^{(e)},
\end{aligned}
\tag{3.8}
$$

where the vector potential is given by

$$
\mathbf{A}^{(e)}(\mathbf{r}) = \int \boldsymbol{\Phi}^{(e)}(\mathbf{r}-\mathbf{r}')\mathbf{J}^{(e)}(\mathbf{r}')d\mathbf{r}' \,, \tag{3.9}
$$

and

$$
\begin{aligned}
\boldsymbol{\Phi}^{(e)}(\mathbf{r}-\mathbf{r}') &= -j\omega\mu_0 \frac{e^{-jk_0|\mathbf{r}-\mathbf{r}'|}}{4\pi|\mathbf{r}-\mathbf{r}'|} \\
&= \frac{-j\omega\mu_0}{2\pi}\int_0^\infty \frac{e^{-\lambda_0|z-z'|}}{2\lambda_0}J_0(rl)l\,dl \,.
\end{aligned}
\tag{3.10}
$$

In the Bessel transform, 3.10, $r = [(x-x')^2 + (y-y')^2]^{1/2}$.

The "layered-space" contribution to the total field is given by

$$
\begin{aligned}
E_x^{(s)}(\mathbf{r}) &= F_x^{(1)} + \frac{1}{k_0^2}\frac{\partial^2}{\partial x^2}F_x + \frac{1}{k_0^2}\frac{\partial^2}{\partial x \partial y}F_y + \frac{1}{k_0^2}\frac{\partial^2}{\partial x \partial z}F_z \\
E_y^{(s)}(\mathbf{r}) &= \frac{1}{k_0^2}\frac{\partial^2}{\partial y \partial x}F_x + F_y^{(1)} + \frac{1}{k_0^2}\frac{\partial^2}{\partial y^2}F_y + \frac{1}{k_0^2}\frac{\partial^2}{\partial y \partial z}F_z \\
E_z^{(s)}(\mathbf{r}) &= \frac{1}{k_0^2}\frac{\partial^2}{\partial z \partial x}F_{zx} + \frac{1}{k_0^2}\frac{\partial^2}{\partial z \partial y}F_{zy} + F_z + \frac{1}{k_0^2}\frac{\partial^2}{\partial z^2}F_z,
\end{aligned}
\tag{3.11}
$$

where

$$
\begin{aligned}
F_x^{(1)}(\mathbf{r}) &= \int d\mathbf{r}' G_{xx}^{(1)}(x-x', y-y'; z, z')J_x(\mathbf{r}') \\
F_x(\mathbf{r}) &= \int d\mathbf{r}' G_{xx}^{(2)}(x-x', y-y'; z, z')J_x(\mathbf{r}') \\
F_y(\mathbf{r}) &= \int d\mathbf{r}' G_{xx}^{(2)}(x-x', y-y'; z, z')J_y(\mathbf{r}') \\
F_z(\mathbf{r}) &= \int d\mathbf{r}' G_{xz}(x-x', y-y'; z, z')J_z(\mathbf{r}') \\
F_y^{(1)}(\mathbf{r}) &= \int d\mathbf{r}' G_{xx}^{(1)}(x-x', y-y'; z, z')J_y(\mathbf{r}')
\end{aligned}
$$

$$
\begin{aligned}
F_{zx}^{(a)}(\mathbf{r}) &= \int d\mathbf{r}' G_{xz}^{(a)}(x-x', y-y'; z, z') J_x(\mathbf{r}') \\
F_{zx}^{(b)}(\mathbf{r}) &= -\int d\mathbf{r}' G_{xz}^{(b)}(x-x', y-y'; z, z') J_x(\mathbf{r}') \\
F_{zy}^{(a)}(\mathbf{r}) &= \int d\mathbf{r}' G_{xz}^{(a)}(x-x', y-y'; z, z') J_y(\mathbf{r}') \\
F_{zy}^{(b)}(\mathbf{r}) &= -\int d\mathbf{r}' G_{xz}^{(b)}(x-x', y-y'; z, z') J_y(\mathbf{r}') \,. \qquad (3.12)
\end{aligned}
$$

The unique kernels that appear in (3.12) are given by

$$
\begin{aligned}
G_{xx}^{(1)} &= \frac{-j\omega\mu_0}{2\pi} \int_0^\infty \left[\frac{c(z'-z)}{2\lambda_0} e^{-2\lambda_0 T} G_{1,-1}^{(E)} + \frac{e^{\lambda_0(z+z')}}{2\lambda_0} e^{-2\lambda_0 z_0} G_1^{(E)} \right. \\
&\quad \left. + \frac{e^{-\lambda_0(z+z')}}{2\lambda_0} e^{2\lambda_0 z_{-1}} G_{-1}^{(E)} \right] J_0(rl) l dl \\
G_{xx}^{(2)} &= \frac{-j\omega\mu_0}{2\pi} \int_0^\infty \left[\frac{c(z'-z)}{2\lambda_0} e^{-2\lambda_0 T} \left(G_{1,-1}^{(E)} \frac{k_0^2}{l^2} + G_{1,-1}^{(M)} \frac{\lambda_0^2}{l^2} \right) \right. \\
&\quad + \frac{e^{\lambda_0(z+z')}}{2\lambda_0} e^{-2\lambda_0 z_0} \left(G_1^{(E)} \frac{k_0^2}{l^2} + G_1^{(M)} \frac{\lambda_0^2}{l^2} \right) \\
&\quad \left. + \frac{e^{-\lambda_0(z+z')}}{2\lambda_0} e^{2\lambda_0 z_{-1}} \left(G_{-1}^{(E)} \frac{k_0^2}{l^2} + G_{-1}^{(M)} \frac{\lambda_0^2}{l^2} \right) \right] J_0(rl) l dl \\
G_{xz} &= \frac{-j\omega\mu_0}{2\pi} \int_0^\infty \left[\frac{c(z'-z)}{2\lambda_0} e^{-2\lambda_0 T} G_{1,-1}^{(M)} - \frac{e^{\lambda_0(z+z')}}{2\lambda_0} e^{-2\lambda_0 z_0} G_1^{(M)} \right. \\
&\quad \left. - \frac{e^{-\lambda_0(z+z')}}{2\lambda_0} e^{2\lambda_0 z_{-1}} G_{-1}^{(M)} \right] J_0(rl) l dl. \qquad (3.13)
\end{aligned}
$$

As before, $r = [(x-x')^2 + (y-y')^2]^{1/2}$ in the Bessel transform. The results of (3.11)–(3.13) follow from (2.54), upon replacing multiplication by $-jk_x$, $-jk_y$, and λ_0 by $\partial/\partial x$, $\partial/\partial y$, $\partial/\partial z$, respectively. In the expression for $\tilde{G}_{zz}$ we use $k_t^2 = \lambda_0^2 + k_0^2$.

The integro-differential equation to which we will apply the method of moments is simply gotten by equating the total electric field, $\mathbf{J}(\mathbf{r})/\sigma_a$, to the sum of the incident field, due to the coil and the infinite-space and layered-space scattered fields:

$$
E_x^{(i)}(\mathbf{r}) = \frac{J_x(\mathbf{r})}{\sigma_a(\mathbf{r})} - E_x^{(0)}(\mathbf{r})[\mathbf{J}] - E_x^{(s)}(\mathbf{r})[\mathbf{J}]
$$

$$E_y^{(i)}(\mathbf{r}) = \frac{J_y(\mathbf{r})}{\sigma_a(\mathbf{r})} - E_y^{(0)}(\mathbf{r})[\mathbf{J}] - E_y^{(s)}(\mathbf{r})[\mathbf{J}]$$

$$E_z^{(i)}(\mathbf{r}) = \frac{J_z(\mathbf{r})}{\sigma_a(\mathbf{r})} - E_z^{(0)}(\mathbf{r})[\mathbf{J}] - E_z^{(s)}(\mathbf{r})[\mathbf{J}] , \tag{3.14}$$

where $\sigma_a(\mathbf{r}) = j\omega(\hat{\varepsilon}(\mathbf{r}) - \hat{\varepsilon}_h)$ is the anomalous conductivity.

3.3 Vector Form of the Integral Equation

The equations of the model, (3.8)–(3.14), can be rewritten to bring out the vector nature more clearly. In deriving the new equations we repeatedly make use of integration-by-parts and then use the fact that the current distribution is limited in space. For example, consider the term $\partial A_x^{(e)}(\mathbf{r})/\partial x$:

$$\begin{aligned}\frac{\partial A_x^{(e)}(\mathbf{r})}{\partial x} &= \frac{\partial}{\partial x}\int \Phi^{(e)}(\mathbf{r}-\mathbf{r}')J_x^{(e)}(\mathbf{r}')d\mathbf{r}' \\ &= \int \frac{\partial}{\partial x}\Phi^{(e)}(\mathbf{r}-\mathbf{r}')J_x^{(e)}(\mathbf{r}')d\mathbf{r}' \\ &= \int -\frac{\partial}{\partial x'}\Phi^{(e)}(\mathbf{r}-\mathbf{r}')J_x^{(e)}(\mathbf{r}')d\mathbf{r}' \\ &= \int \Phi^{(e)}(\mathbf{r}-\mathbf{r}')\frac{\partial J_x^{(e)}(\mathbf{r}')}{\partial x'}d\mathbf{r}' .\end{aligned} \tag{3.15}$$

Following this pattern we derive two other important results for z-derivatives:

$$\begin{aligned}\frac{\partial}{\partial z}\int d\mathbf{r}' G_{xz}^{(a)}(x-x',y-y',z-z')J_z(\mathbf{r}') &= \int d\mathbf{r}' G_{xz}^{(a)}(x-x',y-y',z-z')\frac{\partial J_z(\mathbf{r}')}{\partial z'} \\ \frac{\partial}{\partial z}\int d\mathbf{r}' G_{xz}^{(b)}(x-x',y-y',z+z')J_z(\mathbf{r}') &= -\int d\mathbf{r}' G_{xz}^{(b)}(x-x',y-y',z+z')\frac{\partial J_z(\mathbf{r}')}{\partial z'} .\end{aligned} \tag{3.16}$$

Applying integration-by-parts to (3.8) allows us to rewrite the infinite-space contribution as a single vector equation:

$$\mathbf{E}^{(0)}(\mathbf{r})[\mathbf{J}^{(e)}] = \int \Phi^{(e)}(\mathbf{r}-\mathbf{r}')\mathbf{J}^{(e)}(\mathbf{r}')d\mathbf{r}' + \frac{1}{k_0^2}\nabla\int \Phi^{(e)}(\mathbf{r}-\mathbf{r}')\nabla'\cdot\mathbf{J}^{(e)}(\mathbf{r}')d\mathbf{r}' . \tag{3.17}$$

In a similar manner, we can rewrite each of the component-equations in (3.11) to get the following vector equation for the layered-space contribution:

$$\begin{aligned}
\mathbf{E}^{(s)}(\mathbf{r})[\mathbf{J}^{(e)}] = &\int d\mathbf{r}' G_{xx}^{(1)(a)}(x-x',y-y',z-z')\mathbf{J}_t^{(e)}(\mathbf{r}') \\
&+\mathbf{a}_z \int d\mathbf{r}' G_{xz}^{(a)}(x-x',y-y',z-z')J_z^{(e)}(\mathbf{r}') \\
&+\int d\mathbf{r}' G_{xx}^{(1)(b)}(x-x',y-y',z+z')\mathbf{J}_t^{(e)}(\mathbf{r}') \\
&+\mathbf{a}_z \int d\mathbf{r}' G_{xz}^{(b)}(x-x',y-y',z+z')J_z^{(e)}(\mathbf{r}') \\
&+\nabla_t \int d\mathbf{r}' G_{xx}^{(2)(a)'}(x-x',y-y',z-z')\nabla_t' \cdot \mathbf{J}_t^{(e)}(\mathbf{r}') \\
&+\nabla_t \int d\mathbf{r}' G_{xx}^{(2)(b)'}(x-x',y-y',z+z')\nabla_t' \cdot \mathbf{J}_t^{(e)}(\mathbf{r}') \\
&+\frac{1}{k_0^2}\nabla \int d\mathbf{r}' G_{xz}^{(a)}(x-x',y-y',z-z')\nabla' \cdot \mathbf{J}^{(e)}(\mathbf{r}') \\
&-\frac{1}{k_0^2}\nabla \int d\mathbf{r}' G_{xz}^{(b)}(x-x',y-y',z+z')\nabla' \cdot \mathbf{J}^{(e)}(\mathbf{r}') \,, \qquad (3.18)
\end{aligned}$$

where $\nabla_t = \mathbf{a}_x \frac{\partial}{\partial x} + \mathbf{a}_y \frac{\partial}{\partial y}$ is the transverse-gradient operator; $\nabla_t \cdot \mathbf{J} = \frac{\partial J_x}{\partial x} + \frac{\partial J_y}{\partial y}$ is the transverse divergence; $\mathbf{J}_t = \mathbf{a}_x J_x + \mathbf{a}_y J_y$, and

$$\begin{aligned}
G_{xx}^{(2)(a)'}(x-x',y-y',z-z') &= \frac{-j\omega\mu_0}{2\pi}\int_0^\infty \left[\frac{c(z'-z)}{2\lambda_0}e^{-2\lambda_0 T}\frac{G_{1,-1}^{(E)}-G_{1,-1}^{(M)}}{l^2}\right] J_0(rl)l\,dl \\
G_{xx}^{(2)(b)'}(x-x',y-y',z+z') &= \frac{-j\omega\mu_0}{2\pi}\int_0^\infty \left[\frac{e^{\lambda_0(z+z')}}{2\lambda_0}e^{-2\lambda_0 z_0}\frac{G_1^{(E)}-G_1^{(M)}}{l^2}\right. \\
&\quad \left. +\frac{e^{-\lambda_0(z+z')}}{2\lambda_0}e^{2\lambda_0 z_{-1}}\frac{G_{-1}^{(E)}-G_{-1}^{(M)}}{l^2}\right] J_0(rl)l\,dl.
\end{aligned}$$

(3.19)

The remaining expressions are given in (3.13).

3.4 The Volume-Integral Equations in Terms of Amperian Currents

The development of the Green's dyad in Chap. 2 assumed that the unknowns were the dual currents, $\mathbf{J}_m$ and $\mathbf{J}_e$. By "dual" we mean that one appears in the Faraday–Maxwell law (the first equation in (2.2)) and the other in the Ampere–Maxwell law (the second equation in (2.2)). This works well in presenting the theory in a coherent and consistent manner, but it also means that there will be "dual" Green's dyads for electric–electric and magnetic–magnetic interactions, as shown in Chap. 2. It is advantageous from a computational perspective to be able to use as much code as possible when computing the various dyadic components, and in order to accomplish this we find that writing the anomalous magnetic currents as "Amperian currents" in Ampere's law is beneficial.

We start with Maxwell's equations

$$\begin{aligned} \nabla\times\mathbf{E} &= -j\omega\mathbf{B} \\ \nabla\times\mathbf{H} &= j\omega\mathbf{D}+\mathbf{J}^{(e)} . \end{aligned} \tag{3.20}$$

Now $\mathbf{H}=\mathbf{B}/\mu(\mathbf{r})=\mathbf{B}/\mu_h+\mathbf{B}/\mu(\mathbf{r})-\mathbf{B}/\mu_h=\mathbf{B}/\mu_h-\mathbf{M}_a$, where μ_h is the host permeability and $\mathbf{M}_a$ is the anomalous magnetization vector. Thus the second of Maxwell's equations may be written

$$\nabla\times\mathbf{B}/\mu_h = j\omega\mathbf{D}+\mathbf{J}^{(e)}+\nabla\times\mathbf{M}_a , \tag{3.21}$$

which makes clear that the Amperian current, $\nabla\times\mathbf{M}_a$, is an equivalent anomalous electric current that arises because of the departures of the magnetic permeability of the workpiece from the host permeability, μ_h. $\mathbf{J}^{(e)}$, on the other hand, is an electric current that includes the anomalous current that arises due to differences in electrical conductivity; $\mathbf{J}^{(e)}=\sigma_h\mathbf{E}+(\sigma(\mathbf{r})-\sigma_h)\mathbf{E}=\sigma_h\mathbf{E}+\mathbf{J}_a$.

Even though the Amperian current is electrical, because it appears as a source term in the second Maxwell equation (Ampere's law), we will refer to it as $\mathbf{J}^{(m)}$, to remind us that it is of magnetic origin, and to distinguish it from $\mathbf{J}^{(e)}$ (which now stands for the anomalous electric current, $\mathbf{J}_a$). The important point, however, is that because the Amperian current behaves as an electrical current, we need only use electric–electric Green functions in the new formulation of the problem, as we now show.

The coupled volume-integral equations for $\mathbf{J}^{(e)}$ and $\mathbf{J}^{(m)}$ are

$$\begin{aligned} \mathbf{E}^{(i)}(\mathbf{r}) &= \frac{\mathbf{J}^{(e)}(\mathbf{r})}{\sigma_a(\mathbf{r})} - \mathbf{E}^{(0)}(\mathbf{r})\left[\mathbf{J}^{(e)}\right] - \mathbf{E}^{(s)}(\mathbf{r})\left[\mathbf{J}^{(e)}\right] - \mathbf{E}^{(0)}(\mathbf{r})\left[\mathbf{J}^{(m)}\right] - \mathbf{E}^{(s)}(\mathbf{r})\left[\mathbf{J}^{(m)}\right] \\ \mathbf{B}^{(i)}(\mathbf{r}) &= \frac{\mu(\mathbf{r})\mu_h}{\mu(\mathbf{r})-\mu_h}\mathbf{M}_a + \frac{1}{j\omega}\nabla\times\mathbf{E}^{(0)}(\mathbf{r})\left[\mathbf{J}^{(e)}\right] + \frac{1}{j\omega}\nabla\times\mathbf{E}^{(s)}(\mathbf{r})\left[\mathbf{J}^{(e)}\right] \\ &\quad + \frac{1}{j\omega}\nabla\times\mathbf{E}^{(0)}(\mathbf{r})\left[\mathbf{J}^{(m)}\right] + \frac{1}{j\omega}\nabla\times\mathbf{E}^{(s)}(\mathbf{r})\left[\mathbf{J}^{(m)}\right] . \end{aligned} \tag{3.22}$$

In arriving at the second equation, we have used the fact that $\mathbf{B} = -(1/j\omega)\nabla \times \mathbf{E}$, and $\mathbf{M}_a = ((\mu(\mathbf{r}) - \mu_h)/\mu(\mathbf{r})\mu_h)\mathbf{B}$.

The first part of the first equation in (3.22) is the usual electric–electric model, (3.14). The remaining parts must be determined, and in order to do this we will use the form of the equations given in Sect. 3.3. Because $\mathbf{J}^{(m)}$ has zero divergence, we can write

$$
\begin{aligned}
\mathbf{E}^{(0)}(\mathbf{r})[\mathbf{J}^{(m)}] &= \int \Phi^{(e)}(\mathbf{r}-\mathbf{r}')\mathbf{J}^{(m)}(\mathbf{r}')d\mathbf{r}' \\
\mathbf{E}^{(s)}(\mathbf{r})[\mathbf{J}^{(m)}] &= \int d\mathbf{r}' G_{xx}^{(1)(a)}(x-x',y-y',z-z')\mathbf{J}_t^{(m)}(\mathbf{r}') \\
&\quad +\mathbf{a}_z \int d\mathbf{r}' G_{xz}^{(a)}(x-x',y-y',z-z')J_z^{(m)}(\mathbf{r}') \\
&\quad +\int d\mathbf{r}' G_{xx}^{(1)(b)}(x-x',y-y',z+z')\mathbf{J}_t^{(m)}(\mathbf{r}') \\
&\quad +\mathbf{a}_z \int d\mathbf{r}' G_{xz}^{(b)}(x-x',y-y',z+z')J_z^{(m)}(\mathbf{r}') \\
&\quad +\nabla_t \int d\mathbf{r}' G_{xx}^{(2)(a)'}(x-x',y-y',z-z')\nabla_t' \cdot \mathbf{J}_t^{(m)}(\mathbf{r}') \\
&\quad +\nabla_t \int d\mathbf{r}' G_{xx}^{(2)(b)'}(x-x',y-y',z+z')\nabla_t' \cdot \mathbf{J}_t^{(m)}(\mathbf{r}')\,, \qquad (3.23)
\end{aligned}
$$

where $\nabla_t = \mathbf{a}_x\frac{\partial}{\partial x} + \mathbf{a}_y\frac{\partial}{\partial y}$; $\nabla_t \cdot \mathbf{J} = \frac{\partial J_x}{\partial x} + \frac{\partial J_y}{\partial y}$; $\mathbf{J}_t = \mathbf{a}_x J_x + \mathbf{a}_y J_y$, and

$$
G_{xx}^{(1)(a)}(x-x',y-y',z-z') = \frac{-j\omega\mu_0}{2\pi}\int_0^\infty \left[\frac{c(z'-z)}{2\lambda_0}e^{-2\lambda_0 T}G_{1,-1}^{(E)}\right] J_0(rl)l\,dl
$$

$$
\begin{aligned}
G_{xx}^{(1)(b)}(x-x',y-y',z+z') = \frac{-j\omega\mu_0}{2\pi}\int_0^\infty &\left[\frac{e^{\lambda_0(z+z')}}{2\lambda_0}e^{-2\lambda_0 z_0}G_1^{(E)}\right. \\
&\left. + \frac{e^{-\lambda_0(z+z')}}{2\lambda_0}e^{2\lambda_0 z_{-1}}G_{-1}^{(E)}\right] J_0(rl)l\,dl
\end{aligned}
$$

$$
G_{xx}^{(2)(a)'}(x-x',y-y',z-z') = \frac{-j\omega\mu_0}{2\pi}\int_0^\infty \left[\frac{c(z'-z)}{2\lambda_0}e^{-2\lambda_0 T}\frac{G_{1,-1}^{(E)} - G_{1,-1}^{(M)}}{l^2}\right] J_0(rl)l\,dl
$$

$$
\begin{aligned}
G_{xx}^{(2)(b)'}(x-x',y-y',z+z') = \frac{-j\omega\mu_0}{2\pi}\int_0^\infty &\left[\frac{e^{\lambda_0(z+z')}}{2\lambda_0}e^{-2\lambda_0 z_0}\frac{G_1^{(E)} - G_1^{(M)}}{l^2}\right. \\
&\left. +\frac{e^{-\lambda_0(z+z')}}{2\lambda_0}e^{2\lambda_0 z_{-1}}\frac{G_{-1}^{(E)} - G_{-1}^{(M)}}{l^2}\right] J_0(rl)l\,dl
\end{aligned}
$$

$$G_{xz}^{(a)}(x-x',y-y',z-z') = \frac{-j\omega\mu_0}{2\pi}\int_0^\infty \left[\frac{c(z'-z)}{2\lambda_0}e^{-2\lambda_0 T}G_{1,-1}^{(M)}\right]J_0(rl)ldl$$

$$G_{xz}^{(b)}(x-x',y-y',z+z') = \frac{-j\omega\mu_0}{2\pi}\int_0^\infty \left[-\frac{e^{\lambda_0(z+z')}}{2\lambda_0}e^{-2\lambda_0 z_0}G_1^{(M)}\right.$$

$$\left. - \frac{e^{-\lambda_0(z+z')}}{2\lambda_0}e^{2\lambda_0 z_{-1}}G_{-1}^{(M)}\right]J_0(rl)ldl\,. \tag{3.24}$$

Furthermore,

$$\begin{aligned}
\nabla\times\mathbf{E}^{(0)}(\mathbf{r})\left[\mathbf{J}^{(e)}\right] &= \nabla\times\int \boldsymbol{\Phi}^{(e)}(\mathbf{r}-\mathbf{r}')\mathbf{J}^{(e)}(\mathbf{r}')d\mathbf{r}' \\
\nabla\times\mathbf{E}^{(s)}(\mathbf{r})\left[\mathbf{J}^{(e)}\right] &= \nabla\times\int d\mathbf{r}'G_{xx}^{(1)(a)}(x-x',y-y',z-z')\mathbf{J}_t^{(e)}(\mathbf{r}') \\
&\quad +\nabla\times\mathbf{a}_z\int d\mathbf{r}'G_{xz}^{(a)}(x-x',y-y',z-z')J_z^{(e)}(\mathbf{r}') \\
&\quad +\nabla\times\int d\mathbf{r}'G_{xx}^{(1)(b)}(x-x',y-y',z+z')\mathbf{J}_t^{(e)}(\mathbf{r}') \\
&\quad +\nabla\times\mathbf{a}_z\int d\mathbf{r}'G_{xz}^{(b)}(x-x',y-y',z+z')J_z^{(e)}(\mathbf{r}') \\
&\quad +\nabla\times\nabla_t\int d\mathbf{r}'G_{xx}^{(2)(a)'}(x-x',y-y',z-z')\nabla_t'\cdot\mathbf{J}_t^{(e)}(\mathbf{r}') \\
&\quad +\nabla\times\nabla_t\int d\mathbf{r}'G_{xx}^{(2)(b)'}(x-x',y-y',z+z')\nabla_t'\cdot\mathbf{J}_t^{(e)}(\mathbf{r}')\,,
\end{aligned} \tag{3.25}$$

with the same expressions holding for $\nabla\times\mathbf{E}^{(0)}(\mathbf{r})\left[\mathbf{J}^{(m)}\right]$, $\nabla\times\mathbf{E}^{(s)}(\mathbf{r})\left[\mathbf{J}^{(m)}\right]$.

Chapter 4
Discretization via the Galerkin Method of Moments

We will discretize (3.22) by employing Galerkin's method, which uses the same vector functions for expansion and testing. The spatial derivatives that cause problems will be removed by the testing process. In order to test these derivatives, we introduce special vector expansion functions, called "facet elements" and "edge elements," that comprise products of pulse and tent functions.

Facet elements have been called "volumetric rooftop" functions by Catedra et al. [36]. Volumetric rooftop functions have also been used in [37, 38]. These functions are a generalization of two-dimensional rooftop functions that have been used in problems involving scattering from two-dimensional structures and three-dimensional surfaces [30,32,33].

Facet elements and edge elements are a subset of a more general class of spline-generated basis-functions that are based upon higher-order convolutions of the unit pulse

$$\pi(x) = \begin{cases} 1, & \text{if } 0 \leq x < 1 \\ 0, & \text{otherwise.} \end{cases} \tag{4.1}$$

The reader is invited to study [31, 39] for a more complete development of the subject.

4.1 Expansion of the Anomalous Currents

We introduce facet elements oriented in the x-, y-, and z-directions, such that the derivative with respect to x, y, and z is bounded. This ensures that the divergence of the vector field is bounded, and for this reason facet elements are often referred to

H.A. Sabbagh et al., *Computational Electromagnetics and Model-Based Inversion*, Scientific Computation, DOI 10.1007/978-1-4419-8429-6_4,

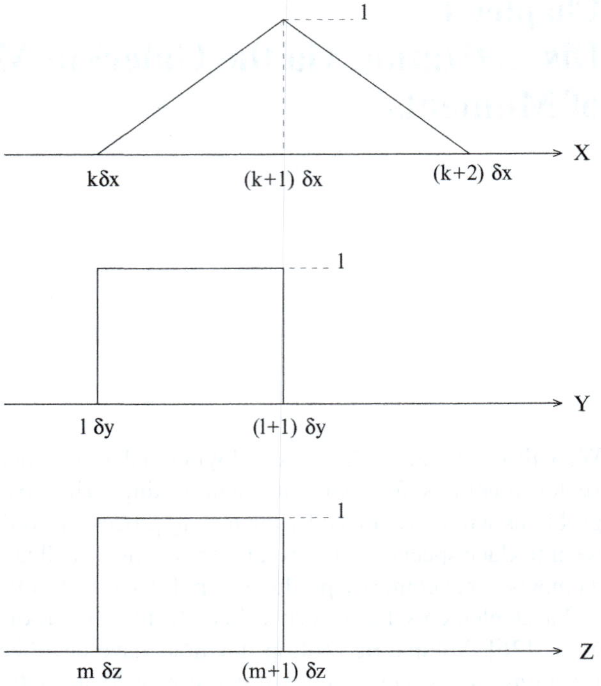

Fig. 4.1 Showing the location of the tent and pulse functions for the facet element $T^{(x)}_{klm}(x,y,z) = \pi_{2k}(x/\delta x)\pi_{1l}(y/\delta y)\pi_{1m}(z/\delta z)$

as "divergence-conforming." We write $T^{(q)(e)}_{klm}(x,y,z)$ for the facet element oriented in the qth direction. The expressions for $T^{(q)(e)}_{klm}$ are:

$$
\begin{aligned}
T^{(x)(e)}_{klm}(\mathbf{r}) &= \pi_{2k}(x/\delta x)\pi_{1l}(y/\delta y)\pi_{1m}(z/\delta z)\\
T^{(y)(e)}_{klm}(\mathbf{r}) &= \pi_{1k}(x/\delta x)\pi_{2l}(y/\delta y)\pi_{1m}(z/\delta z)\\
T^{(z)(e)}_{klm}(\mathbf{r}) &= \pi_{1k}(x/\delta x)\pi_{1l}(y/\delta y)\pi_{2m}(z/\delta z) \qquad (k,l,m) = (0,0,0),\ldots,(N_x,N_y,N_z),
\end{aligned}
\tag{4.2}
$$

where $\pi_{1m}(y/\delta y)$ is the mth unit pulse function and $\pi_{2k}(x/\delta x)$ is the kth tent function, which is the convolution of $\pi_{1k}(x/\delta x)$ with itself. The $T^{(q)(e)}(\mathbf{r})_{klm}$ are called "facet elements," because the qth element is constant over the qth facet of the klmth cell.

Figure 4.1 shows the position of the tent and pulse functions for the facet element $T^{(x)(e)}_{klm}(x,y,z)$.

The support of $T^{(x)(e)}_{klm}(x,y,z)$ is shown in Fig. 4.2. We assume that the conductivity is constant, with the value

$$\sigma_{\text{cell}} = \sigma_{\max} + V_c(\sigma_{\min} - \sigma_{\max}), \tag{4.3}$$

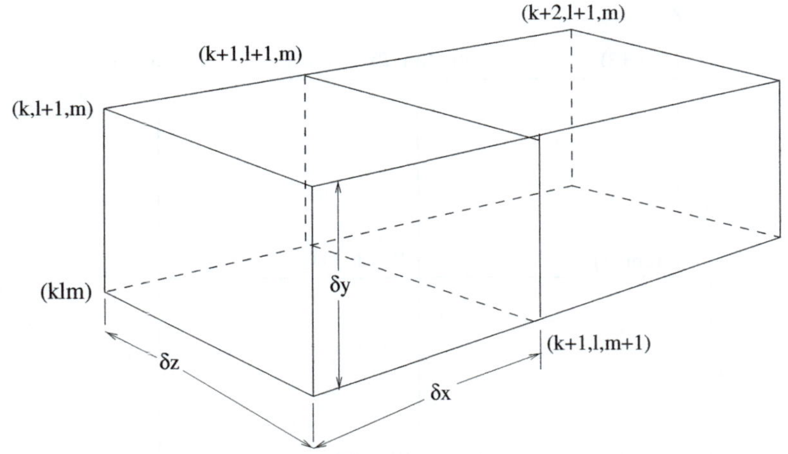

Fig. 4.2 Support of $T_{klm}^{(x)(e)}(x,y,z)$. The conductivity is assumed to be constant within each cell of dimension $\delta x \times \delta y \times \delta z$

within each cell of dimension $\delta x \times \delta y \times \delta z$. $\sigma_{\max}$ and $\sigma_{\min}$ are, respectively, the maximum and minimum conductivities in the problem and V_c is the *conductivity volume-fraction.*

We expand the anomalous electric current vector in terms of the facet elements as

$$J_x^{(e)}(\mathbf{r}) = \sum_{KLM} J_{KLM}^{(x)} T_{KLM}^{(x)}(\mathbf{r})$$
$$J_y^{(e)}(\mathbf{r}) = \sum_{KLM} J_{KLM}^{(y)} T_{KLM}^{(y)}(\mathbf{r})$$
$$J_z^{(e)}(\mathbf{r}) = \sum_{KLM} J_{KLM}^{(z)} T_{KLM}^{(z)}(\mathbf{r}), \tag{4.4}$$

and will then use these same basis-functions for testing the integral equations.

Because $\mathbf{J}^{(m)}(\mathbf{r}) = \nabla \times \mathbf{M}_a(\mathbf{r})$, we expand $\mathbf{M}_a(\mathbf{r})$ in "curl-conforming" edge elements, which have the required differentiability of the curl operation

$$M_x(\mathbf{r}) = \sum_{KLM} M_{KLM}^{(x)} T_{KLM}^{(x)(m)}(\mathbf{r})$$
$$M_y(\mathbf{r}) = \sum_{KLM} M_{KLM}^{(y)} T_{KLM}^{(y)(m)}(\mathbf{r})$$
$$M_z(\mathbf{r}) = \sum_{KLM} M_{KLM}^{(z)} T_{KLM}^{(z)(m)}(\mathbf{r}), \tag{4.5}$$

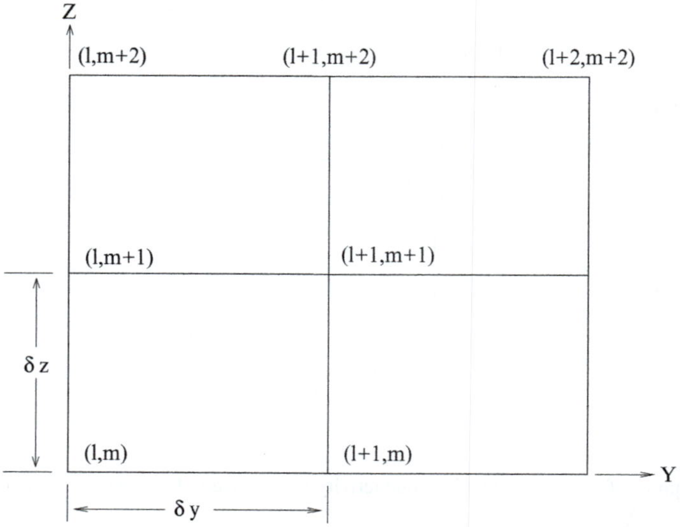

Fig. 4.3 Support of $T_{klm}^{(x)(m)}(x,y,z)$ in (y,z)-space. The support extends one cell in the x-direction, normal to the page. The permeability is assumed to be constant within each cell of dimension $\delta x \times \delta y \times \delta z$

where

$$\begin{aligned} T_{KLM}^{(x)(m)}(\mathbf{r}) &= \pi_{1K}(x)\pi_{2L}(y)\pi_{2M}(z) \\ T_{KLM}^{(y)(m)}(\mathbf{r}) &= \pi_{2K}(x)\pi_{1L}(y)\pi_{2M}(z) \\ T_{KLM}^{(z)(m)}(\mathbf{r}) &= \pi_{2K}(x)\pi_{2L}(y)\pi_{1M}(z). \end{aligned} \tag{4.6}$$

These functions are called edge elements because the expansion coefficient, $M_{KLM}^{(x)}$, is the (constant) value of M_x along the x-directed edge, $(y = (L+1)\delta y, z = (M+1)\delta z)$. There are similar interpretations for $M_{KLM}^{(y)}$ and $M_{KLM}^{(z)}$.

The support of $T_{klm}^{(x)(m)}(x,y,z)$ is shown in Fig. 4.3. We assume that the magnetic permeability is constant, with the value

$$\mu_{\text{cell}} = \mu_{\max} + V_p(\mu_{\min} - \mu_{\max}), \tag{4.7}$$

within each cell of dimension $\delta x \times \delta y \times \delta z$. $\mu_{\max}$ and $\mu_{\min}$ are, respectively, the maximum and minimum permeabilities in the problem and V_p is the *permeability volume-fraction.*

The components of the magnetic current vector are given by

$$\begin{aligned} J_x^{(m)} &= \frac{\partial M_z}{\partial y} - \frac{\partial M_y}{\partial z} \\ &= \sum_{KLM} \left[M_{KLM}^{(z)} \pi_{2K}(x)\pi_{2L}'(y)\pi_{1M}(z) - M_{KLM}^{(y)} \pi_{2K}(x)\pi_{1L}(y)\pi_{2M}'(z) \right] \end{aligned}$$

$$
\begin{aligned}
J_y^{(m)} &= \frac{\partial M_x}{\partial z} - \frac{\partial M_z}{\partial x} \\
&= \sum_{KLM} \left[M_{KLM}^{(x)} \pi_{1K}(x)\pi_{2L}(y)\pi'_{2M}(z) - M_{KLM}^{(z)} \pi'_{2K}(x)\pi_{2L}(y)\pi_{1M}(z) \right] \\
J_z^{(m)} &= \frac{\partial M_y}{\partial x} - \frac{\partial M_x}{\partial y} \\
&= \sum_{KLM} \left[M_{KLM}^{(y)} \pi'_{2K}(x)\pi_{1L}(y)\pi_{2M}(z) - M_{KLM}^{(x)} \pi_{1K}(x)\pi'_{2L}(y)\pi_{2M}(z) \right] \\
\nabla_t \cdot \mathbf{J}_t^{(m)} &= \frac{\partial^2 M_x}{\partial y \partial z} - \frac{\partial^2 M_y}{\partial x \partial z} \\
&= \sum_{KLM} \left[M_{KLM}^{(x)} \pi_{1K}(x)\pi'_{2L}(y)\pi'_{2M}(z) - M_{KLM}^{(y)} \pi'_{2K}(x)\pi_{1L}(y)\pi'_{2M}(z) \right], \qquad (4.8)
\end{aligned}
$$

where the last term is the transverse divergence of the Amperian current.

4.2 Testing the Integral Equations

The first step in discretizing the integral equations, (3.22), is to substitute the expansions, (4.4) and (4.8), for the anomalous currents into the equations and then to "test" the resulting equation. By "testing" we mean taking moments, which is done by multiplying a functional equation by a test function and then integrating over space. When the test function is the same as the expansion function for the unknowns, the method is called the Galerkin variant of the method of moments. We will test each component of the electric equations of (3.22) by the corresponding facet function, $T_{klm}^{(q)(e)}(\mathbf{r})$, and each component of the magnetic equations by the corresponding edge function, $T_{klm}^{(q)(m)}(\mathbf{r})$. The procedure is straightforward, but quite lengthy, so we will show only the results. The terms of (3.22) that involve spatial derivatives are "mollified" by using the two vector identities that are defined in Appendix A.1.

The discretized electric equation is:

$$
\begin{bmatrix} \mathbf{E}^{(ix)} \\ \mathbf{E}^{(iy)} \\ \mathbf{E}^{(iz)} \end{bmatrix} = \begin{bmatrix} \mathbf{Q}^{(x)} & \mathbf{0} & \mathbf{0} \\ \mathbf{0} & \mathbf{Q}^{(y)} & \mathbf{0} \\ \mathbf{0} & \mathbf{0} & \mathbf{Q}^{(z)} \end{bmatrix}^{(ee)} \begin{bmatrix} \mathbf{J}^{(x)} \\ \mathbf{J}^{(y)} \\ \mathbf{J}^{(z)} \end{bmatrix}
- \begin{bmatrix} \mathbf{G}_{(0)}^{(xx)} & \mathbf{G}_{(0)}^{(xy)} & \mathbf{G}_{(0)}^{(xz)} \\ \mathbf{G}_{(0)}^{(yx)} & \mathbf{G}_{(0)}^{(yy)} & \mathbf{G}_{(0)}^{(yz)} \\ \mathbf{G}_{(0)}^{(zx)} & \mathbf{G}_{(0)}^{(zy)} & \mathbf{G}_{(0)}^{(zz)} \end{bmatrix}^{(ee)} \begin{bmatrix} \mathbf{J}^{(x)} \\ \mathbf{J}^{(y)} \\ \mathbf{J}^{(z)} \end{bmatrix}
$$

$$
-\begin{bmatrix} \mathbf{G}_{(a)}^{(xx)} & \mathbf{G}_{(a)}^{(xy)} & \mathbf{G}_{(a)}^{(xz)} \\ \mathbf{G}_{(a)}^{(yx)} & \mathbf{G}_{(a)}^{(yy)} & \mathbf{G}_{(a)}^{(yz)} \\ \mathbf{G}_{(a)}^{(zx)} & \mathbf{G}_{(a)}^{(zy)} & \mathbf{G}_{(a)}^{(zz)} \end{bmatrix}^{(ee)} \begin{bmatrix} \mathbf{J}^{(x)} \\ \mathbf{J}^{(y)} \\ \mathbf{J}^{(z)} \end{bmatrix}
$$

$$
-\begin{bmatrix} \mathbf{G}_{(b)}^{(xx)} & \mathbf{G}_{(b)}^{(xy)} & \mathbf{G}_{(b)}^{(xz)} \\ \mathbf{G}_{(b)}^{(yx)} & \mathbf{G}_{(b)}^{(yy)} & \mathbf{G}_{(b)}^{(yz)} \\ \mathbf{G}_{(b)}^{(zx)} & \mathbf{G}_{(b)}^{(zy)} & \mathbf{G}_{(b)}^{(zz)} \end{bmatrix}^{(ee)} \begin{bmatrix} \mathbf{J}^{(x)} \\ \mathbf{J}^{(y)} \\ \mathbf{J}^{(z)} \end{bmatrix}
$$

$$
-\begin{bmatrix} \mathbf{G}_{(0)} \; \mathbf{G}_{(a)} \; \mathbf{G}_{(b)} \end{bmatrix}^{(em)} \begin{bmatrix} \mathbf{M}^{(x)} \\ \mathbf{M}^{(y)} \\ \mathbf{M}^{(z)} \end{bmatrix}, \tag{4.9}
$$

where the **Q**'s are tri-diagonal matrices, the $\mathbf{G}_{(0)}$'s the infinite-space matrices, the $\mathbf{G}_{(a)}$'s the convolutional layered-space matrices, and the $\mathbf{G}_{(b)}$'s the correlational layered-space matrices. The infinite-space matrices are convolutional, also. The superscript (ee) denotes electric–electric matrices and (em) denotes electric–magnetic matrices. The **J**'s are the unknown electric currents and the **M**'s are the unknown magnetic polarization vectors. The last block in (4.9) is simply a shorthand representation of the three blocks above it, except that it represents electric–magnetic interactions.

The discretized magnetic equation is similar to (4.9) and is given by

$$
\begin{bmatrix} \mathbf{B}^{(ix)} \\ \mathbf{B}^{(iy)} \\ \mathbf{B}^{(iz)} \end{bmatrix} = \begin{bmatrix} \mathbf{Q}^{(x)} & \mathbf{0} & \mathbf{0} \\ \mathbf{0} & \mathbf{Q}^{(y)} & \mathbf{0} \\ \mathbf{0} & \mathbf{0} & \mathbf{Q}^{(z)} \end{bmatrix}^{(mm)} \begin{bmatrix} \mathbf{M}^{(x)} \\ \mathbf{M}^{(y)} \\ \mathbf{M}^{(z)} \end{bmatrix}
$$

$$
+\begin{bmatrix} \mathbf{G}_{(0)}^{(xx)} & \mathbf{G}_{(0)}^{(xy)} & \mathbf{G}_{(0)}^{(xz)} \\ \mathbf{G}_{(0)}^{(yx)} & \mathbf{G}_{(0)}^{(yy)} & \mathbf{G}_{(0)}^{(yz)} \\ \mathbf{G}_{(0)}^{(zx)} & \mathbf{G}_{(0)}^{(zy)} & \mathbf{G}_{(0)}^{(zz)} \end{bmatrix}^{(mm)} \begin{bmatrix} \mathbf{M}^{(x)} \\ \mathbf{M}^{(y)} \\ \mathbf{M}^{(z)} \end{bmatrix}
$$

$$
+\begin{bmatrix} \mathbf{G}_{(a)}^{(xx)} & \mathbf{G}_{(a)}^{(xy)} & \mathbf{G}_{(a)}^{(xz)} \\ \mathbf{G}_{(a)}^{(yx)} & \mathbf{G}_{(a)}^{(yy)} & \mathbf{G}_{(a)}^{(yz)} \\ \mathbf{G}_{(a)}^{(zx)} & \mathbf{G}_{(a)}^{(zy)} & \mathbf{G}_{(a)}^{(zz)} \end{bmatrix}^{(mm)} \begin{bmatrix} \mathbf{M}^{(x)} \\ \mathbf{M}^{(y)} \\ \mathbf{M}^{(z)} \end{bmatrix}
$$

$$
+\begin{bmatrix} \mathbf{G}_{(b)}^{(xx)} & \mathbf{G}_{(b)}^{(xy)} & \mathbf{G}_{(b)}^{(xz)} \\ \mathbf{G}_{(b)}^{(yx)} & \mathbf{G}_{(b)}^{(yy)} & \mathbf{G}_{(b)}^{(yz)} \\ \mathbf{G}_{(b)}^{(zx)} & \mathbf{G}_{(b)}^{(zy)} & \mathbf{G}_{(b)}^{(zz)} \end{bmatrix}^{(mm)} \begin{bmatrix} \mathbf{M}^{(x)} \\ \mathbf{M}^{(y)} \\ \mathbf{M}^{(z)} \end{bmatrix}
$$

$$
+\begin{bmatrix} \mathbf{G}_{(0)} \; \mathbf{G}_{(a)} \; \mathbf{G}_{(b)} \end{bmatrix}^{(me)} \begin{bmatrix} \mathbf{J}^{(x)} \\ \mathbf{J}^{(y)} \\ \mathbf{J}^{(z)} \end{bmatrix}, \tag{4.10}
$$

where **B** is the incident magnetic flux density due to the coil, the superscript (mm) stands for magnetic–magnetic interactions, and (me) stands for magnetic–electric interactions. The magnetic–magnetic **Q** matrices are a little more complicated than the electric–electric ones.

4.3 Solution via Conjugate Gradients

The system of equations, (4.9) and (4.10), that is produced by the method of moments contains a dense matrix, as opposed to the sparse matrices that are produced by finite-element or finite-difference techniques. Hence, it is necessary to develop efficient means of solving the system if the volume-integral method is to be viable in solving realistic three-dimensional problems. When the number of unknowns is $\lesssim 3{,}000$, we can use a direct matrix-decomposition solver, such as the LU-decomposition [41], but for large problems we use the conjugate-gradient algorithm [42].

The conjugate-gradient algorithm, being an iterative method, requires many matrix-vector multiplications. In order for this method to be useful, therefore, there must be an efficient method for computing these products. Fortunately, because we have formulated the volume-integral equations on a regular grid, there exists a very efficient numerical scheme for computing vector-matrix products.

4.3.1 *Efficient Computation of Convolutions and Correlations*

The triple sums that appear in (4.9) and (4.10) consist of three-dimensional convolutions, or two-dimensional convolutions and one-dimensional correlations, which means that we can use three-dimensional FFTs to compute them.

The appropriate theorems (in one dimension) that relate discrete Fourier transforms and convolutions and correlations are ($\Longleftrightarrow$ denotes a discrete transform-pair):

$$
\begin{aligned}
\text{If} \qquad g(j) &\Longleftrightarrow G(n) \\
h(j) &\Longleftrightarrow H(n) \\
\text{Then } \frac{1}{N}\sum_{k=0}^{N-1} g(j+k)h(k) &= \frac{1}{N}\sum_{k=0}^{N-1} g(k)h(k-j) \\
&\Longleftrightarrow G(n)H(-n) \\
&= G(n)H(N-n) \\
\frac{1}{N}\sum_{k=0}^{N-1} g(k)h(j+k) &= \frac{1}{N}\sum_{k=0}^{N-1} g(k-j)h(k) \\
&\Longleftrightarrow G(-n)H(n) \\
&= G(N-n)H(n) \\
\frac{1}{N}\sum_{k=0}^{N-1} g(k)h(j-k) &= \frac{1}{N}\sum_{k=0}^{N-1} g(j-k)h(k) \\
&\Longleftrightarrow G(n)H(n),
\end{aligned}
\tag{4.11}
$$

where $j = 0,\ldots,N-1$, $n = 0,\ldots,N-1$ in all of these. Several points should be made: first note that correlation summing is not commutable and that one must use negative frequencies in the discrete Fourier transform (which, of course, introduces the term $N-n$).

Let's look at the matrix structure of convolution and correlation sums and see how we can use FFT techniques to compute them. We'll work in one dimension. Consider the following convolution, which is written as a vector-matrix equation:

$$\begin{bmatrix} y_0 \\ y_1 \\ y_2 \\ y_3 \end{bmatrix} = \begin{bmatrix} m_0 & m_{-1} & m_{-2} & m_{-3} \\ m_1 & m_0 & m_{-1} & m_{-2} \\ m_2 & m_1 & m_0 & m_{-1} \\ m_3 & m_2 & m_1 & m_0 \end{bmatrix} \begin{bmatrix} x_0 \\ x_1 \\ x_2 \\ x_3 \end{bmatrix}. \tag{4.12}$$

Rewrite this in the expanded form (padding with zeros to get a power of two) in order to achieve a circulant-matrix:

$$\begin{bmatrix} y_0 \\ y_1 \\ y_2 \\ y_3 \\ * \\ * \\ * \\ * \end{bmatrix} = \begin{bmatrix} m_0 & m_{-1} & m_{-2} & m_{-3} & 0 & m_3 & m_2 & m_1 \\ m_1 & m_0 & m_{-1} & m_{-2} & m_{-3} & 0 & m_3 & m_2 \\ m_2 & m_1 & m_0 & m_{-1} & m_{-2} & m_{-3} & 0 & m_3 \\ m_3 & m_2 & m_1 & m_0 & m_{-1} & m_{-2} & m_{-3} & 0 \\ 0 & m_3 & m_2 & m_1 & m_0 & m_{-1} & m_{-2} & m_{-3} \\ m_{-3} & 0 & m_3 & m_2 & m_1 & m_0 & m_{-1} & m_{-2} \\ m_{-2} & m_{-3} & 0 & m_3 & m_2 & m_1 & m_0 & m_{-1} \\ m_{-1} & m_{-2} & m_{-3} & 0 & m_3 & m_2 & m_1 & m_0 \end{bmatrix} \begin{bmatrix} x_0 \\ x_1 \\ x_2 \\ x_3 \\ 0 \\ 0 \\ 0 \\ 0 \end{bmatrix}, \tag{4.13}$$

where the $*$ denotes a discarded entry. Hence, the sequences to be FFT'd are: $(m_0, m_1, m_2, m_3, 0, m_{-3}, m_{-2}, m_{-1})$ and $(x_0, x_1, x_2, x_3, 0, 0, 0, 0)$, and the output sequence is $(y_0, y_1, y_2, y_3, *, *, *, *)$. The order of the entries in the sequences is very important.

Now, for correlations:

$$\begin{bmatrix} y_0 \\ y_1 \\ y_2 \\ y_3 \end{bmatrix} = \begin{bmatrix} m_0 & m_1 & m_2 & m_3 \\ m_1 & m_2 & m_3 & m_4 \\ m_2 & m_3 & m_4 & m_5 \\ m_3 & m_4 & m_5 & m_6 \end{bmatrix} \begin{bmatrix} x_0 \\ x_1 \\ x_2 \\ x_3 \end{bmatrix}. \tag{4.14}$$

Rewrite this in the expanded form (padding with zeros to get a power of two) in order to achieve a circulant-matrix:

$$\begin{bmatrix} y_0 \\ y_1 \\ y_2 \\ y_3 \\ * \\ * \\ * \\ * \end{bmatrix} = \begin{bmatrix} m_0 & m_1 & m_2 & m_3 & m_4 & m_5 & m_6 & 0 \\ m_1 & m_2 & m_3 & m_4 & m_5 & m_6 & 0 & m_0 \\ m_2 & m_3 & m_4 & m_5 & m_6 & 0 & m_0 & m_1 \\ m_3 & m_4 & m_5 & m_6 & 0 & m_0 & m_1 & m_2 \\ m_4 & m_5 & m_6 & 0 & m_0 & m_1 & m_2 & m_3 \\ m_5 & m_6 & 0 & m_0 & m_1 & m_2 & m_3 & m_4 \\ m_6 & 0 & m_0 & m_1 & m_2 & m_3 & m_4 & m_5 \\ 0 & m_0 & m_1 & m_2 & m_3 & m_4 & m_5 & m_6 \end{bmatrix} \begin{bmatrix} x_0 \\ x_1 \\ x_2 \\ x_3 \\ 0 \\ 0 \\ 0 \\ 0 \end{bmatrix}, \tag{4.15}$$

where the $*$ denotes a discarded entry. Hence, the sequences to be FFT'd are: $(m_0, m_1, m_2, m_3, m_4, m_5, m_6, 0)$ and $(x_0, x_1, x_2, x_3, 0, 0, 0, 0)$, and the output sequence is $(y_0, y_1, y_2, y_3, *, *, *, *)$. The order of the entries in the sequences is very important, and also don't forget to negate the frequencies in the transform of the x-sequence.

To summarize: we expand the original data, padding with zeros as necessary to get a circulant matrix, and then take FFTs.

4.3.2 The Conjugate-Gradient Algorithm

Let us write (4.9) and (4.10) as the operator equation

$$Y = \mathscr{A}X, \tag{4.16}$$

where Y denotes the known left-hand side and X denotes the vector of unknown currents. The conjugate gradient (CG) algorithm starts with an initial guess, X_0, from which we compute $R_0 = Y - \mathscr{A}X_0$, $P_1 = Q_0 = \mathscr{A}^* R_0$, where $\mathscr{A}^*$ is the adjoint operator that corresponds to the conjugate-transpose of the matrix blocks in (4.9) and (4.10). In addition, we have a convergence parameter, ε. Then for $k = 1, \ldots,$ if $Test = \|R_k\|/\|Y\| < \varepsilon$, stop; X_k is the optimal solution of (4.16). Otherwise, update X_k by the following steps:

$$\begin{aligned} S_k &= \mathscr{A}P_k \\ a_k &= \frac{\|Q_{k-1}\|^2}{\|S_k\|^2} \\ X_k &= X_{k-1} + a_k P_k \\ R_k &= R_{k-1} - a_k S_k \\ Q_k &= \mathscr{A}^* R_k \\ b_k &= \frac{\|Q_k\|^2}{\|Q_{k-1}\|^2} \\ P_{k+1} &= Q_k + b_k P_k. \end{aligned} \tag{4.17}$$

The convolution and correlation operations that are a part of $\mathscr{A}$ and $\mathscr{A}^*$ are evaluated by using the FFT, as described in the preceding section. This, together with the fact that the storage requirements are reasonably modest, is the reason why the conjugate gradient algorithm becomes attractive for solving potentially large problems in our model.

4.4 Comments and Conclusions

We have formulated the volume-integral approach in terms of the Galerkin variant of the method of moments, in which the unknown anomalous currents and the testing functions are expressed in terms of basis functions that are defined on a regular grid. This results in operators that have very special structures; they are either three-dimensional convolutions or two-dimensional convolutions and one-dimensional correlations, which means that we can use three-dimensional FFTs to accelerate the matrix-vector operations occurring within a conjugate-gradient search algorithm. The use of a highly irregular mesh in the finite-element technique does not allow a similar advantage in the solution process. In the remainder of this text, we will show how this formulation produces extremely efficient solutions of complex problems.

We have not gone into certain technical details, such as comparing operation counts for a direct matrix-vector multiply versus an FFT-assisted operation. Recent texts, such as [43–45], deal with these issues in more generality, while [46] deals with other "fast" algorithms for computational electromagnetics.

Appendix

A.1 Two Theorems of Vector Analysis

Let the testing vector function, $\mathbf{B}(\mathbf{r})$, have a finite support, then the curl-operator can be transferred from one vector function to another within an integral:

$$\begin{aligned}\int \mathbf{B}\cdot\nabla\times\mathbf{A}d\mathbf{r} &= \int \mathbf{A}\cdot\nabla\times\mathbf{B}d\mathbf{r} + \int \nabla\cdot(\mathbf{A}\times\mathbf{B})d\mathbf{r} \\ &= \int \mathbf{A}\cdot\nabla\times\mathbf{B}d\mathbf{r} + \oint (\mathbf{A}\times\mathbf{B})\cdot d\mathbf{S} \\ &= \int \mathbf{A}\cdot\nabla\times\mathbf{B}d\mathbf{r}, \qquad (4.18)\end{aligned}$$

where the surface-integral in the second equality appears as a result of Gauss' divergence theorem. The surface over which this integral is evaluated extends beyond the support of $\mathbf{B}(\mathbf{r})$.

Similarly, we have a transformation of a gradient operator to a divergence operator:

$$\int \mathbf{B} \cdot \nabla V d\mathbf{r} = \int (\nabla \cdot \mathbf{B} V) d\mathbf{r} - \int V \nabla \cdot \mathbf{B} d\mathbf{r}$$
$$= -\int V \nabla \cdot \mathbf{B} d\mathbf{r}.$$

Chapter 5
Computing Network Immittance Functions from Field Calculations

VIC-3D® computes impedances from field calculations by using the *reaction* concept. We will follow Harrington [47, pp. 116–120], in developing this concept.

Harrington defines reaction as

$$[a,b] = \int\int\int \mathbf{E}(a) \cdot \mathbf{J}^{(b)} dV \ , \tag{5.1}$$

where $\mathbf{E}^{(a)}$ is the field due to source a and $\mathbf{J}^{(b)}$ is source b. The reciprocity theorem states that

$$[a,b] = [b,a] \ ; \tag{5.2}$$

that is, the reaction of field a on source b is equal to the reaction of field b on source a.

Harrington also shows that the driving-point and transfer impedances of a linear network can be defined in terms of reactions by

$$z_{ij} = -\frac{[j,i]}{I_i I_j} = -\frac{[i,j]}{I_i I_j} \ . \tag{5.3}$$

This is a way of relating circuit-theoretic ideas, such as z_{ij}, I_i, I_j, with field-theoretic concepts, such as fields, distributed currents, and reaction.

In order to gain further insight into the relationship between field-theoretic and circuit models for eddy-current probes, we employ inductively coupled-circuit theory.

H.A. Sabbagh et al., *Computational Electromagnetics and Model-Based Inversion*, Scientific Computation, DOI 10.1007/978-1-4419-8429-6_5,

5.1 The Classical Bistatic Arrangement

In Fig. 5.1 we show the classical bistatic arrangement. This arrangement subsumes the usual "driver-pickup" and "remote-field" configurations. The transmitter in Fig. 5.1 is excited, and the receiver drives an infinite-impedance load. Hence, the equivalent circuit is as shown in the lower part of Fig. 5.1.

The circuit equations are

$$\begin{aligned} V_1 &= j\omega L_1 I_1 + j\omega M_{13} I_3 \\ 0 &= j\omega M_{13} I_1 + (R_3 + j\omega L_3) I_3 \, . \end{aligned} \tag{5.4}$$

I_1 is the actual current flowing in the exciting (transmitter) coil and I_3 is the distributed anomalous current due to the flaw. I_1 is a circuit current and I_3 a distributed current.

From the second equation in (5.4) we have

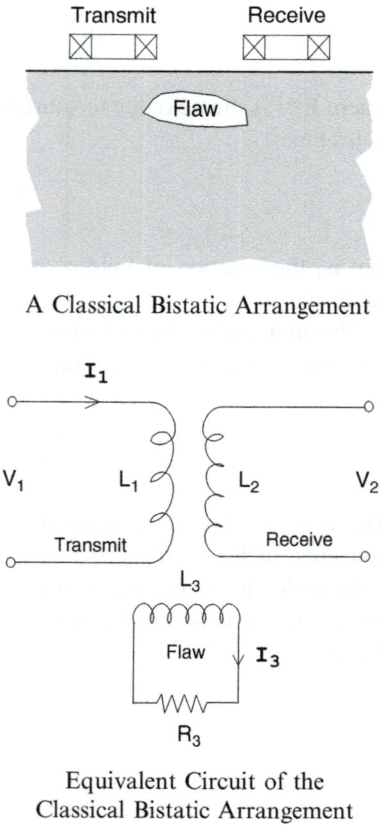

Fig. 5.1 A classical bistatic sensor arrangement, *upper*, and its equivalent circuit, *lower*

$$I_3 = -\frac{j\omega M_{13}}{R_3 + j\omega L_3} I_1 \,. \tag{5.5}$$

Thus, the distributed current is due solely to the exciting current, I_1.

The output voltage, V_0, is given by

$$\begin{aligned} V_0 &= j\omega M_{12} I_1 + j\omega M_{23} I_3 \\ &= j\omega M_{12} I_1 - \frac{[2,3]}{I_2} \,, \end{aligned} \tag{5.6}$$

where we have once again used (5.3) in writing the second term.

The transfer impedance, Z_{01}, is defined to be

$$\begin{aligned} Z_{01} &= \frac{V_0}{I_1} \\ &= j\omega M_{12} - \frac{[2,3]}{I_1 I_2} \\ &= z_{12} - \frac{[2,3]}{I_1 I_2} \,. \end{aligned} \tag{5.7}$$

Now, z_{12} represents the *direct coupling* between the transmitter and receiver; as such it does not take into account the presence of the flaw, which is represented by the distributed current, I_3, but, rather, represents the background against which the flaw must be detected. In remote-field inspection, the transmitter and receiver are sufficiently far apart (remote) that $z_{12} \approx 0$; that is, the direct coupling is practically zero.

We can look at this another way, as well. Because z_{12} has nothing to do with the flaw, we can subtract its effect initially, even if the bistatic arrangement is not remote-field. Hence, we get for the impedance change that is due solely to the flaw:

$$\Delta Z = -\frac{[2,3]}{I_1 I_2} \,, \tag{5.8}$$

where

$$[2,3] = \int\!\!\int\!\!\int_{\text{flaw}} \mathbf{E}^{(2)} \cdot \mathbf{J}^{(3)} dV \,. \tag{5.9}$$

Note that I_1 is the current that actually drives the transmitter, but I_2 is an apparent circuit current that drives the receiver, when the receiver is treated as a transmitter, as well. (It is a feature of reciprocity that it transforms transmitters into receivers and receivers into transmitters.)

Because I_1 and I_2 are circuit currents, they are scalars that can be set to unity. If we normalize I_1 and I_2 to unity, then $\mathbf{E}^{(2)}$ in (5.9) is the incident field within the flaw, due to one ampere in the receiver coil, when the receiver acts like a transmitter. Note that $\mathbf{J}^{(3)}$ is independent of $\mathbf{E}^{(2)}$, because there is only one source for $\mathbf{J}^{(3)}$, namely the one-ampere current in the true transmitter, coil 1.

VIC-3D® computes the reaction [2,3] by means of a scalar product of two discrete vectors. This can be seen by substituting the current expansion, (4.4), into (5.9):

$$
\begin{aligned}
[2,3] &= \iiint_{\text{flaw}} \mathbf{E}^{(2)} \cdot \mathbf{J}^{(3)} dV \\
&= \sum_{KLM} \Big[J_{KLM}^{(3)(x)} \iiint_{\text{flaw}} E^{(2)(x)}(\mathbf{r}) T^{(x)(e)}(\mathbf{r}) dV \\
&\quad + J_{KLM}^{(3)(y)} \iiint_{\text{flaw}} E^{(2)(y)}(\mathbf{r}) T^{(y)(e)}(\mathbf{r}) dV \\
&\quad + J_{KLM}^{(3)(z)} \iiint_{\text{flaw}} E^{(2)(z)}(\mathbf{r}) T^{(z)(e)}(\mathbf{r}) dV \Big] \\
&= \sum_{KLM} \Big[J_{KLM}^{(3)(x)} E_{KLM}^{(2)(x)} + J_{KLM}^{(3)(y)} E_{KLM}^{(2)(y)} + J_{KLM}^{(3)(z)} E_{KLM}^{(2)(z)} \Big] ,
\end{aligned}
\tag{5.10}
$$

where $\mathbf{E}_{KLM}^{(2)}$ is the *KLM*th moment of $\mathbf{E}^{(2)}$. This expansion is extremely important and will appear throughout this text.

5.2 The Differential Probe

The equivalent circuit for a differential-probe system is shown in Fig. 5.2. The exciting coil is driven by the ac source, while the two sense coils are connected

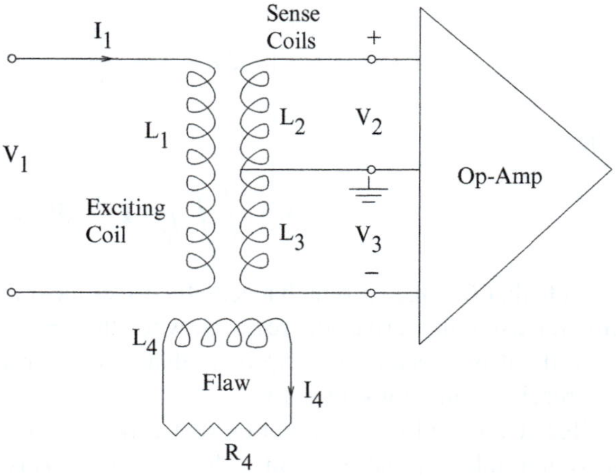

Fig. 5.2 Equivalent circuit for the differential probe

to an infinite-impedance operational amplifier. This means that neither sense coil carries a current. The system equations are

$$\begin{aligned} V_1 &= j\omega L_1 I_1 + j\omega M_{14} I_4 \\ 0 &= j\omega M_{14} I_1 + (R_4 + j\omega L_4) I_4 \ , \end{aligned} \tag{5.11}$$

where I_1 is the actual current flowing in the exciting coil and I_4 represents the effects of the distributed anomalous electric current that flows within the flaw. As before, I_1 is a circuit-current and I_4 is a distributed current. The second equation in (5.11) shows that the distributed current is due only to the exciting current:

$$I_4 = -\frac{j\omega M_{14}}{R_4 + j\omega L_4} I_1 \ . \tag{5.12}$$

The output voltage, $V_0 = V_2 - V_3$, is given by

$$V_0 = V_2 - V_3 = j\omega(M_{12} - M_{13}) I_1 + j\omega(M_{24} - M_{34}) I_4 \ . \tag{5.13}$$

Now, because of the symmetrical placement of the two sense-coils with respect to the exciting coil, we have $M_{12} = M_{13}$. This holds regardless of the presence of the flaw, because the flaw effects are included in the second term in (5.13). Hence, (5.13) becomes

$$\begin{aligned} V_0 &= j\omega(M_{24} - M_{34}) I_4 \\ &= \frac{[3,4]}{I_3} - \frac{[2,4]}{I_2} \ . \end{aligned} \tag{5.14}$$

We call the ratio, $V_0/I_1 = Z_{01}$, the open-circuit transfer impedance of the linear network consisting of the coupled circuits of Fig. 5.2. In terms of the reactions, then, we have

$$Z_{01} = \frac{[3,4]}{I_1 I_3} - \frac{[2,4]}{I_1 I_2} \ . \tag{5.15}$$

I_1 is the actual current in the exciting-coil, whereas I_2 and I_3 are fictitious currents in the two sense-coils, when these coils are treated as transmitters when applying reciprocity theory. If $I_1 = I_2 = I_3 = 1$, then (5.15) is expressed solely in terms of reactions

$$Z_{01} = ([3,4] - [2,4]) \ . \tag{5.16}$$

Each of these reaction terms is interpreted as in the classical bistatic arrangement discussed earlier.

5.3 Impedance of Ferrite-Core Probes

We calculate the driving-point impedance first. This is the impedance seen at the terminals of the probe in the absence of the flaw; that is, $I_4 = 0$ in (5.11). This impedance consists of two parts, the first being the contribution of the coil and the second the contribution of the ferrite core. We will compute each of these contributions by using the reaction principle.

The reaction of field $\mathbf{E}^{(1)}$ on source $\mathbf{J}_e^{(1)}$ is

$$[1,1] = \int \mathbf{J}_e^{(1)} \cdot \mathbf{E}^{(1)} dV \, . \tag{5.17}$$

The source with superscript 1 is the primary source due to the exciting coil. If I_c is the current in the exciting coil, then the driving-point impedance (or self-impedance) seen by the coil is

$$Z = -\frac{[1,1]}{I_c^2} \, . \tag{5.18}$$

If we normalize the excitation to be $I_c = 1$, then

$$\begin{aligned} Z &= -\int \mathbf{J}_e \cdot \mathbf{E}^{(i)} dV \\ &= -\int (\mathbf{J}_c + \nabla \times \mathbf{M}_C) \cdot \mathbf{E}^{(i)} dV \\ &= -\int \mathbf{J}_c \cdot \mathbf{E}^{(i)} dV - \int \mathbf{M}_C \cdot \nabla \times \mathbf{E}^{(i)} dV \\ &= -\int \mathbf{J}_c \cdot \mathbf{E}^{(i)} dV + \int \mathbf{M}_C \cdot j\omega \mathbf{B}^{(i)} dV \, , \end{aligned} \tag{5.19}$$

where we have replaced the superscript 1 by (i) to denote incident fields. The transference of the curl operator in going from the second to the third equation is valid for $\mathbf{M}$ with finite support (see Chap. 4, Appendix A.1). $\mathbf{J}_c$ is the current density in the coil and $\mathbf{M}_C$ is the magnetization of the core.

Upon substituting the expansions for the magnetic solution vectors, (4.5), into (5.19), we get

$$\begin{aligned} Z = &-\int \mathbf{J}_c \cdot \mathbf{E}^{(i)} \\ &+ j\omega \sum_{KLM} \left(M_{KLM}^{(x)} B_{KLM}^{(ix)} + M_{KLM}^{(y)} B_{KLM}^{(iy)} + M_{KLM}^{(z)} B_{KLM}^{(iz)} \right), \end{aligned} \tag{5.20}$$

where $\mathbf{B}_{KLM}^{(i)}$ is the KLMth moment of $\mathbf{B}^{(i)}$. The first term is the contribution of the coil to the driving-point impedance and is computed in Appendix A.1; the second is the contribution of the core. The scalar product in this term is reminiscent of (5.10).

5.4 Computation of Impedance Changes due to the Presence of a Flaw

Here, the crucial thing that must be computed is the reaction between the anomalous current produced by the flaw and the incident field of a coil. In the case of the differential ferrite-core probe, we want to compute [2,4] and [3,4] of (5.16).

The reaction of field, $\mathbf{E}^{(2)}$, on source, $\mathbf{J}_e^{(1)}$, is

$$[1,2] = \int \mathbf{J}_e^{(1)} \cdot \mathbf{E}^{(2)} dV \, . \tag{5.21}$$

The source with superscript 1 is the primary source due to the exciting coil and superscript 2 denotes scattered fields (and their sources) due to the flaw. If I_c is the current in the exciting coil, then the change in impedance due to the flaw, as seen by the coil is

$$\Delta Z = -\frac{[1,2]}{I_c^2} = -\frac{[2,1]}{I_c^2} \, , \tag{5.22}$$

where we have used the reciprocity theorem. If we normalize the excitation to be $I_c = 1$, then

$$\begin{aligned} \Delta Z &= -\int \mathbf{J}_e \cdot \mathbf{E}^{(i)} dV \\ &= -\int (\mathbf{J} + \nabla \times \mathbf{M}) \cdot \mathbf{E}^{(i)} dV \\ &= -\int \mathbf{J} \cdot \mathbf{E}^{(i)} dV - \int \mathbf{M} \cdot \nabla \times \mathbf{E}^{(i)} dV \\ &= -\int \mathbf{J} \cdot \mathbf{E}^{(i)} dV + \int \mathbf{M} \cdot j\omega \mathbf{B}^{(i)} dV \, , \end{aligned} \tag{5.23}$$

where we have dropped the superscript 2 and replaced the superscript 1 by (i) to denote incident fields. The transference of the curl operator in going from the second to the third equation is valid for $\mathbf{M}$ with finite support (see Chap. 4, Appendix A.1).

Upon substituting the expansion for the magnetic solution vectors, (4.5), and the corresponding one for the electric current, (4.4), into (5.23), we get

$$\begin{aligned} \Delta Z = -\sum_{KLM} \Big[& J_{KLM}^{(x)} E_{KLM}^{(ix)} + J_{KLM}^{(y)} E_{KLM}^{(iy)} + J_{KLM}^{(z)} E_{KLM}^{(iz)} \\ & - j\omega \left(M_{KLM}^{(x)} B_{KLM}^{(ix)} + M_{KLM}^{(y)} B_{KLM}^{(iy)} + M_{KLM}^{(z)} B_{KLM}^{(iz)} \right) \Big] \, . \end{aligned} \tag{5.24}$$

This is a sum of scalar products of the electric current and magnetic polarization solution vectors with the incident electric field and magnetic flux-density moment vectors.

Appendix

A.1 Calculation of Circular Coil Impedance

The incident field due to a circular coil lying in region 0 is given by

$$\mathbf{E}_0^{(i)}(r,z) = -j\omega\mu_0 2\pi \mathbf{a}_\phi \left[\int_{\text{coil}} \mathbf{G}_{00}(r,z;r',z')\cdot\mathbf{J}_c(r',z')r'dr'dz'\right], \tag{5.25}$$

where

$$\mathbf{G}_{00}(r,z;r',z') = \mathbf{a}_\phi \frac{1}{2\pi}\int_0^\infty \frac{e^{-\lambda_0|z-z'|}+R_0 e^{-\lambda_0(z+z')}}{2\lambda_0} J_1(rl)J_1(r'l)l\,dl\,. \tag{5.26}$$

R_0 is the reflection coefficient from the top surface of the workpiece and is calculated using the ideas presented in Chap. 2.

We assume that the coil has a rectangular cross-section and carries a current, $I_c = 1A$, that is uniformly distributed over this cross-section with N_c turns per square meter ($\mathbf{J}_c = N_c\mathbf{a}_\phi$). The integrals with respect to (r',z') over the coil are evaluated first:

$$\int_{\text{coil}} \mathbf{G}_{00}(r,z;r',z')\cdot\mathbf{J}_c(r',z')r'dr'dz' =$$

$$\frac{N_c}{2\pi}\int_0^\infty \frac{\mathscr{I}(r_1 l, r_2 l)}{l}\left[\frac{F_2(z_1,z_2,z)+R_0 e^{-\lambda_0 z}F_1(z_1,z_2)}{2\lambda_0}\right] J_1(rl)dl\,, \tag{5.27}$$

where (r_1,r_2) are the inner and outer radii of the coil and (z_1,z_2) the bottom and top coordinates of the coil.

Then

$$\begin{aligned}
Z_c &= -\int \mathbf{J}_c\cdot\mathbf{E}^{(i)}dV \\
&= -2\pi N_c \int\int E^{(i)}(r,z)r\,dr\,dz \\
&= j\omega\mu_0(2\pi)^2 N_c \int_{\text{coil}} r\,dr\,dz \int_{\text{coil}} \mathbf{G}_{00}(r,z;r',z')\cdot\mathbf{J}_c(r',z')r'dr'dz' \\
&= j\omega\mu_0 2\pi N_c^2 \int_{\text{coil}} r\,dr\,dz \left\{\int_0^\infty \frac{\mathscr{I}(r_1 l, r_2 l)}{l}\right. \\
&\quad \times \left.\left[\frac{F_2(z_1,z_2,z)+R_0 e^{-\lambda_0 z}F_1(z_1,z_2)}{2\lambda_0}\right] J_1(rl)dl\right\} \\
&= j\omega\mu_0 2\pi N_c^2 \int_0^\infty \frac{\mathscr{I}^2(r_1 l, r_2 l)}{l^3}\left[\frac{F_3(z_1,z_2;z_1,z_2)+R_0 F_1^2(z_1,z_2)}{2\lambda_0}\right] dl\,.
\end{aligned} \tag{5.28}$$

The functions $F_1(z_1,z_2)$, $F_2(z_1,z_2,z)$, $F_3(z_1,z_2;z_1,z_2)$ and $\mathscr{I}(r_1 l, r_2 l)$ are defined by [27]

$$
\begin{aligned}
F_1(a,b) &= \int_a^b e^{-\lambda_0 z} dz \\
&= \frac{e^{-\lambda_0 a} - e^{-\lambda_0 b}}{\lambda_0} \\
F_2(a,b,c) &= \int_a^b e^{-\lambda_0 |c-z|} dz \\
&= \frac{e^{-\lambda_0 (a-c)} - e^{-\lambda_0 (b-c)}}{\lambda_0}, \quad c \leq a \\
&= \frac{2 - e^{-\lambda_0 (c-a)} - e^{-\lambda_0 (b-c)}}{\lambda_0}, \quad a \leq c \leq b \\
&= \frac{e^{-\lambda_0 (c-b)} - e^{-\lambda_0 (c-a)}}{\lambda_0}, \quad b \leq c \\
F_3(z_1,z_2,z_i^{(-)},z_i^{(+)}) &= \int_{z_1}^{z_2} dz' \int_{z_i^{(-)}}^{z_i^{(+)}} dz e^{-\lambda_0 |z-z'|} \\
&= I_1 + I_2 + I_3 + I_4 \\
I_1 &= -2x_1/\lambda_0 - e^{-\lambda_0 x_1}/\lambda_0^2, x_1 > 0, x_1 = z_i^{(+)} - z_2 \\
&= -e^{\lambda_0 x_1}/\lambda_0^2, x_1 < 0 \\
I_2 &= -2x_2/\lambda_0 - e^{-\lambda_0 x_2}/\lambda_0^2, x_2 > 0, x_2 = z_i^{(-)} - z_1 \\
&= -e^{\lambda_0 x_2}/\lambda_0^2, x_2 < 0 \\
I_3 &= 2x_3/\lambda_0 + e^{-\lambda_0 x_3}/\lambda_0^2, x_3 > 0, x_3 = z_i^{(-)} - z_2 \\
&= e^{\lambda_0 x_3}/\lambda_0^2, x_3 < 0, \\
I_4 &= 2x_4/\lambda_0 + e^{-\lambda_0 x_4}/\lambda_0^2, x_4 > 0, x_4 = z_i^{(+)} - z_1 \\
&= e^{\lambda_0 x_4}/\lambda_0^2, x_4 < 0 \\
\mathscr{I}(a,b) &= \int_a^b z J_1(z) dz \,, \qquad (5.29)
\end{aligned}
$$

and $J_1(r)$ is the Bessel function of the first kind, order 1.

A.2 A Coupled-Circuit Model of the Volume-Integral Equation

Maxwell's second equation defines the three current systems that we are interested in:

$$\begin{aligned}\nabla \times \mathbf{H} &= j\omega\mathbf{D} + \mathbf{J}_c + \sigma(\mathbf{r})\mathbf{E} \\ &= j\omega\mathbf{D} + \mathbf{J}_c + \sigma_h\mathbf{E} + (\sigma(\mathbf{r}) - \sigma_h)\mathbf{E} \\ &= j\omega\mathbf{D} + \mathbf{J}_c + \mathbf{J}_h + \mathbf{J}_a \,,\end{aligned} \tag{5.30}$$

where $j\omega\mathbf{D}$ is the displacement current, $\mathbf{J}_c$ the coil current, $\mathbf{J}_h$ the host current, and $\mathbf{J}_a$ the anomalous current due to the anomalous conductivity, $\sigma(\mathbf{r}) - \sigma_h$.

Setting aside consideration of the displacement current, the three remaining current systems are going to be modeled by the three coupled circuits of Fig. 5.3 in order to get a simpler interpretation of the volume-integral equation. $M_{ch} = M_{hc}$ is the mutual inductance between the coil and host, $M_{ca} = M_{ac}$ is the mutual inductance between the coil and anomaly, and $M_{ha} = M_{ah}$ is the mutual inductance between the host and anomaly.

The loop equations (Kirchoff's Voltage Law) for this system are

$$\begin{aligned}V_0 &= I_c(R_c + j\omega L_c) + j\omega M_{ch}I_h + j\omega M_{ca}I_a \\ 0 &= j\omega M_{ch}I_c + (R_h + j\omega L_h)I_h + j\omega M_{ha}I_a \\ 0 &= j\omega M_{ca}I_c + j\omega M_{ha}I_h + (R_a + j\omega L_a)I_a \,.\end{aligned} \tag{5.31}$$

The circuit elements, R and L, should not be taken literally. They represent such things as energy loss and storage, which occur in both lumped circuits (such as this one) and distributed fields.

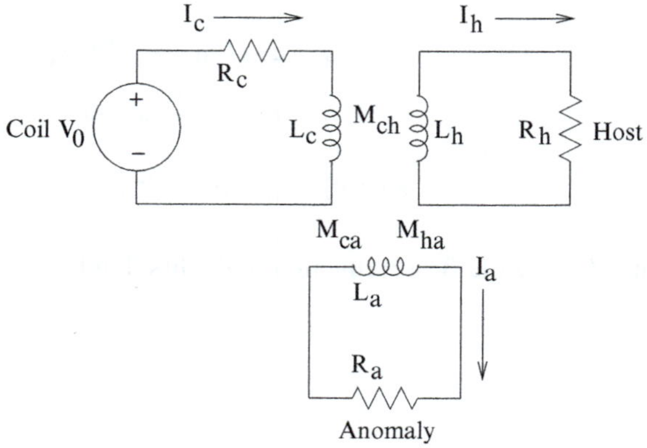

Fig. 5.3 A coupled-circuit model showing the three current systems, I_c, I_h, and I_a

If we eliminate I_h from the bottom two equations in (5.31), we get an equation between the coil current and the anomalous current:

$$0 = j\omega M_{ca} I_c + \frac{\omega^2 M_{ch} M_{ha} I_c}{R_h + j\omega L_h} + \frac{\omega^2 M_{ha}^2 I_a}{R_h + j\omega L_h} + (R_a + j\omega L_a) I_a \,, \tag{5.32}$$

which can be rewritten as

$$-j\omega M_{ca} I_c - \frac{\omega^2 M_{ch} M_{ha} I_c}{R_h + j\omega L_h} = (R_a + j\omega L_a) I_a + \frac{\omega^2 M_{ha}^2 I_a}{R_h + j\omega L_h}$$

or

$$\mathbf{E}_{\mathrm{inc}} = \frac{\mathbf{J}_a}{\sigma_a} + \mathbf{E}(\mathbf{r})\left[\mathbf{J}_a\right] \,. \tag{5.33}$$

From this equation it is clear that the incident field and the "integral operator functional" depend upon the host parameters, as we knew. If the host is freespace, we are tempted to let $R_h \to \infty$, which would make the host conduction current, I_h, vanish. We must replace R_h in this case by a capacitor to simulate the host displacement current, thereby maintaining the presence of the integral operator functional in the last equation of (5.33). Equation (5.33) illustrates the role played by the mutual inductances in modeling "action at a distance" terms, such as the incident field and integral operator. We also see that if $R_a \to \infty$ (which corresponds to $\sigma_a \to 0$), then the anomalous current also vanishes, $I_a \to 0$. The second of (5.31) also makes clear that the host current, I_h, depends upon both the coil current and the anomalous current, I_a. The host current is not simply due to I_c alone, unless the anomaly is not excited.

We can go further and compute the driving-point impedance of the coil. It is a straightforward procedure to eliminate I_h and I_a in favor of I_c in the bottom two equations of (5.31) and then substitute this result into the first equation to get a relation between the driving voltage, V_0, and the loaded coil current, I_c. From this we obtain the driving-point impedance as

$$\begin{aligned} Z_{\mathrm{in}} = \frac{V_0}{I_c} &= R_c + j\omega L_c + \frac{\omega^2 M_{ch}^2 (R_a + j\omega L_a) - j\omega^3 M_{ch} M_{ha} M_{ca}}{(R_h + j\omega L_h)(R_a + j\omega L_a) + \omega^2 M_{ha}^2} \\ &+ \frac{\omega^2 M_{ca}^2 (R_h + j\omega L_h) - j\omega^3 M_{ch} M_{ha} M_{ca}}{(R_h + j\omega L_h)(R_a + j\omega L_a) + \omega^2 M_{ha}^2} \\ &= R_c + j\omega L_c + \frac{\omega^2 M_{ch}^2}{R_h + j\omega L_h} \\ &+ \frac{\omega^2 M_{ca}^2 (R_h + j\omega L_h)^2 - j2\omega^3 M_{ch} M_{ca} M_{ha} (R_h + j\omega L_h) - \omega^4 M_{ch}^2 M_{ha}^2}{(R_h + j\omega L_h)\left[(R_h + j\omega L_h)(R_a + j\omega L_a) + \omega^2 M_{ha}^2\right]} \,. \end{aligned} \tag{5.34}$$

The first two terms in (5.34) are the free-space coil driving-point impedance and the third is the additional term due to the host in the absence of an anomaly (the "unflawed host"). This term vanishes when the coil is well removed from the workpiece, so that $M_{ch} \to 0$.

The final term is the change caused by the presence of an anomaly. It vanishes under two distinct conditions: $R_a \to \infty$, which, as we have seen before, corresponds to the absence of the flaw, and M_{ca}, $M_{ha} \to 0$. This condition corresponds to the vanishing of $\mathbf{E}_{\text{inc}}$ and the integral operator in (5.33), thereby forcing the anomalous current, I_a, to vanish. This occurs when the coil is well removed from the anomaly.

Let's take another look at (5.34). Each of the last two terms can be interpreted as an "$\mathbf{E}_{\text{inc}} \cdot \mathbf{J}$" with an appropriate $\mathbf{E}_{\text{inc}}$ and $\mathbf{J}$. If $I_a = 0$, then the solution of the second equation in (5.31) for I_h is $I_h = -j\omega M_{ch} I_c/(R_h + j\omega L_h)$, and when this is multiplied by $E_{\text{inc}} = -j\omega M_{ch} I_c$ we get the second term in (5.34), after negation and setting $I_c = 1$. Hence, we interpret I_h as being the anomalous or "scattering" current when $I_a = 0$, and $-j\omega M_{ch} I_c$ as the incident field *acting in free space*.

The more usual case in the hierarchy is $I_a \neq 0$, which means that (5.33) applies. If we solve this equation for I_a, and then multiply that result by the left-hand side, we get the third term in (5.34), after negation and setting $I_c = 1$. Again, we have consistency with the theory of calculating changes in impedance, but now E_{inc} must account for the presence of the host current, I_h, as well as the coil current, I_c. The environment of each of the scattering currents is different, and that determines how E_{inc} is to be computed. Note that E_{inc} in (5.33) contains two terms, the first being the direct contribution of the coil acting in free-space and the second the effect of the coil acting through the host current, I_h.

A.2.1 Frequency-Response Loci

We can now derive some very useful results for analyzing and interpreting model results based on the frequency response of the measured impedance. Return to (5.34) and set $M_{ca} = M_{ha} = 0$, which corresponds to the unflawed host. Throughout this discussion we will work with $Z' = Z_{\text{in}} - R_c$, which is done because R_c has nothing to do with magnetic coupling of the coil to the host. Furthermore, we will work with $Z'/\omega L_c$, which is the impedance normalized to the free-space reactance of the coil. It is, of course, a dimensionless quantity.

We have, after rationalizing the third (coupling) term in (5.34),

$$\begin{aligned}\frac{Z'}{\omega L_c} &= \frac{\omega M_{ch}^2 R_h}{L_c(R_h^2 + \omega^2 L_h^2)} + j\left(1 - \frac{\omega^2 M_{ch}^2 L_h/L_c}{R_h^2 + \omega^2 L_h^2}\right) \\ &= \frac{\omega k^2 L_h R_h}{R_h^2 + \omega^2 L_h^2} + j\left(\frac{R_h^2 + \omega^2 L_h^2 - \omega^2 k^2 L_h^2}{R_h^2 + \omega^2 L_h^2}\right), \qquad (5.35)\end{aligned}$$

where we have introduced the coupling-coefficient, $k^2 = M_{ch}/L_c L_h$ [108, page 399].

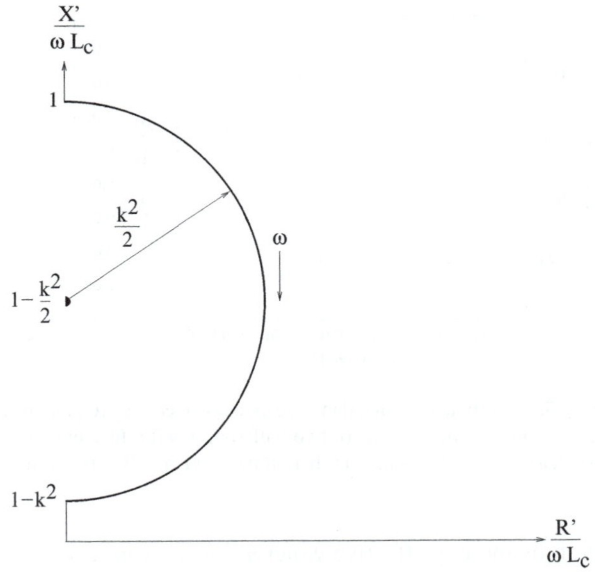

Fig. 5.4 Illustrating the semicircle frequency locus of the normalized impedance when the circuit elements, R_h, L_h, and k are frequency independent. This is the "ideal" Förster plot

It will be advantageous to shift the origin of the coordinate system for the normalized impedance to $j(1-k^2/2)$:

$$\begin{aligned}\frac{Z'}{\omega L_c} - j\left(1-\frac{k^2}{2}\right) &= \frac{\omega k^2 L_h R_h}{R_h^2+\omega^2 L_h^2} + j\left[\frac{R_h^2+\omega^2 L_h^2 - \omega^2 k^2 L_h^2}{R_h^2+\omega^2 L_h^2} - \left(1-\frac{k^2}{2}\right)\right] \\ &= \frac{k^2}{2}\left[\frac{2\omega L_h R_h}{R_h^2+\omega^2 L_h^2} + j\frac{R_h^2-\omega^2 L_h^2}{R_h^2+\omega^2 L_h^2}\right], \end{aligned} \tag{5.36}$$

and from this it is easily shown that

$$\left|\frac{Z'}{\omega L_c} - j\left(1-\frac{k^2}{2}\right)\right|^2 = \frac{k^4}{4}\,. \tag{5.37}$$

This is an extremely important result and is the basis for interpreting both model and experimental results. If the circuit parameters, R_h, L_h, and k are all frequency independent, as in an "ideal" lumped circuit, then the locus of impedance vs. frequency is a semicircle, the "ideal" Förster plot shown in Fig. 5.4. The diameter of this semicircle directly measures the coupling coefficient, k^2. The terminal points of the diameter correspond to $\omega = 0$ and $\omega \to \infty$, where $R'/\omega L_c$ vanishes.

Of course, field calculations of eddy-current phenomena do not lead to "ideal" lumped circuit elements. For example, we know that at high frequencies $R_h \sim \sqrt{\omega}$ because of the skin effect.[1] This means that $R'/\omega L_c$ in (5.35) dies out more slowly

[1] See [109, pp. 286–303] for a thorough discussion of skin effect on circuit elements.

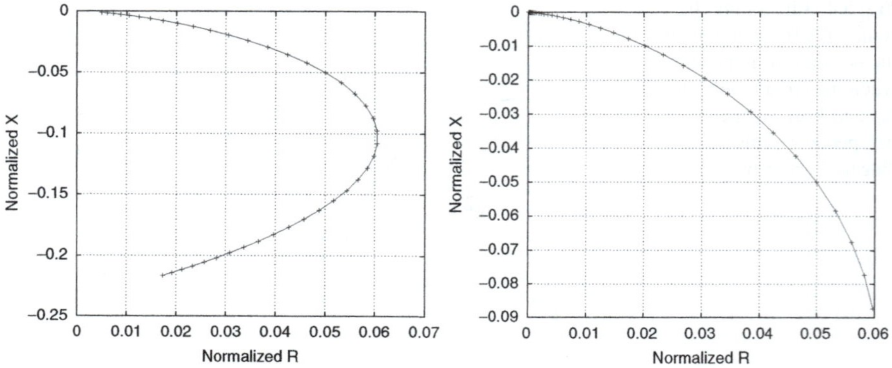

Fig. 5.5 Normalized impedance responses over the frequency range of 100 Hz to 1 MHz, for a given coil and lift-off, but for two half-spaces with different conductivities. The half-space on the left has $\sigma = 3 \times 10^7$ S/m, and that on the right $\sigma = 3 \times 10^5$ S/m

with ω, thereby effectively increasing k^2, where we now interpret this parameter to be the distance between the terminal points of the locus on the $X'/\omega L_c$ axis, as in the ideal case, even though this locus is no longer semicircular. We can anticipate the increase in k^2 due to skin effect on physical grounds; at higher frequencies, the current in the host material is forced closer and closer to the upper boundary (the "skin"), thereby increasing the coupling between it and the coil current.

Figure 5.5 shows a model calculation of the normalized impedance function for a coil over a half-space whose conductivity is $\sigma = 3 \times 10^7$ S/m. The frequency range is 100 Hz to 1 MHz, with the locus of normalized impedance going clockwise. We clearly see that the normalized impedance response is strongly elongated in the reactance component. We can take the chord length along this direction from $\omega = 0$ to $\omega \to \infty$ and conclude that the "effective" $k^2 \approx 0.23$.

When we redo the calculation with the same probe over the same frequency range, but with a half-space whose conductivity is $\sigma = 3 \times 10^5$, we get the result shown in the right-hand side of Fig. 5.5. What is interesting about these two plots is that the loci of normalized impedances lie on the same curve. S.N. Vernon [110] was the first to observe that, given a probe, data collected from materials of very different resistivities fall on the same curve that she called a "universal impedance diagram." She did this empirically; Bowler et al. [111] confirmed this result using model calculations based on the precursor of **VIC-3D®** [25].

It is important to realize, however, that despite the fact that the impedance curve loci are identical over the entire frequency range of $0 < f < \infty$, the responses for each frequency are considerably different. Compare, for example, that at 1 MHz, the left-hand curve has virtually closed on the imaginary axis, whereas the right-hand curve has barely reached the point of maximum normalized R'.

Chapter 6
Advanced Probe Models Based on Magnetic Dipoles and Ferrite Cores

6.1 Modeling Nonstandard Probes

By a "nonstandard probe," we mean one that is not the usual air-core circular coil with a square cross-section, whose axis is normal to a plane surface, as in the pancake coil of Appendix A.1 of Chap. 5, or with its axis coinciding with the axis of a tube, as in the bobbin coils of Chap. 9. In this chapter, we develop a theory that allows us to efficiently discretize the current density in such probes, thereby allowing us to use transfer matrices (these will be discussed in Chap. 10) and similar mathematical constructs for solving problems.

In particular, this chapter deals with probes that are characterized by their magnetic-dipole moments, which means that all equivalent electric currents are solenoidal (divergence-free). These models allow us to calculate inductive effects, only, which, of course, is a main concern in eddy-current modeling. In the next chapter, we analyze probes that are driven by electric dipoles, which allows us to calculate capacitive effects as well as inductive effects. This allows the probe models to be used over a greater frequency range.

6.2 The Incident Field Due to a Ferrite-Core Probe

We only need to compute the incident electric field at a flaw, because the incident magnetic field can be expressed in terms of the electric field, as we know.

The Green function that we have worked with previously was that for which the source and field points were in the same layer of the workpiece. Now, upon referring to Fig. 2.1 of Chap. 2, we need the Green function, $\mathbf{G}^{(\pm q0)}$, in which the source point is in region 0, as before, but the field point is in the qth layer above region 0, or the qth layer below the region. The first situation carries the label $+q0$ and the second $-q0$.

H.A. Sabbagh et al., *Computational Electromagnetics and Model-Based Inversion*, Scientific Computation, DOI 10.1007/978-1-4419-8429-6_6,

We compute the Green functions using the same four-vector algebraic approach that was used in Chap. 2, and then the resulting volume-integral differential equation is derived exactly as is explained in Chap. 3. Because the Amperian currents, which are of magnetic origin, are actually equivalent electric currents [because they appear in the second of Maxwell's equations (Ampere's law)], we only need to compute the electric–electric dyadic Green functions. The resulting differentio-integral equations are

$$\begin{aligned}
E_x(\mathbf{r}) &= \int G^{(q0)(1)}(x-x',y-y';z,z')J_x(\mathbf{r}')d\mathbf{r}' \\
&\quad +\frac{\partial}{\partial x}\int G^{(q0)(2)}(x-x',y-y';z,z')\nabla_t'\cdot\mathbf{J}_t(\mathbf{r}')d\mathbf{r}' \\
&\quad +\frac{1}{k_0^2}\frac{\partial}{\partial x}\int G^{(q0)(3)}(x-x',y-y';z,z')\nabla'\cdot\mathbf{J}(\mathbf{r}')d\mathbf{r}' \\
E_y(\mathbf{r}) &= \int G^{(q0)(1)}(x-x',y-y';z,z')J_y(\mathbf{r}')d\mathbf{r}' \\
&\quad +\frac{\partial}{\partial y}\int G^{(q0)(2)}(x-x',y-y';z,z')\nabla_t'\cdot\mathbf{J}_t(\mathbf{r}')d\mathbf{r}' \\
&\quad +\frac{1}{k_0^2}\frac{\partial}{\partial y}\int G^{(q0)(3)}(x-x',y-y';z,z')\nabla'\cdot\mathbf{J}(\mathbf{r}')d\mathbf{r}' \\
E_z(\mathbf{r}) &= \int G^{(q0)(4)}(x-x',y-y';z,z')J_z(\mathbf{r}')d\mathbf{r}' \\
&\quad +\frac{1}{k_0^2}\frac{\partial}{\partial z}\int G^{(q0)(5)}(x-x',y-y';z,z')\nabla'\cdot\mathbf{J}(\mathbf{r}')d\mathbf{r}' \,, \qquad (6.1)
\end{aligned}$$

which holds for $\mathbf{r}$ in region q. If $\mathbf{r}$ is in region $-q$, then replace $(q0)$ by $(-q0)$.

The result of (6.1) is quite general and could be used, for example, in finding the interaction between a flaw in region 0 and one in region $\pm q$, as in the Spatial Decomposition Algorithm of Chap. 10. The Amperian current produced by the core of such a probe has zero-divergence, as does the current in the exciting coil of the probe. Hence, the last term in each of the equations in (6.1) vanishes. The electric currents that flow in the various coils of the probe lie in the transverse, (x,y), plane, because of the orientation of the coils. This means that $\nabla_t'\cdot\mathbf{J}(\mathbf{r}')=0$ for these currents.

Upon calling the coil current, $\mathbf{J}^{(c)}(\mathbf{r})$, and the current of magnetic origin (the Amperian current) $\mathbf{J}^{(m)}(\mathbf{r})$, then (6.1) becomes

$$\begin{aligned}
E_x(\mathbf{r}) &= \int G^{(q0)(1)}(x-x',y-y';z,z')J_x^{(c)}(\mathbf{r}')d\mathbf{r}' \\
&\quad +\int G^{(q0)(1)}(x-x',y-y';z,z')J_x^{(m)}(\mathbf{r}')d\mathbf{r}' \\
&\quad +\frac{\partial}{\partial x}\int G^{(q0)(2)}(x-x',y-y';z,z')\nabla_t'\cdot\mathbf{J}_t^{(m)}(\mathbf{r}')d\mathbf{r}'
\end{aligned}$$

$$E_y(\mathbf{r}) = \int G^{(q0)(1)}(x-x',y-y';z,z')J_y^{(c)}(\mathbf{r}')d\mathbf{r}'$$
$$+\int G^{(q0)(1)}(x-x',y-y';z,z')J_y^{(m)}(\mathbf{r}')d\mathbf{r}'$$
$$+\frac{\partial}{\partial y}\int G^{(q0)(2)}(x-x',y-y';z,z')\nabla_t'\cdot\mathbf{J}_t^{(m)}(\mathbf{r}')d\mathbf{r}'$$
$$E_z(\mathbf{r}) = \int G^{(q0)(4)}(x-x',y-y';z,z')J_z^{(m)}(\mathbf{r}')d\mathbf{r}' . \tag{6.2}$$

The kernels in (6.1) and (6.2) are computed using the same four-vector approach of Chap. 2, but because of their complexity will not be shown here.

Equation (6.2) is still quite rigorous, within the context of the nature of the currents, $\mathbf{J}^{(c,m)}(\mathbf{r})$. We shall see shortly that is possible for common eddy-current problems to eliminate the $E_z(\mathbf{r})$ term, also. This occurs when the coil region is air, and the field point is in a highly-conducting region, which is the usual combination in eddy-current problems. The contributions of the coil current in the x- and y-equation of (6.2) are precisely the classical terms produced by circular coils of rectangular cross-section that **VIC-3D®** already computes. Of course, if the coils are noncircular, but, say, are rectangular in the (x,y)-plane, then we must evaluate the contributions in the same manner as in the case of the magnetic currents. See [92] for a benchmark validation of this model.

6.3 A Solenoidal Current Model

At the typical frequencies of interest in eddy-current NDE, currents that flow in probes can be assumed to be divergenceless: $\nabla\cdot\mathbf{J}_c = 0$, which implies that we can write $\mathbf{J}_c = \nabla\times\mathbf{M}$, with the magnetization vector, $\mathbf{M}$, expanded in edge elements, as in (4.5) and (4.6) of Chap. 4. We cannot determine $\mathbf{M}$ uniquely from the curl equation without specifying another constraint condition. Helmholtz' theorem [48] uses knowledge of $\nabla\cdot\mathbf{M}$ as this additional condition, but we will simply set $M_x = 0$. This will limit slightly our flexibility in defining arbitrary current sources but will suffice for our purposes.

Thus, we rewrite expansion (4.8) for the current density in the following form:

$$\mathbf{J}_c(x,y,z) = \sum_{KLM} M_{KLM}^{(z)}\left[T_{KLM,y}^{(z)(m)}(\mathbf{r})\mathbf{a}_x - T_{KLM,x}^{(z)(m)}(\mathbf{r})\mathbf{a}_y\right]$$
$$+\sum_{KLM} M_{KLM}^{(y)}\left[T_{KLM,x}^{(y)(m)}(\mathbf{r})\mathbf{a}_z - T_{KLM,z}^{(y)(m)}(\mathbf{r})\mathbf{a}_x\right], \tag{6.3}$$

where $T_{KLM,q}^{(p)(m)}$ denotes the qth spatial derivative of the pth-oriented edge element. The parameters, δ_x, δ_y, δ_z, refer to the source grid, not the flaw grid.

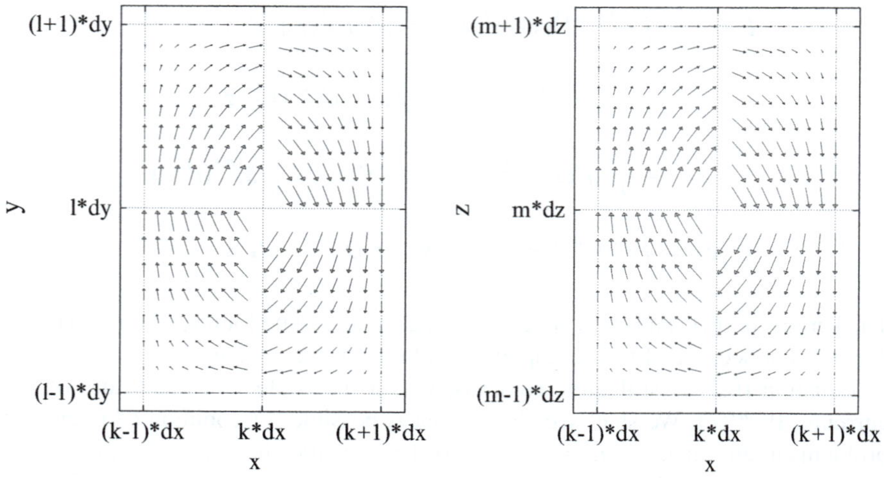

Fig. 6.1 Circulations. *Left*: $T^{(z)(m)}_{klm,x}(\mathbf{r})\mathbf{a}_y - T^{(z)(m)}_{klm,y}(\mathbf{r})\mathbf{a}_x$, in the (x,y)-plane, about the edge joining the points (x_k, y_l, z_{m-1}) and (x_k, y_l, z_m); *right*: $T^{(y)(m)}_{klm,x}(\mathbf{r})\mathbf{a}_z - T^{(y)(m)}_{klm,z}(\mathbf{r})\mathbf{a}_x$, in the (x,z)-plane, about the edge joining the points (x_k, y_{l-1}, z_m) and (x_k, y_l, z_m)

The physical interpretation of the basis vectors is: $T^{(z)(m)}_{klm,x}(\mathbf{r})\mathbf{a}_y - T^{(z)(m)}_{klm,y}(\mathbf{r})\mathbf{a}_x$ is a circulation in the x–y plane, about the edge joining the points (x_k, y_l, z_{m-1}) and (x_k, y_l, z_m). $T^{(y)(m)}_{klm,x}(\mathbf{r})\mathbf{a}_z - T^{(y)(m)}_{klm,z}(\mathbf{r})\mathbf{a}_x$ is a circulation in the x–z plane, about the edge joining the points (x_k, y_{l-1}, z_m) and (x_k, y_l, z_m). The circulation is confined to four adjacent cells sharing the edge on the axis of circulation. These circulations are plotted in Fig. 6.1.

The solution of (6.3) for the expansion coefficients is given by

$$M^{(y)}_{klm} = \frac{1}{\delta y}\int_{y_{l-1}}^{y_l}\int_{x_0}^{x_k} J^{(z)}_c(x,y,z_m)dxdy$$

$$M^{(z)}_{klm} = -\frac{1}{\delta_z}\int_{z_{m-1}}^{z_m}\int_{x_0}^{x_k} J^{(y)}_c(x,y_l,z)dxdz\,, \tag{6.4}$$

where x_0 lies outside the source region. Now that we have $\mathbf{M}_{KLM} = \left[0,\, M^{(y)}_{KLM},\, M^{(z)}_{KLM}\right]$, we can use the transfer matrices of (10.4) to calculate incident fields on flaws.

6.4 Response of a Rectangular Coil When Rotated Relative to a Crack

We will apply the theory of nonstandard probes to model an experimental benchmark test, thereby validating the modeling algorithm that was developed in the last section. R.J. Ditchburn and S.K. Burke, writing in NDT&E International 38 (2005), pp. 690–700, show that the maximum response of an infinite crack to a rectangular coil is maximum when the long axis of the coil is parallel to the crack. They conjecture that the opposite orientation dependence holds for small cracks. Here, we demonstrate that their conjecture holds for the case of the coil shown in Fig. 6.2, and the half-penny model of the fatigue crack shown in Figs. 6.3 and 6.4. The host material is Inconel 718, which is nonmagnetic.

In Fig. 6.5, we show the driving-point impedance of the coil as it is rotated through ninety degrees while it remains centered over the crack. The zero angle corresponds to the case in which the x-axis of the coil coincides with the x-axis of the crack.

6.4.1 *The Anomalous (Scattering) Currents*

In order to get a good physical understanding of the response of eddy-current probes, it is useful to compute and plot the currents within the anomaly, as well as fields produced by those currents. We will consider here only the anomalous currents.

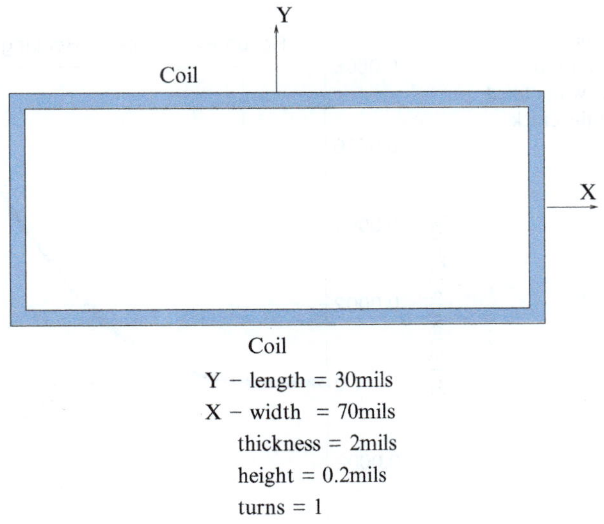

Fig. 6.2 A rectangular coil

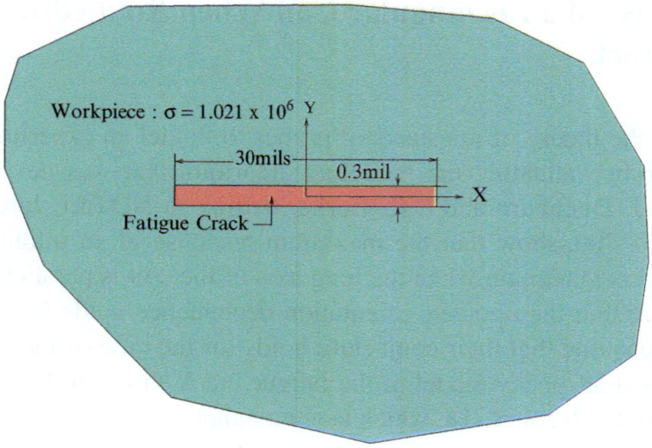

Fig. 6.3 Top view of a half-penny fatigue crack

Fig. 6.4 Side view of the same half-penny fatigue crack

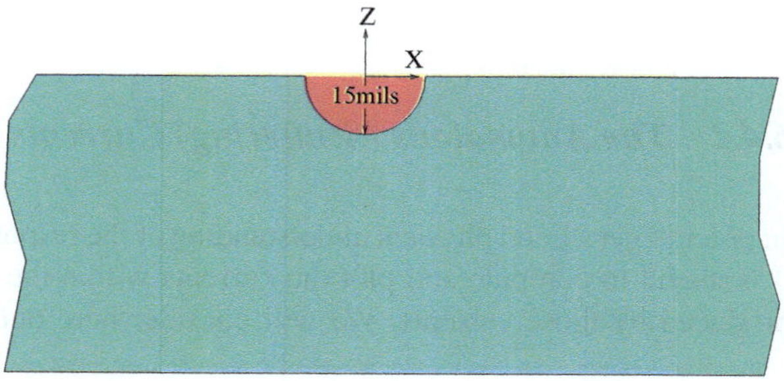

Fig. 6.5 Driving-point impedance of the transmit coil as it is rotates while fixed over the center of the crack

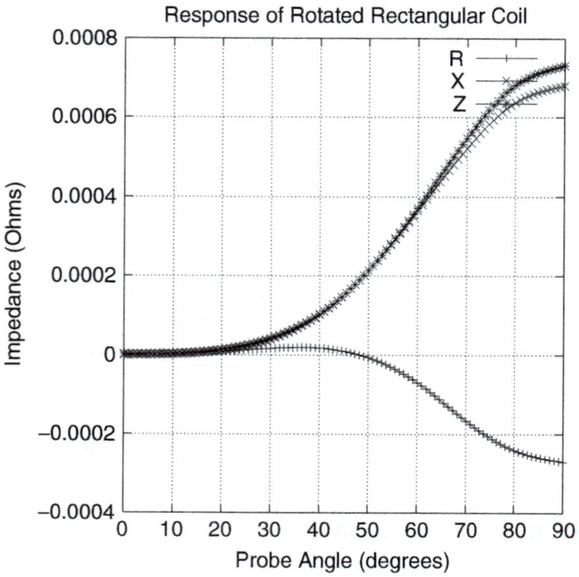

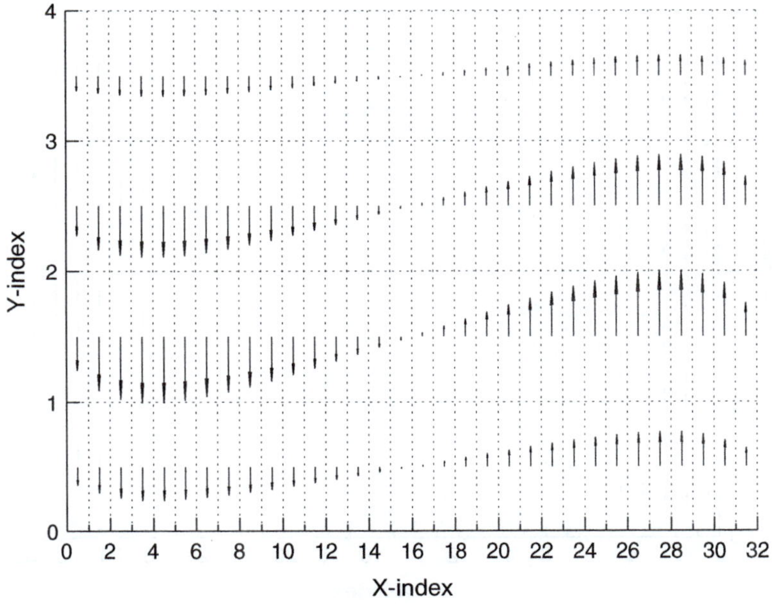

Fig. 6.6 Projections onto the *xy*-plane of the currents in the top layer of the anomaly: real component at 0° multiplied by 2.0×10^{-9}

Figures 6.6–6.9 illustrate the projections onto the *xy*-plane of the anomalous (scattering) currents in the top layer of the crack. The coil is oriented at either 0° or 90° relative to the crack, as explained above. In order to facilitate the interpretation of these figures, we have converted J_x into units of cell width and J_y into units of cell length. By doing this, a current that flows toward the corner of a cell will be plotted as a vector that points toward the corner of the cell. It is clear, now, why the response at 90° is much larger than at 0°; the anomalous currents are an order of magnitude larger with the 90° orientation than with the 0°. Furthermore, it should be clear that the currents within the very narrow crack are non-solenoidal; i.e., they have a nonzero divergence. In extreme cases of very thin, very long cracks this can cause problems with the convergence of the conjugate-gradient solver unless special preconditioners are used, and the development of such preconditioners for volume-integral equations is an active research area. Finally, we note that the currents are predominately oriented in the *y*-direction, as one might expect for a long, thin crack with the shape shown in Fig. 6.3.

In order to eliminate the distortion due to the large cell aspect-ratio, we plot the current in a single cell so that the plotted cell has its true aspect ratio. Figures 6.10–6.13 are the result.

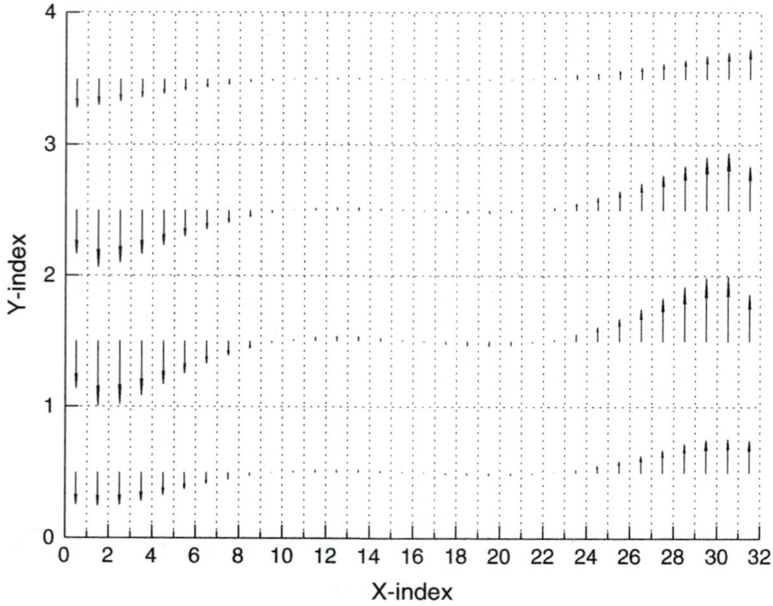

Fig. 6.7 Projections onto the *xy*-plane of the currents in the top layer of the anomaly: imaginary component at 0° multiplied by 2.0×10^{-8}

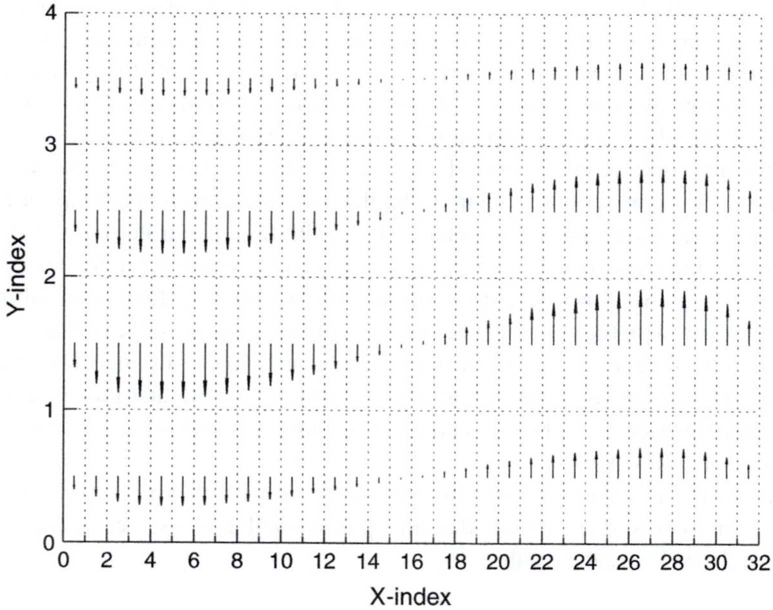

Fig. 6.8 Projections onto the *xy*-plane of the currents in the top layer of the anomaly: real component at 90° multiplied by 2.0×10^{-10}

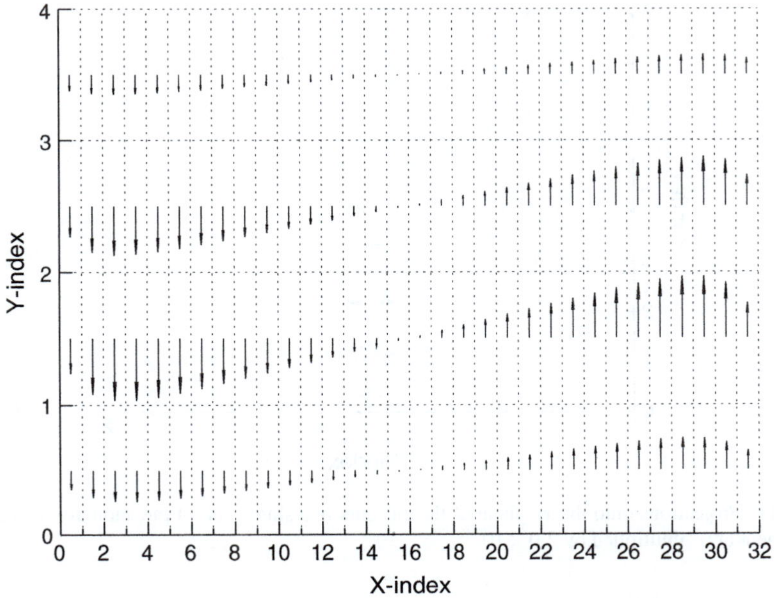

Fig. 6.9 Projections onto the *xy*-plane of the currents in the top layer of the anomaly: imaginary component at 90° multiplied by 1.0×10^{-10}

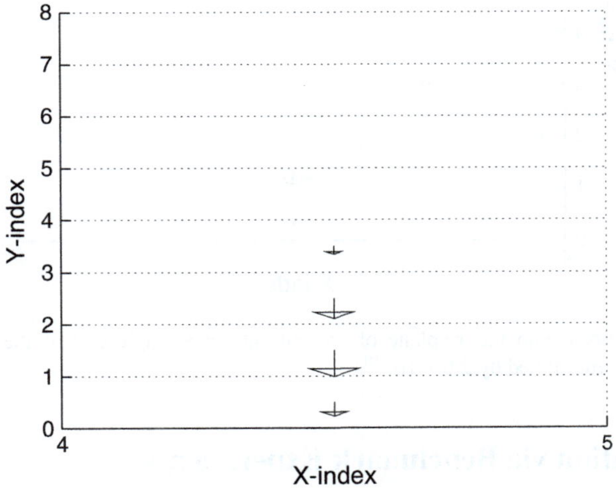

Fig. 6.10 Projections onto the *xy*-plane of the currents in a single cell of the anomaly: real component at 0° multiplied by 2.0×10^{-9}

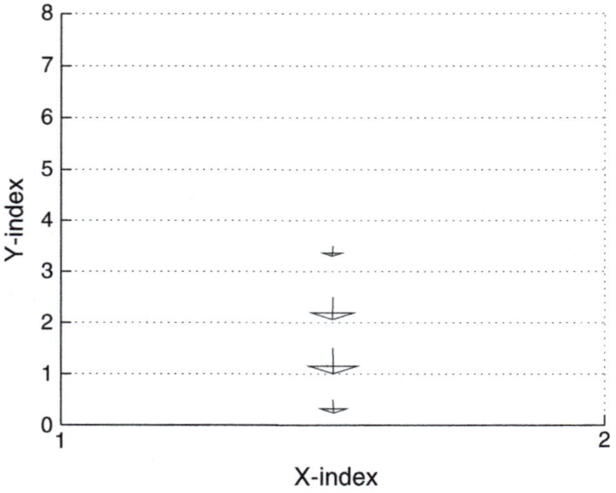

Fig. 6.11 Projections onto the *xy*-plane of the currents in a single cell of the anomaly: imaginary component at 0° multiplied by 2.0×10^{-8}

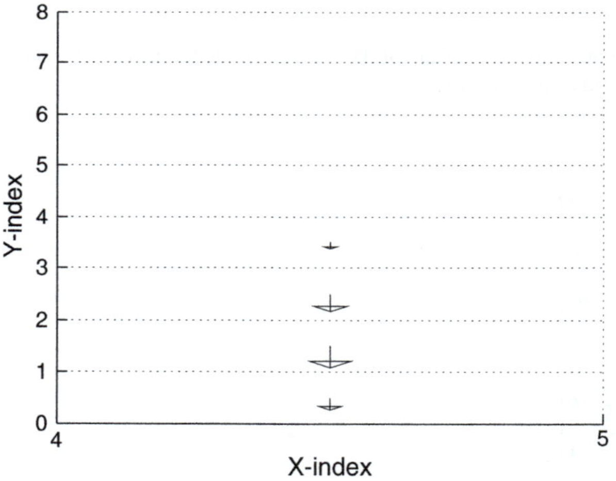

Fig. 6.12 Projections onto the *xy*-plane of the currents in a single cell of the anomaly: real component at 90° multiplied by 2.0×10^{-10}

6.5 Validation via Benchmark Experiments

We will use the theory of this chapter to model several benchmark experiments that are described in the Ditchburn–Burke (D–B) paper referred to above. In particular, we will show model results for several experiments that involve the 50-turn rectangular spiral coil whose aspect ratio is 2:1. We use the same dimensions as in the paper: width = 18.52 mm, length = 36.68 mm, thickness = 8.71 mm,

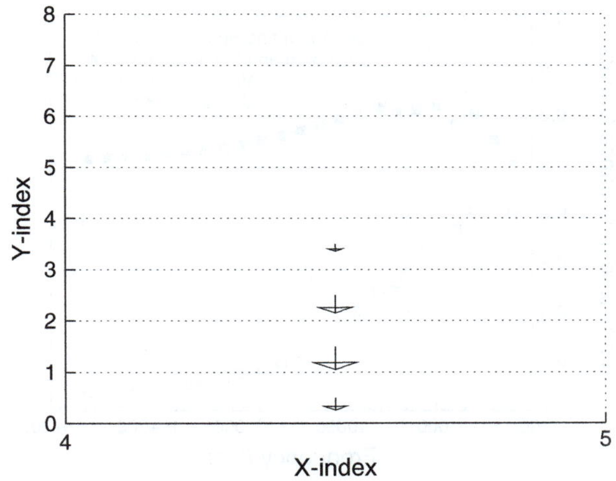

Fig. 6.13 Projections onto the xy-plane of the currents in a single cell of the anomaly: imaginary component at 90° multiplied by 1.0×10^{-10}

height = 0.01 mm. This value of height is the same as the thickness of the individual tracks of copper forming each spiral turn, $\approx 10\,\mu$m. With these values, we calculate the freespace inductance of the coil to be $L_0 = 48.33\,\mu$H, whereas the measured value is $50.25\,\mu$H, the error being less than 4 %.

Next, we model experimental results for the coil over defect-free aluminum and steel plates. The conductivity of aluminum is 1.653×10^7 S/m, and for steel, 3.472×10^6 S/m. The relative magnetic permeability of the steel plate is 85. In both cases, we used the same "fitted lift-off" in the paper: for aluminum, 0.12 mm, and for steel, 0.05 mm.

The results for the aluminum plate are shown in Fig. 6.14 and for steel in Fig. 6.15.

6.6 Modeling a Differential-Receive Ferrite-Core Probe

The differential-receive ferrite-core probe, often called a "split-core differential" or "split-D" probe, is used to detect flaws in bolt-holes in aircraft structures or jet engine disks. We demonstrate a model result for a split-core differential probe interacting with a flaw in an aluminum substrate. Figure 6.16 shows a model split-D probe. Note the presence of the "split" ferrite cores and D-shaped receive coils, which give the probe its name. The response produced by this probe is the transfer impedance between the transmit (or driver) coil and the series-connected receive coils.

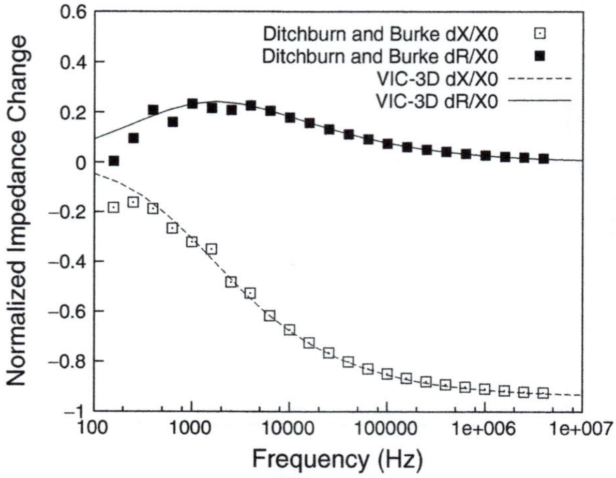

Fig. 6.14 **VIC-3D®** model results for the D–B experiment of the coil over a defect-free host of aluminum

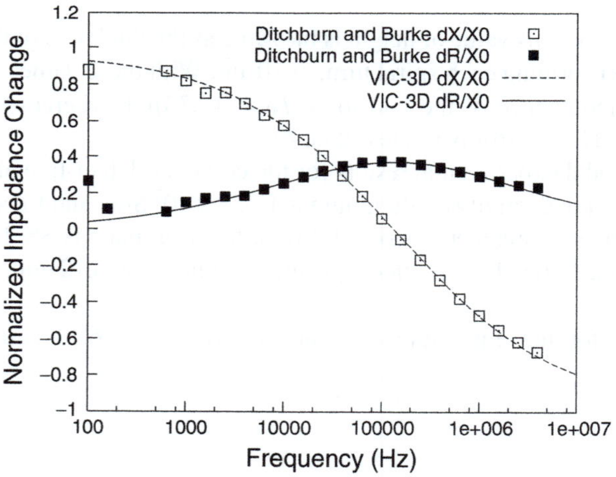

Fig. 6.15 **VIC-3D®** model results for the D–B experiment of the coil over a defect-free host of steel

6.6.1 Sizing Surface Cracks: Experimental Test Cases

We'll apply a model of the split-D probe to the problem of characterizing surface cracks within bolt holes in a benchmark test case. Figure 6.17 illustrates the test cases, and Fig. 6.18 shows a photomicrograph of an actual through-wall fatigue crack of the type that we are going to size. The bottom part of Fig. 6.18 shows the EDM notch that is used for calibrating the data. The eddy-current instrument that

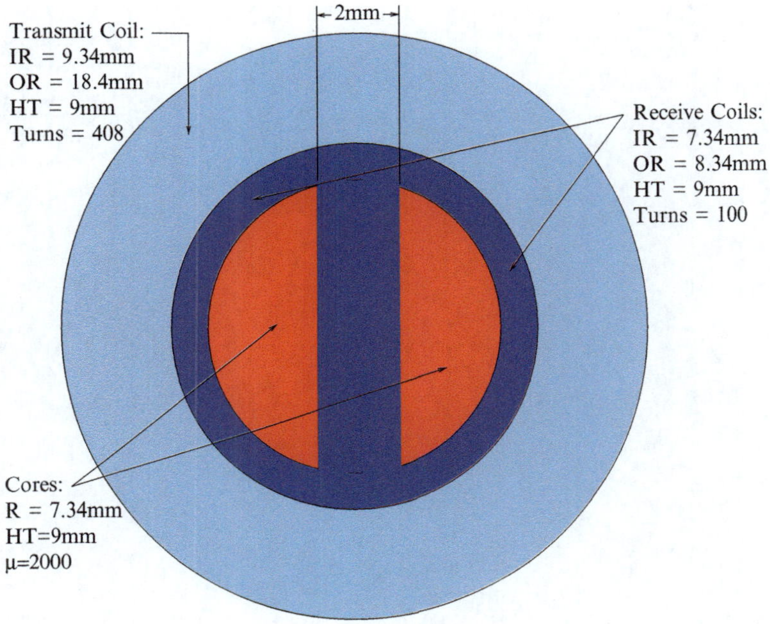

Fig. 6.16 The split-D coil configuration

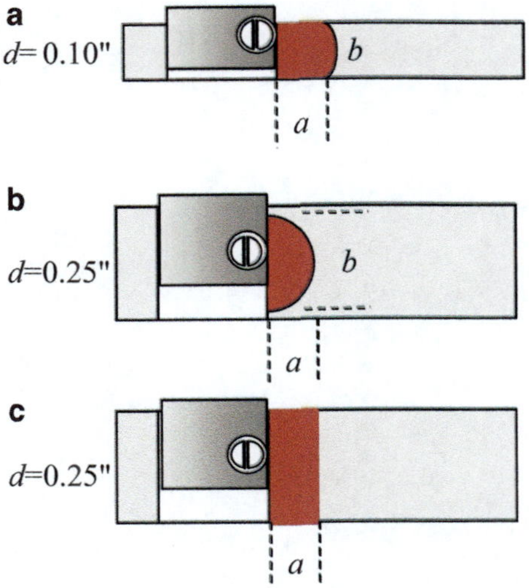

Fig. 6.17 Illustrating experimental test cases for sizing surface fatigue cracks using a split-D probe within a bolt hole in an aluminum substrate. The probe (small white circle) is embedded in a mandrel, which is then rotated azimuthally about a vertical axis that is not shown. Simultaneously, the mandrel is lifted vertically, thereby generating a two-dimensional raster scan. The top figure (**a**) illustrates the cross-section of a through-wall crack in a 0.100″-thick host, the middle figure (**b**) illustrates a mid-bore crack in a 0.250″-thick host, and the bottom figure (**c**) illustrates the EDM notch in a 0.250″-thick host that is used to calibrate the test setup (see Fig. 6.18)

Fig. 6.18 Illustrating a photomicrograph of an actual fatigue crack extending from the surface of a bolt hole within an aluminum substrate (*top*), together with an EDM notch that is used for calibrating the entire test setup (*bottom*). The "nominal" width of the crack is 0.007 mm (0.28 mils) and the width of the notch is 6.1 mils

is used in the tests measures "instrument volts," which are then scaled to impedances for further processing by **VIC-3D®** using the β-transformation scheme described in Sect. 15.2.

6.6.2 An Interpolation Algorithm and Inverse Problem

In order to reduce the burden of computing the response of very thin, long cracks, we will resort to a "surrogate" interpolation algorithm. The theory behind the algorithm is based on the following formula for the change in impedance of a probe due to scattering from an anomaly, as determined by fundamental electromagnetic theory (recall Chap. 5):

$$Z_a = \int\int\int_{V_a} \mathbf{E}_0 \cdot \mathbf{J}_a dV \;, \tag{6.5}$$

where we assume that the probe is excited with a current of one ampere. $\mathbf{E}_0$ is the incident field due to the probe in the absence of the anomaly, $\mathbf{J}_a$ is the anomalous (scattering) current, and V_a is the volume of the anomaly. $\mathbf{E}_0$ and $\mathbf{J}_a$ are both bounded, with bounded derivatives, so that the integral in (6.5) vanishes to polynomial order in V_a as $V_a \to 0$. This certainly includes the case where the cross-section of the anomaly remains fixed and the width, $W \to 0$.

Thus, we can derive an interpolation scheme in W that takes advantage of this fact. Assuming that W is "small enough," we can postulate the second-order approximation, $Z_a = aW + bW^2$. If we compute values of Z_a at $W = 0.125$ and 0.25 mil, calling them $Z_{0.125}$ and $Z_{0.25}$, respectively, then it is easy to show that $a = 16Z_{0.125} - 4Z_{0.25}$, $b = -64Z_{0.125} + 32Z_{0.25}$, which yields the final result

$$Z_a = (16W - 64W^2)Z_{0.125} - (4W - 32W^2)Z_{0.25} \;. \tag{6.6}$$

NLSE uses higher-order spline interpolations to perform similar surrogate duties when solving inverse problems.[1] What is novel in our next example is to use interpolation to solve the forward problem, using the value of W that NLSE gives us.

We will solve an inverse problem to reconstruct the width of an 0.018″-deep through-wall crack. We assume the length to be 0.100″, as in Fig. 6.17a, and we will use a block model of the crack. In order to simplify the calculations, however, we will assume that the crack lies in a half-space, as in Figs. 6.3 and 6.4. Figure 6.19 shows the blending functions, $Z_{0.125}$ and $Z_{0.25}$, for the interpolating grid with nodes at $W = 0$, 0.125, 0.250 mil. This allows a quadratic interpolation in NLSE. The scaled input data for this crack are shown in Fig. 6.20.[2]

---

[1]This will be fully explained in Chap. 12.

[2]These data were supplied by David Forsyth and Mark Keiser of Texas Research Institute Austin.

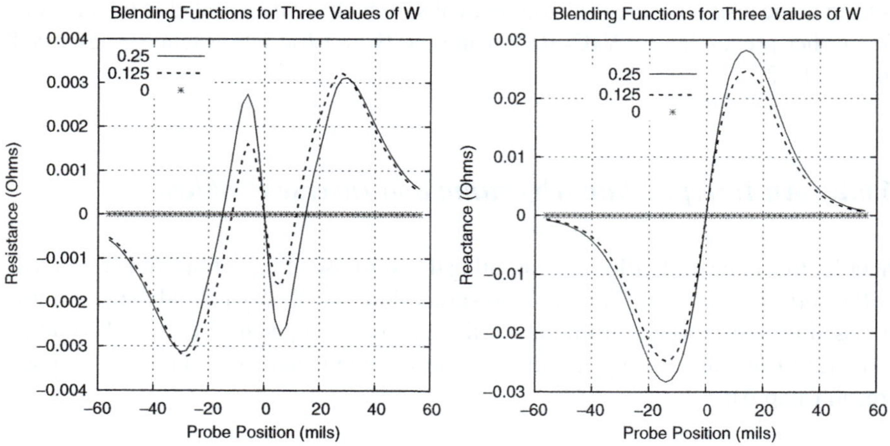

Fig. 6.19 Blending functions for the interpolation table with nodes at $W = 0$, 0.125 and 0.25 mil. The functions associated with the latter two nodes are the $Z_{0.125}$ and $Z_{0.25}$ of (6.6), respectively

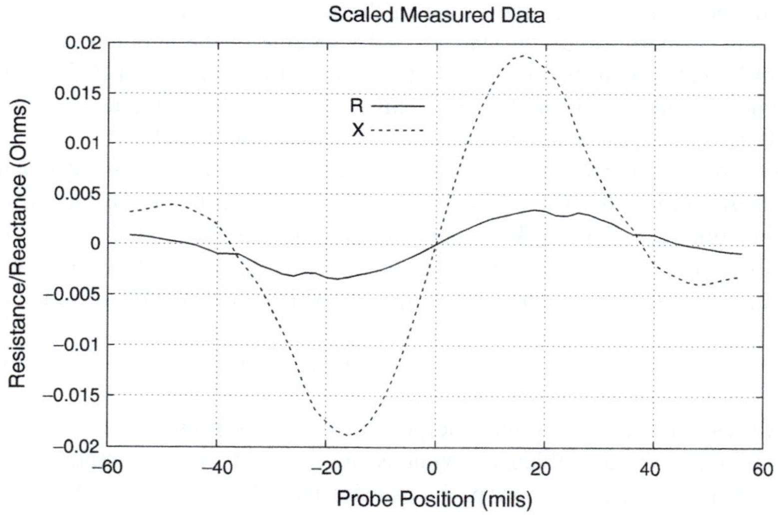

Fig. 6.20 Showing the scaled measured data taken by the split-D probe. These data will be submitted to NLSE for inversion to determine W

The result of the inversion is $W = 0.0793$, and the model results that are obtained when using the quadratic interpolation formula derived in (6.6) with $W = 0.0793$ are shown in Fig. 6.21, along with the scaled input data. Clearly, there is a good fit in reactance. The model resistance shows a higher spatial frequency content (more wiggles) than the data, but this is probably due to uncertainties in measuring the resistance. In addition, there remains a slight "shoulder" in the scaled reactance data that the model does not reproduce, and this may be due in part to the measurements,

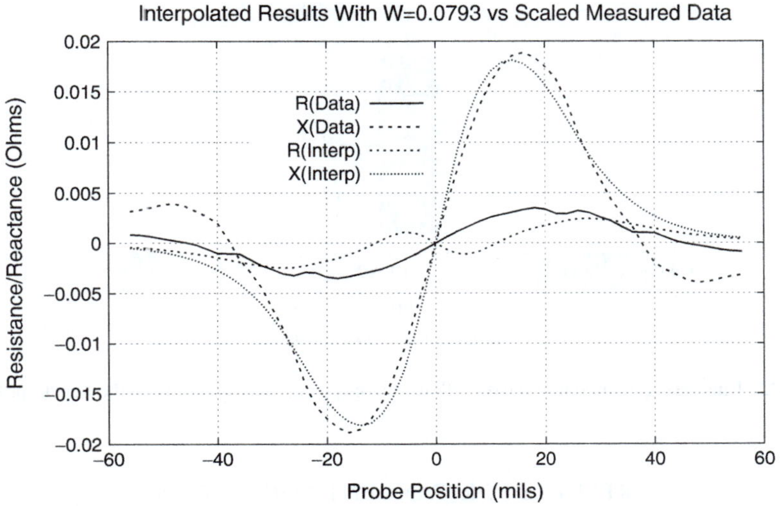

Fig. 6.21 Comparing the model result with $W = 0.0793$ mil and the scaled data input to NLSE

as well as to an imperfect model that does not include the effects of the finite edges of the workpiece.

The ability to detect and size a fatigue crack whose width is of the order of 0.05–0.5 mil is enhanced through the use of a split-D probe, rather than a single absolute coil. The split-D probe is essentially a mathematical differentiator, and taking the "derivative" of the response past a crack will give a signal that is sensitive to the width of the crack. More complex applications of surrogate models in computational electromagnetics are given in [9].

6.7 Validation Test: Tangent Coil Over a Crack in a Thin Plate

One of the important new features in modeling probes is the ability to rotate coils in a variety of directions, allowing us, for example, to model circular tangent coils. Figure 6.22 illustrates a tangent coil over a flawed workpiece, comprising a through-wall crack in a thin plate. Burke and Rose have used this arrangement to produce benchmark data which we have used to validate the rotation feature (see [49]).

Figure 6.23 shows a comparison of the benchmark data and a model calculation, when the tangent coil is scanned over the slot at 2 kHz, and Fig. 6.24 compares the benchmark data and a calculation of a frequency scan of the tangent coil, when it is located over the center of the slot. The frequency range extends from 251 Hz to 10 kHz.

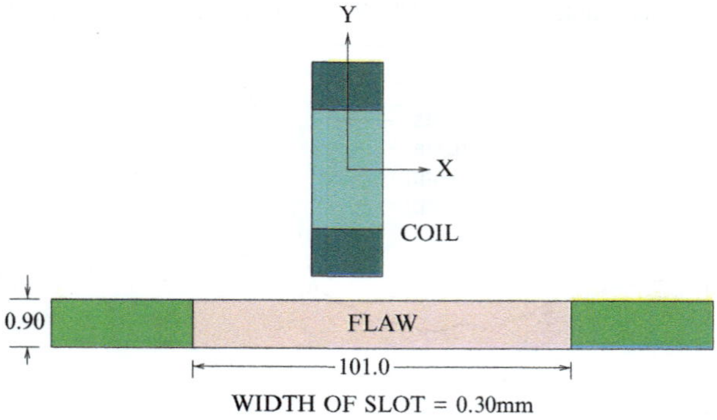

Fig. 6.22 Illustrating a tangent coil over a flawed workpiece, corresponding to the benchmark test of [49]

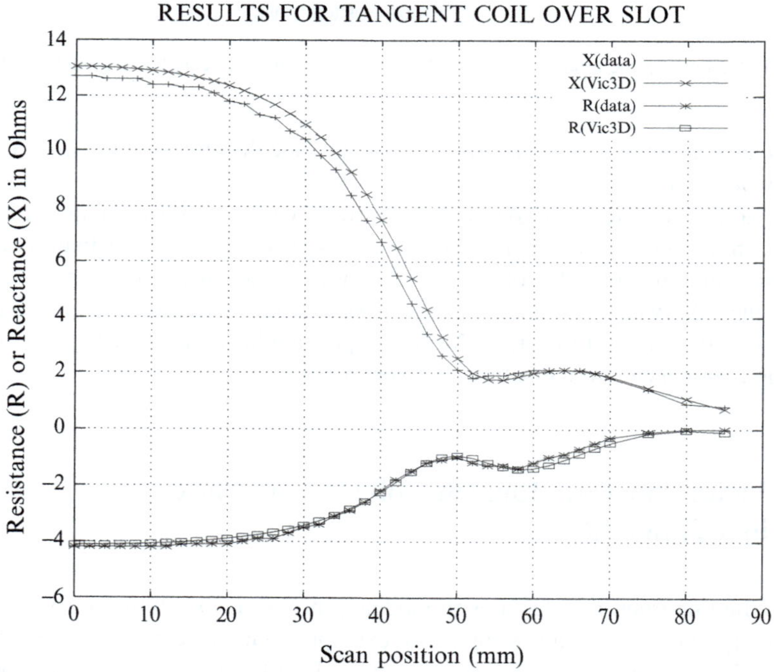

Fig. 6.23 Comparison of calculations with the position-scan benchmark data of [49]

Clearly, the magnetic-dipole model that we have developed in this chapter is able to model the interaction of a tangent coil with a through-wall slot over a broad frequency range, even for a slot as large as 101 mm. The results shown here were obtained using a modest flaw grid of $128 \times 2 \times 4$ cells, and a very coarse grid of $2 \times 4 \times 4$ cells for the probe.

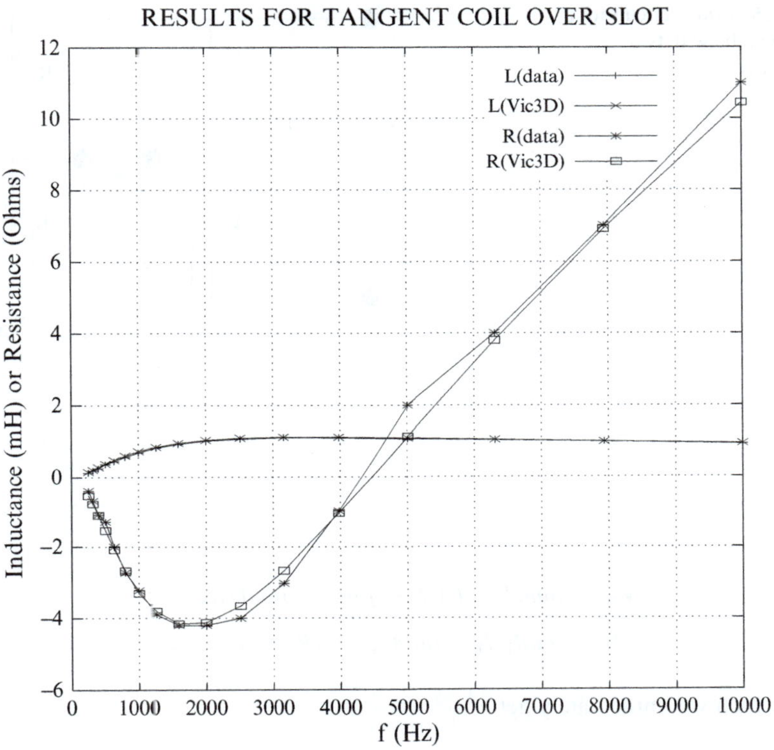

Fig. 6.24 Comparison of calculations with the frequency-scan benchmark data of [49]

Appendix

A.1 Analysis of a Bridge Circuit for a Split-D Probe

The split-D coil configuration shown in Fig. 6.16 comprises one transmit and two receive coils. Another common split-D probe contains no separate transmit coil but has two identical coils connected in a bridge circuit that serves as the driver of the two coils.

Figure 6.25 shows a standard bridge circuit that drives the split-D coils, labeled L_1 and L_2, in the presence of a flaw, shown as an inductor, L_3, and resistor, R_3. The inductance of the two coils is L; M is the mutual inductance between these two coils, and M_{13}, M_{23}, are, respectively, the mutual inductance between L_3 and L_1 and L_3 and L_2. The bridge is open-circuited, and its output voltage is V_0.

The equations for the circuit of Fig. 6.25 are

$$V_{\text{in}} = I_1(R_0 + j\omega L) + j\omega M I_2 + j\omega M_{13} I_3$$

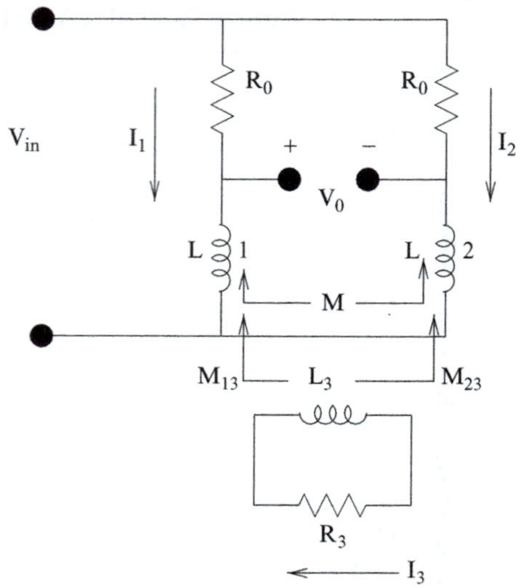

Fig. 6.25 Bridge circuit for a split-D probe with two transmit coils

$$
\begin{aligned}
V_{\text{in}} &= j\omega M I_1 + I_2(R_0 + j\omega L) + j\omega M_{23} I_3 \\
0 &= j\omega M_{13} I_1 + j\omega M_{23} I_2 + (R_3 + j\omega L_3) I_3 \,,
\end{aligned}
\tag{6.7}
$$

from which we immediately get

$$
I_3 = -\frac{j\omega M_{13} I_1 + j\omega M_{23} I_2}{R_3 + j\omega L_3} \,.
$$

When this is substituted into (6.7), we have

$$
\begin{aligned}
V_{\text{in}} &= I_1 \left[R_0 + j\omega L + \frac{\omega^2 M_{13}^2}{R_3 + j\omega L_3} \right] + I_2 \left[j\omega M + \frac{\omega^2 M_{13} M_{23}}{R_3 + j\omega L_3} \right] \\
V_{\text{in}} &= I_1 \left[j\omega M + \frac{\omega^2 M_{13} M_{23}}{R_3 + j\omega L_3} \right] + I_2 \left[R_0 + j\omega L + \frac{\omega^2 M_{23}^2}{R_3 + j\omega L_3} \right] \\
V_0 &= I_1 \left[j\omega L + \frac{\omega^2 M_{13}^2}{R_3 + j\omega L_3} - \frac{\omega^2 M_{13} M_{23}}{R_3 + j\omega L_3} - j\omega M \right] \\
&\quad + I_2 \left[j\omega M - j\omega L + \frac{\omega^2 M_{13} M_{23}}{R_3 + j\omega L_3} - \frac{\omega^2 M_{23}^2}{R_3 + j\omega L_3} \right] .
\end{aligned}
\tag{6.8}
$$

Now, if coil 1 and coil 2 are both considered to be transmit coils, then **VIC-3D®** considers them to be in series carrying one ampere. Hence, $I_1 = I_2 = 1$ in the expression for V_0 in (6.8), from which follows the result

$$V_0 = Z_{\text{out}} = \frac{\omega^2 (M_{13}^2 - M_{23}^2)}{R_2 + j\omega L_3} \, . \tag{6.9}$$

One way to guarantee that the circuit of Fig. 6.25 will behave as a current source such that $I_1 \approx I_2$ is to make the balance resistance, R_0, much greater than any of the other impedance elements over the frequency range of interest. In this case, $I_1 \approx I_2 \approx V_{\text{in}}/R_0$.

M_{13} and M_{23} depend upon the scan position of the probe relative to the flaw. All other parameters are independent of the probe position. It is clear from (6.9) that if the flaw is symmetrically placed with respect to the probe, then $M_{13} = M_{23}$, which means that $V_0 = 0$. Furthermore, because the probe is symmetrical, it follows that V_0 will be an antisymmetrical function of the probe position if the flaw is symmetrical with respect to the probe position. This gives the usual antisymmetrical response that we expect of a differential probe.

Chapter 7
Advanced Probe Models Based on Electric Dipoles

7.1 Introduction

Eddy-current measurements of conductivity spectra are needed to gauge the near-surface residual stress in surface-treated turbine engine components for life extension, where conductivity and stress are related via the piezoresistive effect. Broadband (0.1–100 MHz) planar spiral-coil eddy-current probes are currently being developed using photolithography for the purpose of efficiently measuring the conductivity spectra for subsequent inversion to residual stress versus depth profiles.

The key problem is that the probes do not perform well at high frequencies (>30 MHz) due to stray capacitive effects, which begin to dominate as the resonant frequency of the probe is approached. This parasitic effect is mainly due to capacitive coupling between adjacent traces of the coil, which makes it especially difficult to predict in terms of sensitivity. At high frequencies, the lift-off sensitivity of current spiral coils becomes unstable above about 25 MHz. There is a need, therefore, to develop models to account for these problematic parasitic effects and allow for probe optimization in the design stage, rather than through the tedious and inefficient trial-and-error approach. The key parameters of interest to the probe designer are coil diameter, trace width, and trace spacing. The probes could be single-layered or multilayered, circular or rectangular, shielded or unshielded.

In this chapter, we will apply the volume-integral algorithm to develop a computer code to design spiral-coil eddy-current probes for optimal broadband performance. This code will allow the designer to predict probe performance as a function of frequency.

7.2 Development of the Electric-Dipole Mathematical Model

The current version of **VIC-3D®** assumes that all applied current sources can be modeled as magnetic dipoles. This is reasonable because the currents are assumed to be solenoidal (zero divergence) and flow in parallel planes. Thus, the sources can

H.A. Sabbagh et al., *Computational Electromagnetics and Model-Based Inversion*, Scientific Computation, DOI 10.1007/978-1-4419-8429-6_7,

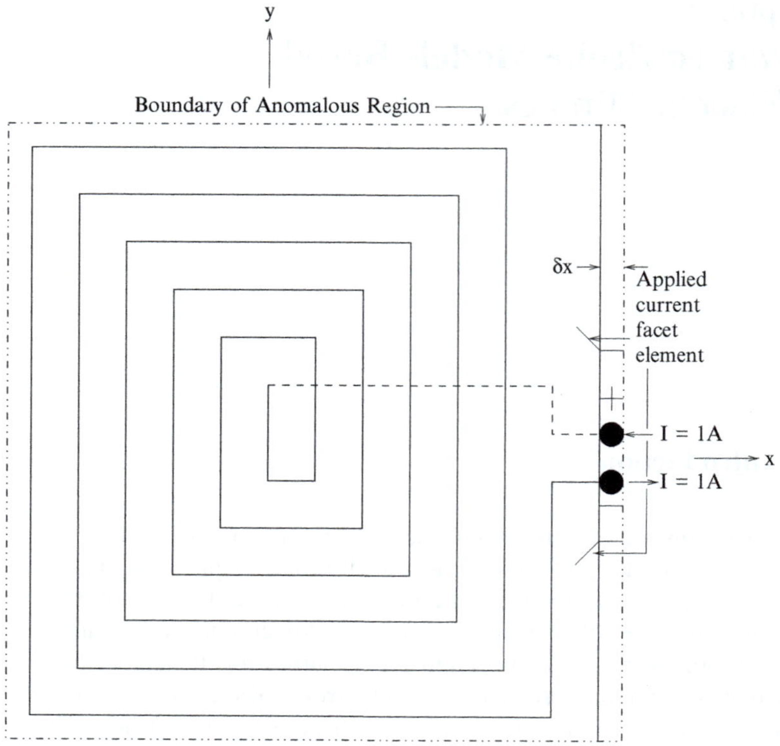

Fig. 7.1 A printed-circuit rectangular-spiral coil lying within an "anomalous region." The electric dipole produced by the driving-point current excites the metallic traces of the coil. The convention in electric-circuit theory is that current enters the positive terminal of a passive load and leaves through the negative terminal

be defined by a single magnetic dipole that is oriented normal to the planes. The field produced by these sources, then, are incident upon the anomalous region of the problem. This model works well for typical eddy-current problems but allows only the computation of the inductance of the source. The solenoidal nature of the current precludes any possibility of computing the driving-point capacitance of the coil. In order to compute capacitance, and therefore the resonant frequency of the coil, we must relax the assumption of solenoidal current-flow and assume that the excitation of the coil is an electric dipole.

We are going to apply our volume-integral algorithm to the metallic traces contained within the anomalous region shown in Fig. 7.1. These traces are the "anomaly" of the problem (much like a ferrite core in a conventional eddy-current probe), and the driving-point terminals form an electric dipole that will drive the coil.

The discretized electric equation is:

$$\begin{aligned}
\begin{bmatrix} \mathbf{E}^{(ix)} \\ \mathbf{E}^{(iy)} \\ \mathbf{E}^{(iz)} \end{bmatrix} &= \begin{bmatrix} \mathbf{Q}^{(x)} & \mathbf{0} & \mathbf{0} \\ \mathbf{0} & \mathbf{Q}^{(y)} & \mathbf{0} \\ \mathbf{0} & \mathbf{0} & \mathbf{Q}^{(z)} \end{bmatrix}^{(ee)} \begin{bmatrix} \mathbf{J}^{(x)} \\ \mathbf{J}^{(y)} \\ \mathbf{J}^{(z)} \end{bmatrix} \\
&- \begin{bmatrix} \mathbf{G}_{(0)}^{(xx)} & \mathbf{G}_{(0)}^{(xy)} & \mathbf{G}_{(0)}^{(xz)} \\ \mathbf{G}_{(0)}^{(yx)} & \mathbf{G}_{(0)}^{(yy)} & \mathbf{G}_{(0)}^{(yz)} \\ \mathbf{G}_{(0)}^{(zx)} & \mathbf{G}_{(0)}^{(zy)} & \mathbf{G}_{(0)}^{(zz)} \end{bmatrix}^{(ee)} \begin{bmatrix} \mathbf{J}^{(x)} \\ \mathbf{J}^{(y)} \\ \mathbf{J}^{(z)} \end{bmatrix} \\
&- \begin{bmatrix} \mathbf{G}_{(a)}^{(xx)} & \mathbf{G}_{(a)}^{(xy)} & \mathbf{G}_{(a)}^{(xz)} \\ \mathbf{G}_{(a)}^{(yx)} & \mathbf{G}_{(a)}^{(yy)} & \mathbf{G}_{(a)}^{(yz)} \\ \mathbf{G}_{(a)}^{(zx)} & \mathbf{G}_{(a)}^{(zy)} & \mathbf{G}_{(a)}^{(zz)} \end{bmatrix}^{(ee)} \begin{bmatrix} \mathbf{J}^{(x)} \\ \mathbf{J}^{(y)} \\ \mathbf{J}^{(z)} \end{bmatrix} \\
&- \begin{bmatrix} \mathbf{G}_{(b)}^{(xx)} & \mathbf{G}_{(b)}^{(xy)} & \mathbf{G}_{(b)}^{(xz)} \\ \mathbf{G}_{(b)}^{(yx)} & \mathbf{G}_{(b)}^{(yy)} & \mathbf{G}_{(b)}^{(yz)} \\ \mathbf{G}_{(b)}^{(zx)} & \mathbf{G}_{(b)}^{(zy)} & \mathbf{G}_{(b)}^{(zz)} \end{bmatrix}^{(ee)} \begin{bmatrix} \mathbf{J}^{(x)} \\ \mathbf{J}^{(y)} \\ \mathbf{J}^{(z)} \end{bmatrix} \\
&- \begin{bmatrix} \mathbf{G}_{(0)} & \mathbf{G}_{(a)} & \mathbf{G}_{(b)} \end{bmatrix}^{(em)} \begin{bmatrix} \mathbf{M}^{(x)} \\ \mathbf{M}^{(y)} \\ \mathbf{M}^{(z)} \end{bmatrix}, \qquad (7.1)
\end{aligned}$$

where the $\mathbf{Q}$'s are tri-diagonal matrices, the $\mathbf{G}_{(0)}$'s the infinite-space matrices, the $\mathbf{G}_{(a)}$'s the convolutional layered-space matrices, and the $\mathbf{G}_{(b)}$'s the correlational layered-space matrices. The infinite-space matrices are convolutional, also. The superscript (ee) denotes electric–electric matrices and (em) denotes electric–magnetic matrices. The $\mathbf{J}$'s are the unknown electric currents and the $\mathbf{M}$'s are the unknown magnetic polarization vectors. The last block in (7.1) is simply a shorthand representation of the three blocks above it, except that it represents electric–magnetic interactions.

From here on we will ignore magnetic interactions and will combine the various $\mathbf{G}$-matrices into one. The incident electric field moments on the left-hand are due, of course, to any independent sources, currents, or charges, as in the usual case of a separate probe coil. We compute an effective incident electric field-moment due to the dipole in the following way. Assume that the positive terminal in Fig. 7.1 carries 1A of current at cell $k^+l^+m^+$ and that the negative terminal carries 1A of current at cell $k^-l^-m^-$. Then in the configuration shown in Fig. 7.1, we have $J^{(x)}_{k^+l^+m^+} = -1/\delta y\delta z, J^{(y)}_{k^+l^+m^+} = J^{(z)}_{k^+l^+m^+} = 0$, and $J^{(x)}_{k^-l^-m^-} = 1/\delta y\delta z, J^{(y)}_{k^-l^-m^-} = J^{(z)}_{k^-l^-m^-} = 0$, where $\delta x,\ \delta y,\ \delta z$ are the dimensions of a cell in the anomalous region. Then, from (7.1) we have for the incident field moments

$$E^{(ix)}_{klm} = \left[-G^{(xx)}_{klm,k^+l^+m^+} + G^{(xx)}_{klm,k^-l^-m^-}\right] \times \frac{1}{\delta y \delta z}$$

$$E^{(iy)}_{klm} = \left[-G^{(yx)}_{klm,k^+l^+m^+} + G^{(yx)}_{klm,k^-l^-m^-}\right] \times \frac{1}{\delta y \delta z}$$

$$E^{(iz)}_{klm} = \left[-G^{(zx)}_{klm,k^+l^+m^+} + G^{(zx)}_{klm,k^-l^-m^-}\right] \times \frac{1}{\delta y \delta z} . \tag{7.2}$$

This electric-dipole moment vector can be thought of as the "constraint electric field" that arises because of the unit-constrained current at $k^+l^+m^+$ and $k^-l^-m^-$.

Now that we have transformed the known current-density vectors at $k^+l^+m^+$ and $k^-l^-m^-$ into the incident electric-field moments (this is akin to using the **G**-matrices as "transfer matrices"), we are left with an anomalous current of zero at these cells. After all, a current at a location is either known or unknown (anomalous), but not both. Thus, if the current is known, then the anomalous current must be zero. The way to force the anomalous current to vanish is to make the cells at $k^+l^+m^+$, $k^-l^-m^-$ to contain only host material. Hence, with this proviso, the problem can be solved completely as a standard problem for any anomalous region, using (7.2) as the incident field. In this case, the anomalous region of Fig. 7.1 will contain the source dipole.

The position of the applied current facet elements (plus and minus) are included within the grid for the anomalous region. The magnitude of the current density at the boundary of the cell for the + terminal is $-1/\delta y \delta z$, and the magnitude of the current density for the − terminal is $1/\delta y \delta z$.

Once we have the **J**'s, we can compute the change in driving-point impedance seen at the terminals of the unit-current source in the usual way as the dot-product of the **J**'s with the incident field moments:

$$\Delta Z = -\Sigma_{klm}\left[J^{(x)}_{klm}E^{(ix)}_{klm} + J^{(y)}_{klm}E^{(iy)}_{klm} + J^{(z)}_{klm}E^{(iz)}_{klm}\right]$$

$$= \frac{1}{\delta y \delta z}\Sigma_{klm}\left\{J^{(x)}_{klm}\left[-G^{(xx)}_{klm,k^+l^+m^+} + G^{(xx)}_{klm,k^-l^-m^-}\right]\right.$$

$$+J^{(y)}_{klm}\left[-G^{(yx)}_{klm,k^+l^+m^+} + G^{(yx)}_{klm,k^-l^-m^-}\right]$$

$$\left.+J^{(z)}_{klm}\left[-G^{(zx)}_{klm,k^+l^+m^+} + G^{(zx)}_{klm,k^-l^-m^-}\right]\right\} \tag{7.3}$$

The unloaded driving-point impedance of the dipole is given by taking the "reaction" of the driving-point current with itself, i.e., by taking the dot product of the driving-point current with the electric field moment produced by that current:

$$Z^{(0)} = -\left(\frac{1}{\delta y \delta z}\right)^2 \left[G^{(xx)}_{k^+l^+m^+,k^+l^+m^+} + G^{(xx)}_{k^-l^-m^-,k^-l^-m^-} - G^{(xx)}_{k^+l^+m^+,k^-l^-m^-}\right.$$

$$\left.-G^{(xx)}_{k^-l^-m^-,k^+l^+m^+}\right] . \tag{7.4}$$

The driving-point impedance of the loaded dipole is the sum of (7.3) and (7.4).

7.3 An Example

Consider the $8 \times 8 \times 2$ mm grid with $8 \times 8 \times 2$ cells shown in Fig. 7.2. If the bottom layer comprises only blue (host) cells, then the current is constrained to flow only in the top layer, which means that there will be no z-directed currents at all within the grid.

The current enters cell 3, which makes it the positive terminal, and leaves cell 1, making it the negative terminal. Thus, cells 1 and 3 make up an electric dipole that will excite the remaining cells. In order to determine the excitation source, we constrain the expansion coefficient for the x-directed current at the facet between cells 3 and 4 to be $-1 \times 10^6\,\mathrm{A/m^2}$ and for the x-directed current at the facet between cells 1 and 2 to be $+1 \times 10^6\,\mathrm{A/m^2}$. Clearly, the current in cells 1 and 2 and 3 and 4 is constrained to flow only in the x-direction.

At frequencies that are low enough that this body is electrically large, we can model this structure by the electrical equivalent circuit shown in Fig. 7.3. The capacitor, C_0, is associated with the stored electric energy within the gap between cells 1 and 3 of the grid, the resistor is associated with energy loss in the metal, and the inductance is associated with the stored magnetic energy within the area enclosed by the coil.

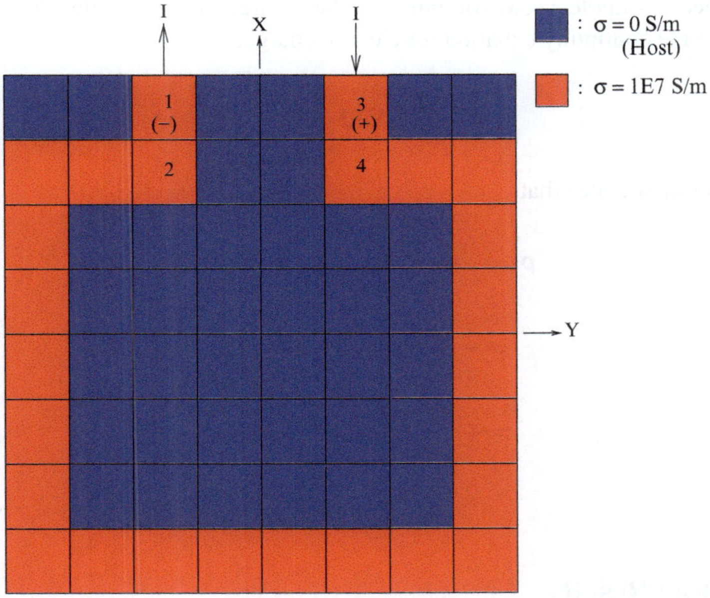

Fig. 7.2 Illustrating a square coil on an $8 \times 8 \times 2$ grid

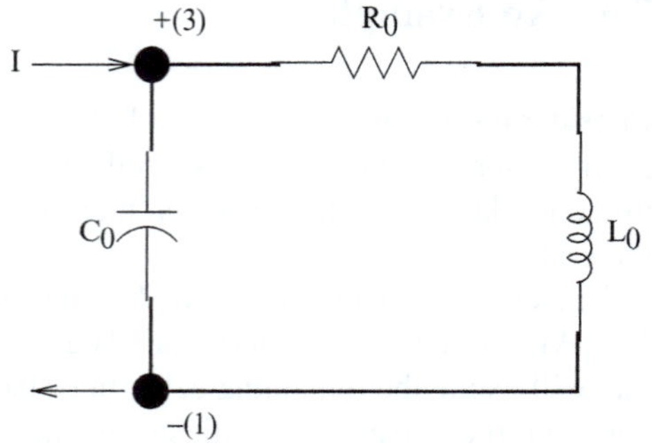

Fig. 7.3 An equivalent electric circuit for the structure shown in Fig. 7.2 at low frequencies

7.4 Another Example

Consider the same problem as before, but this time let the excitation current lie in the y-direction, as shown in Fig. 7.4. In this case, the excitation currents lie in the y-direction, and we show that the sources are limited to a single cell. Hence, for the $(-)$ terminal, the current is a "ramp" function, and in the $(+)$ terminal, it is a "slide" function. Thus, in computing the incident field vectors in (7.2), we would use only the appropriate "ramp" or "slide" portions of the matrix elements.

The electric-dipole vector oriented in the y-direction is a little clearer now, because of the continuity equation for electric charge:

$$\frac{\partial \rho}{\partial t} + \nabla \cdot \mathbf{J} = 0\,, \tag{7.5}$$

from which we deduce that

$$\begin{aligned}\rho &= -\frac{1}{j\omega}\nabla\cdot\mathbf{J}\\ &= -\frac{10^9}{j\omega}\ \mathrm{Q/m^3}\ (-)\ \text{terminal}\\ &= +\frac{10^9}{j\omega}\ \mathrm{Q/m^3}\ (+)\ \text{terminal.}\end{aligned} \tag{7.6}$$

7.5 Initial Results

Figures 7.5–7.11 show the results of the dipole calculation for the example of Fig. 7.2, but using ramp functions at the terminals (cells 1 and 3) instead of tent functions. The tent function that spans cells 1 and 2, as well as the tent function that

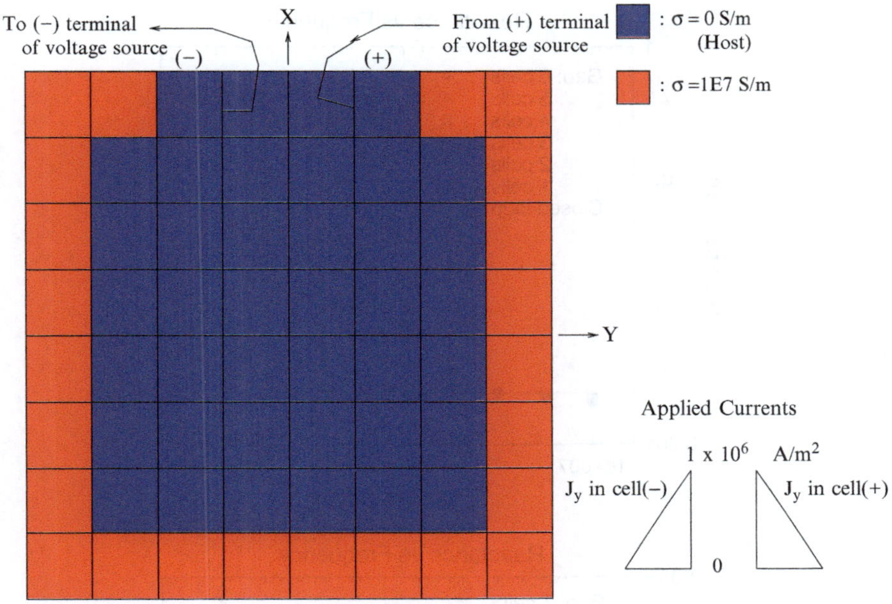

Fig. 7.4 Illustrating a square coil on an $8 \times 8 \times 2$ grid. The excitation current lies in the *y*-direction this time, and the nature of the excitation currents within their cells is shown

spans cells 3 and 4 are treated as unknowns to be computed. These are "freespace" results—there is no workpiece, and the grid has $8 \times 8 \times 2$ cells. The frequency of the broad resonance increases as the gap increases, which suggests that the associated capacitance decreases (see the discussion of Fig. 7.12). The sharp resonances, on the other hand, do not change frequency with the gap, which suggests that they may be transmission-line effects associated with the length of the loop.

Figure 7.12 is a plot of the terminal capacitance as a function of the distance between the terminals, as deduced from the frequency of the zero-crossing of the broad resonance in the reactance functions. This result is reasonable, because we expect this capacitance to decrease with an increased separation of the terminals. In addition, this figure includes the capacitance of the "naked terminals" as a function of the gap between the terminals. These capacitances are computed from the reactances (not shown) of the naked terminals at 10^8 Hz. The latter capacitances are smaller than the former since they include only the ramp functions of the terminal and not the tent functions that complete the terminal posts.

We can do a simple analysis of the circuit of Fig. 7.3 and estimate the parameters from the frequency responses of the driving-point impedances that we have just computed. The driving-point impedance is given by

$$Z_{\text{in}} = \frac{1}{j\omega C_0 + \dfrac{1}{R_0 + j\omega L_0}}$$

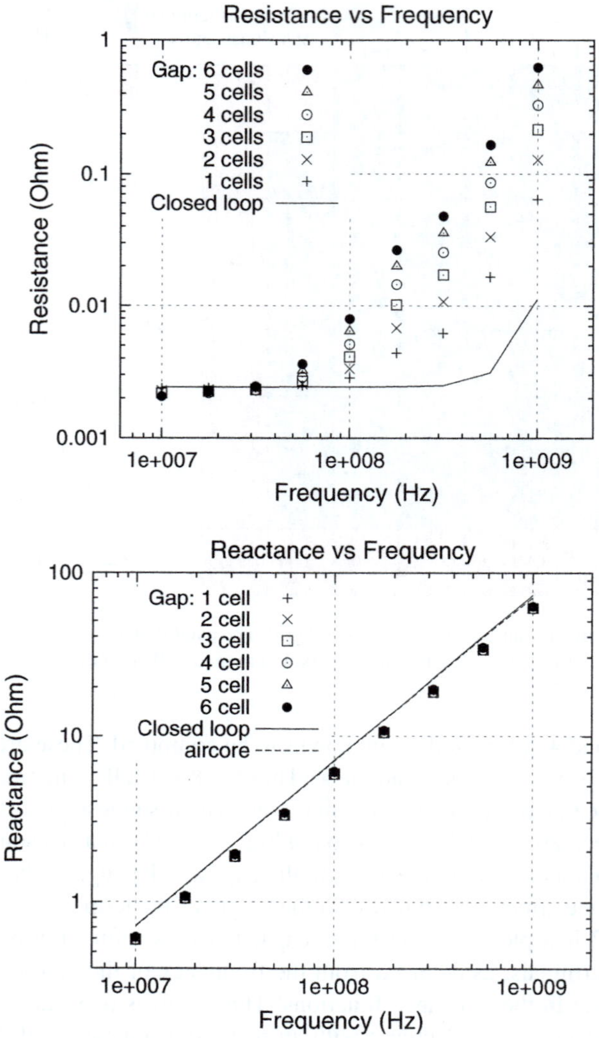

Fig. 7.5 Results of the dipole model calculation in freespace at frequencies below the first resonance. The curve labeled "aircore" is the result of **VIC-3D**®'s standard aircore calculation for a single-loop coil enclosing the same area. *Top*: resistance. *Bottom*: reactance

$$
\begin{aligned}
&= \frac{R_0 + j\omega L_0}{1 - \omega^2 L_0 C_0 + j\omega R_0 C_0} \\
&= \frac{R_0 + j\omega(L_0[1 - \omega^2 L_0 C_0] - R_0^2 C_0)}{(1 - \omega^2 L_0 C_0)^2 + (\omega R_0 C_0)^2} \,. \qquad (7.7)
\end{aligned}
$$

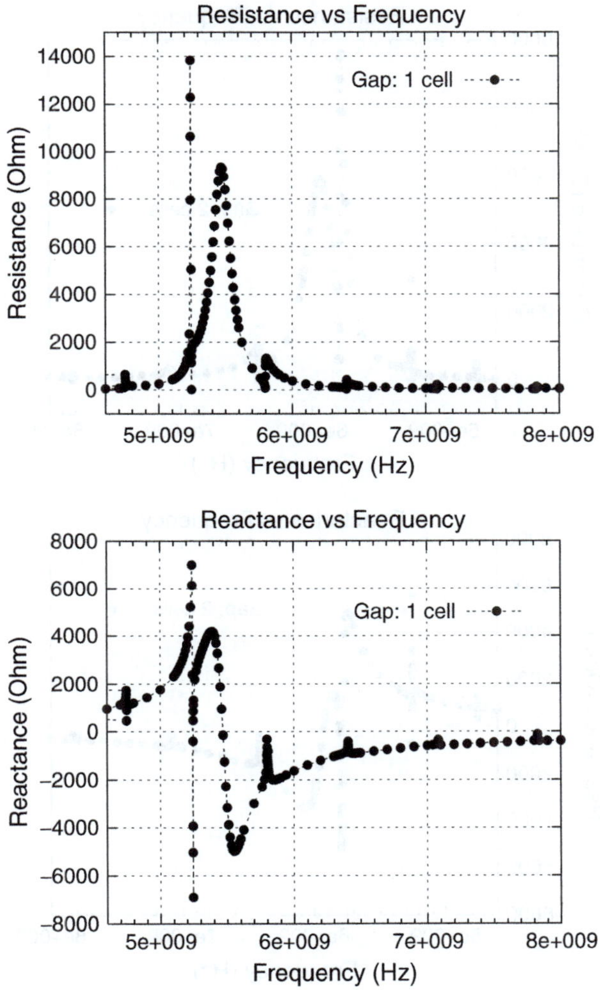

Fig. 7.6 Results for the $8 \times 8 \times 2$ grid, one-cell-gap (1 mm) model. *Top*: resistance. *Bottom*: reactance

Resonance occurs at that frequency for which the total reactance vanishes, or $L_0(1 - \omega_R^2 L_0 C_0) = R_0^2 C_0$, from which

$$\omega_R = \sqrt{\frac{1}{L_0 C_0}}\sqrt{\frac{L_0 - C_0 R_0^2}{L_0}}$$
$$\approx \sqrt{\frac{1}{L_0 C_0}}, \tag{7.8}$$

for $R_0^2 C_0 \ll L_0$.

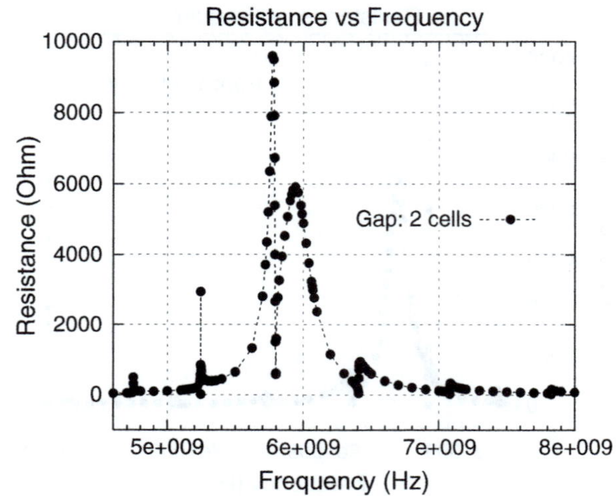

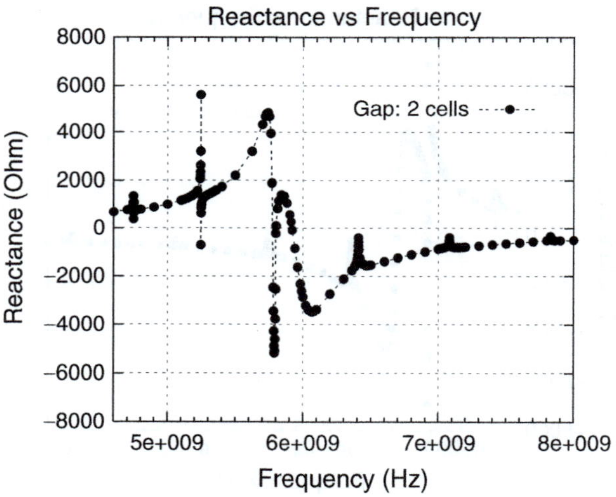

Fig. 7.7 Results for the $8 \times 8 \times 2$ grid, two-cell-gap (2 mm) model. *Top*: resistance. *Bottom*: reactance

We can estimate R_0 as follows: the coil conductivity is 1×10^7 S/m, and the cells are cubes of 1 mm on a side. Thus, a path of N cells will have a resistance of

$$\begin{aligned} R_0 &= \rho \times l/A \\ &= 1 \times 10^{-7} \times N \times 1 \times 10^{-3}/(1 \times 10^{-6}) \\ &= N \times 1 \times 10^{-4}\Omega\,, \end{aligned} \tag{7.9}$$

from which we develop the following table (Table 7.1).

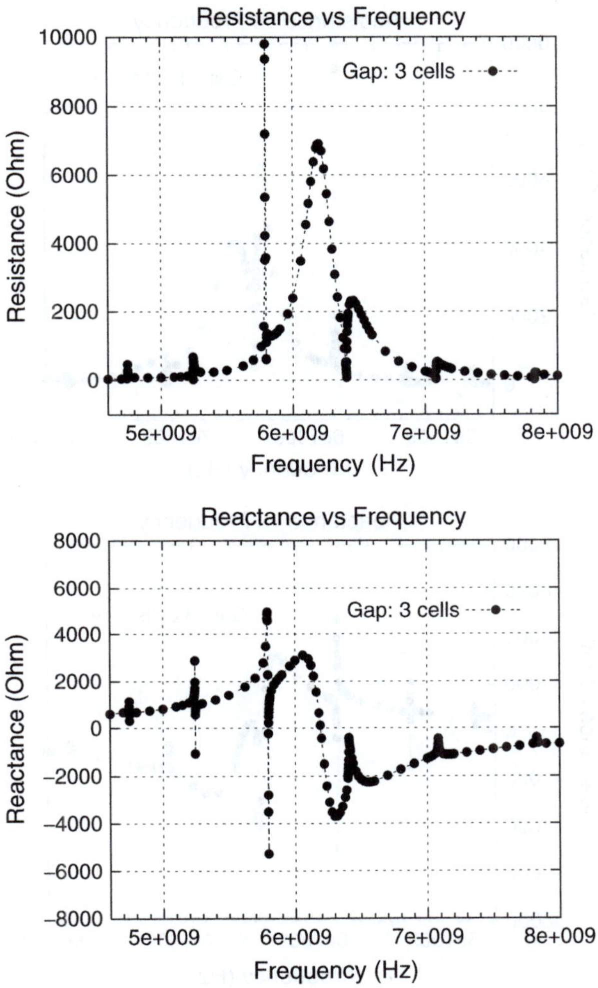

Fig. 7.8 Results for the $8 \times 8 \times 2$ grid, three-cell-gap (3 mm) model. *Top*: resistance. *Bottom*: reactance

These results are in good agreement with the low-frequency resistances plotted in Figs. 7.5 and 7.16.

We can estimate L_0 from the reactance at 1×10^8 Hz in Fig. 7.5: $L_0 = X_L/(2\pi \times 1 \times 10^8)$ (Table 7.2).

Finally, we estimate the capacitance from the resonant frequencies as shown in (7.8). These are the "Full Circuit" capacitance values that are plotted in Fig. 7.12 (Table 7.3).

Figure 7.13 shows the change in impedance when a 1 mm plate is placed below the loop, and Figs. 7.14 and 7.15 show the results of a lift-off scan at 10 MHz. The plotted results in the latter two figures are the change due to the plate.

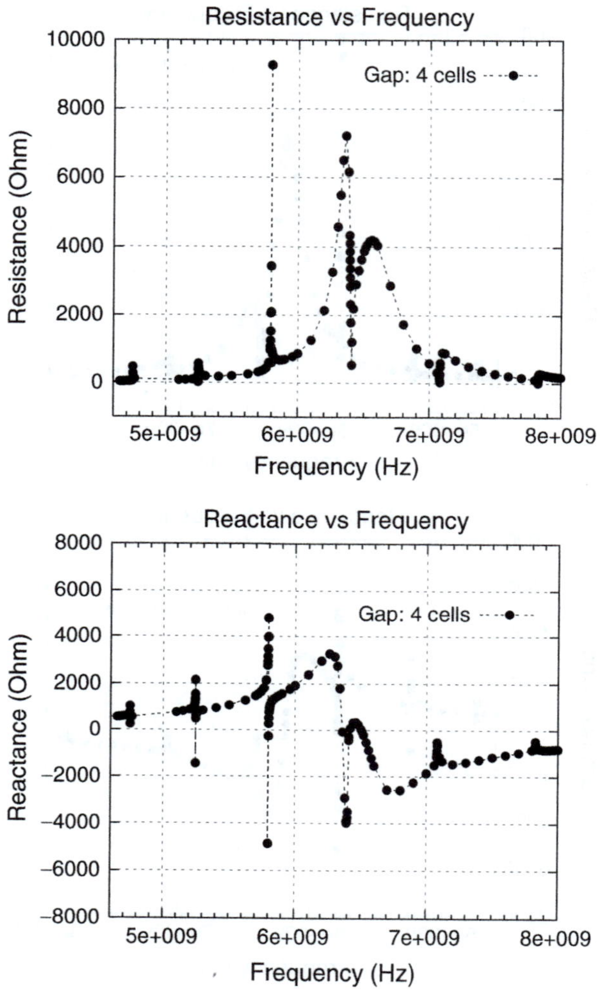

Fig. 7.9 Results for the $8 \times 8 \times 2$ grid, four-cell-gap (4 mm) model. *Top*: resistance. *Bottom*: reactance

7.6 Results With a Grid Of $16 \times 16 \times 4$ Cells

Figure 7.16 shows the low-frequency response of the refined grid, which should be compared with the corresponding results of Fig. 7.5. Finally, we show in Fig. 7.17 the impedance response of the "naked" terminal-pair.

The result for the "naked" terminal pair is interesting. It represents a resistor, whose resistance varies with frequency, in series with an object whose log-log plot varies linearly (for the most part), with a negative slope. Thus, this object's

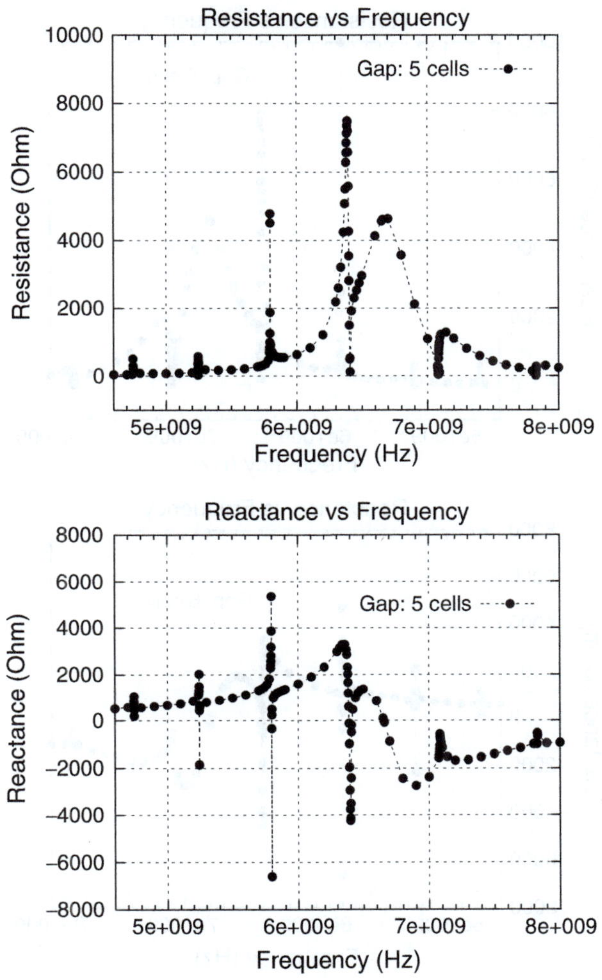

Fig. 7.10 Results for the $8 \times 8 \times 2$ grid, five-cell-gap (5 mm) model. *Top*: resistance. *Bottom*: reactance

frequency response is a simple negative-power law that obviously represents a capacitor. In order to compute the capacitance of this object, we take the logarithm of the usual expression, $X_C = 1/(2\pi \times f \times C)$, and get $\log X_C = -\log 2\pi - \log f - \log C$. Now, the reactance (not shown) of the naked terminals is $X_C = 5.06 \times 10^4\,\Omega$ when $f = 1 \times 10^8$ Hz, so that $\log C = -13.489$, or $C = 3.24 \times 10^{-14}$ which is in good agreement with the corresponding one-cell results of Fig. 7.12. The lack of smoothness in the resistance plot of Fig. 7.17 is due to lack of precision in the calculations.

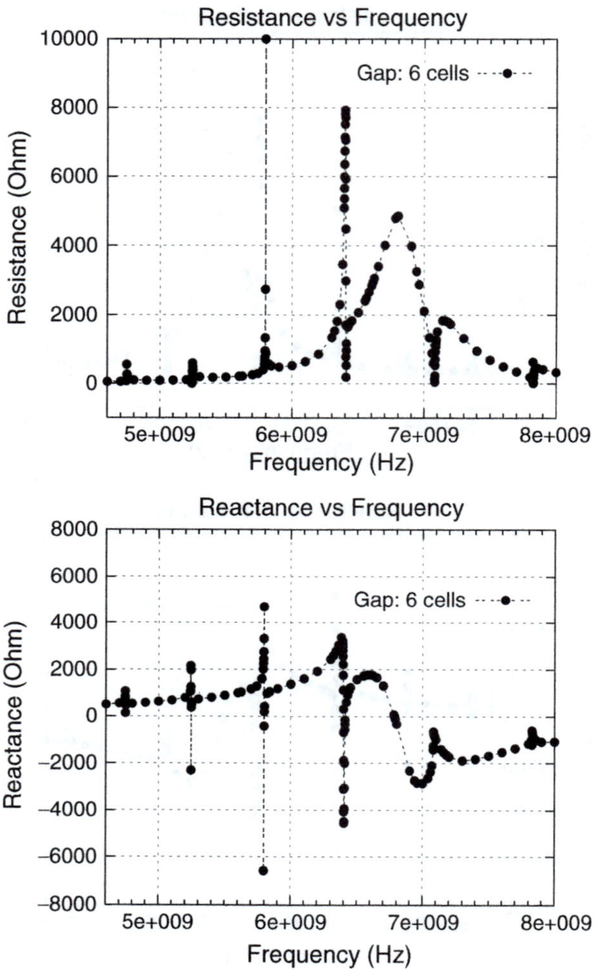

Fig. 7.11 Results for the $8 \times 8 \times 2$ grid, six-cell-gap (6 mm) model. *Top*: resistance. *Bottom*: reactance

7.7 Computation of the Divergence of the Currents

Figure 7.18 shows the divergence of the current in the loop with a 1-cell (1 mm) gap at 10 MHz, and Fig. 7.19 shows the same thing at 1 GHZ. In both of these figures, J_x flows out of the terminal at $0 < x < 0.001$, $0.002 < y < 0.003$ and into the terminal at $0 < x < 0.001$, $0.004 < y < 0.005$. The units are A/m^3. Figure 7.20 shows the divergence of the current in the closed-loop at 10 MHz. The two driving tent functions span the cells in the region $0.002 < x < 0.005$, $0 < y < 0.001$ and flow in the positive x-direction. The units are A/m^3.

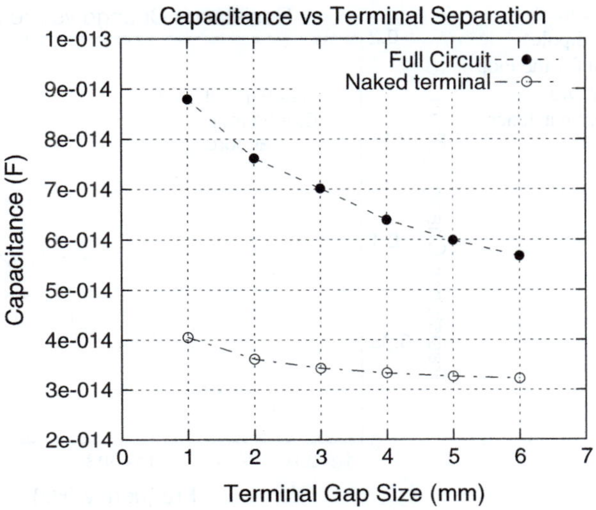

Fig. 7.12 Showing the "loaded" and "naked" terminal capacitance as a function of the distance between the terminals

Table 7.1 Estimating R_0

Test	N (cells)	R_0 (Ω)
1	27	2.7×10^{-3}
2	26	2.6×10^{-3}
3	25	2.5×10^{-3}
4	24	2.4×10^{-3}
5	23	2.3×10^{-3}
6	22	2.2×10^{-3}

Table 7.2 Estimating L_0

Test	X_L (Ω)	L_0 (H)
1	6.078	9.678×10^{-9}
2	5.962	9.494×10^{-9}
3	5.918	9.424×10^{-9}
4	5.897	9.390×10^{-9}
5	5.990	9.538×10^{-9}
6	6.100	9.713×10^{-9}

Table 7.3 Estimating C_0

Test	f_R (Hz)	L_0 (H)	C_0 (F)
1	5.46×10^9	9.678×10^{-9}	8.79×10^{-14}
2	5.92×10^9	9.494×10^{-9}	7.62×10^{-14}
3	6.19×10^9	9.424×10^{-9}	7.02×10^{-14}
4	6.50×10^9	9.390×10^{-9}	6.39×10^{-14}
5	6.66×10^9	9.538×10^{-9}	5.99×10^{-14}
6	6.78×10^9	9.713×10^{-9}	5.68×10^{-14}

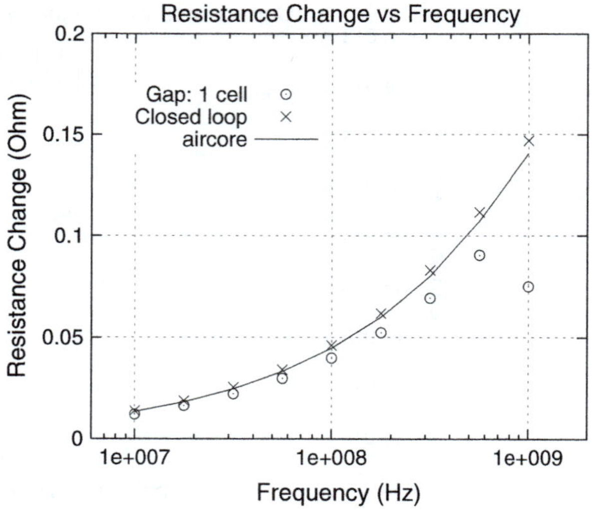

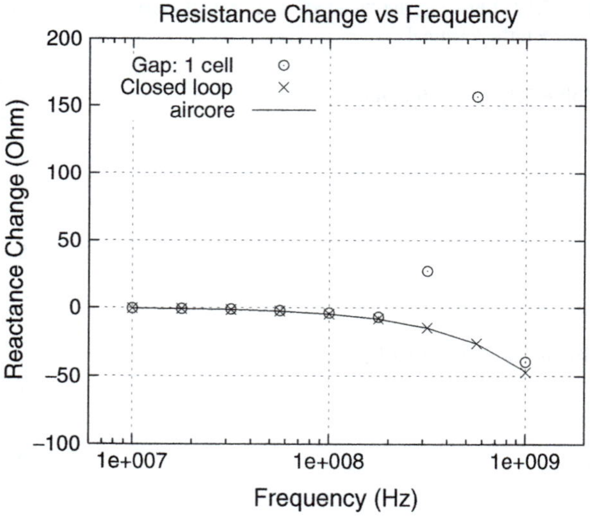

Fig. 7.13 Change in impedance of the dipole model calculation for the loop over a 1 mm plate. *Top*: resistance. *Bottom*: reactance

Note that the 1-mm gap coil and the closed-loop coil have comparable divergence values at 10 MHz, but the response of the 1-mm gap coil at 1 GHz is orders-of-magnitude larger, as we would expect. At this frequency resonance effects, as manifested by the increased capacitive reactance, begin to assert themselves, and we always associate capacitive effects with charge buildup, which is the same as divergence of the current.

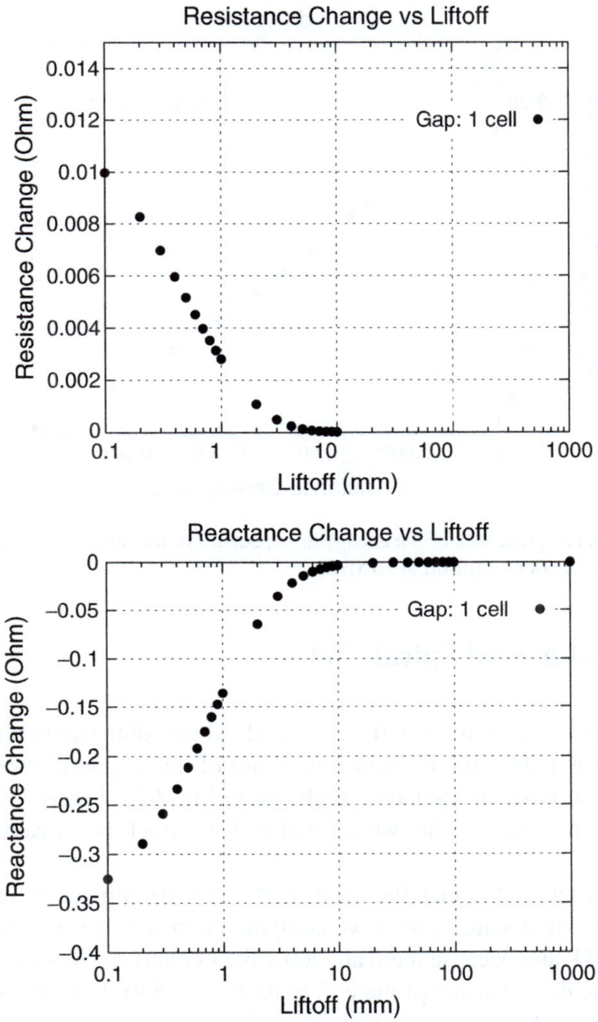

Fig. 7.14 Change in impedance of the dipole model calculation over a 1 mm plate as a function of lift-off at 10 MHz. *Top*: resistance. *Bottom*: reactance

7.8 Three-Turn Spiral Coil

Figure 7.21 shows a three-turn spiral coil, whose results are displayed in the next few figures. Figure 7.22 shows the frequency scan of this coil in freespace, and Fig. 7.23 shows these same results below the resonance around 2 GHz. Finally, Figs. 7.24 and 7.25 illustrate the results of the dipole model of the three-turn spiral coil when a 1 mm plate is placed below the coil.

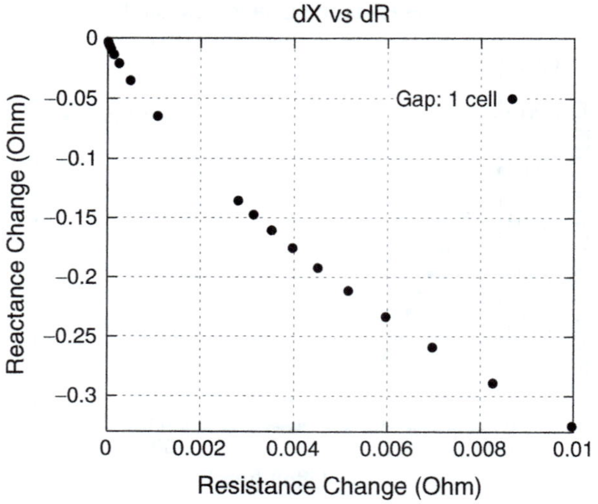

Fig. 7.15 Impedance-plane plot of the change in impedance of the dipole model calculation over a 1 mm plate as a function of lift-off at 10 MHz

7.9 A Two-Layered Spiral Coil

Figure 7.26 shows a two-layered spiral coil whose dimensions are $0.8\,\mathrm{mm} \times 0.8\,\mathrm{mm} \times 0.15\,\mathrm{mm}$. The low-frequency response of this coil is shown in Fig. 7.27, and the response through resonance is shown in Fig. 7.28. Figure 7.29 shows the high-frequency response of the two-layered coil, in which two weaker resonances appear.

The conductivity of the metallic traces is $\sigma = 1 \times 10^7$ S/m, as before. Given the dimensions that were stated above, we compute the resistance of each voxel to be $R_{\mathrm{voxel}} = 0.002\,\Omega$, and because there are 260 voxels connected in series in Fig. 7.26, we compute the dc resistance of the coil to be $R_0 = 0.520\,\Omega$, which is in excellent agreement with the low-frequency resistance shown in Fig. 7.27.

Using the results of Figs. 7.27 and 7.28, we can estimate the other elements of the equivalent circuit of Fig. 7.3. From Fig. 7.27 we estimate the inductance to be $L_0 = 1.59 \times 10^{-8}$H, and from Fig. 7.28 we estimate the resonant frequency to be 6.4×10^9Hz, or $\omega_0 = 40.2 \times 10^9$rad/sec. Using these results we compute $C_0 = 1/(L_0 \times \omega_0^2) = 3.89 \times 10^{-14}$Fd. For a high-Q parallel resonant circuit of this type, the peak resistance at resonance is given by $R_{\mathrm{res}} = L_0/(C_0 R_0) = 7.86 \times 10^5\,\Omega$, which is in excellent agreement with the result shown in Fig. 7.28.

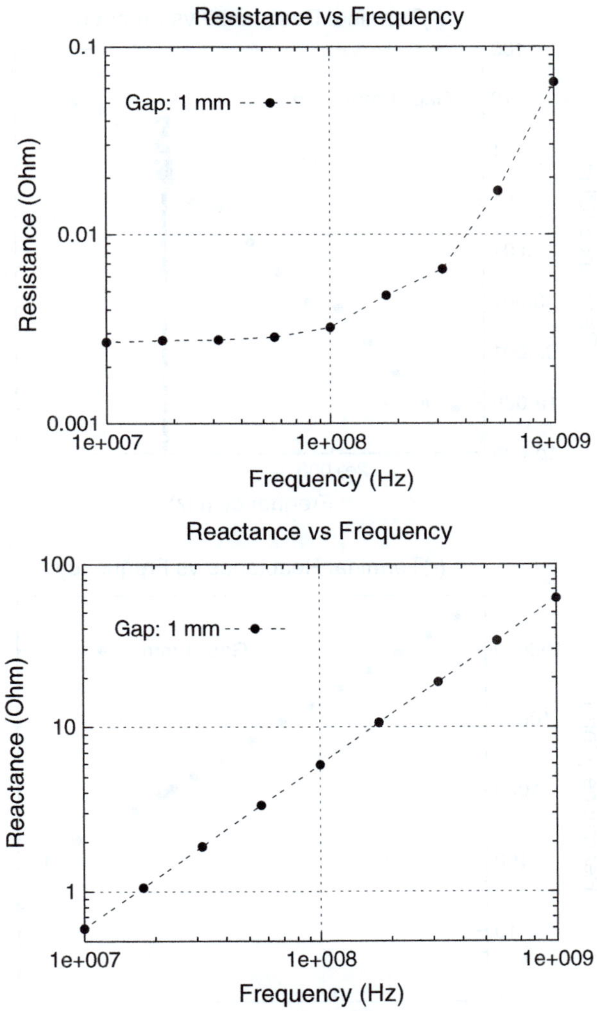

Fig. 7.16 Results of the $16 \times 16 \times 4$-cell 1-mm gap model at low frequencies. *Top*: resistance. *Bottom*: reactance

7.10 A Model of a THz Transmitter on GaAs

In [93], the authors describe a high-efficiency transmitter structure for terahertz (THz) frequencies. The antenna consists of two $10\,\mu$m-wide metal lines deposited on semi-insulating GaAs, with a separation of $100\,\mu$m. Two metal tabs extend out from these lines toward each other. Figure 7.30 illustrates the structure. We have modeled this system by using a $16 \times 16 \times 2$-grid ($160\,\mu\text{m} \times 160\,\mu\text{m} \times 20\,\mu\text{m}$) placed on a 1 mm thick plate having the nominal parameters for GaAs, $\sigma = 100$ S/m and $\varepsilon_{\text{rel}} = 10.9$ [94].

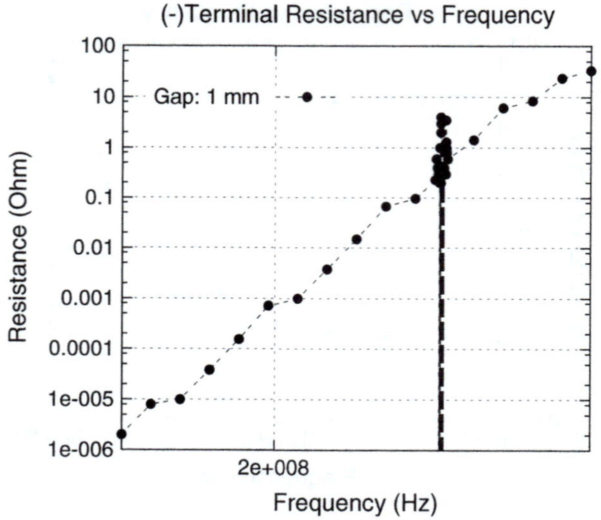

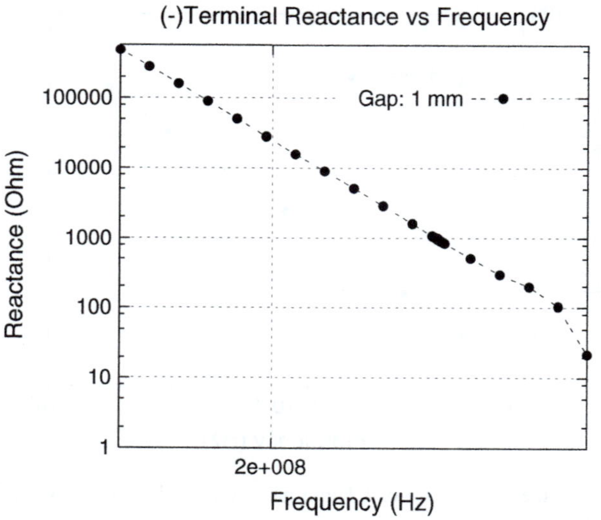

Fig. 7.17 Results for the "naked" driving-point terminal response for the $16 \times 16 \times 4$-cell 1-mm gap model. *Top*: resistance. *Bottom*: reactance

The results of the impedance calculation are shown in Figs. 7.31 and 7.32. It is clear from the high-frequency results shown in the bottom of Fig. 7.32 that the impedance response can be modeled by the series-parallel circuit of Fig. 7.33 up to a frequency of about 8×10^{11} Hz. We can estimate the parameters of Fig. 7.33 by using the data shown in Figs. 7.31 and 7.32.

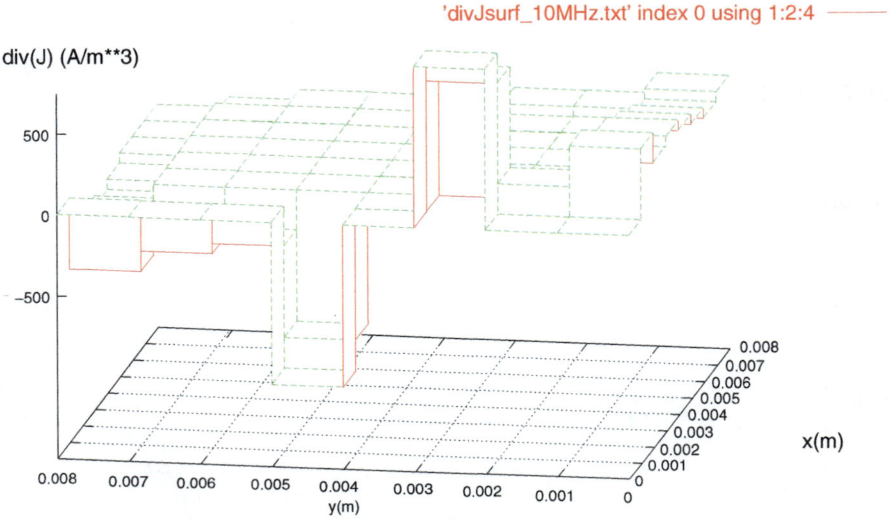

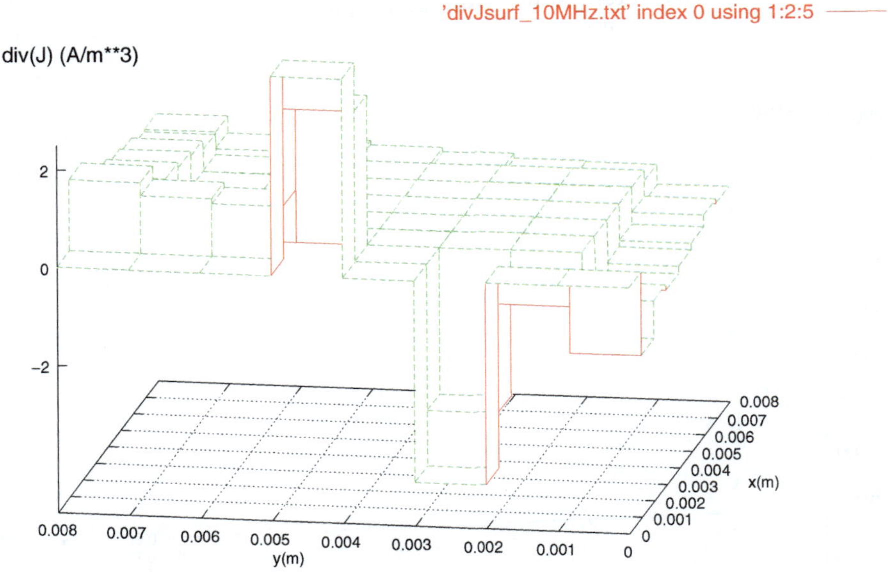

Fig. 7.18 Divergence of the current of the 1-mm gap model at 10 MHz. *Top*: real. *Bottom*: imaginary

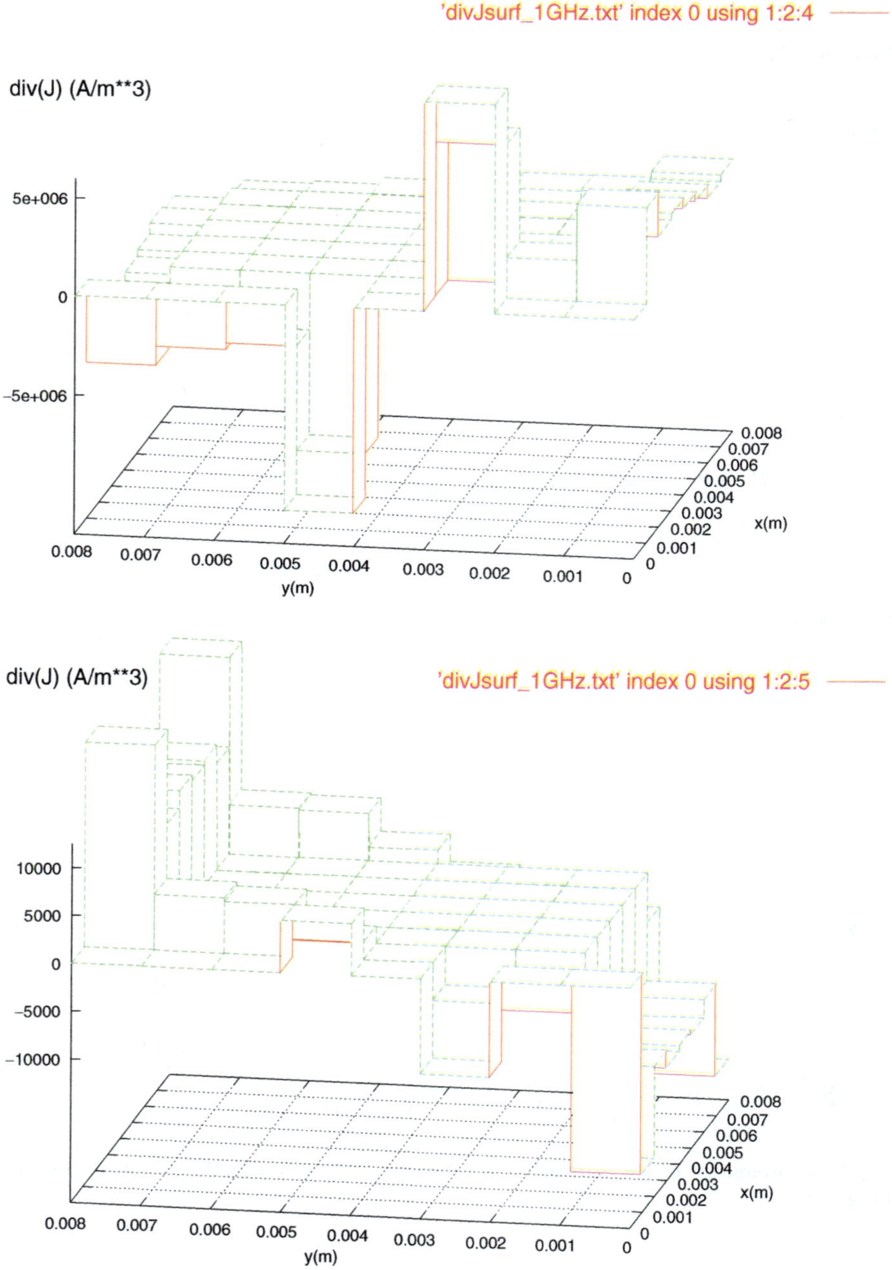

Fig. 7.19 Divergence of the current of the 1-mm gap model at 1 GHz. *Top*: real. *Bottom*: imaginary

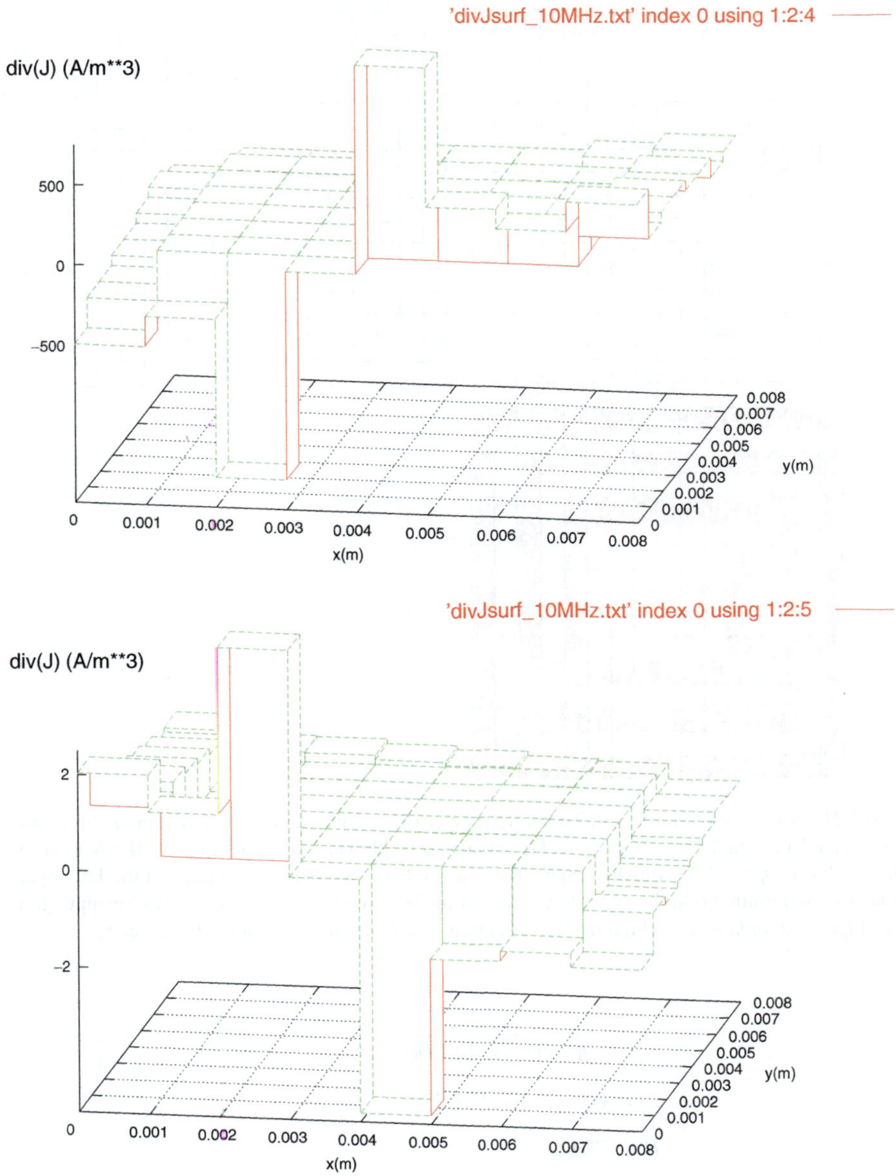

Fig. 7.20 Divergence of the current of the closed-loop model at 10 MHz. *Top*: real. *Bottom*: imaginary

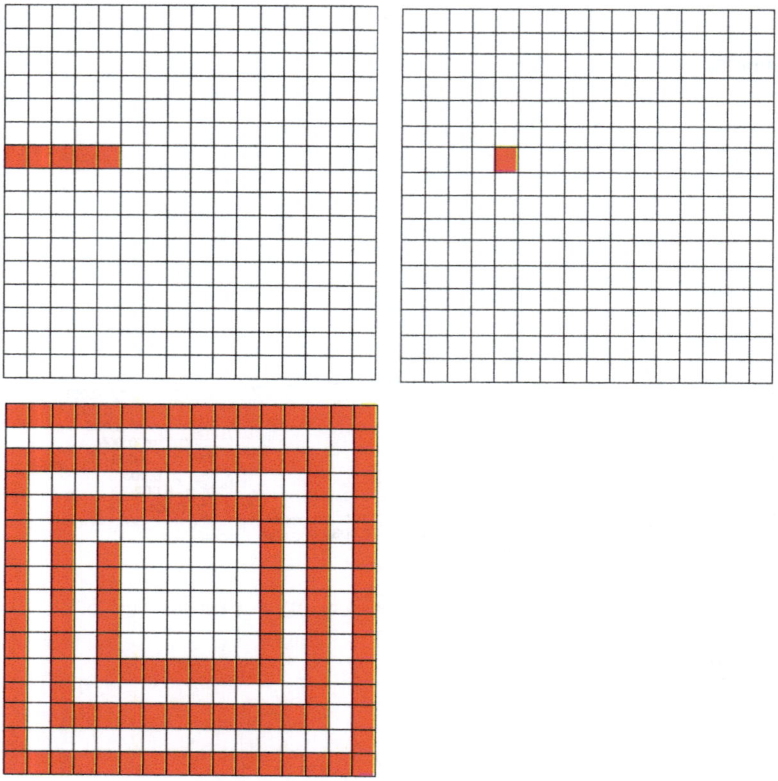

Fig. 7.21 A three-turn spiral coil. The bottom layer of the three-layer system contains the spiral coil (*red*). The middle layer *(upper right)* contains a single *via* (*red*) that connects the *bottom* to the *top* layer *(upper left)*, which contains the lead (*red*) to the negative terminal on the left edge. The positive terminal is at the upper left corner of the bottom-most layer. This makes the input gap one-layer (the *middle* layer) high in the z-direction, and five cells wide along the left edge

$$
\begin{aligned}
C_0 &\approx \frac{1}{2\pi \times 10^9 \times 24{,}000} \\
&= 6.63 \times 10^{-15}\,\mathrm{Fd} \\
R &\approx 70\Omega \\
L &\approx \frac{1}{(2\pi \times 3 \times 10^{11})^2 \times 6.63 \times 10^{-15}} \\
&= 4.25 \times 10^{-11}\,\mathrm{H} \\
C_1 &\approx \frac{1}{(2\pi \times 5 \times 10^{11})^2 \times 4.25 \times 10^{-11}} \\
&= 2.38 \times 10^{-15}\,\mathrm{Fd}\,. \qquad (7.10)
\end{aligned}
$$

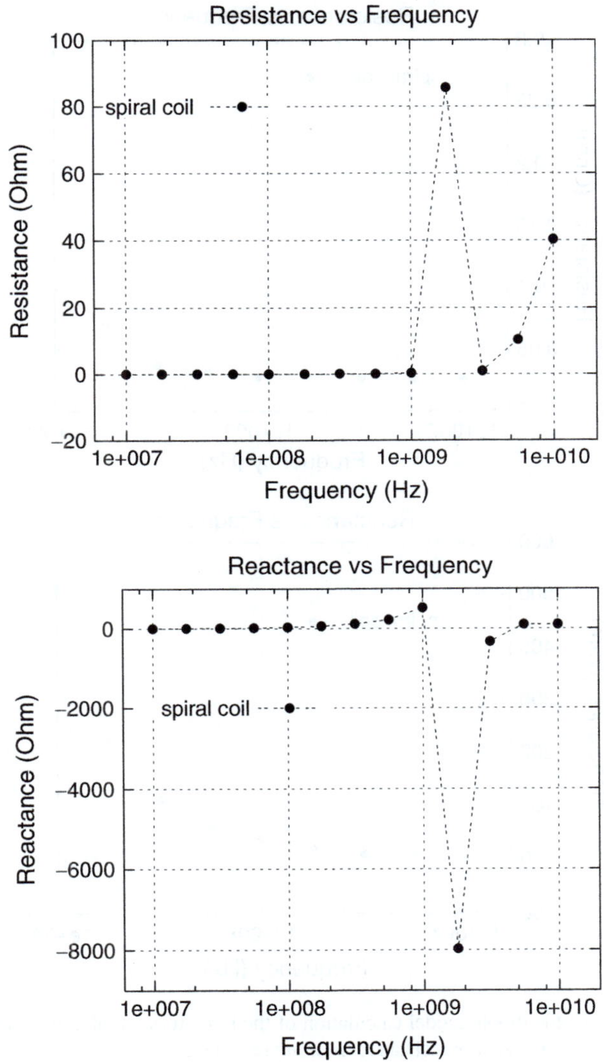

Fig. 7.22 Frequency scan of the three-turn spiral coil of Fig. 7.21 in freespace. *Top*: resistance. *Bottom*: reactance

We argue that because $C_1 \ll C_0$, the shunt capacitor is an open circuit at the lower frequencies, which means that the overall impedance response will be that of a series *RLC* resonant circuit. That is the condition shown with the resonance at about 3×10^{11} Hz. Immediately beyond the series resonance, the series branch is inductive, and resonates with the shunt capacitor, C_1. The typical characteristic of a parallel resonant circuit, in which the resistance rises significantly in the vicinity of the reactance cross-over point, is clearly shown, with the parallel resonant frequency

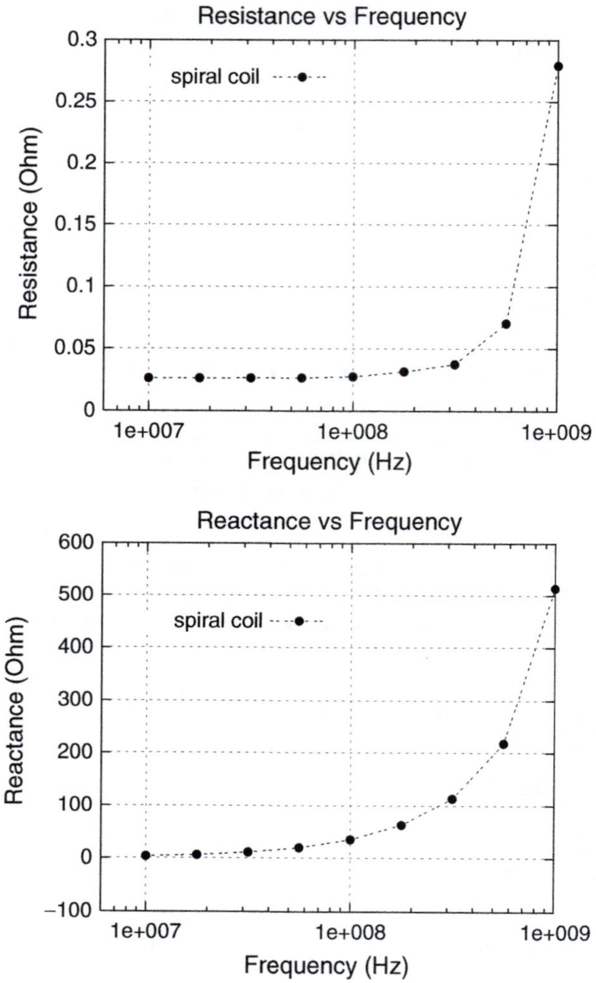

Fig. 7.23 Results of the dipole model calculation of the three-turn spiral coil in freespace below the resonance around 2 GHz. *Top*: resistance. *Bottom*: reactance

being approximately 5×10^{11} Hz. The wiggles beyond 8×10^{11} are attributed to parasitic elements that are not easily represented in a simple lumped equivalent circuit. This response is contrary to that shown in the earlier models, in which the parallel resonance occurs first. The reason for that is that the simple closed-loop circuit of the earlier models has a significant inductance, which dominates the low-frequency response, contrary to the THz transmitter, which is simply a pair of open terminals.

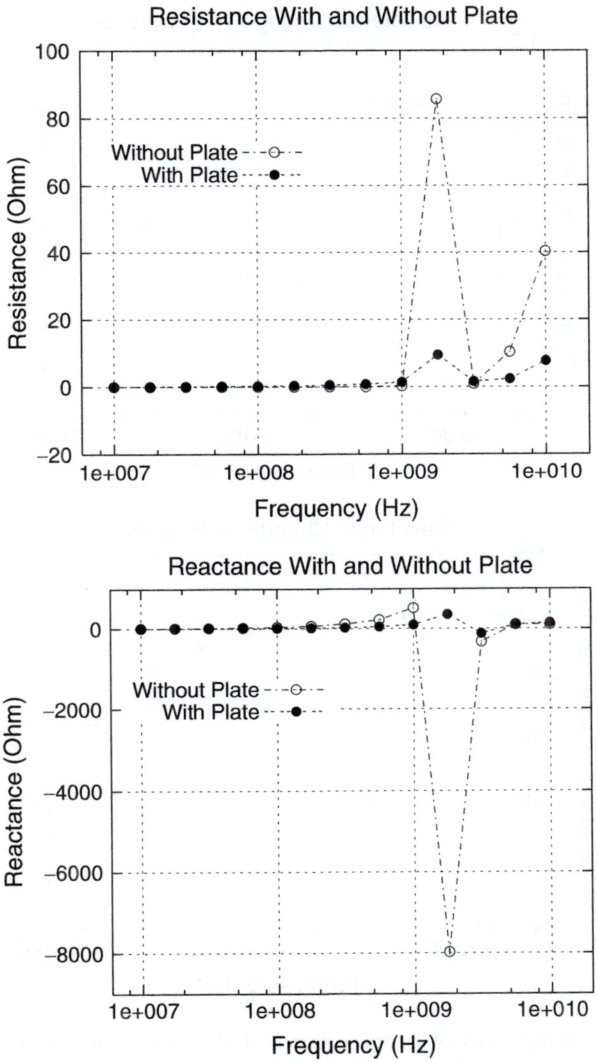

Fig. 7.24 Impedance of the dipole model calculation of the three-turn spiral coil in freespace and over a 1 mm plate. *Top*: resistance. *Bottom*: reactance

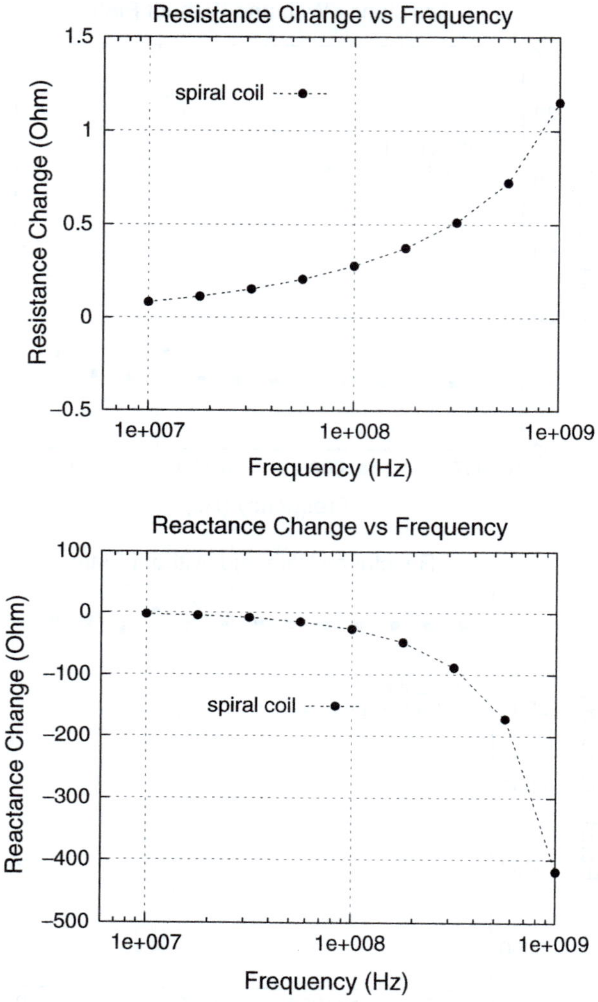

Fig. 7.25 Change in impedance of the dipole model calculation of the three-turn spiral coil over a 1 mm plate at low frequencies. *Top*: resistance. *Bottom*: reactance

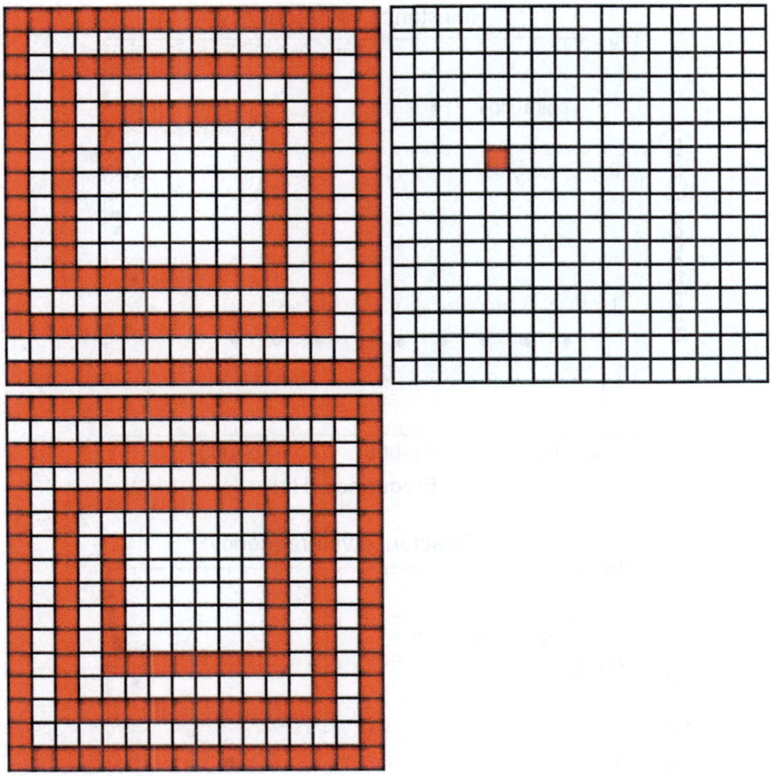

Fig. 7.26 A two-layered spiral coil

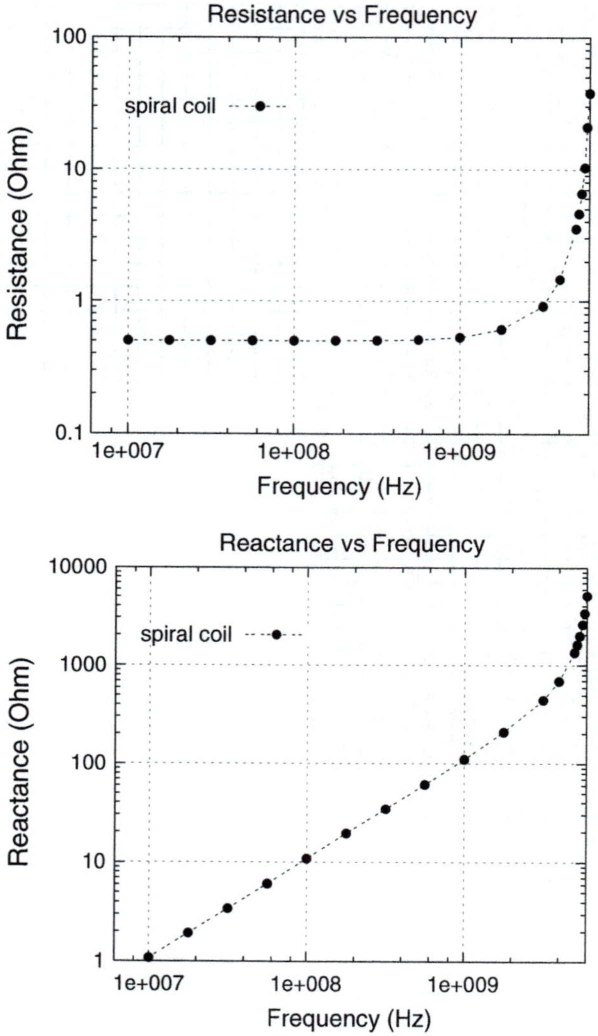

Fig. 7.27 Results of the dipole model calculation of the two-layered spiral coil in freespace below resonance. *Top*: resistance. *Bottom*: reactance

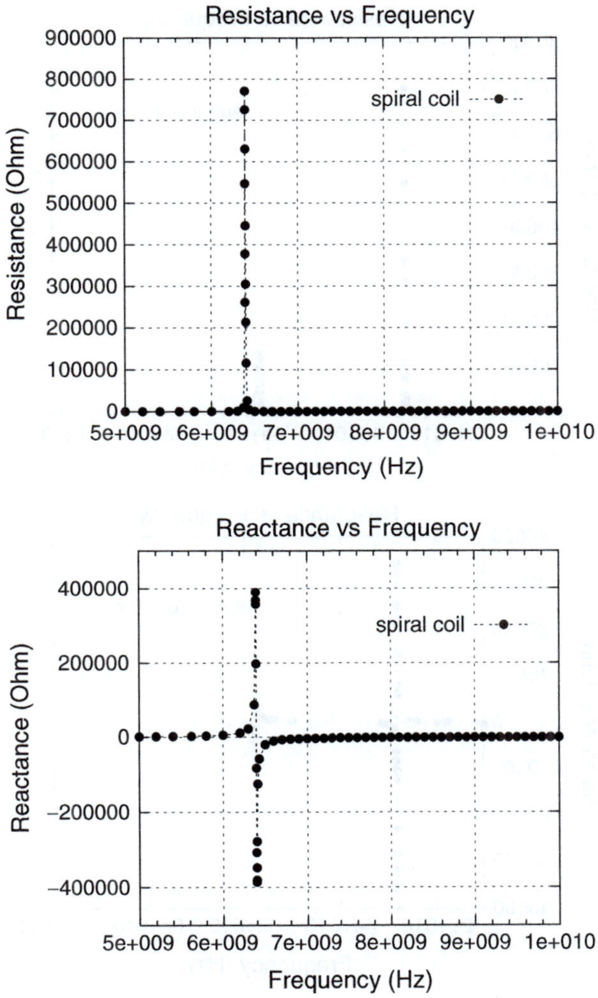

Fig. 7.28 Results of the dipole model calculation of the two-layered spiral coil in freespace through resonance. *Top*: resistance. *Bottom*: reactance

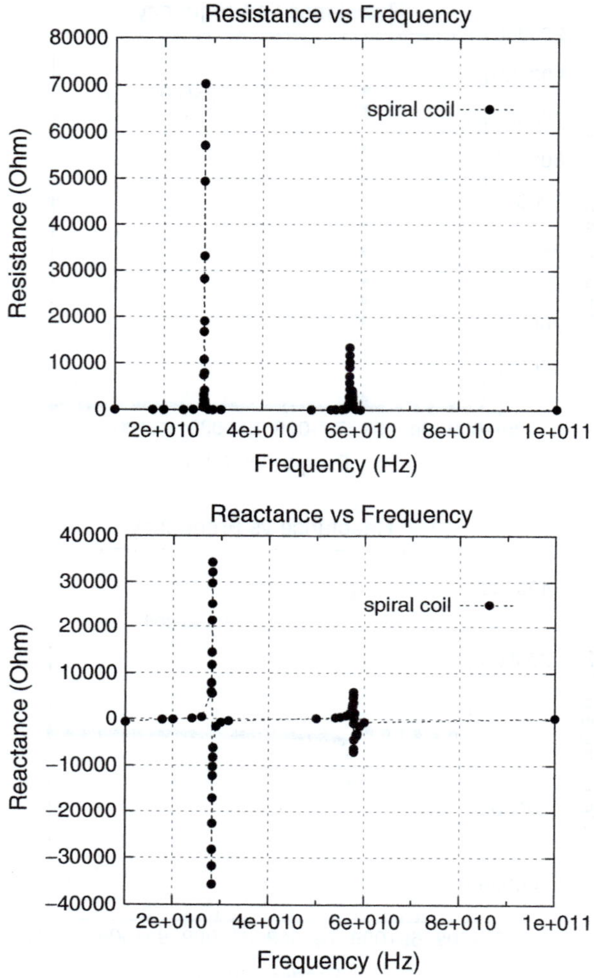

Fig. 7.29 Results of the dipole model calculation of the two-layered spiral coil in freespace at higher frequencies; two weaker resonances appear. *Top*: resistance. *Bottom*: reactance

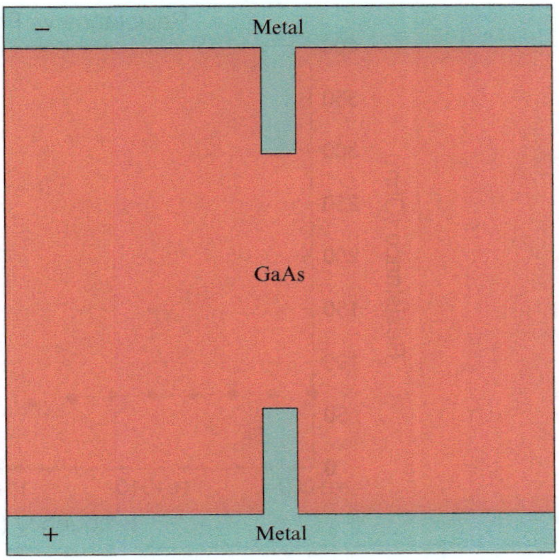

Fig. 7.30 Sketch of a THz transmitter (after Fig. 3 of [93])

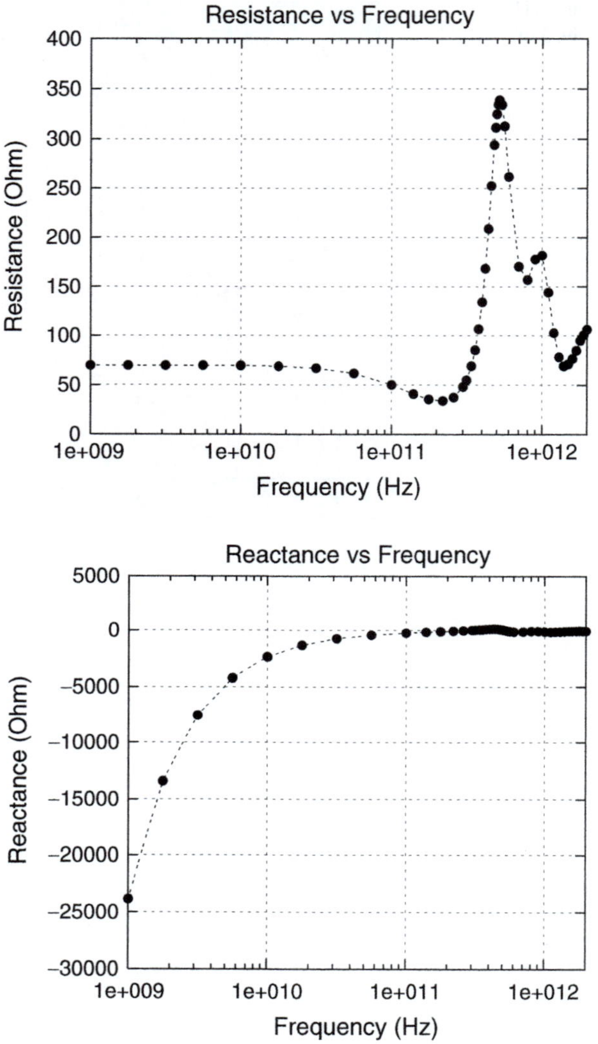

Fig. 7.31 Results of the dipole-model calculation for the THz transmitter of Fig. 3 of [93]. *Top*: resistance. *Bottom*: reactance

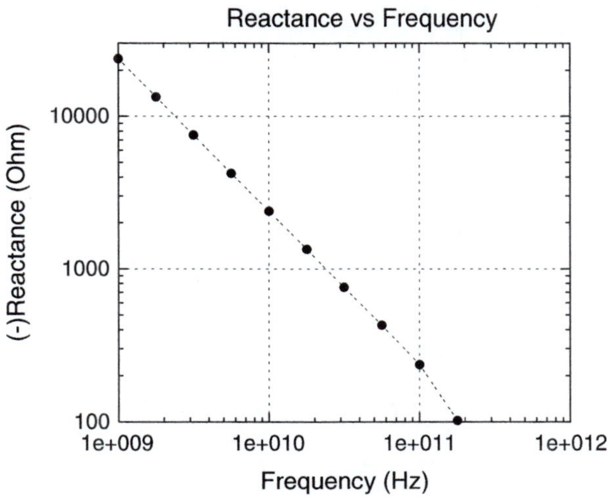

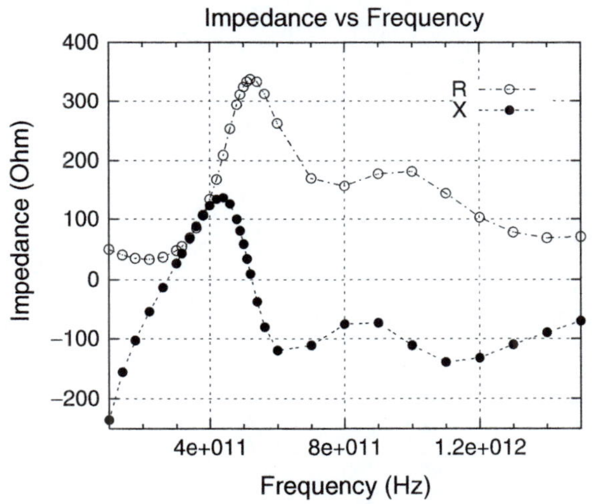

Fig. 7.32 Results of the dipole-model calculation for the THz transmitter of Fig. 3 of [93]. *Top*: low-frequency reactance. *Bottom*: high-frequency impedance

Fig. 7.33 A series-parallel circuit whose impedance is equivalent to that of Fig. 7.32 up to about 8×10^{11} Hz

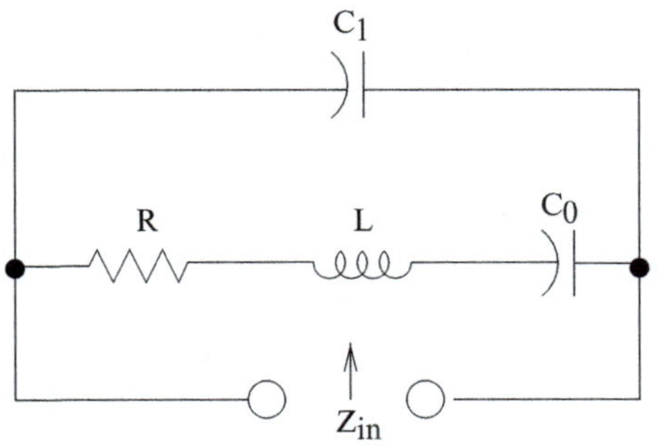

Chapter 8
Planar and Conforming Arrays of Probes

In the last chapter, we developed a volume-integral approach to the analysis of printed-circuit probes. Such probes are finding increased use in planar and conformable arrays, because of their small size and the ability to be manufactured in various shapes. Arrays are typically composed of a set of transmitter coils and receiver coils, which allows a variety of multiplexing schemes to generate responses with given features (see [96]). For example, one can generate a two-dimensional "scan" with a suitable array, without moving it. This can be very effective when doing voxel-based inversions, as will be discussed in a later volume of this work.

In this chapter we will discuss a model for a conforming array that consists of a periodic arrangement of transmit–receive coils in the azimuthal direction around a center axis (a "necklace"), which is used for the inspection of steam generator and heat-exchanger tubing in nuclear power plants. In the second part of the chapter we will describe a planar array of printed-circuit coils that has been developed to acquire data for a voxel-based inversion algorithm.

Throughout this chapter we implicitly make use of the reciprocity theorem of Chap. 5 as it is applied to bistatic arrays comprising a transmit and receive coil. Thus the impedances that we compute are transfer impedances.

8.1 Modeling a Circular Array

The ZETEC X-Probe is used for the rapid inspection of steam generator tubes in nuclear power plants.[1] It comprises a periodic azimuthal array of sensors, whose fundamental basis cell is shown in Fig. 8.1 (we have chosen the dimensions shown

---

[1]We are indebted to Prof. Sung-Jin Song of Sungkyunkwan University for bringing this probe design to our attention.

H.A. Sabbagh et al., *Computational Electromagnetics and Model-Based Inversion*, Scientific Computation, DOI 10.1007/978-1-4419-8429-6_8,

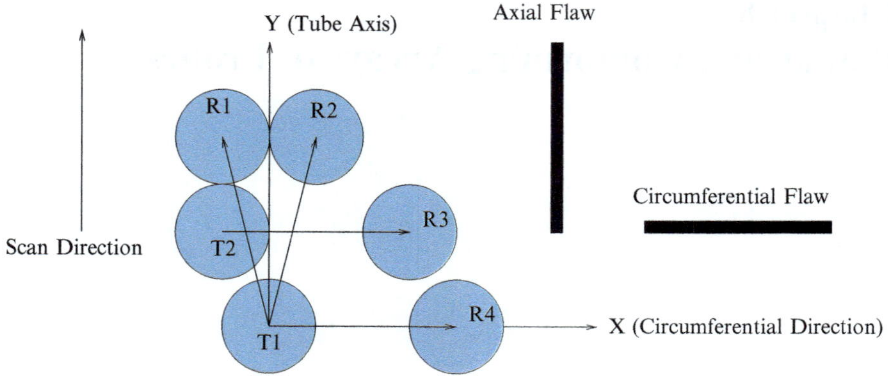

Coordinates(mm): T1(0,0)
T2(−2.5,5)
R1(−2.5,10)
R2(2.5,10)
R3(7.5,5)
R4(10,0)

Fig. 8.1 Showing the basis cell of the X-Probe coil arrangement. "T" stands for a transmit coil and "R" for a receive coil

in this figure). The array is designed for the inspection of steam generator tubes, and operates by being pulled through the tube in the axial direction. The configuration comprising T1R1 and T1R2 is sensitive to axial flaws, whereas T2R3 and T1R4 are sensitive to circumferential flaws. "T" stands for a transmit coil, and "R" for a receive coil.

We have created a model for the interaction of the basis cell with a flaw of dimensions 0.5 mm × 3 mm × 1.5 mm (width × length × height), located in an aluminum half-space with conductivity $\sigma = 2.801 \times 10^7$ S/m and excitation frequency of 100 kHz. The flaw is oriented in either the axial direction (the Y-axis in Fig. 8.1) or the circumferential direction (the X-axis in Fig. 8.1). The long dimension of the flaw determines its orientation.

We show the responses of T1R1 in the axial and circumferential directions and T1R4 in the circumferential direction in Fig. 8.2. The response of T1R2 is identical with that of T1R1 when the flaw is symmetrically placed with respect to the Y -axis, so that we show only the T1R1 response. Clearly, T1R1 is sensitive to axial flaws, while being relatively insensitive to circumferential flaws, while T1R4 is sensitive to circumferential flaws, exactly as advertised. It is interesting to note that the configuration T1R4 relative to the circumferential flaw is precisely the "parallel scan" configuration discussed in [95].

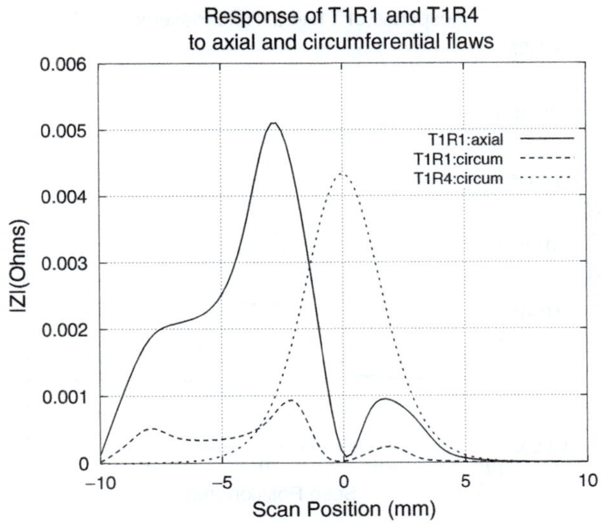

Fig. 8.2 Comparing the responses of T1R1 and T1R4 to axial and circumferential cracks

Table 8.1 Parameters for solid models

Solid model	A (mm)	B (mm)	Depth (mm)
Block	0.6125	0.6125	1.5
Hole	0.69	0.69	1.5
Cone	1.2	1.2	1.5
Semi-ellipsoid	0.846	0.846	1.5

A and B are transverse parameters

8.1.1 Modeling With the T1R4 Configuration

We'll do some modeling with the T1R4 configuration using various solid models shown in Table 8.1. These solids all have equal volumes, are surface-breaking, and have the same maximum depth, 1.5 mm, into the host. The A and B parameters determine the extent of the solid in the transverse (X,Y) plane. Our first exercise is to determine the sensitivity of the T1R4 configuration to the X-coordinate of the hole as the probe is scanned in the Y-direction. The X-coordinates of the hole are 1, 2, 3, 4, and 5 mm. Recall that the center of T1 is $X = 0$ mm and the center of R4 is $X = 10$ mm, which means that a hole location of 5 mm is at the center of the configuration. We do not need to go beyond 5 mm because of the obvious mirror symmetry about this point.

We contrast two conditions for the experiment; the first is with a host of aluminum, $\sigma_{\text{al}} = 2.801 \times 10^7$ S/m, and the second with a host of stainless steel, $\sigma_{\text{ss}} = 1 \times 10^6$ S/m. Both problems are run at a frequency of 100 kHz. The results are shown in Fig. 8.3. Two things, in particular, stand out in these results. In the

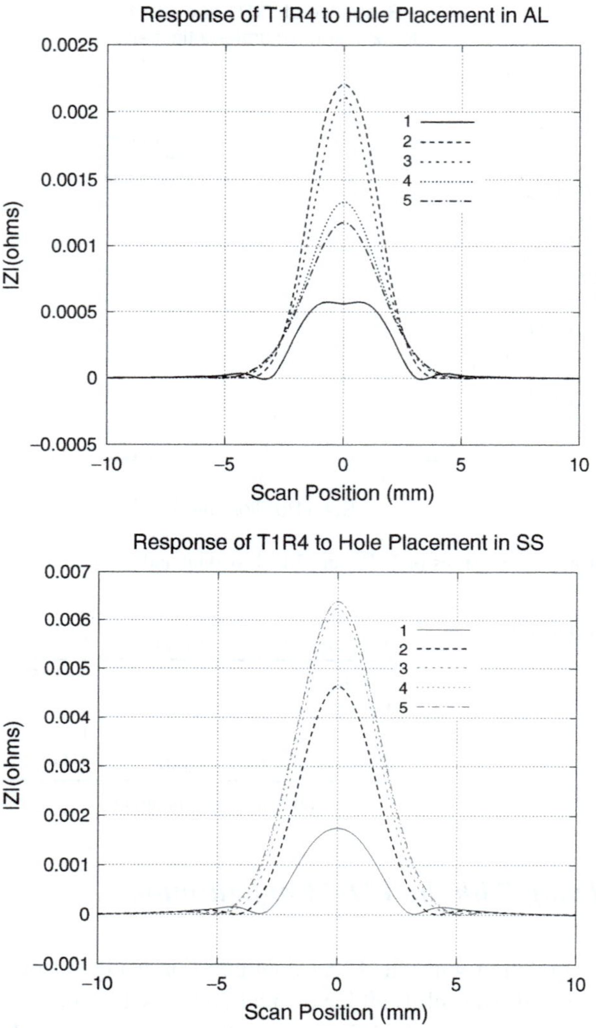

Fig. 8.3 Response of the T1R4 configuration to the X-position of a hole as the probe is scanned in the Y-direction. The coordinates of the hole are 1, 2, 3, 4, and 5 mm, and the frequency is 100 kHz. *Top*: aluminum substrate. *Bottom*: stainless-steel substrate

first place, the response in the aluminum substrate is not a monotonic function of the location of the hole, whereas it is in the stainless-steel substrate (locations 4 and 5 are virtually identical). This means that the parameters that define the mutual inductive coupling between the hole and either coil depend upon the electromagnetic properties of the structure as well as the frequency.

The second point is that the response is much larger in the stainless-steel substrate than in the aluminum. The reason for this is that the skin depth, which defines the

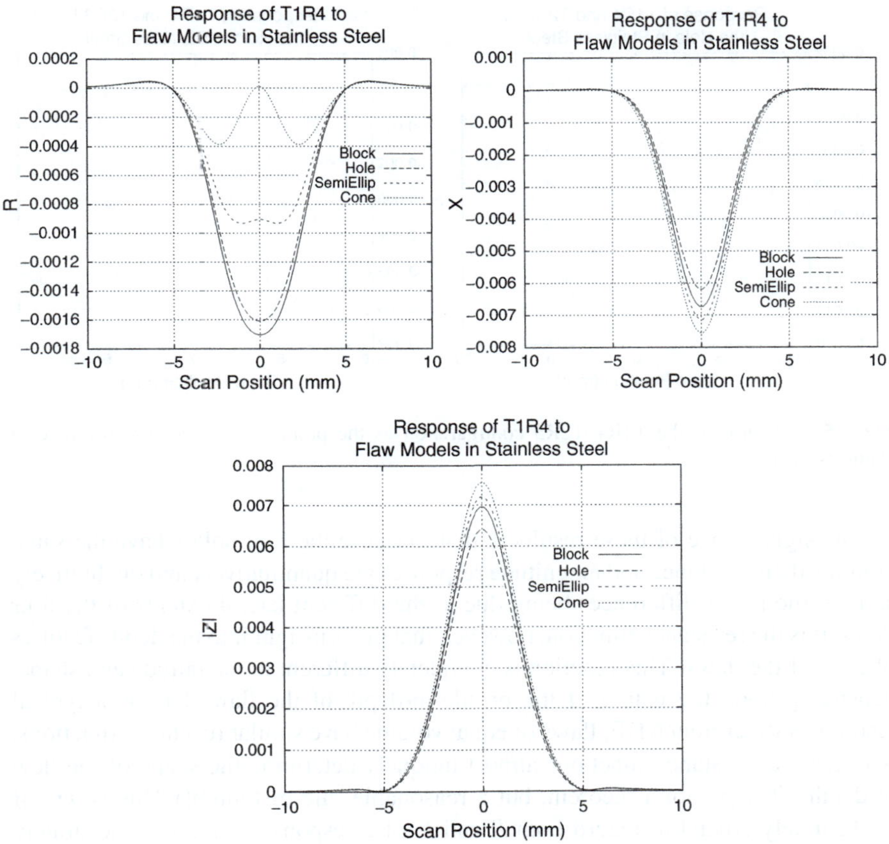

Fig. 8.4 Response of each of the four solid models to the T1R4 configuration, when each is centered at $X = 5, Y = 0$

penetration extent of the coil fields in any direction, is much smaller in aluminum than in stainless steel. This means that more of the field of either the transmit or receive coil in stainless steel interacts with the hole than in aluminum.

Models of the type that we have just computed are useful in making probability-of-detection estimates. For example, by choosing frequencies correctly for a given host material, so that the response with respect to flaw position is monotonic, as in the bottom part of Fig. 8.3, we can choose the spacing between the transmit and receive coils so that the response will have a given minimum value with respect to flaw position. This allows one to estimate the probability of detection for the given transmit–receive configuration.

The second exercise is to compute the response of the T1R4 configuration to each of the solid models of Table 8.1, when each is located midway between T1 and R4, i.e., at $X = 5$, $Y = 0$, in stainless steel, and the excitation frequency is 100 kHz. The responses are plotted in Fig. 8.4.

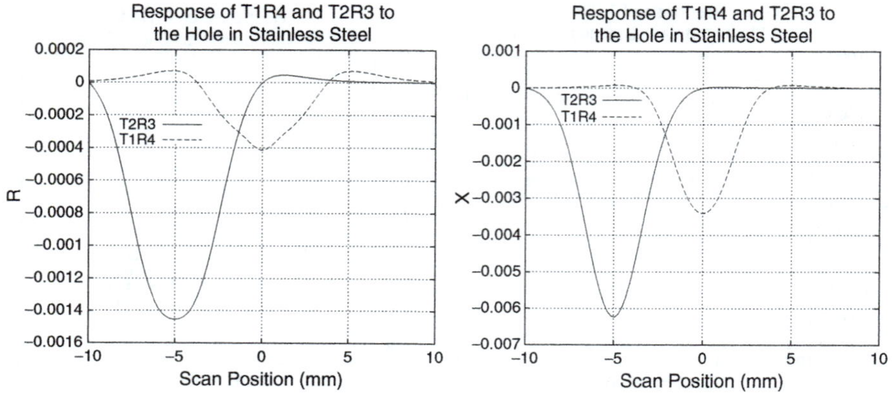

Fig. 8.5 Response of the T1R4–T2R3 configuration as the probe is scanned past the hole in stainless steel

The significance of these results is that, because the four solids have the same volume, their reactance and magnitude responses are quantitatively and qualitatively similar, the major differences being due to the different lateral extents of the four flaws. It is the resistance function, however, that gives insight into the depth-features of each of the flaws. This function is completely different in magnitude and shape, depending upon the nature of the profile-in-depth of the flaw. This is a typical result in eddy-current NDE; flaws of equal volume have similar reactance functions, whereas the resistance functions almost uniquely determine the shape of the flaw in depth. (This is not a theorem, but a reasonable rule of thumb!) This is why it is absolutely crucial to record both R and X of a response if one is to accurately reconstruct the flaw from impedance data. This holds true for any measurement technique; two "channels" are required, not just magnitude, say, or reactance.

We can think of the cavity of the flaw as storing magnetic energy, so that, if all conditions are equal, flaws of equal volume will store equal energy, and since the time-rate-of-change of magnetic energy determines the inductive reactance of a system, we can understand that the reactance function of equal-volume flaws will be equal, or nearly so.

8.1.2 Modeling with the T1R4–T2R3 Configuration

Consider the response of the combined T1R4–T2R3 configuration as the probe is scanned past the hole of Table 8.1, which is located at $(X = 1.6\,\text{mm}, Y = 0\,\text{mm})$ in stainless steel. The results are shown in Fig. 8.5. Because the T2R3 dipole is located at $Y = 5$, its response occurs before the response of the T1R4 dipole as the combined probe is scanned from $Y = -10$ to 10 mm. The advantage of the joint configuration is that one dipole, in this case, T2R3, captures more of the hole than the other, thereby giving an improved signal response.

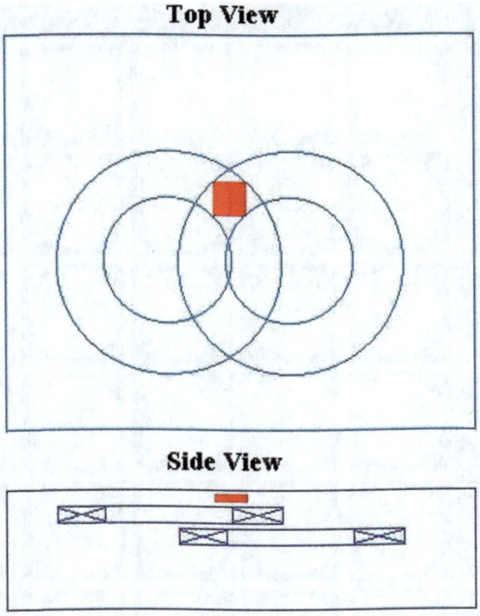

Fig. 8.6 Sketch of the T/R-array with two large drive-coils and an 8 × 8 array of receive coils in the *shaded rectangle*

8.2 Modeling a Planar T/R Array

Figure 8.6 illustrates a planar T/R array that is designed to produce benchmark data to validate voxel-based inversion algorithms. The algorithms are not of importance here; we intend to develop an approach to characterizing such arrays through techniques that have been developed for analyzing electrical (and microwave) N-port networks. The 8 × 8 array of printed-circuit receive-sensors is shown in Fig. 8.7.

The freespace response (reactance) of the sensor when drive 1 is excited at 1 MHz is shown in Table 8.2 and the response when drive 2 is excited at 1 MHz is shown in Table 8.3. The response at any other frequency is directly proportional to the frequency. In particular note the symmetries of the responses.

8.3 N-Port Theory of the T/R Array

The transmit coils are driven by the A-channel of an HP3577A network analyzer, and the receive coils are connected to the B-channel of the analyzer, as shown in Fig. 8.8. The electronic commutator at the transmit ports sequentially connects the two transmit coils, represented as Ports A and B, to the A-channel of the network analyzer. While connected to either of the transmit ports, the electronic commutator will sequentially connect each of the receive coils, represented as Ports 1 through 16 (for the 4 × 4 array), to the B-channel of the analyzer. Thus, the operation reduces to measuring a series of scattering coefficients for a simple 2-port network.

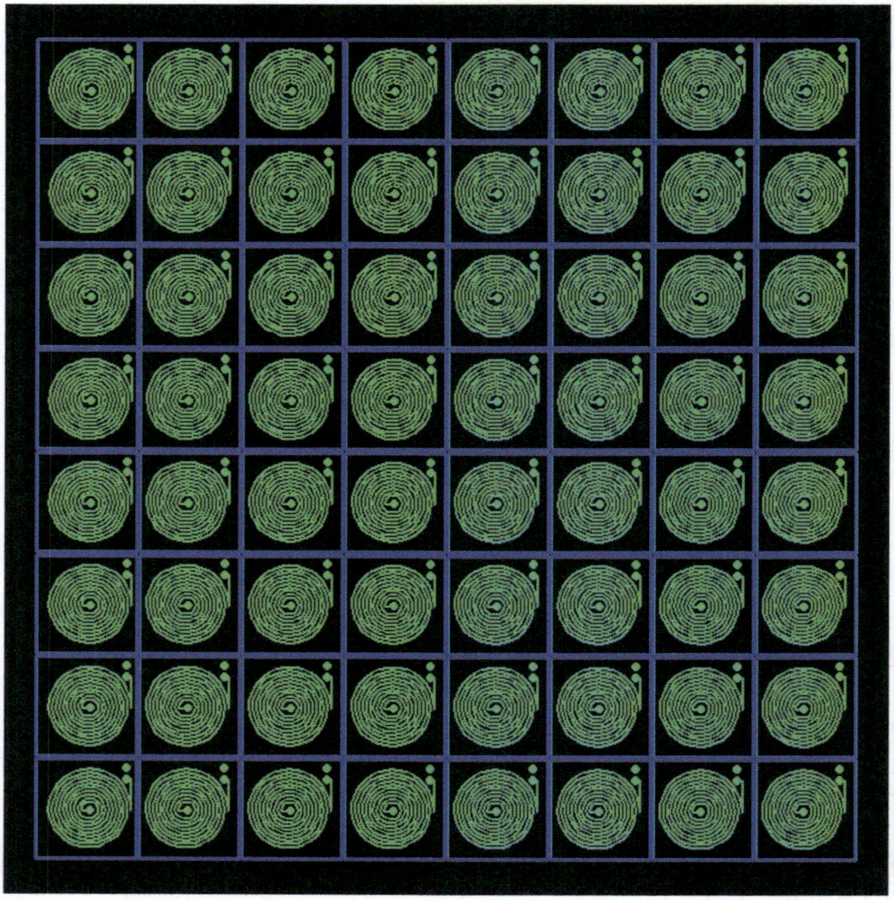

Fig. 8.7 Illustrating the 8×8 array of receive-sensors

For example, when transmit coil A is connected to the analyzer, we will measure the sequence of scattering coefficients, S_{AA}, S_{Ai}, S_{iA}, S_{ii}, for $i = 1,\ldots,16$. With these results, we can calculate the set of open-circuit transfer impedances from port A to port i via

$$Z_{Ai} = \frac{2Z_0 S_{Ai}}{1 - S_{AA} - S_{ii} + (S_{AA}S_{ii} - S_{Ai}S_{iA})}, \quad i = 1,\ldots,16, \tag{8.1}$$

where Z_0 is the characteristic impedance, $50\,\Omega$, of the transmission lines that connect the coils to the network analyzer channels. This relation is derived from the application of scattering theory to electrical networks. See, e.g., N. Balabanian and T. Bickart, *Electrical Network Theory*, J. Wiley & Sons, Inc., New York, 1969, Chap. 8, or H.J. Carlin and A.B. Giordano, *Network Theory*, Prentice-Hall, Inc.,

Table 8.2 Freespace response (reactance) of the sensor when drive 1 is excited at 1MHz

Row index	Column index 1	2	3	4	5	6	7	8
1	0.086667	0.07897	0.067643	0.054604	0.035038	0.016082	−0.011478	−0.049784
2	0.11716	0.10924	0.097852	0.082588	0.063816	0.042137	0.019594	−0.011792
3	0.14929	0.13967	0.12581	0.10896	0.089246	0.066876	0.041743	0.015422
4	0.18371	0.17076	0.15412	0.13412	0.11258	0.089016	0.063428	0.034018
5	0.22034	0.20369	0.18303	0.15891	0.13368	0.10812	0.081338	0.053026
6	0.26323	0.23871	0.21134	0.1824	0.15324	0.12421	0.09614	0.066032
7	0.31381	0.275	0.23754	0.20234	0.16917	0.13792	0.10758	0.076681
8	0.37296	0.31155	0.2607	0.21799	0.18134	0.14681	0.11491	0.084233

Table 8.3 Freespace response (reactance) of the sensor when drive 2 is excited at 1 MHz

Row index	Column index 1	2	3	4	5	6	7	8
1	−0.048081	−0.011022	0.015848	0.035748	0.054344	0.067393	0.078287	0.085688
2	−0.01108	0.019723	0.042402	0.064251	0.082398	0.096835	0.109	0.11698
3	0.015891	0.042336	0.0674	0.089226	0.10909	0.12534	0.13897	0.14869
4	0.035562	0.064129	0.08916	0.11344	0.13414	0.15422	0.17029	0.18254
5	0.053813	0.082191	0.10893	0.13406	0.15904	0.18267	0.20341	0.21926
6	0.067035	0.096412	0.12518	0.15409	0.18256	0.21131	0.23876	0.26188
7	0.078053	0.10861	0.13872	0.16998	0.20314	0.23853	0.27545	0.31296
8	0.085347	0.11685	0.14826	0.1822	0.21882	0.26143	0.31258	0.37377

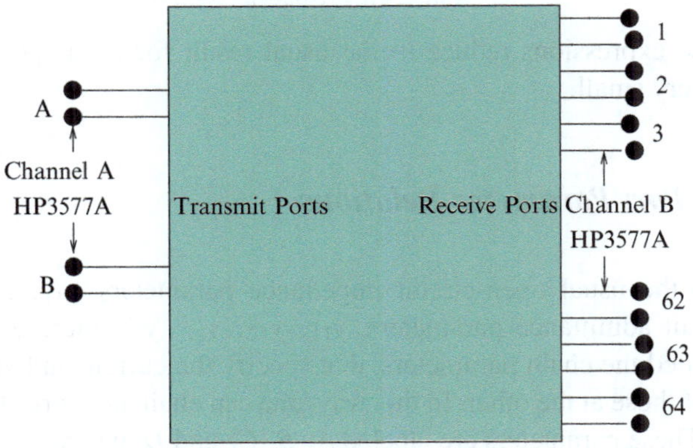

Fig. 8.8 Illustrating an N-port system representation of the T/R array and its connection to the HP3577A network analyzer

Englewood Cliffs, 1964, Chap. 4. For a reciprocal network of the type that we have, $S_{Ai} = S_{iA}$. Open-circuit transfer impedances are the input data for voxel-based inversion algorithms.

Table 8.4 Relationship between two-port parameters

Open-circuit impedance parameters		Short-circuit admittance parameters		Chain parameters	
z_{11}	z_{12}	$\frac{y_{22}}{\lvert y\rvert}$	$\frac{-y_{12}}{\lvert y\rvert}$	$\frac{A}{C}$	$\frac{AD-BC}{C}$
z_{21}	z_{22}	$\frac{-y_{21}}{\lvert y\rvert}$	$\frac{y_{11}}{\lvert y\rvert}$	$\frac{1}{C}$	$\frac{D}{C}$
$\frac{z_{22}}{\lvert z\rvert}$	$\frac{-z_{12}}{\lvert z\rvert}$	y_{11}	y_{12}	$\frac{D}{B}$	$\frac{-(AD-BC)}{B}$
$\frac{-z_{21}}{\lvert z\rvert}$	$\frac{z_{11}}{\lvert z\rvert}$	y_{21}	y_{22}	$\frac{-1}{B}$	$\frac{A}{B}$
$\frac{z_{11}}{z_{21}}$	$\frac{\lvert z\rvert}{z_{21}}$	$\frac{-y_{22}}{y_{21}}$	$\frac{-1}{y_{21}}$	A	B
$\frac{1}{z_{21}}$	$\frac{z_{22}}{z_{21}}$	$\frac{-\lvert y\rvert}{y_{21}}$	$\frac{-y_{11}}{y_{21}}$	C	D

For completeness, we'll write the open-circuit driving-point impedances at ports A and i that correspond to (8.1):

$$Z_{AA} = \frac{Z_0(1+S_{AA})(1-S_{ii})+S_{Ai}S_{iA}}{1-S_{AA}-S_{ii}+(S_{AA}S_{ii}-S_{Ai}S_{iA})}$$

$$Z_{ii} = \frac{Z_0(1-S_{AA})(1+S_{ii})+S_{Ai}S_{iA}}{1-S_{AA}-S_{ii}+(S_{AA}S_{ii}-S_{Ai}S_{iA})}\,. \tag{8.2}$$

Note that these expressions reduce to the usual result for a one-port structure if $S_{Ai} = S_{iA}$ are very small.

8.3.1 Two-Port Parameter Relations

In addition to the usual open-circuit impedance parameters, z_{11}, z_{12}, z_{21}, z_{22}, and short-circuit admittance parameters, y_{11}, y_{12}, y_{21}, y_{22}, there exists a set of parameters, called the chain parameters, that specify the current and voltage at one port in terms of those at the other. In this way, one can chain networks together very conveniently. These parameters are labeled A, B, C, and D, where

$$V_1 = AV_2 - BI_2$$

$$I_1 = CV_2 - DI_2\,. \tag{8.3}$$

Table 8.4 gives the relationship between the three sets of two-port parameters.[2] The vertical bars denote the determinant of the corresponding 2×2 matrix. It is a

[2] Taken from N. Balabanian and T. Bickart, *Electrical Network Theory*, J. Wiley & Sons, Inc., New York, 1969, Chap. 3.

simple matter to verify that the determinant of the chain matrix, $AD - BC = \frac{z_{12}}{z_{21}} = \frac{y_{12}}{y_{21}} = 1$, for reciprocal two-ports. Hence, for such networks the inverse chain matrix always exists.

Two independent two-ports can be chained together to produce a composite third two-port simply by multiplying their corresponding chain matrices:

$$\begin{bmatrix} A_3 & B_3 \\ C_3 & D_3 \end{bmatrix} = \begin{bmatrix} A_1 & B_1 \\ C_1 & D_1 \end{bmatrix} \times \begin{bmatrix} A_2 & B_2 \\ C_2 & D_2 \end{bmatrix} . \tag{8.4}$$

Let the matrix, $\mathbf{M}_3 = \mathbf{M}_1 \times \mathbf{M}_2$, represent the composite effect of the "model" driver-receive coil two-port, $\mathbf{M}_2$, and the remaining "parasitic" effects of the connections within the T/R array, $\mathbf{M}_1$. Then after measuring the scattering coefficients of the composite system in, say, freespace, by means of the HP3577A, and using (8.1) and (8.2) to compute the open-circuit impedance parameters, followed by application of the appropriate entries in Table 8.4, we have $\mathbf{M}_3$. We compute $\mathbf{M}_2$ by means of the **VIC-3D®** model, and when we right-multiply $\mathbf{M}_3$ by $\mathbf{M}_2^{-1}$, which is guaranteed to exist, we have the desired parasitic chain matrix:

$$\mathbf{M}_1 = \mathbf{M}_3 \times \mathbf{M}_2^{-1} . \tag{8.5}$$

If, then, we assume that the parasitic effects are unchanged whether the T/R array is in freespace or over a host material, we simply left-multiply the new measured composite chain matrix by $\mathbf{M}_1^{-1}$ to get the new model chain matrix that corresponds to the current test configuration. Using the top two lines of Table 8.4 allows us to transform from the A, B, C, D parameters of the model to the open-circuit impedance parameters that **VIC-3D®** uses.

If, as is likely, the parasitic effects change when the T/R array is over a host, we redo the procedure described above by replacing freespace by an unflawed version of the host, or a metal whose conductivity is close to that of the host.

8.3.2 Calculation of Freespace Chain Matrix at 1 MHz

There are 128 chain matrices for the two drive coils and 64 receive coils. We'll compute a few of them here. First, we note that the freespace driving-point impedance of either drive coil is $Z_D = j562.64\,\Omega$ at 1 MHz, and the driving-point impedance of each receive coil is $Z_R = j5.7559\,\Omega$ in freespace at 1 MHz. These are model values computed by **VIC-3D®**. The entries in Tables 8.2 and 8.3 are the open-circuit transfer impedances between the appropriate drive coil and each of the receive coils, also calculated by **VIC-3D®**.

We'll label each receive coil by its row-column indices, and each drive coil by either A or B, with A corresponding to drive 1, and B to drive 2. Thus, our notation

Table 8.5 DC resistance (Ω) of several receive coils

Connector	7	7	7	7	9	9	9	9
Pins	5	4	3	2	5	4	3	2
	2.7	2.9	3.2	3.4	2.7	3.0	3.2	3.4

Table 8.6 Measured freespace scattering parameters at 1MHz taken with the HP3577A

		Connector	7	7	7	7
		Pins	5	4	3	2
S_{11}	Mag	dBm	−0.133	−0.133	−0.133	−0.133
	Phase	degrees	−3.957	−3.992	−4.016	−4.044
S_{21}	Mag	dBm	−62.4	−62.6	−60.3	−59.4
	Phase	degrees	162	156	159	158
S_{12}	Mag	dBm	−62.4	−62.3	−60.2	−59.5
	Phase	degrees	162	157	159	158
S_{22}	Mag	dBm	−1.101	−1.199	−1.280	−1.370
	Phase	degrees	160.70	0160.70	0160.50	0160.44

yields $Z_{AA} = Z_{BB} = j562.64$, $Z_{ij,ij} = j5.7559$, for all ij, ij, and, for example, $Z_{A,45} = Z_{45,A} = j0.11258$. With this notation, we'll compute the chain matrix for the coupled pair $(A,45)$ using the relationships of Table 8.4:

$$\begin{bmatrix} A & B \\ C & D \end{bmatrix}_{(A,45)} = \begin{bmatrix} \dfrac{562.64}{0.11258} & \dfrac{-562.64 \times 5.7559 + (0.11258)^2}{j0.11258} \\[2ex] \dfrac{1}{j0.11258} & \dfrac{5.7559}{0.11258} \end{bmatrix}$$

$$= \begin{bmatrix} 4997.69 & j28766.09 \\ -j8.88257 & 51.1272 \end{bmatrix}. \tag{8.6}$$

8.3.3 Application to Measured Data at 1 MHz

The dc resistance of the two drive coils are 14.0 Ω (yellow–orange) and 15.3 Ω (red–brown). The dc resistance for several receive coils are listed in Table 8.5.

Table 8.6 shows measured freespace scattering parameters taken at 1 MHz with the HP3577A, and Table 8.7 gives the scattering parameters expressed as complex numbers.

Using (8.1) and (8.2) and Table 8.7, we compute the open-circuit driving-point and transfer impedances at 1 MHz and list the results in Table 8.8. It is clear from these results that the driver coil is operating beyond resonance because of the large negative (capacitive) reactance in S_{11}, as well as a large resistance component.

Table 8.7 Measured freespace scattering parameters at 1 MHz expressed as complex numbers

Connector	7	7	7	7
Pins	5	4	3	2
S_{11}	0.9848	0.9848	0.9848	0.9848
	$0.9976 - j0.069$	$0.9976 - 0.0696$	$0.9975 - j0.070$	$0.9975 - j0.0705$
S_{21}	7.5858×10^{-4}	7.4131×10^{-4}	9.661×10^{-4}	10.715×10^{-4}
	$-0.951 + j0.3090$	$-0.9135 + j0.4067$	$-0.9336 + j0.3584$	$-0.9272 + j0.3746$
S_{12}	7.5858×10^{-4}	7.4131×10^{-4}	9.661×10^{-4}	10.715×10^{-4}
	$-0.951 + j0.3090$	$-0.9135 + j0.4067$	$-0.9336 + j0.3584$	$-0.9272 + j0.3746$
S_{22}	0.8809	0.8711	0.8630	0.8541
	$-0.9438 + j0.3305$	$-0.9438 + j0.3305$	$-0.9426 + j0.3338$	$-0.9423 + j0.3348$

Table 8.8 Computed Z-parameters at 1 MHz

Pin	Z_{11}	$Z_{12} = Z_{21}$	Z_{22}
5	0.306552D + 03,	−0.561103D − 01,	0.325738D + 01,
	−0.137948D + 04	0.580112D + 00	0.846631D + 01
4	0.300809D + 03,	0.642640D − 02,	0.354377D + 01,
	−0.136906D + 04	0.567862D + 00	0.845987D + 01
3	0.298757D + 03,	−0.308957D − 01,	0.378596D + 01,
	−0.136126D + 04	0.738724D + 00	0.854385D + 01
2	0.294125D + 03,	−0.182279D − 01,	0.405050D + 01,
	−0.135279D + 04	0.818309D + 00	0.856359D + 01

The first component listed is the resistance, and the second the reactance (Smatrix.f90)

This has an effect on the other parameters. For example, from Tables 8.2 and 8.3, we expect the reactance of Z_{12} to have much smaller values than those shown in Table 8.8.

By using the data of Table 8.8 in the two-port relations of Table 8.4, we compute the ABCD-matrices for the array at 1 MHz. The results are shown in Table 8.9. The integer at the left is the pin number, and the complex number beneath each matrix is the determinant of the matrix. Note that each determinant is equal to the theoretical value of unity.

8.3.4 Application of Two-Port Theory to Measured Data

Now, we'll apply the chain-matrix algorithm described in (8.4) to the measured data. In this example, we'll let $\mathbf{M}_3$ correspond to the pin 5 measured data and $\mathbf{M}_2$ correspond to the Drive 2(1,1) model data. When we apply (8.4) to these matrices, we compute $\mathbf{M}_1$ to be as shown in Table 8.10. When we substitute the Z-parameter data for $\mathbf{M}_1$ in Table 8.10 into the generic T-network 2-port of Fig. 8.9, we obtain the equivalent 2-port T-network for the parasitic structure shown in Fig. 8.10. Though we do not need to show the parasitic network, since we are not trying to synthesize it (in fact, we're trying to eliminate it!), it is interesting to note that it is physically realizable with passive elements.

Table 8.9 ABCD matrices for the array based upon the HP3577A measurements at 1 MHz

2	−0.166034D+04	−0.322446D+03	−0.396387D+04	−0.155253D+05
	−0.272074D−01	−0.122143D+01	0.103496D+02	−0.518038D+01
	0.10000D+01	0.11595D−11		
3	−0.185639D+04	−0.326783D+03	−0.423618D+04	−0.170986D+05
	−0.565164D−01	−0.135132D+01	0.113315D+02	−0.559892D+01
	0.10000D+01	−0.15228D−11		
4	−0.240460D+04	−0.556934D+03	−0.380976D+04	−0.223168D+05
	0.199263D−01	−0.176077D+01	0.149665D+02	−0.607118D+01
	0.10000D+01	−0.22027D−12		
5	−0.240655D+04	−0.295666D+03	−0.533580D+04	−0.213383D+05
	−0.165187D+00	−0.170783D+01	0.139209D+02	−0.696156D+01
	0.10000D+01	0.18492D−11		

The integer at the left denotes the pin, and the complex number beneath each array is the value of the determinant of the matrix (ABCD_matrix_HP3577A.f90)

Table 8.10 Data associated with $\mathbf{M}_1$ (system_matrix.f90)

ABCD matrix elements of M1			
−0.155705D+06	0.146371D+06	0.823540D+08	0.876060D+08
−0.125013D+03	−0.850823D+02	−0.478708D+05	0.703375D+05
Z-parameter matrix elements of M1			
0.306623D+03	−0.137952D+04	−0.546689D−02	0.372069D−02
0.419358D−03	−0.562640D+03		
Determinant of M1			
0.999999D+00	0.413507D−06		
ABCD matrix elements of M1inv			
−0.478708D+05	0.703376D+05	−0.823541D+08	−0.876060D+08
0.125014D+03	0.850824D+02	−0.155705D+06	0.146371D+06
Consistency check: M1inv x M3 = M2			
−0.117019D+05	0.483884D−02	−0.278519D−01	−0.673551D+05
0.860027D−05	0.207983D+02	−0.119713D+03	0.495023D−04

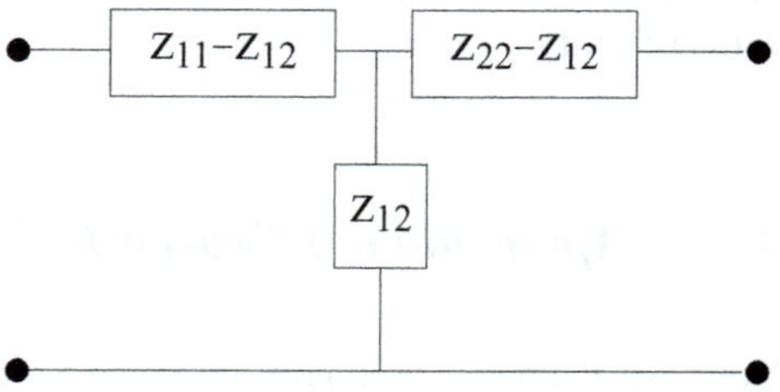

Fig. 8.9 Showing a 2-port T-network constructed from knowledge of Z_{11}, $Z_{12} = Z_{21}$, and Z_{22}

Figure 8.11 shows the networks associated with $\mathbf{M}_1$ and $\mathbf{M}_2$ coupled together between the new ports 1 and 2. The coupled network has the same Z-parameters as originally tabulated for Pin 5 in Table 8.8. (We'll leave this as an exercise for the reader.) This shows the consistency between the algebra of ABCD-matrices and simple network theory.

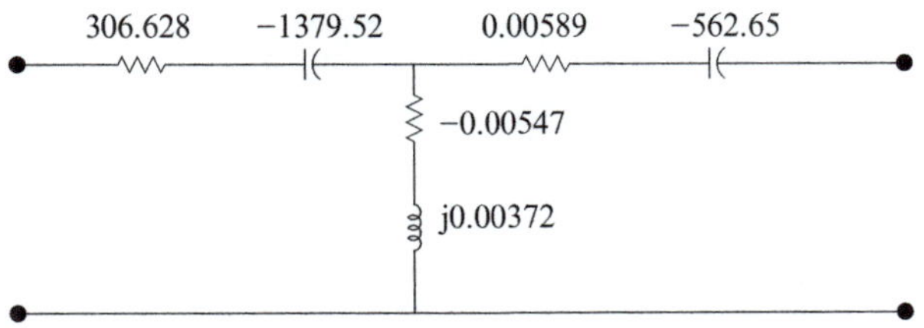

Fig. 8.10 Showing the equivalent T-network 2-port for the parasitic structure

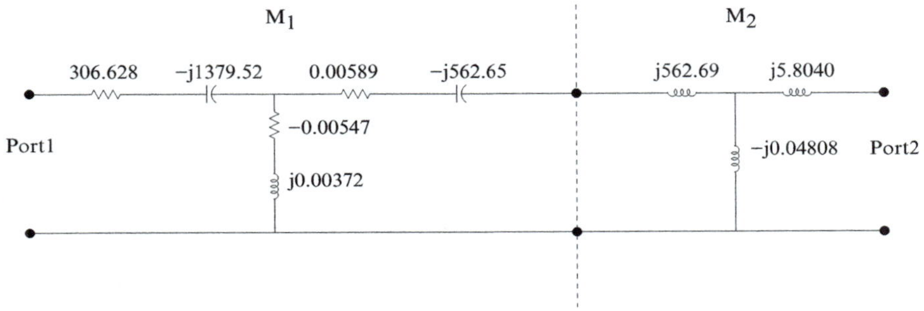

Fig. 8.11 Showing the coupled networks that correspond to $\mathbf{M}_1$ (the parasitic elements) and $\mathbf{M}_2$ (the model data for Drive 2(1,1)). The composite network extending from Port1 to Port2 has the same Z-parameters as shown in Table 8.8 for Pin 5

8.3.5 Modeling N-Port Array on an Aluminum Host: $\sigma_{\mathbf{Al}} = 2.801 \times 10^7$ S/m

Let $\mathbf{M}_3 = \mathbf{M}_1 \times \mathbf{M}_2$, where

$$\begin{aligned}\mathbf{M}_1 &= \text{“parasitic” effects measured on unflawed Al workpiece}\\ \mathbf{M}_2 &= \text{model results on unflawed Al workpiece computed by VIC3D(c)}\\ \mathbf{M}_3 &= \text{measured data on unflawed Al workpiece}\end{aligned} \tag{8.7}$$

and let

$$\begin{aligned}\mathbf{M}_1 &= \text{“parasitic” effects measured on flawed Al workpiece (same as above)}\\ \mathbf{M}_2' &= \text{model results on flawed Al workpiece}\\ \mathbf{M}_3' &= \text{measured data on flawed Al workpiece}\end{aligned} \tag{8.8}$$

Compute $\mathbf{M}_1 = \mathbf{M}_3 \times \mathbf{M}_2^{-1}$ from (8.7) and $\mathbf{M}_2' = \mathbf{M}_1^{-1} \times \mathbf{M}_3'$ from (8.8). The anomalous data for inversion are contained in $\mathbf{M}_2' - \mathbf{M}_2$. Actually, all we want is $\delta Z_{12} = \dfrac{1}{C_2'} - \dfrac{1}{C_2}$, where C_2 and C_2' are entries in the ABCD-matrices for $\mathbf{M}_2$ and $\mathbf{M}_2'$, respectively.

Chapter 9
Multilayered Media with Cylindrical Geometries

9.1 Introduction

In this chapter, we describe the mathematical development of a general axisymmetric model for **VIC-3D**®. This model is capable of analyzing tubes with tube supports and roll-expanded transition zones. Features such as magnetite and sludge, are included, and materials may be either ferromagnetic or nonmagnetic. The model described in this chapter will include only differential (or absolute) bobbin coils.

Flaws, or anomalies, can be of three types: (1) axisymmetric (such as circumferential rings, tube supports, roll-expanded transition zones), (2) thin axially-oriented cracks, and (3) user-defined flaws, such as intergranular attack (IGA) or corrosion pits. The incident fields due to bobbin coils are computed using the axisymmetric theory developed in this chapter.

9.2 Some Typical Problems in Steam Generator Tubing

There are a number of rather complicated geometries that appear in the inspection of steam generator tubing by means of eddy currents. Figures 9.1–9.3 illustrate several of them, and Fig. 9.4 illustrates a number of different flaws that must be modeled. The model must not only contend with these geometries, but it must deal with ferromagnetic bodies, as well.

9.3 Coupled Ferromagnetic Integral Equations

It is possible to account for ferromagnetic effects by introducing an anomalous magnetic current, together with magnetic–magnetic, magnetic–electric, and electric–magnetic Green functions. For this problem, which is axisymmetric, and

H.A. Sabbagh et al., *Computational Electromagnetics and Model-Based Inversion*, Scientific Computation, DOI 10.1007/978-1-4419-8429-6_9,

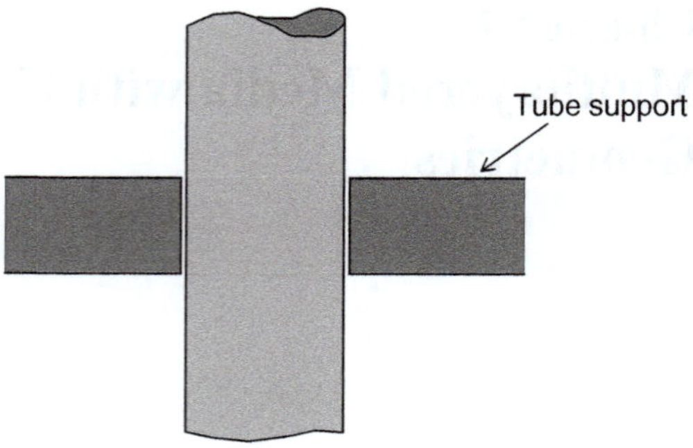

Fig. 9.1 Illustrating a steam generator tube with a tube support

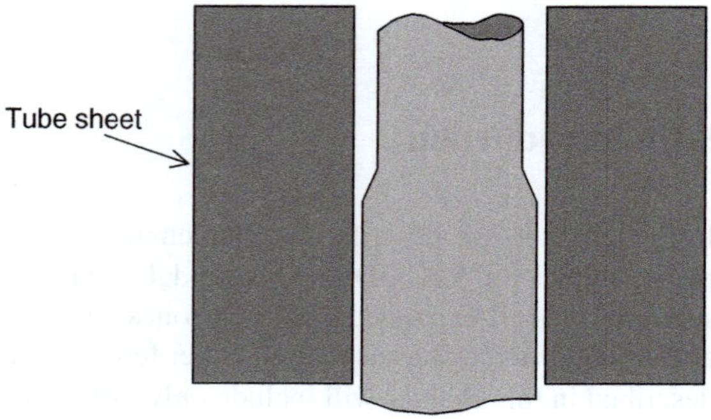

Fig. 9.2 Illustrating a roll-expanded steam generator tube

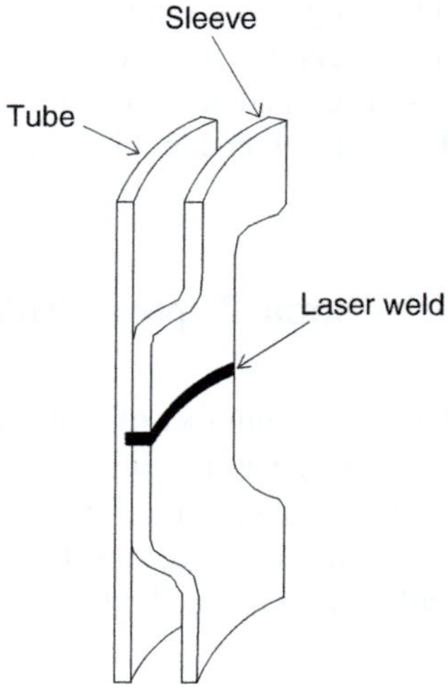

Fig. 9.3 Illustrating a laser-welded sleeve in a steam generator tube

therefore solenoidal in the electric variables, it will be easier to use an "all electric" model for the coupled system, following the same ideas that we developed in the context of Amperian currents in Chap. 3. The electric–electric Green function is simple, and because all electric variables are solenoidal, the matrices should be well conditioned for all frequencies.

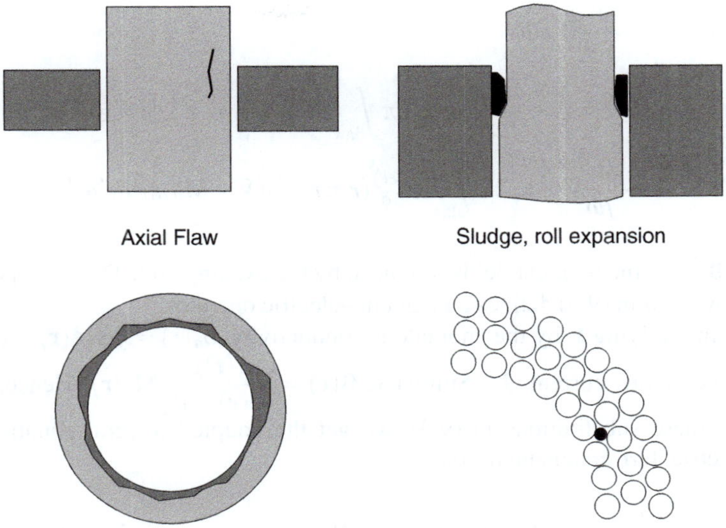

Fig. 9.4 Illustrating a variety of anomalies ("flaws") in steam generator tubes

We start with Maxwell's equations

$$\begin{aligned} \nabla \times \mathbf{E} &= -j\omega\mathbf{B} \\ \nabla \times \mathbf{H} &= j\omega\mathbf{D} + \mathbf{J}^{(e)} . \end{aligned} \tag{9.1}$$

Now $\mathbf{H} = \mathbf{B}/\mu(\mathbf{r}) = \mathbf{B}/\mu_h + \mathbf{B}/\mu(\mathbf{r}) - \mathbf{B}/\mu_h = \mathbf{B}/\mu_h - \mathbf{M}_a$, where μ_h is the host permeability and $\mathbf{M}_a$ is the anomalous magnetization vector. Thus the second of Maxwell's equations may be written

$$\nabla \times (\mathbf{B}/\mu_h) = j\omega\mathbf{D} + \mathbf{J}^{(e)} + \nabla \times \mathbf{M}_a , \tag{9.2}$$

which makes clear that the Amperian current, $\nabla \times \mathbf{M}_a$, is an equivalent anomalous electric current that arises because of the departures of the magnetic permeability of the workpiece from the host permeability, μ_h. $\mathbf{J}^{(e)}$, on the other hand, is an electric current that includes the anomalous current that arises due to differences in electrical conductivity; $\mathbf{J}^{(e)} = \sigma_h\mathbf{E} + (\sigma(\mathbf{r}) - \sigma_h)\mathbf{E} = \sigma_h\mathbf{E} + \mathbf{J}_a$.

The problem is axisymmetric, which means that $\mathbf{E}$, $\mathbf{J}_a$, and $\nabla \times \mathbf{M}_a$ have only a nonzero ϕ-component, whereas $\mathbf{M}_a$ has z- and r-components. Hence, we have as the solutions of (9.1) and (9.2)

$$\begin{aligned} \mathbf{E}(r,z) &= \mathbf{E}^{(i)}(r,z) + \mathbf{a}_\phi 2\pi \int_{\text{flaw}} G_{\phi\phi}^{(ee)}(r,z;r',z') J_a(r',z') r' dr' dz' \\ &\quad + \mathbf{a}_\phi 2\pi \int_{\text{flaw}} G_{\phi\phi}^{(ee)}(r,z;r',z') (\nabla \times \mathbf{M}_a)_\phi r' dr' dz' \end{aligned}$$

$$\begin{aligned}
\mathbf{B}(r,z) &= -\frac{1}{j\omega}\nabla\times\mathbf{E} \\
&= \mathbf{B}^{(i)}(r,z) - \frac{1}{j\omega}\nabla\times\mathbf{a}_\phi 2\pi\int_{\text{flaw}} G_{\phi\phi}^{(ee)}(r,z;r',z')J_a(r',z')r'dr'dz' \\
&\quad -\frac{1}{j\omega}\nabla\times\mathbf{a}_\phi 2\pi\int_{\text{flaw}} G_{\phi\phi}^{(ee)}(r,z;r',z')(\nabla\times\mathbf{M}_a)_\phi r'dr'dz' \,. \qquad (9.3)
\end{aligned}$$

$\mathbf{E}^{(i)}$ and $\mathbf{B}^{(i)}$ are the incident fields produced by the exciting coil. The curl operation divided by $-j\omega$ in (9.3) defines a magnetic–electric operator.

Upon multiplying $\mathbf{E}$ by the anomalous conductivity, $\sigma_a(r) = j\omega(\hat{\varepsilon}(\mathbf{r}) - \hat{\varepsilon}_h)$, we get the anomalous current, $\mathbf{J}_a$. Similarly, $\mathbf{B}(\mathbf{r}) = \dfrac{\mu(\mathbf{r})\mu_h}{\mu(\mathbf{r})-\mu_h}\mathbf{M}_a(\mathbf{r})$. Hence, when we make these substitutions in (9.3), we get the coupled integral equations for axisymmetric, ferromagnetic problems:

$$\begin{aligned}
\frac{J_a(r,z)}{\sigma_a(r,z)} &= E^{(i)}(r,z) + 2\pi\int_{\text{flaw}} G_{\phi\phi}^{(ee)}(r,z;r',z')J_a(r',z')r'dr'dz' \\
&\quad + 2\pi\int_{\text{flaw}} G_{\phi\phi}^{(ee)}(r,z;r',z')(\nabla\times\mathbf{M}_a)_\phi r'dr'dz' \\
\frac{\mu(r,z)\mu_h}{\mu(r,z)-\mu_h}\mathbf{M}_a(r,z) &= \mathbf{B}^{(i)}(r,z) - \frac{1}{j\omega}\nabla\times\mathbf{a}_\phi 2\pi\int_{\text{flaw}} G_{\phi\phi}^{(ee)}(r,z;r',z')J_a(r',z')r'dr'dz' \\
&\quad -\frac{1}{j\omega}\nabla\times\mathbf{a}_\phi 2\pi\int_{\text{flaw}} G_{\phi\phi}^{(ee)}(r,z;r',z')(\nabla\times\mathbf{M}_a)_\phi r'dr'dz'. \qquad (9.4)
\end{aligned}$$

We have dropped the vector components from the electrical variables; the scalar equation in (9.4) is the ϕ-component of the electric field integral equation.

9.4 Discretization: Method of Moments

Because $\mathbf{M}(r,z) = M_r(r,z)\mathbf{a}_\rho + M_z(r,z)\mathbf{a}_z$, it follows that $(\nabla\times\mathbf{M})_\phi = \partial M_r/\partial z - \partial M_z/\partial r$. Therefore, in order for $\mathbf{M}$ to belong to $H(curl)$, the space of vector functions with bounded curls, it follows that M_r must be everywhere differentiable with respect to z, and M_z must be everywhere differentiable with respect to r. An expansion of the following form will satisfy these criteria:

$$\begin{aligned}
M_a^{(r)}(r,z) &= \sum_{lm} M_{lm}^{(r)}\pi_{1l}(r)\pi_{2m}(z) \\
M_a^{(z)}(r,z) &= \sum_{lm} M_{lm}^{(z)}\pi_{2l}(r)\pi_{1m}(z)\,, \qquad (9.5)
\end{aligned}$$

where π_{1m} is the unit pulse function, which starts at the mth node, and π_{2m} is the triangular function, which starts at the mth node. $J_a(r,z)$, on the other hand, can be expanded in pulse functions

$$J_a(r,z) = \sum_{lm} J_{lm}\pi_{1l}(r)\pi_{1m}(z) \ , \tag{9.6}$$

because there are no derivatives appearing in (9.4)(a), except for $\nabla \times \mathbf{M}_a$.

We proceed to discretize (9.4), assuming that $\nu(r,z) = \dfrac{\mu(r,z)\mu_h}{\mu(r,z)-\mu_h}$ and $\sigma_a(r,z)$ are constant in each cell. The result, after taking moments of (9.4)(a) is

$$E^{(i)}_{lm} = \frac{r^{(+)}_l + r^{(-)}_l}{2}\frac{\delta r\delta z}{\sigma_{lm}}J_{lm} - \sum_{LM} G^{(ee)}_{lm,LM}J_{LM} - \sum_{LM} G^{(em)(r)}_{lm,LM}M^{(r)}_{LM} + \sum_{LM} G^{(em)(z)}_{lm,LM}M^{(z)}_{LM} \tag{9.7}$$

where

$$\begin{aligned}
E^{(i)}_{lm} &= \int\!\!\int E^{(i)}(r,z)\pi_{1l}(r)\pi_{1m}(z)rdrdz \\
G^{(ee)}_{lm,LM} &= 2\pi\int\!\!\int rdrdz\pi_{1l}(r)\pi_{1m}(z)\int\!\!\int G(r,z;r',z')\pi_{1L}(r')\pi_{1M}(z')r'dr'dz' \\
G^{(em)(r)}_{lm,LM} &= 2\pi\int\!\!\int rdrdz\pi_{1l}(r)\pi_{1m}(z)\int\!\!\int G(r,z;r',z')\pi_{1L}(r')\pi'_{2M}(z')r'dr'dz' \\
&= \frac{G^{(ee)}_{lm,LM} - G^{(ee)}_{lm,LM+1}}{\delta z} \\
G^{(em)(z)}_{lm,LM} &= 2\pi\int\!\!\int rdrdz\pi_{1l}(r)\pi_{1m}(z)\int\!\!\int G(r,z;r',z')\pi'_{2L}(r')\pi_{1M}(z')r'dr'dz' \\
&= \frac{G^{(ee)}_{lm,LM} - G^{(ee)}_{lm,L+1M}}{\delta r} \ .
\end{aligned} \tag{9.8}$$

The discretized version of (9.4)(b) is

$$\begin{aligned}
B^{(i)(r)}_{lm} &= \delta r\frac{r^{(+)}_l + r^{(-)}_l}{2}\sum_M Q^{(r)}_{mM}M^{(r)}_{lM} + \sum_{LM} G^{(me)(r)}_{lm,LM}J_{LM} + \sum_{LM} G^{(mm)(rr)}_{lm,LM}M^{(r)}_{LM} \\
&\quad - \sum_{LM} G^{(mm)(rz)}_{lm,LM}M^{(z)}_{LM} \\
B^{(i)(z)}_{lm} &= \delta z\sum_L Q^{(z)}_{lL}M^{(z)}_{Lm} - \sum_{LM} G^{(me)(z)}_{lm,LM}J_{LM} - \sum_{LM} G^{(mm)(zr)}_{lm,LM}M^{(r)}_{LM} + \sum_{LM} G^{(mm)(zz)}_{lm,LM}M^{(z)}_{LM} \ ,
\end{aligned} \tag{9.9}$$

where the tri-diagonal matrices are symmetric and have the following nonzero entries:

$$\begin{aligned}
Q^{(r)}_{mM} &= \int \nu(r_l^{(-)}, z)\pi_{2M}(z)\pi_{2m}(z)dz \\
&= \frac{\delta z}{6}\begin{cases} \nu_{lm} & \text{if } M = m-1 \\ 2(\nu_{lm}+\nu_{lm+1}) & \text{if } M = m \\ \nu_{lm+1} & \text{if } M = m+1 \end{cases} \\
Q^{(z)}_{lL} &= \int \nu(r, m\delta z)\pi_{2L}(r)\pi_{2l}(r)rdr \\
&= \delta r^2 \begin{cases} \nu_{lm}\left(\frac{1}{12}+\frac{l+r_0/\delta r}{6}\right) & \text{if } L = l-1 \\ \nu_{lm}\left(\frac{1}{4}+\frac{l+r_0/\delta r}{3}\right)+\nu_{l+1,m}\left(\frac{10}{24}+\frac{l+r_0/\delta r}{3}\right) & \text{if } L = l \\ \nu_{l+1,m}\left(\frac{1}{4}+\frac{l+r_0/\delta r}{6}\right) & \text{if } L = l+1 \end{cases}
\end{aligned} \tag{9.10}$$

where r_0 is the radial starting point of the grid. $r_l^{(-)} = r_0 + l\delta r$ and $r_l^{(+)} = r_0 + (l+1)\delta r$. The first entry in each of these matrices is the lower diagonal, the second the main diagonal, and the third the upper diagonal.

The other matrices are given by

$$\begin{aligned}
B^{(i)(r)}_{lm} &= \int\int \mathbf{B}^{(i)}(r,z)\cdot\mathbf{a}_\rho\pi_{1l}(r)\pi_{2m}(z)rdrdz \\
&= -\frac{1}{j\omega}\frac{E^{(i)}_{lm}-E^{(i)}_{lm+1}}{\delta z} \\
B^{(i)(z)}_{lm} &= \int\int \mathbf{B}^{(i)}(r,z)\cdot\mathbf{a}_z\pi_{2l}(r)\pi_{1m}(z)rdrdz \\
&= \frac{1}{j\omega}\frac{E^{(i)}_{lm}-E^{(i)}_{l+1m}}{\delta r} \\
G^{(me)(r)}_{lm,LM} &= \frac{2\pi}{j\omega}\int\int rdrdz\pi_{1l}(r)\pi'_{2m}(z)\int\int G(r,z;r',z')\pi_{1L}(r')\pi_{1M}(z')r'dr'dz' \\
&= \frac{1}{j\omega}\frac{G^{(ee)}_{lm,LM}-G^{(ee)}_{lm+1,LM}}{\delta z}
\end{aligned}$$

$$
\begin{aligned}
G^{(mm)(rr)}_{lm,LM} &= \frac{2\pi}{j\omega}\int\int r dr dz \pi_{1l}(r)\pi'_{2m}(z)\int\int G(r,z;r',z')\pi_{1L}(r')\pi'_{2M}(z')r'dr'dz' \\
&= \frac{G^{(ee)}_{lm,LM} - G^{(ee)}_{lm+1,LM} - G^{(ee)}_{lm,LM+1} + G^{(ee)}_{lm+1,LM+1}}{j\omega\delta z^2} \\
G^{(mm)(rz)}_{lm,LM} &= \frac{2\pi}{j\omega}\int\int r dr dz \pi_{1l}(r)\pi'_{2m}(z)\int\int G(r,z;r',z')\pi'_{2L}(r')\pi_{1M}(z')r'dr'dz' \\
&= \frac{G^{(ee)}_{lm,LM} - G^{(ee)}_{lm+1,LM} - G^{(ee)}_{lm,L+1M} + G^{(ee)}_{lm+1,L+1M}}{j\omega\delta r\delta z} \\
G^{(me)(z)}_{lm,LM} &= \frac{2\pi}{j\omega}\int\int r dr dz \pi'_{2l}(r)\pi_{1m}(z)\int\int G(r,z;r',z')\pi_{1L}(r')\pi_{1M}(z')r'dr'dz' \\
&= \frac{1}{j\omega}\frac{G^{(ee)}_{lm,LM} - G^{(ee)}_{l+1m,LM}}{\delta r} \\
G^{(mm)(zr)}_{lm,LM} &= \frac{2\pi}{j\omega}\int\int r dr dz \pi'_{2l}(r)\pi_{1m}(z)\int\int G(r,z;r',z')\pi_{1L}(r')\pi'_{2M}(z')r'dr'dz' \\
&= \frac{G^{(ee)}_{lm,LM} - G^{(ee)}_{l+1m,LM} - G^{(ee)}_{lm,LM+1} + G^{(ee)}_{l+1m,LM+1}}{j\omega\delta r\delta z} = G^{(mm)(rz)}_{LM,lm} \\
G^{(mm)(zz)}_{lm,LM} &= \frac{2\pi}{j\omega}\int\int r dr dz \pi'_{2l}(r)\pi_{1m}(z)\int\int G(r,z;r',z')\pi'_{2L}(r')\pi_{1M}(z')r'dr'dz' \\
&= \frac{G^{(ee)}_{lm,LM} - G^{(ee)}_{l+1m,LM} - G^{(ee)}_{lm,L+1M} + G^{(ee)}_{l+1m,L+1M}}{j\omega\delta r^2} .
\end{aligned}
\tag{9.11}
$$

9.5 Calculation of Matrices

Keep in mind that the Green function that appears in the integrands of (9.8) and (9.11) is really $G^{(ee)}_{\phi\phi}(r,z;r',z')$. We will derive expressions for this function later, but for the present we will use the results.

Refer to Fig. 9.5, which shows an axisymmetric layered tube with concentric layers. A coaxial filamentary current source of unit strength is located at $r = r'$ in the layer bounded by $r_{s-1} < r' < r_{s'-1}$. The wave parameter in the ith layer is given by $\alpha_i = (k_i^2 - h^2)^{1/2}$, where h is the one-dimensional spatial Fourier transform variable, and $k_i^2 = \omega^2\mu_i\hat{\varepsilon}_i = \omega^2\mu_i\varepsilon_i - j\omega\mu_i\sigma_i$. The eigenvectors, $\mathbf{V}_{1m}, \mathbf{V}_{2m}, \mathbf{V}_{3m}, \mathbf{V}_{4m}$, shown in Fig. 9.5 are defined in Appendix A.1.

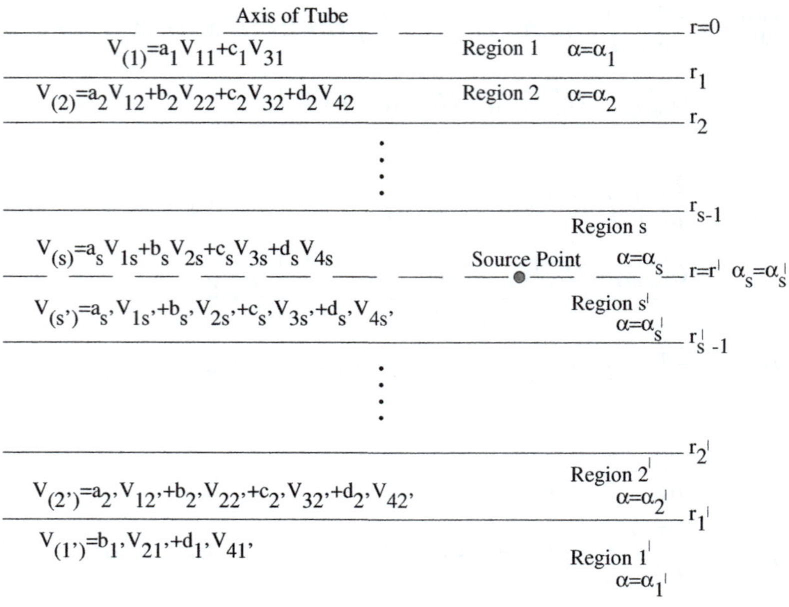

Fig. 9.5 An axisymmetric layered tube with concentric layers

We introduce the following two coefficients that define a ϕ-directed filamentary electric current source:

$$c_1' = \frac{\pi\alpha_s}{j2} \cdot \frac{U_{12}(s')e^{-j\alpha_s r_{s'-1}} J_1(\alpha_s r') + U_{22}(s')e^{-j\alpha_s \Delta_s} e^{j\alpha_s r_{s-1}} H_1^{(2)}(\alpha_s r')}{e^{-j2\alpha_s\Delta_s} U_{22}(s')V_{11}(s) - U_{12}(s')V_{21}(s)}$$

$$d_{1'}' = \frac{\pi\alpha_s}{j2} \cdot \frac{V_{11}(s)e^{-j\alpha_s\Delta_s} e^{-j\alpha_s r_{s'-1}} J_1(\alpha_s r') + V_{21}(s)e^{j\alpha_s r_{s-1}} H_1^{(2)}(\alpha_s r')}{e^{-j2\alpha_s\Delta_s} U_{22}(s')V_{11}(s) - U_{12}(s')V_{21}(s)}, \quad (9.12)$$

where $\Delta_s = r_{s'-1} - r_{s-1}$. These coefficients are derived in Appendix A.1.

The electric field in the various regions is defined in terms of these parameters, as follows:
In region m ($r_{m-1} < r < r_m$), above the source region:

$$\tilde{E}_\phi(r,r') = -j\omega\frac{\mu_m}{\alpha_m} c_1' \left[V_{11}(m)e^{-j\alpha_m(r_m - r)} j_1(\alpha_m r) + V_{21}(m)e^{j\alpha_m(r_m - r)} h_1^{(2)}(\alpha_m r) \right], \quad (9.13)$$

and in region m' ($r_{m'} < r < r_{m'-1}$), below the source region:

$$\tilde{E}_\phi(r,r') = -j\omega\frac{\mu_{m'}}{\alpha_{m'}} d_{1'}' \left[U_{12}(m')e^{-j\alpha_{m'}(r_{m'} - r)} j_1(\alpha_{m'} r) + U_{22}(m')e^{j\alpha_{m'}(r_{m'} - r)} h_1^{(2)}(\alpha_{m'} r) \right]. \quad (9.14)$$

In the source region, $(r_{s-1} < r, r' < r_{s'-1})$:

$$\tilde{E}_\phi(r,r') = -j\omega\frac{\mu_s}{\alpha_s}d'_{1'}\left[U_{12}(s')e^{-j\alpha_s(r_{s-1}-r)}j_1(\alpha_s r) + U_{22}(s')e^{j\alpha_s(r_{s-1}-r)}h_1^{(2)}(\alpha_s r)\right]$$

$$r > r'$$

$$= -j\omega\frac{\mu_s}{\alpha_s}c'_1\left[V_{11}(s)e^{-j\alpha_s(r_{s'-1}-r)}j_1(\alpha_s r) + V_{21}(s)e^{j\alpha_s(r_{s'-1}-r)}h_1^{(2)}(\alpha_s r)\right]$$

$$r < r'\,. \tag{9.15}$$

As an interesting application of (9.15), if we assume that there is only one region that occupies all space, namely the source region, then (9.15) yields

$$\tilde{E}_\phi(r,r') = -\pi\frac{\omega\mu_s}{2}H_1^{(2)}(\alpha_s r')J_1(\alpha_s r)\ ,\ r < r'$$

$$\tilde{E}_\phi(r,r') = -\pi\frac{\omega\mu_s}{2}H_1^{(2)}(\alpha_s r)J_1(\alpha_s r')\ ,\ r' < r\,. \tag{9.16}$$

Alternatively, we can treat (9.16) as an "incident Green function," which is then scattered from the various layers, producing the total function in (9.15).

The spatial-domain Green function, in which the field point, r, is in region m, m', s, and the source point in region s is given by the inverse Fourier transform of the corresponding expression in (9.13)–(9.15)

$$G(r,z;r',z') = \frac{1}{(2\pi)^2}\int_{-\infty}^{\infty}\tilde{E}(r,r')e^{jh(z-z')}dh\,. \tag{9.17}$$

The $G(r,z;r',z')$ that appear in (9.8) and (9.11) have r, r' in the same (source) region, so it corresponds to (9.15) being substituted into (9.17).

The U's and V's are transfer coefficients that will be defined recursively shortly. Note that when $s = 1$, i.e., when there are no layers above the source point in Fig. 9.5, then $V_{11}(1) = 1$, $V_{21}(1) = 0$; when $s' = 1'$, i.e., when there are no layers below the source point, then $U_{22}(1') = 1$, $U_{12}(1') = 0$.

It will be more convenient to compute the infinite-space Green function by using

$$G(r,z;r',z') = -\frac{j\omega\mu_s}{2\pi}\int_0^{\infty}\frac{e^{-\alpha_s|z-z'|}}{2\alpha_s}J_1(r'l)J_1(rl)l\,dl\ , \tag{9.18}$$

which is the same as the free-space Green function that is used in the ferrite-core module. Equations (9.17) and (9.18) produce the same result for a filamentary current loop in infinite space, even though they correspond to different coordinate systems.

When computing matrix elements that involve $G^{(ee)}$ in the source region, it will be advantageous to separate the incident, or infinite-space, term, (9.16), from the layered-space terms, as in Chap. 2. Because (9.15) includes both terms, we subtract (9.16) from (9.15) to get the layered-space terms.

We start by rewriting (9.12) in a slightly more familiar form that involves reflection coefficients:

$$c'_1 = \frac{\pi\alpha_s}{j2}\,\frac{\left(\dfrac{U_{12}(s')}{U_{22}(s')V_{11}(s)}\right)e^{j2\alpha_s\Delta_s}e^{-j\alpha_s r_{s'-1}}J_1(\alpha_s r') + \left(\dfrac{e^{j\alpha_s\Delta_s}}{V_{11}(s)}\right)e^{j\alpha_s r_{s-1}}H_1^{(2)}(\alpha_s r')}{1-R_{s',s}e^{j2\alpha_s\Delta_s}}$$

$$d'_{1'} = \frac{\pi\alpha_s}{j2}\,\frac{\left(\dfrac{e^{j\alpha_s\Delta_s}}{U_{22}(s')}\right)e^{-j\alpha_s r_{s'-1}}J_1(\alpha_s r') + \left(\dfrac{V_{21}(s)}{U_{22}(s')V_{11}(s)}\right)e^{j2\alpha_s\Delta_s}e^{j\alpha_s r_{s-1}}H_1^{(2)}(\alpha_s r')}{1-R_{s',s}e^{j2\alpha_s\Delta_s}}, \tag{9.19}$$

where the double reflection coefficient is defined by

$$R_{s',s} = \frac{U_{12}(s')V_{21}(s)}{U_{22}(s')V_{11}(s)}\,. \tag{9.20}$$

Next, substitute (9.19) into (9.15):

$$\begin{aligned}
\tilde{E}_\phi(r,r') &= -j\omega\mu_s\left(\frac{\pi}{j2}\right)\cdot \\
&\quad\left[\frac{\left(\dfrac{e^{j\alpha_s\Delta_s}}{U_{22}(s')}\right)e^{-j\alpha_s r_{s'-1}}J_1(\alpha_s r') + \left(\dfrac{V_{21}(s)}{U_{22}(s')V_{11}(s)}\right)e^{j2\alpha_s\Delta_s}e^{j\alpha_s r_{s-1}}H_1^{(2)}(\alpha_s r')}{1-R_{s',s}e^{j2\alpha_s\Delta_s}}\right]\cdot \\
&\quad\left[U_{12}(s')e^{-j\alpha_s(r_{s-1}-r)}j_1(\alpha_s r) + U_{22}(s')e^{j\alpha_s(r_{s-1}-r)}h_1^{(2)}(\alpha_s r)\right] \\
&\quad r > r' \\
&= -j\omega\mu_s\left(\frac{\pi}{j2}\right)\cdot \\
&\quad\left[\frac{\left(\dfrac{U_{12}(s')}{U_{22}(s')V_{11}(s)}\right)e^{j2\alpha_s\Delta_s}e^{-j\alpha_s r_{s'-1}}J_1(\alpha_s r') + \left(\dfrac{e^{j\alpha_s\Delta_s}}{V_{11}(s)}\right)e^{j\alpha_s r_{s-1}}H_1^{(2)}(\alpha_s r')}{1-R_{s',s}e^{j2\alpha_s\Delta_s}}\right]\cdot \\
&\quad\left[V_{11}(s)e^{-j\alpha_s(r_{s'-1}-r)}j_1(\alpha_s r) + V_{21}(s)e^{j\alpha_s(r_{s'-1}-r)}h_1^{(2)}(\alpha_s r)\right] \\
&\quad r < r'.
\end{aligned} \tag{9.21}$$

Upon expanding and simplifying we get

$$\tilde{E}_{\phi}(r,r') = -j\omega\mu_s\left(\frac{\pi}{j2}\right)\left[\frac{\left(\frac{U_{12}(s')}{U_{22}(s')}\right)e^{-j\alpha_s(2r_{s-1}-(r+r'))}j_1(\alpha_s r')j_1(\alpha_s r)}{1-R_{s',s}e^{j2\alpha_s\Delta_s}}\right.$$

$$+\frac{H_1^{(2)}(\alpha_s r)J_1(\alpha_s r')+R_{s',s}e^{j2\alpha_s\Delta_s}H_1^{(2)}(\alpha_s r')J_1(\alpha_s r)}{1-R_{s',s}e^{j2\alpha_s\Delta_s}}$$

$$\left.+\frac{\left(\frac{V_{21}(s)}{V_{11}(s)}\right)e^{j\alpha_s(2r_{s'-1}-(r+r'))}h_1^{(2)}(\alpha_s r')h_1^{(2)}(\alpha_s r)}{1-R_{s',s}e^{j2\alpha_s\Delta_s}}\right]$$

$$r > r'$$

$$= -j\omega\mu_s\left(\frac{\pi}{j2}\right)\left[\frac{\left(\frac{U_{12}(s')}{U_{22}(s')}\right)e^{-j\alpha_s(2r_{s-1}-(r+r'))}j_1(\alpha_s r')j_1(\alpha_s r)}{1-R_{s',s}e^{j2\alpha_s\Delta_s}}\right.$$

$$+\frac{H_1^{(2)}(\alpha_s r')J_1(\alpha_s r)+R_{s',s}e^{j2\alpha_s\Delta_s}H_1^{(2)}(\alpha_s r)J_1(\alpha_s r')}{1-R_{s',s}e^{j2\alpha_s\Delta_s}}$$

$$\left.+\frac{\left(\frac{V_{21}(s)}{V_{11}(s)}\right)e^{j\alpha_s(2r_{s'-1}-(r+r'))}h_1^{(2)}(\alpha_s r')h_1^{(2)}(\alpha_s r)}{1-R_{s',s}e^{j2\alpha_s\Delta_s}}\right]$$

$$r' > r \tag{9.22}$$

Call $\frac{U_{12}(s')}{U_{22}(s')} = \rho_{s'}e^{-j2\alpha_s\Delta_s}$, $\frac{V_{21}(s)}{V_{11}(s)} = \rho_s e^{-j2\alpha_s\Delta_s}$. Then, clearly, $R_{s',s} = \rho_{s'}\rho_s e^{-j4\alpha_s\Delta_s}$. $\rho_{s'}$ corresponds to reflections from the surface $r = r_{s'-1}$, and ρ_s corresponds to reflections from the surface $r = r_{s-1}$.

Finally, subtract (9.16) from (9.22) to get the layered-space fields in the region $r_{s-1} < r, r' < r_{s'-1}$:

$$\tilde{E}_{\phi}^{(ls)} = -j\omega\mu_s\left(\frac{\pi}{j2}\right)\left[\frac{\rho_{s'}e^{-j\alpha_s(2r_{s'-1}-(r+r'))}j_1(\alpha_s r')j_1(\alpha_s r)+\rho_s e^{j\alpha_s(2r_{s-1}-(r+r'))}h_1^{(2)}(\alpha_s r')h_1^{(2)}(\alpha_s r)}{1-\rho_{s',s}e^{-j2\alpha_s\Delta_s}}\right.$$

$$\left.+\frac{\rho_{s',s}e^{-j2\alpha_s\Delta_s}\left(H_1^{(2)}(\alpha_s r')J_1(\alpha_s r)+H_1^{(2)}(\alpha_s r)J_1(\alpha_s r')\right)}{1-\rho_{s',s}e^{-j2\alpha_s\Delta_s}}\right]. \tag{9.23}$$

The layered-space field is continuous, along with its derivatives at $r = r'$. We can also replace the Hankel and Bessel functions with their scaled versions in (9.23):

$$H_1^{(2)}(\alpha_s r')J_1(\alpha_s r) + H_1^{(2)}(\alpha_s r)J_1(\alpha_s r') = e^{-j\alpha_s(r'-r)}h_1^{(2)}(\alpha_s r')j_1(\alpha_s r) + e^{j\alpha_s(r'-r)}h_1^{(2)}(\alpha_s r)j_1(\alpha_s r').$$

Note the similarity between these results and the (a) and (b) terms (convolution and correlation) in Chap. 2. Of course, because the scaled Bessel and Hankel functions are not independent of (r,r'), the results of (9.23) are not rigorously convolutions and correlations in (r,r'). The expression (9.23), however, does make clear the presence of incident and reflected fields at the layers, r and r', just as in Chap. 2.

In a similar manner, we can derive expressions for the transmitted fields above and below the source region. In region m $(r_{m-1} < r < r_m)$ above the source region, we have, upon substituting (9.12) into (9.13):

$$\tilde{E}_\phi(r,r') = -j\omega\mu_m\left(\frac{\alpha_s}{\alpha_m}\right)\left(\frac{\pi}{j2}\right) \times \left[\frac{\rho_{s'}e^{-j\alpha_s\Delta_s}e^{-j\alpha_s(r_{s'-1}-r')}j_1(\alpha_s r') + e^{j\alpha_s(r_{s-1}-r')}h_1^{(2)}(\alpha_s r')}{1-\rho_{s',s}e^{-j2\alpha_s\Delta_s}}\right] \times \left[\frac{V_{21}(m)}{V_{11}(s)e^{-j\alpha_s\Delta_s}}e^{j\alpha_m(r_m-r)}h_1^{(2)}(\alpha_m r) + \frac{V_{11}(m)}{V_{11}(s)e^{-j\alpha_s\Delta_s}}e^{-j\alpha_m(r_m-r)}j_1(\alpha_m r)\right], \tag{9.24}$$

and in region m' $(r_{m'} < r < r_{m'-1})$ below the source region:

$$\tilde{E}_\phi(r,r') = -j\omega\mu_{m'}\left(\frac{\alpha_s}{\alpha_{m'}}\right)\left(\frac{\pi}{j2}\right) \times \left[\frac{\rho_s e^{-j\alpha_s\Delta_s}e^{j\alpha_s(r_{s-1}-r')}h_1^{(2)}(\alpha_s r') + e^{-j\alpha_s(r_{s'-1}-r')}j_1(\alpha_s r')}{1-\rho_{s',s}e^{-j2\alpha_s\Delta_s}}\right] \times \left[\frac{U_{12}(m')}{U_{22}(s')e^{-j\alpha_s\Delta_s}}e^{-j\alpha_{m'}(r_{m'}-r)}j_1(\alpha_{m'}r) + \frac{U_{22}(m')}{U_{22}(s')e^{-j\alpha_s\Delta_s}}e^{j\alpha_{m'}(r_{m'}-r)}h_1^{(2)}(\alpha_{m'}r)\right]. \tag{9.25}$$

By writing

$$\frac{V_{21}(m)}{V_{11}(s)} = \frac{V_{11}(m)}{V_{11}(s)} \cdot \frac{V_{21}(m)}{V_{11}(m)}$$

$$= \frac{V_{11}(m)}{V_{11}(s)} R_m$$

$$\frac{U_{12}(m')}{U_{22}(s')} = \frac{U_{22}(m')}{U_{22}(s')} \cdot \frac{U_{12}(m')}{U_{22}(m')}$$

$$= \frac{U_{22}(m')}{U_{22}(s')} R_{m'} \,, \tag{9.26}$$

then we can rewrite (9.24) and (9.25) as

$$\tilde{E}_\phi(r,r') = -j\omega\mu_m \left(\frac{\alpha_s}{\alpha_m}\right)\left(\frac{\pi}{j2}\right)\left(\frac{V_{11}(m)}{V_{11}(s)e^{-j\alpha_s\Delta_s}}\right)$$
$$\times \left[\frac{\rho_{s'} e^{-j\alpha_s\Delta_s} e^{-j\alpha_s(r_{s'-1}-r')} j_1(\alpha_s r') + e^{j\alpha_s(r_{s-1}-r')} h_1^{(2)}(\alpha_s r')}{1-\rho_{s',s} e^{-j2\alpha_s\Delta_s}}\right]$$
$$\times \left[R_m e^{j\alpha_m(r_m-r)} h_1^{(2)}(\alpha_m r) + e^{-j\alpha_m(r_m-r)} j_1(\alpha_m r)\right], \tag{9.27}$$

in region m $(r_{m-1} < r < r_m)$ above the source region, and

$$\tilde{E}_\phi(r,r') = -j\omega\mu_{m'} \left(\frac{\alpha_s}{\alpha_{m'}}\right)\left(\frac{\pi}{j2}\right)\left(\frac{U_{22}(m')}{U_{22}(s')e^{-j\alpha_s\Delta_s}}\right)$$
$$\times \left[\frac{\rho_s e^{-j\alpha_s\Delta_s} e^{j\alpha_s(r_{s-1}-r')} h_1^{(2)}(\alpha_s r') + e^{-j\alpha_s(r_{s'-1}-r')} j_1(\alpha_s r')}{1-\rho_{s',s} e^{-j2\alpha_s\Delta_s}}\right]$$
$$\times \left[R_{m'} e^{-j\alpha_{m'}(r_{m'}-r)} j_1(\alpha_{m'} r) + e^{j\alpha_{m'}(r_{m'}-r)} h_1^{(2)}(\alpha_{m'} r)\right], \tag{9.28}$$

in region m' $(r_{m'} < r < r_{m'-1})$ below the source region. We will return to a discussion of the computation of the reflection coefficients (the R's) shortly, but right now we'll get into details of the computation of the matrix elements.

Consider

$$G^{(ee)}_{lm,LM} = 2\pi \int\int r\,dr\,dz\,\pi_{1l}(r)\pi_{1m}(z) \int\int G(r,z;r',z')\pi_{1L}(r')\pi_{1M}(z')r'dr'dz'$$

where

$$G(r,z;r',z') = G^{(0)}(r,z;r',z') + G^{(s)}(r,z;r',z') \,. \tag{9.29}$$

$G^{(0)}(r,z;r',z')$ is the infinite-space Green function and $G^{(s)}(r,z;r',z')$ is the layered-space Green function:

$$G^{(0)}(r,z;r',z') = -\frac{j\omega\mu_s}{2\pi}\int_0^\infty \frac{e^{-\alpha_s|z-z'|}}{2\alpha_s}J_1(r'l)J_1(rl)l\,dl$$
$$G^{(s)}(r,z;r',z') = -\frac{\omega\mu_s}{8\pi}\int_{-\infty}^{\infty} e^{jh(z-z')}$$
$$\times\left[\frac{\rho_{s'}e^{-j\alpha_s(2r_{s'-1}-(r+r'))}j_1(\alpha_s r')j_1(\alpha_s r)+\rho_s e^{j\alpha_s(2r_{s-1}-(r+r'))}h_1^{(2)}(\alpha_s r')h_1^{(2)}(\alpha_s r)}{1-\rho_{s',s}e^{-j2\alpha_s\Delta_s}}\right.$$
$$\left.+\frac{\rho_{s',s}e^{-j2\alpha_s\Delta_s}\left(e^{-j\alpha_s(r'-r)}h_1^{(2)}(\alpha_s r')j_1(\alpha_s r)+e^{j\alpha_s(r'-r)}h_1^{(2)}(\alpha_s r)j_1(\alpha_s r')\right)}{1-\rho_{s',s}e^{-j2\alpha_s\Delta_s}}\right]dh\,. \tag{9.30}$$

Hence, the infinite-space matrix element is:

$$G^{(0)(ee)}_{lm,LM} = -j\omega\mu_s\int_0^\infty \frac{F(m-M)}{2\alpha_s}\frac{\mathscr{I}(L(h\delta r),(L+1)(h\delta r))\cdot\mathscr{I}(l(h\delta r),(l+1)(h\delta r))}{h^3}dh\,, \tag{9.31}$$

where

$$F(m-M) = \delta z^2\begin{cases} e^{\alpha_s\delta z(m-M+1)}\left(\dfrac{1-e^{-\alpha_s\delta z}}{\alpha_s\delta z}\right)^2 & \text{, if } m-M\le -1\\ 2\left(\dfrac{e^{-\alpha_s\delta z}-1+\alpha_s\delta z}{(\alpha_s\delta z)^2}\right) & \text{, if } m-M=0\\ e^{-\alpha_s\delta z(m-M-1)}\left(\dfrac{1-e^{-\alpha_s\delta z}}{\alpha_s\delta z}\right)^2 & \text{, if } m-M\ge 1\end{cases}$$
$$\mathscr{I}(a,b) = \int_a^b zJ_1(z)dz\,. \tag{9.32}$$

Note that $F(m-M)=F(M-m)$; i.e., it is symmetrical in (m,M), and corresponds to the F_3 function in (5.29).

The layered-space matrix element is given by:

$$G^{(s)(ee)}_{lm,LM} = -\frac{\omega\mu_s}{4}\delta z^2\int_{-\infty}^{\infty}\left(\frac{2-e^{-jh\delta z}-e^{jh\delta z}}{h^2\delta z^2}\right)e^{jh(m-M)\delta z}$$
$$\times\left[R^{(1)}(\alpha_s)\frac{\mathscr{I}(l(\alpha_s\delta r),(l+1)(\alpha_s\delta r))\cdot\mathscr{I}(L(\alpha_s\delta r),(L+1)(\alpha_s\delta r))}{\alpha_s^4}\right.$$
$$+R^{(2)}(\alpha_s)\frac{\mathscr{H}(l(\alpha_s\delta r),(l+1)(\alpha_s\delta r))\cdot\mathscr{H}(L(\alpha_s\delta r),(L+1)(\alpha_s\delta r))}{\alpha_s^4}$$

$$+R^{(3)}(\alpha_s)\left(\frac{\mathscr{I}(l(\alpha_s\delta r),(l+1)(\alpha_s\delta r))\cdot\mathscr{H}(L(\alpha_s\delta r),(L+1)(\alpha_s\delta r))}{\alpha_s^4}\right.$$
$$\left.\left.+\frac{\mathscr{H}(l(\alpha_s\delta r),(l+1)(\alpha_s\delta r))\cdot\mathscr{I}(L(\alpha_s\delta r),(L+1)(\alpha_s\delta r))}{\alpha_s^4}\right)\right]dh\,, \tag{9.33}$$

where

$$\begin{aligned}
R^{(1)}(\alpha_s) &= \frac{\rho_{s'}e^{-j\alpha_s 2r_{s'-1}}}{1-\rho_{s',s}e^{-j2\alpha_s\Delta_s}} \\
R^{(2)}(\alpha_s) &= \frac{\rho_s e^{j\alpha_s 2r_{s-1}}}{1-\rho_{s',s}e^{-j2\alpha_s\Delta_s}} \\
R^{(3)}(\alpha_s) &= \frac{\rho_{s',s}e^{-j2\alpha_s\Delta_s}}{1-\rho_{s',s}e^{-j2\alpha_s\Delta_s}} \\
\mathscr{H}(z_1,z_2) &= \int_{z_1}^{z_2} H_1^{(2)}(z)z\,dz.
\end{aligned} \tag{9.34}$$

9.5.1 Bobbin Coil Incident Field and Moments

Now we calculate the field produced by a bobbin coil that extends from $l_0 \le z \le l_1$, $r_0 \le r \le r_1$, and is located in region c. First, we note that the current density flowing within such a coil is N_cI, where N_c is the density of turns (i.e., turns per unit area), and I is the current flowing within the coil. Hence, we have

$$\begin{aligned}
E^{(i)}(r,z) &= 2\pi N_cI\int_{l_0}^{l_1}dz'\int_{r_0}^{r_1}r'dr'G(r,z;r',z') \\
&= \frac{N_cI}{2\pi}\int_{-\infty}^{\infty}e^{jhz}dh\left[\int_{l_0}^{l_1}e^{-jhz'}dz'\int_{r_0}^{r_1}r'dr'\tilde{E}(r,r')\right].
\end{aligned} \tag{9.35}$$

$\tilde{E}(r,r')$ is given by (9.27) if the field point is above region c, and by (9.28) otherwise. Hence,

$$E^{(i)} = \frac{N_cI}{2\pi}\int_{-\infty}^{\infty}\frac{e^{jh(z-l_0)}-e^{jh(z-l_1)}}{jh}\left[\int_{r_0}^{r_1}r'dr'\tilde{E}(r,r')\right]dh\,. \tag{9.36}$$

Upon substituting (9.27) and (9.28) into (9.36), we get

$$E^{(i)}(r,z) = j\omega\mu_m\frac{N_cI}{4}\int_{-\infty}^{\infty}\left(\frac{\alpha_c}{\alpha_m}\right)\left(\frac{V_{11}(m)}{V_{11}(c)e^{-j\alpha_c\Delta_c}}\right)\frac{e^{jh(z-l_0)}-e^{jh(z-l_1)}}{h}$$
$$\times\left[R^{(4)}(\alpha_c)\frac{\mathscr{I}(\alpha_c r_0,\alpha_c r_1)}{\alpha_c^2}+R^{(5)}(\alpha_c)\frac{\mathscr{H}(\alpha_c r_0,\alpha_c r_1)}{\alpha_c^2}\right]$$

$$
\begin{aligned}
&\times \left[R_m e^{j\alpha_m(r_m-r)} h_1^{(2)}(\alpha_m r) + e^{-j\alpha_m(r_m-r)} j_1(\alpha_m r)\right] dh \\
&r_{m-1} < r < r_m \quad \text{(above coil)} \\
&= j\omega\mu_{m'} \frac{N_c I}{4} \int_{-\infty}^{\infty} \left(\frac{\alpha_c}{\alpha_{m'}}\right) \left(\frac{U_{22}(m')}{U_{22}(c')e^{-j\alpha_c\Delta_c}}\right) \frac{e^{jh(z-l_0)} - e^{jh(z-l_1)}}{h} \\
&\times \left[R^{(6)}(\alpha_c) \frac{\mathscr{H}(\alpha_c r_0, \alpha_c r_1)}{\alpha_c^2} + R^{(7)}(\alpha_c) \frac{\mathscr{I}(\alpha_c r_0, \alpha_c r_1)}{\alpha_c^2}\right] \\
&\times \left[R_{m'} e^{-j\alpha_{m'}(r_{m'}-r)} j_1(\alpha_{m'} r) + e^{j\alpha_{m'}(r_{m'}-r)} h_1^{(2)}(\alpha_{m'} r)\right] dh \\
&r_{m'} < r < r_{m'-1} \quad \text{(below coil)},
\end{aligned}
\tag{9.37}
$$

where

$$
\begin{aligned}
R^{(4)}(\alpha_c) &= \frac{\rho_{c'} e^{-j\alpha_c\Delta_c} e^{-j\alpha_c r_{c'-1}}}{1 - \rho_{c',c} e^{-j2\alpha_c\Delta_c}} \\
R^{(5)}(\alpha_c) &= \frac{e^{j\alpha_c r_{c-1}}}{1 - \rho_{c',c} e^{-j2\alpha_c\Delta_c}} \\
R^{(6)}(\alpha_c) &= \frac{\rho_c e^{-j\alpha_c\Delta_c} e^{j\alpha_c r_{c-1}}}{1 - \rho_{c',c} e^{-j2\alpha_c\Delta_c}} \\
R^{(7)}(\alpha_c) &= \frac{e^{-j\alpha_c r_{c'-1}}}{1 - \rho_{c',c} e^{-j2\alpha_c\Delta_c}} \\
\Delta_c &= r_{c'-1} - r_{c-1} \, .
\end{aligned}
\tag{9.38}
$$

The first result in (9.37) would be used for an encircling-coil, because in that case the entire tube is "above" the coil. Similarly, the second result would be used for an internal-coil, because the entire tube is "below" the coil.

Because the dependency on r and r', as well as z and z', is separated in (9.37), it is easy to compute the moments of the incident field, as given in (9.8):

$$
\begin{aligned}
E_{lm}^{(i)} &= \omega\mu_m \frac{N_c I}{4} \int_{-\infty}^{\infty} \left(\frac{\alpha_c}{\alpha_m}\right) \left(\frac{V_{11}(m)}{V_{11}(c)e^{-j\alpha_c\Delta_c}}\right) \\
&\times \left[\frac{e^{-jh(l_0-(m+1)\delta z)} - e^{-jh(l_0-m\delta z)} - e^{-jh(l_1-(m+1)\delta z)} + e^{-jh(l_1-m\delta z)}}{h^2}\right] \\
&\times \left[R^{(4)}(\alpha_c) \frac{\mathscr{I}(\alpha_c r_0, \alpha_c r_1)}{\alpha_c^2} + R^{(5)}(\alpha_c) \frac{\mathscr{H}(\alpha_c r_0, \alpha_c r_1)}{\alpha_c^2}\right] \\
&\times \left[R_m e^{j\alpha_m r_m} \frac{\mathscr{H}(l\alpha_m\delta r, (l+1)\alpha_m\delta r)}{\alpha_m^2} + e^{-j\alpha_m r_m} \frac{\mathscr{I}(l\alpha_m\delta r, (l+1)\alpha_m\delta r)}{\alpha_m^2}\right] dh \\
&r_{m-1} < r < r_m \quad \text{(above coil)}
\end{aligned}
$$

$$
\begin{aligned}
&= \omega\mu_{m'} \frac{N_c I}{4} \int_{-\infty}^{\infty} \left(\frac{\alpha_c}{\alpha_{m'}}\right) \left(\frac{U_{22}(m')}{U_{22}(c')e^{-j\alpha_c\Delta_c}}\right) \\
&\times \left[\frac{e^{-jh(l_0-(m+1)\delta z)} - e^{-jh(l_0-m\delta z)} - e^{-jh(l_1-(m+1)\delta z)} + e^{-jh(l_1-m\delta z)}}{h^2}\right] \\
&\times \left[R^{(6)}(\alpha_c)\frac{\mathscr{H}(\alpha_c r_0, \alpha_c r_1)}{\alpha_c^2} + R^{(7)}(\alpha_c)\frac{\mathscr{I}(\alpha_c r_0, \alpha_c r_1)}{\alpha_c^2}\right] \\
&\times \left[R_{m'} e^{-j\alpha_{m'} r_{m'}} \frac{\mathscr{I}(l\alpha_{m'}\delta r, (l+1)\alpha_{m'}\delta r)}{\alpha_{m'}^2} + e^{j\alpha_{m'} r_{m'}} \frac{\mathscr{H}(l\alpha_{m'}\delta r, (l+1)\alpha_{m'}\delta r)}{\alpha_{m'}^2}\right] dh \\
& r_{m'} < r < r_{m'-1} \quad \text{(below coil)} . \qquad (9.39)
\end{aligned}
$$

9.5.2 Driving-Point Impedance of a Bobbin Coil

The next thing that we must do with the coil is compute its driving-point impedance. We will use the reaction-formula

$$
Z = -\frac{\langle \mathbf{E} \cdot \mathbf{J} \rangle}{I^2} , \qquad (9.40)
$$

where $\mathbf{J}$ is the current density in the coil ($= N_c I \mathbf{a}_\phi$), and I is the total current. Thus,

$$
\begin{aligned}
Z &= -\frac{2\pi N_c I}{I^2} \int\int E(r,z) r dr dz \\
&= -\frac{(2\pi)^2 (N_c I)^2}{I^2} \int r dr \int dz \left\{ \int r' dr' \int dz' \left[G^{(0)}(r,z;r',z') \right.\right. \\
&\quad \left.\left. + G^{(s)}(r,z;r',z') \right] \right\} , \qquad (9.41)
\end{aligned}
$$

where $G^{(0)}(r,z;r',z')$ and $G^{(s)}(r,z;r',z')$ are the infinite-space and layered-space Green functions given in (9.30). Note that the integral over $G^{(0)}$ gives the free-space impedance, $Z^{(0)}$, of the coil, and the integral over $G^{(s)}$ gives the change in impedance, ΔZ, due to the presence of the scattering body (the tube). Hence,

$$
\begin{aligned}
Z &= Z^{(0)} + \Delta Z \\
&= j\omega\mu_c 2\pi N_c^2 \int_0^\infty \frac{F^{(c)}(l_1-l_0)}{2\alpha_c} \cdot \frac{\mathscr{I}^2(r_0 h, r_1 h)}{h^3} dh \\
&+ \omega\mu_c \left(\frac{\pi}{2}\right) N_c^2 \int_{-\infty}^{\infty} \left(\frac{2 - e^{-jh(l_1-l_0)} - e^{jh(l_1-l_0)}}{h^2}\right) \\
&\times \left[R^{(1)}(\alpha_c) \frac{\mathscr{I}^2(\alpha_c r_0, \alpha_c r_1)}{\alpha_c^4} + R^{(2)}(\alpha_c) \frac{\mathscr{H}^2(\alpha_c r_0, \alpha_c r_1)}{\alpha_c^4} \right. \\
&\left. + 2R^{(3)}(\alpha_c) \frac{\mathscr{I}(\alpha_c r_0, \alpha_c r_1)\mathscr{H}(\alpha_c r_0, \alpha_c r_1)}{\alpha_c^4} \right] dh \,, \qquad (9.42)
\end{aligned}
$$

where $F^{(c)}(l_1-l_0) = 2\left[\frac{-1 + e^{-(l_1-l_0)\alpha_c} + \alpha_c(l_1-l_0)}{\alpha_c^2}\right]$ is the specialization to the coil of the F-function of (9.32).

Appendix

A.1 Cylindrical Eigenvectors and the Green Function

We will develop the general theory of eigenvectors in cylindrical coordinates and then complete the development of the Green functions for axisymmetric problems.

Maxwell's equations are

$$
\begin{aligned}
\nabla \times \mathbf{E} &= -j\omega\mu \mathbf{H} - \mathbf{M}^{(i)} \\
\nabla \times \mathbf{H} &= j\omega\hat{\varepsilon}\mathbf{E} + \mathbf{J}^{(i)} \,, \qquad (9.43)
\end{aligned}
$$

where $\mathbf{M}^{(i)}$ and $\mathbf{J}^{(i)}$ are impressed current sources, $\mathbf{M}^{(i)}$ being magnetic, and $\mathbf{J}^{(i)}$ electric.

In cylindrical coordinates, the component forms of (9.43) are

$$\frac{1}{r}\frac{\partial E_z}{\partial \phi} - \frac{\partial E_\phi}{\partial z} = -j\omega\mu H_r - M_r^{(i)}$$

$$\frac{\partial E_r}{\partial z} - \frac{\partial E_z}{\partial r} = -j\omega\mu H_\phi - M_\phi^{(i)}$$

$$\frac{1}{r}\frac{\partial (rE_\phi)}{\partial r} - \frac{1}{r}\frac{\partial E_r}{\partial \phi} = -j\omega\mu H_z - M_z^{(i)}$$

$$\frac{1}{r}\frac{\partial H_z}{\partial \phi} - \frac{\partial H_\phi}{\partial z} = j\omega\hat{\varepsilon} E_r + J_r^{(i)}$$

$$\frac{\partial H_r}{\partial z} - \frac{\partial H_z}{\partial r} = j\omega\hat{\varepsilon} E_\phi + J_\phi^{(i)}$$

$$\frac{1}{r}\frac{\partial (rH_\phi)}{\partial r} - \frac{1}{r}\frac{\partial H_r}{\partial \phi} = j\omega\hat{\varepsilon} E_z + J_z^{(i)} . \tag{9.44}$$

We expand all field variables in a Fourier series in ϕ and a Fourier integral in z:

$$F(r,\phi,z) = \sum_{m=-\infty}^{\infty} e^{jm\phi} \int_{-\infty}^{\infty} \tilde{F}(r,m,h) e^{jhz} dh$$

$$\tilde{F}(r,m,h) = \frac{1}{(2\pi)^2} \int_0^{2\pi} e^{-jm\phi'} d\phi' \int_{-\infty}^{\infty} e^{-jhz'} F(r,\phi',z') dz' , \tag{9.45}$$

where the second equation is the direct transform from physical space to frequency (wave-number) space.

Upon transforming (9.44), we get, after cancelling the common factor, $e^{jn\phi} e^{jhz}$:

$$\frac{jn}{r}\tilde{E}_z - jh\tilde{E}_\phi = -j\omega\mu \tilde{H}_r - \tilde{M}_r^{(i)}$$

$$jh\tilde{E}_r - \frac{d\tilde{E}_z}{dr} = -j\omega\mu \tilde{H}_\phi - \tilde{M}_\phi^{(i)}$$

$$\frac{1}{r}\frac{d(r\tilde{E}_\phi)}{dr} - j\frac{n}{r}\tilde{E}_r = -j\omega\mu \tilde{H}_z - \tilde{M}_z^{(i)}$$

$$\frac{jn}{r}\tilde{H}_z - jh\tilde{H}_\phi = j\omega\hat{\varepsilon}\tilde{E}_r + \tilde{J}_r^{(i)}$$

$$jh\tilde{H}_r - \frac{d\tilde{H}_z}{dr} = j\omega\hat{\varepsilon}\tilde{E}_\phi + \tilde{J}_\phi^{(i)}$$

$$\frac{1}{r}\frac{d(r\tilde{H}_\phi)}{dr} - j\frac{n}{r}\tilde{H}_r = j\omega\hat{\varepsilon}\tilde{E}_z + \tilde{J}_z^{(i)} . \tag{9.46}$$

The first and fourth equations are free of derivatives. Hence, they allow us to eliminate $\tilde{H}_r$ and $\tilde{E}_r$ in favor of the transverse components, $\tilde{H}_z$, $\tilde{H}_\phi$, $\tilde{E}_z$, $\tilde{E}_\phi$:

$$\tilde{H}_r = -\frac{n}{\omega\mu r}\tilde{E}_z + \frac{h}{\omega\mu}\tilde{E}_\phi - \frac{\tilde{M}_r^{(i)}}{j\omega\mu}$$

$$\tilde{E}_r = \frac{n}{\omega\hat{\varepsilon} r}\tilde{H}_z - \frac{h}{\omega\hat{\varepsilon}}\tilde{H}_\phi - \frac{\tilde{J}_r^{(i)}}{j\omega\hat{\varepsilon}} . \qquad (9.47)$$

The remaining equations are

$$jh\tilde{E}_r - \frac{d\tilde{E}_z}{dr} = -j\omega\mu\tilde{H}_\phi - \tilde{M}_\phi^{(i)}$$

$$\frac{1}{r}\frac{d(r\tilde{E}_\phi)}{dr} - j\frac{n}{r}\tilde{E}_r = -j\omega\mu\tilde{H}_z - \tilde{M}_z^{(i)}$$

$$jh\tilde{H}_r - \frac{d\tilde{H}_z}{dr} = j\omega\hat{\varepsilon}\tilde{E}_\phi + \tilde{J}_\phi^{(i)}$$

$$\frac{1}{r}\frac{d(r\tilde{H}_\phi)}{dr} - j\frac{n}{r}\tilde{H}_r = j\omega\hat{\varepsilon}\tilde{E}_z - \tilde{J}_z^{(i)} . \qquad (9.48)$$

Upon eliminating $\tilde{E}_r$ and $\tilde{H}_r$, we get

$$\frac{d\tilde{E}_z}{dr} = \frac{jhn}{\omega\hat{\varepsilon} r}\tilde{H}_z + \left(-\frac{jh^2}{\omega\hat{\varepsilon} r} + \frac{j\omega\mu}{r}\right)(r\tilde{H}_\phi) - \frac{h}{\omega\hat{\varepsilon}}\tilde{J}_r^{(i)} + \tilde{M}_\phi^{(i)}$$

$$\frac{d(r\tilde{E}_\phi)}{dr} = \left(\frac{jn^2}{\omega\hat{\varepsilon} r} - j\omega\mu r\right)\tilde{H}_z - \frac{jnh}{\omega\hat{\varepsilon} r}(r\tilde{H}_\phi) - \frac{n}{\omega\hat{\varepsilon}}\tilde{J}_r^{(i)} - r\tilde{M}_z^{(i)}$$

$$\frac{d\tilde{H}_z}{dr} = -\frac{jhn}{\omega\mu r}\tilde{E}_z + \left(\frac{jh^2}{\omega\mu r} - \frac{j\omega\hat{\varepsilon}}{r}\right)(r\tilde{E}_\phi) - \tilde{J}_\phi^{(i)} - \frac{h}{\omega\mu}\tilde{M}_r^{(i)}$$

$$\frac{d(r\tilde{H}_\phi)}{dr} = \left(-\frac{jn^2}{\omega\mu r} + j\omega\hat{\varepsilon} r\right)\tilde{E}_z + \frac{jnh}{\omega\mu r}(r\tilde{E}_\phi) + r\tilde{J}_z^{(i)} - \frac{n}{\omega\mu}\tilde{M}_r^{(i)} . \qquad (9.49)$$

This system can be written in the following vector-matrix form:

$$\frac{d}{dr}\begin{bmatrix}\tilde{E}_z\\ r\tilde{E}_\phi\\ \tilde{H}_z\\ r\tilde{H}_\phi\end{bmatrix}=\begin{bmatrix}0 & 0 & a_{11}(r) & a_{12}(r)\\ 0 & 0 & a_{21}(r) & a_{22}(r)\\ b_{11}(r) & b_{12}(r) & 0 & 0\\ b_{21}(r) & b_{22}(r) & 0 & 0\end{bmatrix}\begin{bmatrix}\tilde{E}_z\\ r\tilde{E}_\phi\\ \tilde{H}_z\\ r\tilde{H}_\phi\end{bmatrix}$$

$$+\begin{bmatrix}u_{11} & 0 & 0 & 0 & u_{15} & 0\\ u_{21} & 0 & 0 & 0 & 0 & u_{26}\\ 0 & u_{32} & 0 & u_{34} & 0 & 0\\ 0 & 0 & u_{43} & u_{44} & 0 & 0\end{bmatrix}\begin{bmatrix}\tilde{J}_r\\ \tilde{J}_\phi\\ \tilde{J}_z\\ \tilde{M}_r\\ \tilde{M}_\phi\\ \tilde{M}_z\end{bmatrix}, \tag{9.50}$$

where

$$a_{11}=\frac{jhn}{\omega\hat{\varepsilon}r}\quad a_{12}=\frac{-jh^2}{\omega\hat{\varepsilon}r}+\frac{j\omega\mu}{r}\quad a_{21}=\frac{jn^2}{\omega\hat{\varepsilon}r}-j\omega\mu r\quad a_{22}=\frac{-jnh}{\omega\hat{\varepsilon}r}$$

$$b_{11}=\frac{-jhn}{\omega\mu r}\quad b_{12}=\frac{jh^2}{\omega\mu r}-\frac{j\omega\hat{\varepsilon}}{r}\quad b_{21}=\frac{-jn^2}{\omega\mu r}+j\omega\hat{\varepsilon}r\quad b_{22}=\frac{jnh}{\omega\mu r} \tag{9.51}$$

$$u_{11}=\frac{-h}{\omega\hat{\varepsilon}}\qquad u_{15}=1\qquad u_{21}=\frac{-n}{\omega\hat{\varepsilon}}\qquad u_{26}=-r$$

$$u_{32}=-1\qquad u_{34}=\frac{-h}{\omega\mu}\qquad u_{43}=r\quad u_{44}=\frac{-n}{\omega\mu}.$$

In order to compute the eigenvectors of (9.50), we work with the homogeneous system, which we rewrite as $d\mathbf{V}/dr=\mathbf{MV}$. It is straightforward to prove, by appealing to (9.51), that $\mathbf{M}^2=(n^2/r^2+h^2-\omega^2\mu\hat{\varepsilon})\mathbf{I}$, where $\mathbf{I}$ is the identity matrix. Furthermore, we have

$$\frac{d\mathbf{M}}{dr}=-\frac{\mathbf{M}}{r}+\begin{bmatrix}0 & 0 & 0 & 0\\ 0 & 0 & -j2\omega\mu & 0\\ 0 & 0 & 0 & 0\\ j2\omega\hat{\varepsilon} & 0 & 0 & 0\end{bmatrix}.$$

Then

$$\begin{aligned}\frac{d^2\mathbf{V}}{dr^2} &= \frac{d\mathbf{M}}{dr}\mathbf{V} + \mathbf{M}\frac{d\mathbf{V}}{dr} \\ &= \frac{d\mathbf{M}}{dr}\mathbf{V} + \mathbf{M}^2\mathbf{V} \\ &= -\frac{1}{r}\frac{d\mathbf{V}}{dr} + \left(\frac{n^2}{r^2} + h^2 - \omega^2\mu\hat{\varepsilon}\right)\mathbf{V} + \begin{bmatrix} 0 \\ -j2\omega\mu V_3 \\ 0 \\ j2\omega\hat{\varepsilon}V_1 \end{bmatrix}. \end{aligned} \tag{9.52}$$

The solutions for the eigenvector components, $\tilde{E}_z$ and $\tilde{H}_z$, are obtained immediately from $\dfrac{d^2v}{dr^2} + \dfrac{1}{r}\dfrac{dv}{dr} - \left(\dfrac{n^2}{r^2} + h^2 - \omega^2\mu\hat{\varepsilon}\right)v = 0$. This is satisfied by the two linearly-independent functions, $J_n(\alpha r)$, the Bessel function of the first kind, of order n, and $H_n^{(2)}(\alpha r)$, the Hankel function of the second kind, of order n, where $\alpha = (k^2 - h^2)^{1/2}$ and $k^2 = \omega^2\mu\hat{\varepsilon} = \omega^2\mu(\varepsilon + \sigma/j\omega)$.

In order to compute the components, $r\tilde{E}_\phi$ and $r\tilde{H}_\phi$, we do not use (9.52), but return to the fundamental system (9.50). Using the first and third equations, we have $d\tilde{E}_z/dr = a_{11}(r)\tilde{H}_z + a_{12}(r)(r\tilde{H}_\phi)$. Hence

$$\begin{aligned}\tilde{H}_\phi &= \frac{1}{ra_{12}(r)}\left[\frac{d\tilde{E}_z}{dr} - a_{11}(r)\tilde{H}_z\right] \\ &= -\frac{1}{\alpha^2}\left[j\omega\hat{\varepsilon}\frac{d\tilde{E}_z}{dr} + \frac{hn}{r}\tilde{H}_z\right].\end{aligned}$$

By duality, we have

$$\begin{aligned}\tilde{E}_\phi &= \frac{1}{rb_{12}(r)}\left[\frac{d\tilde{H}_z}{dr} - b_{11}(r)\tilde{E}_z\right] \\ &= \frac{1}{\alpha^2}\left[j\omega\mu\frac{d\tilde{H}_z}{dr} - \frac{hn}{r}\tilde{E}_z\right].\end{aligned}$$

Hence, the four eigenvectors of (9.50) are

$$\begin{bmatrix}\tilde{E}_z \\ r\tilde{E}_\phi \\ \tilde{H}_z \\ r\tilde{H}_\phi\end{bmatrix} = \begin{bmatrix} J_n(\alpha r) \\ -\frac{hn}{\alpha^2}J_n(\alpha r) \\ 0 \\ -\frac{j\omega\hat{\varepsilon}}{\alpha}rJ_n'(\alpha r)\end{bmatrix} \begin{bmatrix} H_n^{(2)}(\alpha r) \\ -\frac{hn}{\alpha^2}H_n^{(2)}(\alpha r) \\ 0 \\ -\frac{j\omega\hat{\varepsilon}}{\alpha}rH_n^{(2)\prime}(\alpha r)\end{bmatrix} \begin{bmatrix} 0 \\ \frac{j\omega\mu}{\alpha}rJ_n'(\alpha r) \\ J_n(\alpha r) \\ -\frac{hn}{\alpha^2}J_n(\alpha r)\end{bmatrix} \begin{bmatrix} 0 \\ \frac{j\omega\mu}{\alpha}rH_n^{(2)\prime}(\alpha r) \\ H_n^{(2)}(\alpha r) \\ -\frac{hn}{\alpha^2}H_n^{(2)}(\alpha r)\end{bmatrix}. \tag{9.53}$$

We label the vectors, $\mathbf{V}_1$, $\mathbf{V}_2$, $\mathbf{V}_3$, $\mathbf{V}_4$, respectively, from left to right. It is straightforward to show that these vectors satisfy the homogeneous form of (9.50). The first two eigenmodes are TM (Transverse Magnetic) to z, because $\tilde{H}_z = 0$, and the last two are TE (Transverse Electric) to z, because $\tilde{E}_z = 0$. When we specialize these equations to the axisymmetric case, in which the azimuthal Fourier wavenumber, n, vanishes, we will get another form of TE and TM, namely in the radial direction. This, as we will see, is of great advantage in solving problems with radially layered media.

We now establish the conditions of continuity/discontinuity on $\mathbf{V}$ at singular current sources. Before doing this, we note that the Fourier transform of a unit point-current source located at $\phi = \phi'$, $z = z'$ is given by $\mathscr{F}\{\delta(\phi - \phi')\delta(z - z')\} = \frac{1}{(2\pi)^2} e^{-jn\phi'} e^{-jhz'}$.

Now, if we have a unit vector current source located at (r', ϕ', z'), then the source term in (9.50) becomes

$$\mathbf{U}\mathscr{I}\frac{\delta(r - r')}{r'}\frac{e^{-jn\phi'} e^{-jhz'}}{(2\pi)^2}, \tag{9.54}$$

where $\mathscr{I}$ is a unit vector. Then upon integrating (9.50) an epsilon distance with respect to r across the cylinder $r = r'$, we get

$$\mathbf{V}^{(+)} - \mathbf{V}^{(-)} = \frac{\mathbf{U}}{r'}\mathscr{I}\frac{e^{-jn\phi'} e^{-jhz'}}{(2\pi)^2}, \tag{9.55}$$

where the superscript $(+)$ denotes the limit $r \to r'$ from above and $(-)$ denotes the limit $r \to r'$ from below.

A.1.1 Application to Axisymmetric Systems

The specialization of the preceding to axisymmetric problems requires only that we set the azimuthal wave number, n, to zero. Then the eigenvectors become:

$$\mathbf{V}_1 = \begin{bmatrix} J_0(\alpha r) \\ 0 \\ 0 \\ -j\dfrac{\omega\hat{\varepsilon}}{\alpha} r J_0'(\alpha r) \end{bmatrix}, \mathbf{V}_2 = \begin{bmatrix} H_0^{(2)}(\alpha r) \\ 0 \\ 0 \\ -j\dfrac{\omega\hat{\varepsilon}}{\alpha} r H_0^{(2)\prime}(\alpha r) \end{bmatrix}$$

$$\mathbf{V}_3 = \begin{bmatrix} 0 \\ j\dfrac{\omega\mu}{\alpha} r J_0'(\alpha r) \\ J_0(\alpha r) \\ 0 \end{bmatrix}, \mathbf{V}_4 = \begin{bmatrix} 0 \\ j\dfrac{\omega\mu}{\alpha} r H_0^{(2)\prime}(\alpha r) \\ H_0^{(2)}(\alpha r) \\ 0 \end{bmatrix}. \tag{9.56}$$

When we recall (9.47), we see that $\tilde{H}_r = 0,\ \tilde{E}_r = 0$, for axisymmetric problems, which means that we have TE and TM fields with respect to the radial direction, as well.

In order to compute the Green function, recall Fig. 9.5. We use eigenvectors $\mathbf{V}_1$ and $\mathbf{V}_3$ in the region next to $r=0$, because these two vectors contain $J_0(\alpha_1 r)$, which is continuous at $r=0$. Conversely, $\mathbf{V}_2$ and $\mathbf{V}_4$ contain $H_0^{(2)}(\alpha_{1'}r)$, which, because it represents an out-going cylindrical wave, satisfies the radiation condition at infinity, which includes Region $1'$.

Our strategy is to eliminate $(a_2,b_2,c_2,d_2),\ldots,(a_k,b_k,c_k,d_k)$ in favor of (a_1,c_1), and $(a_{2'},b_{2'},c_{2'},d_{2'})$, $\ldots$, $(a_{k'},b_{k'},c_{k'},d_{k'})$ in favor of $(b_{1'},d_{1'})$, and then use jump conditions at $r=r'$ to determine $(a_1,c_1,b_{1'},d_{1'})$. The intermediate steps are accomplished by equating the four-vectors at each interface.

For example, at $r=r_1$, we have $a_1\mathbf{V}_{11}+c_1\mathbf{V}_{31} = a_2\mathbf{V}_{12}+b_2\mathbf{V}_{22}+c_2\mathbf{V}_{32}+\mathbf{V}_{42}$, which produces the two systems

$$\begin{aligned} a_1 J_0(\alpha_1 r_1) &= a_2 J_0(\alpha_2 r_1) + b_2 H_0^{(2)}(\alpha_2 r_1) \\ a_1\left(\frac{\alpha_2}{\alpha_1}\right)\left(\frac{\hat{\varepsilon}_1}{\hat{\varepsilon}_2}\right) J_0'(\alpha_1 r_1) &= a_2 J_0'(\alpha_2 r_1) + b_2 H_0^{(2)'}(\alpha_2 r_1) \end{aligned}$$

$$\begin{aligned} c_1 J_0(\alpha_1 r_1) &= c_2 J_0(\alpha_2 r_1) + d_2 H_0^{(2)}(\alpha_2 r_1) \\ c_1\left(\frac{\alpha_2}{\alpha_1}\right)\left(\frac{\mu_1}{\mu_2}\right) J_0'(\alpha_1 r_1) &= c_2 J_0'(\alpha_2 r_1) + d_2 H_0^{(2)'}(\alpha_2 r_1)\,. \end{aligned} \tag{9.57}$$

Similarly, at the bottom, $r=r_{1'}$, we have

$$\begin{aligned} b_{1'} H_0^{(2)}(\alpha_{1'} r_{1'}) &= a_{2'} J_0(\alpha_{2'} r_{1'}) + b_{2'} H_0^{(2)}(\alpha_{2'} r_{1'}) \\ b_{1'}\left(\frac{\alpha_{2'}}{\alpha_{1'}}\right)\left(\frac{\hat{\varepsilon}_{1'}}{\hat{\varepsilon}_{2'}}\right) H_0^{(2)'}(\alpha_{1'} r_{1'}) &= a_{2'} J_0'(\alpha_{2'} r_{1'}) + b_{2'} H_0^{(2)'}(\alpha_{2'} r_{1'}) \end{aligned}$$

$$\begin{aligned} d_{1'} H_0^{(2)}(\alpha_{1'} r_{1'}) &= c_{2'} J_0(\alpha_{2'} r_{1'}) + d_{2'} H_0^{(2)}(\alpha_{2'} r_{1'}) \\ d_{1'}\left(\frac{\alpha_{2'}}{\alpha_{1'}}\right)\left(\frac{\mu_{1'}}{\mu_{2'}}\right) H_0^{(2)'}(\alpha_{1'} r_{1'}) &= c_{2'} J_0'(\alpha_{2'} r_{1'}) + d_{2'} H_0^{(2)'}(\alpha_{2'} r_{1'})\,. \end{aligned} \tag{9.58}$$

In general, at $r=r_M$,

$$\begin{aligned} &a_M\mathbf{V}_{1M}+b_M\mathbf{V}_{2M}+c_M\mathbf{V}_{3M}+d_M\mathbf{V}_{4M} \\ &= a_{M+1}\mathbf{V}_{1(M+1)}+b_{M+1}\mathbf{V}_{2(M+1)}+c_{M+1}\mathbf{V}_{3(M+1)}+d_{M+1}\mathbf{V}_{4(M+1)} \end{aligned}$$

produces

$$a_M J_0(\alpha_M r_M) + b_M H_0^{(2)}(\alpha_M r_M) = a_{M+1} J_0(\alpha_{M+1} r_M) + b_{M+1} H_0^{(2)}(\alpha_{M+1} r_M)$$

$$a_M \left(\frac{\alpha_{M+1}}{\alpha_M}\right)\left(\frac{\hat{\varepsilon}_M}{\hat{\varepsilon}_{M+1}}\right) J_0'(\alpha_M r_M) + b_M \left(\frac{\alpha_{M+1}}{\alpha_M}\right)\left(\frac{\hat{\varepsilon}_M}{\hat{\varepsilon}_{M+1}}\right) H_0^{(2)'}(\alpha_M r_M)$$

$$= a_{M+1} J_0'(\alpha_{M+1} r_M) + b_{M+1} H_0^{(2)'}(\alpha_{M+1} r_M)$$

$$c_M J_0(\alpha_M r_M) + d_M H_0^{(2)}(\alpha_M r_M) = c_{M+1} J_0(\alpha_{M+1} r_M) + d_{M+1} H_0^{(2)}(\alpha_{M+1} r_M)$$

$$c_M \left(\frac{\alpha_{M+1}}{\alpha_M}\right)\left(\frac{\mu_M}{\mu_{M+1}}\right) J_0'(\alpha_M r_M) + d_M \left(\frac{\alpha_{M+1}}{\alpha_M}\right)\left(\frac{\mu_M}{\mu_{M+1}}\right) H_0^{(2)'}(\alpha_M r_M)$$

$$= c_{M+1} J_0'(\alpha_{M+1} r_M) + d_{M+1} H_0^{(2)'}(\alpha_{M+1} r_M) \,. \tag{9.59}$$

This system holds also for $r = r_{M'}$.

What about the source interface, $r = r'$? We will use (9.55) (with $n = 0$ since this is an axisymmetric problem). Furthermore, since we need only consider electric sources (because magnetic sources can be derived from electric ones, by virtue of Amperian currents), we have $\tilde{J}_r = 0,\ \tilde{J}_\phi = 1,\ \tilde{J}_z = 0,\ \tilde{M}_r = 0,\ \tilde{M}_\phi = 0, \tilde{M}_z = 0$. Hence, upon invoking (9.50) and (9.51) we have

$$\mathbf{U}\mathscr{I} = \begin{bmatrix} 0 \\ 0 \\ u_{32} \\ 0 \end{bmatrix} = \begin{bmatrix} 0 \\ 0 \\ -1 \\ 0 \end{bmatrix} , \tag{9.60}$$

so that (9.55) becomes

$$\mathbf{V}^{(s')}(r') - \mathbf{V}^{(s)}(r') = \frac{1}{(2\pi)^2} \begin{bmatrix} 0 \\ 0 \\ -1 \\ 0 \end{bmatrix} \frac{e^{-jhz'}}{r'} \,. \tag{9.61}$$

Upon referring, once again, to Fig. 9.5, (9.61) is seen to produce the system

$$(a_{s'} - a_s)\begin{bmatrix} J_0(\alpha_s r') \\ 0 \\ 0 \\ -j\frac{\omega\hat{\varepsilon}_s}{\alpha_s} r' J_0'(\alpha_s r') \end{bmatrix} + (b_{s'} - b_s)\begin{bmatrix} H_0^{(2)}(\alpha_s r') \\ 0 \\ 0 \\ -j\frac{\omega\hat{\varepsilon}_s}{\alpha_s} r' H_0^{(2)'}(\alpha_s r') \end{bmatrix}$$

$$+(c_{s'} - c_s)\begin{bmatrix} 0 \\ j\frac{\omega\mu_s}{\alpha_s} r' J_0'(\alpha_s r') \\ J_0(\alpha_s r') \\ 0 \end{bmatrix} + (d_{s'} - d_s)\begin{bmatrix} 0 \\ j\frac{\omega\mu_s}{\alpha_s} r' H_0^{(2)'}(\alpha_s r') \\ H_0^{(2)}(\alpha_s r') \\ 0 \end{bmatrix}$$

$$= \begin{bmatrix} 0 \\ 0 \\ -\frac{1}{r'} \\ 0 \end{bmatrix} \frac{e^{-jhz'}}{(2\pi)^2}. \tag{9.62}$$

The solution of (9.62) is obtained in a straightforward manner:

$$a_{s'} - a_s = 0$$

$$b_{s'} - b_s = 0$$

$$c_{s'} - c_s = -\frac{H_1^{(2)}(\alpha_s r')\pi\alpha_s}{j2} \frac{e^{-jhz'}}{(2\pi)^2}$$

$$d_{s'} - d_s = \frac{J_1(\alpha_s r')\pi\alpha_s}{j2} \frac{e^{-jhz'}}{(2\pi)^2}. \tag{9.63}$$

In arriving at the final form of (9.63), we made use of the Bessel-function identities, $J_0'(z) = -J_1(z)$, $H_0^{(2)'}(z) = -H_1^{(2)}(z)$, and also of the Wronskian identity, $H_0^{(2)}(z)J_0'(z) - J_0(z)H_0^{(2)'}(z) = j\frac{2}{\pi z}$.

Our problem is considerably simplified by realizing that all a's and b's vanish. This agrees with the fact that $E_z = 0$ and $H_\phi = 0$ in this problem, and the only places that E_z and H_ϕ appear are in the eigenvectors, $\mathbf{V}_1$, $\mathbf{V}_2$, whose excitation coefficients are a and b. $a = 0$, $b = 0$ satisfies (9.59) and (9.63), of course. Hence, all we need to work with are the last two equations in (9.59) and (9.63).

For computational convenience, we introduce the scaled Hankel and Bessel functions, $h_{0,1}(z)$, $j_{0,1}(z)$, through the relations

$$H_{0,1}^{(2)}(z) = e^{-jz} h_{0,1}(z)$$
$$J_{0,1}(z) = e^{jz} j_{0,1}(z) \,. \tag{9.64}$$

Wherever the coefficient c appears in (9.59), it is multiplied by J, and d is multiplied by $H^{(2)}$. Hence, we introduce the scaled coefficients, c', d' through $cJ = c'j$, $dH = d'h$, which yields

$$c' = ce^{jz}$$
$$d' = de^{-jz} \,. \tag{9.65}$$

Now we proceed to solve the scaled version of (9.59)(c,d)

$$c'_M j_0(\alpha_M r_M) + d'_M h_0^{(2)}(\alpha_M r_M) = c'_{M+1} e^{-j\alpha_{M+1}\Delta_{M+1}} j_0(\alpha_{M+1} r_M) + d'_{M+1} e^{j\alpha_{M+1}\Delta_{M+1}} h_0^{(2)}(\alpha_{M+1} r_M)$$

$$c'_M \left(\frac{\alpha_{M+1}}{\alpha_M}\right)\left(\frac{\mu_M}{\mu_{M+1}}\right) j_1(\alpha_M r_M) + d'_M \left(\frac{\alpha_{M+1}}{\alpha_M}\right)\left(\frac{\mu_M}{\mu_{M+1}}\right) h_1^{(2)}(\alpha_M r_M)$$
$$= c'_{M+1} e^{-j\alpha_{M+1}\Delta_{M+1}} j_1(\alpha_{M+1} r_M) + d'_{M+1} e^{j\alpha_{M+1}\Delta_{M+1}} h_1^{(2)}(\alpha_{M+1} r_M) \,, \tag{9.66}$$

where we have used the derivative identities defined below (9.63). We have also defined

$$\Delta_{M+1} = r_{M+1} - r_M$$
$$\Delta_{M'+1} = -(r_{M'+1} - r_{M'}) \,, \tag{9.67}$$

where $\Delta_{M'+1}$ is defined with a negative sign, because $r_{M'+1} < r_{M'}$.

We express (9.66) in matrix notation

$$\begin{bmatrix} c'_{M+1} \\ d'_{M+1} \end{bmatrix} = \begin{bmatrix} T_{11}(M+1,M) & T_{12}(M+1,M) \\ T_{21}(M+1,M) & T_{22}(M+1,M) \end{bmatrix} \begin{bmatrix} c'_M \\ d'_M \end{bmatrix} \,, \tag{9.68}$$

where

$$T_{11}(M+1,M) = \frac{\pi\alpha_{M+1} r_M e^{j\alpha_{M+1}\Delta_{M+1}}}{j2}\Big[h_1^{(2)}(\alpha_{M+1}r_M)j_0(\alpha_M r_M)$$

$$-\,h_0^{(2)}(\alpha_{M+1}r_M)j_1(\alpha_M r_M)\left(\frac{\alpha_{M+1}}{\alpha_M}\right)\left(\frac{\mu_M}{\mu_{M+1}}\right)\Big]$$

$$T_{12}(M+1,M) = \frac{\pi\alpha_{M+1} r_M e^{j\alpha_{M+1}\Delta_{M+1}}}{j2}\Big[h_1^{(2)}(\alpha_{M+1}r_M)h_0^{(2)}(\alpha_M r_M)$$

$$-\,h_0^{(2)}(\alpha_{M+1}r_M)h_1^{(2)}(\alpha_M r_M)\left(\frac{\alpha_{M+1}}{\alpha_M}\right)\left(\frac{\mu_M}{\mu_{M+1}}\right)\Big]$$

$$T_{21}(M+1,M) = \frac{\pi\alpha_{M+1} r_M e^{-j\alpha_{M+1}\Delta_{M+1}}}{j2}\Big[\left(\frac{\alpha_{M+1}}{\alpha_M}\right)\left(\frac{\mu_M}{\mu_{M+1}}\right)j_0(\alpha_{M+1}r_M)j_1(\alpha_M r_M)$$

$$-\,j_1(\alpha_{M+1}r_M)j_0(\alpha_M r_M)\Big]$$

$$T_{22}(M+1,M) = \frac{\pi\alpha_{M+1} r_M e^{-j\alpha_{M+1}\Delta_{M+1}}}{j2}\Big[\left(\frac{\alpha_{M+1}}{\alpha_M}\right)\left(\frac{\mu_M}{\mu_{M+1}}\right)j_0(\alpha_{M+1}r_M)h_1^{(2)}(\alpha_M r_M)$$

$$-\,j_1(\alpha_{M+1}r_M)h_0^{(2)}(\alpha_M r_M)\Big]\,. \tag{9.69}$$

In arriving at the final version of (9.69), we used the fact that the Wronskian for unscaled Bessel functions holds also for scaled Bessel functions; $j_0(z)h_1^{(2)}(z) - j_1(z)h_0^{(2)}(z) = j2/\pi z$.

The T-matrix is called the transition matrix from region M to $M+1$. The boundary condition on (9.68) is $d_1' = 0,\ c_{1'}' = 0$.

The corresponding transition matrix from M' to $M'+1$ (i.e., below the source) is:

$$\begin{bmatrix} c'_{M'+1} \\ d'_{M'+1} \end{bmatrix} = \begin{bmatrix} T_{11}(M'+1,M') & T_{12}(M'+1,M') \\ T_{21}(M'+1,M') & T_{22}(M'+1,M') \end{bmatrix} \begin{bmatrix} c'_{M'} \\ d'_{M'} \end{bmatrix}, \tag{9.70}$$

where

$$T_{11}(M'+1,M') = \frac{\pi\alpha_{M'+1}r_{M'}e^{-j\alpha_{M'+1}\Delta_{M'+1}}}{j2}\left[h_1^{(2)}(\alpha_{M'+1}r_{M'})j_0(\alpha_{M'}r_{M'})\right.$$

$$\left.-h_0^{(2)}(\alpha_{M'+1}r_{M'})j_1(\alpha_{M'}r_{M'})\left(\frac{\alpha_{M'+1}}{\alpha_{M'}}\right)\left(\frac{\mu_{M'}}{\mu_{M'+1}}\right)\right]$$

$$T_{12}(M'+1,M') = \frac{\pi\alpha_{M'+1}r_{M'}e^{-j\alpha_{M'+1}\Delta_{M'+1}}}{j2}\left[h_1^{(2)}(\alpha_{M'+1}r_{M'})h_0^{(2)}(\alpha_{M'}r_{M'})\right.$$

$$\left.-h_0^{(2)}(\alpha_{M'+1}r_{M'})h_1^{(2)}(\alpha_{M'}r_{M'})\left(\frac{\alpha_{M'+1}}{\alpha_{M'}}\right)\left(\frac{\mu_{M'}}{\mu_{M'+1}}\right)\right]$$

$$T_{21}(M'+1,M') = \frac{\pi\alpha_{M'+1}r_{M'}e^{j\alpha_{M'+1}\Delta_{M'+1}}}{j2}\left[\left(\frac{\alpha_{M'+1}}{\alpha_{M'}}\right)\left(\frac{\mu_{M'}}{\mu_{M'+1}}\right)j_0(\alpha_{M'+1}r_{M'})j_1(\alpha_{M'}r_{M'})\right.$$

$$\left.-j_1(\alpha_{M'+1}r_{M'})j_0(\alpha_{M'}r_{M'})\right]$$

$$T_{22}(M'+1,M') = \frac{\pi\alpha_{M'+1}r_{M'}e^{j\alpha_{M'+1}\Delta_{M'+1}}}{j2}\left[\left(\frac{\alpha_{M'+1}}{\alpha_{M'}}\right)\left(\frac{\mu_{M'}}{\mu_{M'+1}}\right)j_0(\alpha_{M'+1}r_{M'})h_1^{(2)}(\alpha_{M'}r_{M'})\right.$$

$$\left.-j_1(\alpha_{M'+1}r_{M'})h_0^{(2)}(\alpha_{M'}r_{M'})\right]. \tag{9.71}$$

In the source region, define the scaled source variables as:

$$c'_s = e^{j\alpha_s r_{s'-1}}c_s\ ,\ c'_{s'} = e^{j\alpha_s r_{s-1}}c_{s'}$$

$$d'_s = e^{-j\alpha_s r_{s'-1}}d_s\ ,\ d'_{s'} = e^{-j\alpha_s r_{s-1}}d_{s'}\ . \tag{9.72}$$

In terms of these scaled coefficients, the source equations, (9.63), become

$$e^{j\alpha_s(r_{s'-1}-r_{s-1})}c'_{s'} - c'_s = -\frac{\pi\alpha_s}{j2}e^{j\alpha_s r_{s'-1}}H_1^{(2)}(\alpha_s r')\frac{e^{-jhz'}}{(2\pi)^2}$$

$$d'_{s'} - e^{j\alpha_s(r_{s'-1}-r_{s-1})}d'_s = \frac{\pi\alpha_s}{j2}e^{-j\alpha_s r_{s-1}}J_1(\alpha_s r')\frac{e^{-jhz'}}{(2\pi)^2}\ . \tag{9.73}$$

Note that all exponentials have exponents that are proportional to the difference of nearby radii. For example, $r_{s'-1} - r_{s-1}$ is the width of the source region, and $e^{j\alpha_s r_{s-1}}H_{0,1}^{(2)}(\alpha_s r') \to e^{-j\alpha_s(r'-r_{s-1})}$, where $r' > r_{s-1}$, and $e^{-j\alpha_s r_{s'-1}}J_{0,1}(\alpha_s r') \to e^{-j\alpha_s(r_{s'-1}-r')}$, where $r_{s'-1} > r'$. Hence, the right-hand sides of (9.73) vanish to exponential order for $\Re\alpha_s \to \infty$, thereby ensuring proper behavior of the scaled coefficients. Note that the unknowns for the entire system of equations are scaled coefficients.

Now

$$\begin{bmatrix} c'_s \\ d'_s \end{bmatrix} = \overline{\overline{T}}(s,s-1)\cdots\overline{\overline{T}}(2,1)\begin{bmatrix} c'_1 \\ 0 \end{bmatrix}$$

$$\begin{bmatrix} c'_{s'} \\ d'_{s'} \end{bmatrix} = \overline{\overline{T}}(s',s'-1)\cdots\overline{\overline{T}}(2',1')\begin{bmatrix} 0 \\ d'_{1'} \end{bmatrix}, \tag{9.74}$$

where $\overline{\overline{T}}$ is a 2×2 matrix, with entries T_{11}, T_{12}, T_{21}, T_{22}, as defined in (9.69) and (9.71).

Call

$$\overline{\overline{T}}(s,s-1)\cdots\overline{\overline{T}}(2,1) = \overline{\overline{V}}(s)$$

$$\overline{\overline{T}}(s',s'-1)\cdots\overline{\overline{T}}(2',1') = \overline{\overline{U}}(s'), \tag{9.75}$$

then

$$c'_s = V_{11}(s)c'_1, d'_s = V_{21}(s)c'_1$$
$$c'_{s'} = U_{12}^{(s')}d'_{1'}, d'_{s'} = U_{22}^{(s')}d'_{1'}\,. \tag{9.76}$$

In general

$$c'_M = V_{11}(M)c'_1, d'_M = V_{21}(M)c'_1$$
$$c'_{M'} = U_{12}(M')d'_{1'}, d'_{M'} = U_{22}(M')d'_{1'}\,. \tag{9.77}$$

Couple (9.76) with (9.73), and call $r_{s'-1} - r_{s-1} = \Delta_s$; the results are then

$$e^{j\alpha_s\Delta_s}U_{12}(s')d'_{1'} - V_{11}(s)c'_1 = -\frac{\pi\alpha_s}{j2}e^{j\alpha_s r_{s'-1}}H_1^{(2)}(\alpha_s r')\frac{e^{-jhz'}}{(2\pi)^2}$$

$$U_{22}(s')d'_{1'} - e^{j\alpha_s\Delta_s}V_{21}(s)c'_1 = \frac{\pi\alpha_s}{j2}e^{-j\alpha_s r_{s-1}}J_1(\alpha_s r')\frac{e^{-jhz'}}{(2\pi)^2}, \tag{9.78}$$

which has for its solution:

$$c_1' = \frac{\pi\alpha_s}{j2} \frac{U_{12}(s')e^{j\alpha_s\Delta_s}e^{-j\alpha_s r_{s-1}}J_1(\alpha_s r') + U_{22}(s')e^{j\alpha_s r_{s'-1}}H_1^{(2)}(\alpha_s r')}{U_{22}(s')V_{11}(s) - e^{j2\alpha_s\Delta_s}U_{12}(s')V_{21(s)}} \frac{e^{-jhz'}}{(2\pi)^2}$$

$$d_{1'}' = \frac{\pi\alpha_s}{j2} \frac{V_{11}(s)e^{-j\alpha_s r_{s-1}}J_1(\alpha_s r') + V_{21}(s)e^{j\alpha_s r_{s'-1}}e^{j\alpha_s\Delta_s}H_1^{(2)}(\alpha_s r')}{U_{22}(s')V_{11}(s) - e^{j2\alpha_s\Delta_s}U_{12}(s')V_{21}(s)} \frac{e^{-jhz'}}{(2\pi)^2} \quad (9.79)$$

This is the same as (9.12), with which we started the calculation of matrices.

Chapter 10
Some Special Topics in Computational Electromagnetics

10.1 Spatial Decomposition Algorithm

"Spatial decomposition" is a generic term for the first step in transforming a functional equation into its discrete version prior to numerical computation. Umashankar [51] introduced the term "spatial decomposition technique" for a method of solving two-dimensional scattering problems from electrically large bodies using boundary-integral equations. As far as we know, we are the first to apply the "spatial decomposition algorithm" to the solution of three-dimensional scattering problems when the scatterer (the "flaw") occupies several regions with different electrical properties using volume-integral equations.

If the flaw extends over two or more layers with different electrical constitutive properties, as in Fig. 10.1, then (4.9) and (4.10) still hold in each layer (with different matrices), but now the incident field-moment vectors, $[\mathbf{E}^{(ix)}, \mathbf{E}^{(iy)}, \mathbf{E}^{(iz)}]$, $[\mathbf{B}^{(ix)}, \mathbf{B}^{(iy)}, \mathbf{B}^{(iz)}]$, depend upon the coil current plus anomalous currents in other layers. This means that we must introduce new Green dyadics; the Green dyadic that we have worked with previously was that for which the source and field points were in the same layer of the workpiece. Now, upon referring to Fig. 2.1 of Chap. 2, we need the Green function, $\mathbf{G}^{(\pm q0)}$, in which the source point is in region 0, as before, but the field point is in the qth layer above region 0, or the qth layer below the region. The first situation carries the label $+q0$, and the second $-q0$.

We compute the Green functions using the same four-vector algebraic approach, (2.15), that was used in Chap. 2, and then the resulting volume-integral differential equation is derived exactly as explained in Chaps. 2 and 3. Because the Amperian currents, which are of magnetic origin, are actually equivalent electric currents [because they appear in the second of Maxwell's equations (Ampere's law)], we only need to compute the electric-electric dyadic Green functions.

The resulting differential volume-integral relations for the electric anomalous current, $\mathbf{J}^{(e)}(\mathbf{r})$, and the electric current of magnetic origin (the Amperian current or simply "magnetic current"), $\mathbf{J}^{(m)}(\mathbf{r})$, are

H.A. Sabbagh et al., *Computational Electromagnetics and Model-Based Inversion*, Scientific Computation, DOI 10.1007/978-1-4419-8429-6_10,

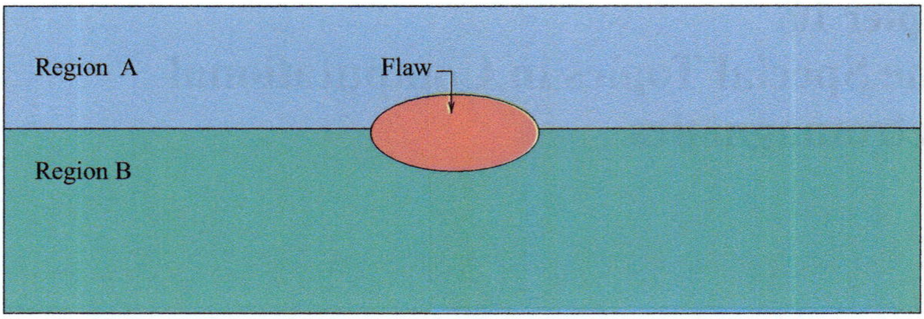

Fig. 10.1 A flaw in multiple layers

$$\begin{aligned}
E_x(\mathbf{r}) = & \int G^{(q0)(1)}(x-x',y-y';z,z')J_x^{(e)}(\mathbf{r}')d\mathbf{r}' \\
& + \int G^{(q0)(1)}(x-x',y-y';z,z')J_x^{(m)}(\mathbf{r}')d\mathbf{r}' \\
& + \frac{\partial}{\partial x}\int G^{(q0)(2)}(x-x',y-y';z,z')\nabla_t'\cdot\mathbf{J}_t^{(e)}(\mathbf{r}')d\mathbf{r}' \\
& + \frac{\partial}{\partial x}\int G^{(q0)(2)}(x-x',y-y';z,z')\nabla_t'\cdot\mathbf{J}_t^{(m)}(\mathbf{r}')d\mathbf{r}' \\
& + \frac{1}{k_0^2}\frac{\partial}{\partial x}\int G^{(q0)(3)}(x-x',y-y';z,z')\nabla'\cdot\mathbf{J}^{(e)}(\mathbf{r}')d\mathbf{r}' \\
E_y(\mathbf{r}) = & \int G^{(q0)(1)}(x-x',y-y';z,z')J_y^{(e)}(\mathbf{r}')d\mathbf{r}' \\
& + \int G^{(q0)(1)}(x-x',y-y';z,z')J_y^{(m)}(\mathbf{r}')d\mathbf{r}' \\
& + \frac{\partial}{\partial y}\int G^{(q0)(2)}(x-x',y-y';z,z')\nabla_t'\cdot\mathbf{J}_t^{(e)}(\mathbf{r}')d\mathbf{r}' \\
& + \frac{\partial}{\partial y}\int G^{(q0)(2)}(x-x',y-y';z,z')\nabla_t'\cdot\mathbf{J}_t^{(m)}(\mathbf{r}')d\mathbf{r}' \\
& + \frac{1}{k_0^2}\frac{\partial}{\partial y}\int G^{(q0)(3)}(x-x',y-y';z,z')\nabla'\cdot\mathbf{J}^{(e)}(\mathbf{r}')d\mathbf{r}' \\
E_z(\mathbf{r}) = & \int G^{(q0)(4)}(x-x',y-y';z,z')J_z^{(e)}(\mathbf{r}')d\mathbf{r}' \\
& + \int G^{(q0)(4)}(x-x',y-y';z,z')J_z^{(m)}(\mathbf{r}')d\mathbf{r}' \\
& + \frac{1}{k_0^2}\frac{\partial}{\partial z}\int G^{(q0)(5)}(x-x',y-y';z,z')\nabla'\cdot\mathbf{J}^{(e)}(\mathbf{r}')d\mathbf{r}' \,, \qquad (10.1)
\end{aligned}$$

which hold for $\mathbf{r}$ in region q and $\mathbf{r}'$ in region 0. The notation for the Z-layering, including the source region, is shown in Fig. 2.1 of Chap. 2. If $\mathbf{r}$ is in region $-q$, then replace $(q0)$ by $(-q0)$. In deriving the final version of (10.1), we used the fact that $\nabla \cdot \mathbf{J}^{(m)}(\mathbf{r}) = 0$.

The kernels that appear in (10.1) are given by

$$G^{(q0)(1)} = -\frac{j\omega\mu_q}{2\pi}\int_0^\infty \frac{\tilde{B}_{q0}}{2\lambda_0}\tilde{G}_{q0}^{(E)}(z,z')J_0(rl)l\,dl$$

$$G^{(q0)(2)} = -\frac{j\omega\mu_q}{2\pi}\int_0^\infty \left[\frac{\tilde{B}_{q0}\tilde{G}_{q0}^{(E)}(z,z') - \dfrac{\mu_0\lambda_q}{\mu_q\lambda_0}\tilde{D}_{q0}\tilde{G}_{q0}^{(M)}(z,z')}{2\lambda_0 l^2}\right] J_0(rl)l\,dl$$

$$G^{(q0)(3)} = -\frac{j\omega\mu_q}{2\pi}\int_0^\infty \frac{\mu_0\lambda_q}{\mu_q\lambda_0}\frac{\tilde{D}_{q0}}{2\lambda_0}\tilde{G}_{q0}^{(M)}(z,z')J_0(rl)l\,dl$$

$$G^{(q0)(4)} = -\frac{j\omega\mu_q}{2\pi}\int_0^\infty \frac{\mu_0}{\mu_q}\frac{\tilde{D}_{q0}}{2\lambda_0}\tilde{G}_{q0}^{(M)\prime}(z,z')J_0(rl)l\,dl$$

$$G^{(q0)(5)} = -\frac{j\omega\mu_q}{2\pi}\int_0^\infty \frac{\mu_0\lambda_0}{\mu_q\lambda_q}\frac{\tilde{D}_{q0}}{2\lambda_0}\tilde{G}_{q0}^{(M)}(z,z')J_0(rl)l\,dl$$

$$G^{(-q0)(1)} = -\frac{j\omega\mu_{-q}}{2\pi}\int_0^\infty \frac{\tilde{A}_{-q0}}{2\lambda_0}\tilde{G}_{-q0}^{(E)}(z,z')J_0(rl)l\,dl$$

$$G^{(-q0)(2)} = -\frac{j\omega\mu_{-q}}{2\pi}\int_0^\infty \left[\frac{\tilde{A}_{-q0}\tilde{G}_{-q0}^{(E)}(z,z') - \dfrac{\mu_0\lambda_{-q}}{\mu_{-q}\lambda_0}\tilde{C}_{-q0}\tilde{G}_{-q0}^{(M)}(z,z')}{2\lambda_0 l^2}\right] J_0(rl)l\,dl$$

$$G^{(-q0)(3)} = -\frac{j\omega\mu_{-q}}{2\pi}\int_0^\infty \frac{\mu_0\lambda_{-q}}{\mu_{-q}\lambda_0}\frac{\tilde{C}_{-q0}}{2\lambda_0}\tilde{G}_{-q0}^{(M)}(z,z')J_0(rl)l\,dl$$

$$G^{(-q0)(4)} = -\frac{j\omega\mu_{-q}}{2\pi}\int_0^\infty \frac{\mu_0}{\mu_{-q}}\frac{\tilde{C}_{-q0}}{2\lambda_0}\tilde{G}_{-q0}^{(M)\prime}(z,z')J_0(rl)l\,dl$$

$$G^{(-q0)(5)} = -\frac{j\omega\mu_{-q}}{2\pi}\int_0^\infty \frac{\mu_0\lambda_0}{\mu_{-q}\lambda_{-q}}\frac{\tilde{C}_{-q0}}{2\lambda_0}\tilde{G}_{-q0}^{(M)}(z,z')J_0(rl)l\,dl\,, \tag{10.2}$$

where $r = [(x-x')^2 + (y-y')^2]^{1/2}$, and the source point, z', is in region 0, and the field point, z, is in region $\pm q$.

The $\tilde{G}$'s that appear in (10.2) are given by the product of terms that involve the source parameters only and those that involve the field parameters:

$$\tilde{G}_{q0}^{(E)}(z,z') = \left[\frac{e^{-\lambda_0(z_0-z')} + R_{0(-)}^{(E)} e^{-\lambda_0(z_0+z'-2z_{-1})}}{1 - R_{0(+)}^{(E)} R_{0(-)}^{(E)} e^{-2\lambda_0\Delta_0}}\right] \left(e^{-\lambda_q(z-z_q)} + R_{q(+)}^{(E)} e^{\lambda_q(z-z_q)}\right)$$

$$\tilde{G}_{q0}^{(M)}(z,z') = \left[\frac{e^{-\lambda_0(z_0-z')} + R_{0(-)}^{(M)} e^{-\lambda_0(z_0+z'-2z_{-1})}}{1 - R_{0(+)}^{(M)} R_{0(-)}^{(M)} e^{-2\lambda_0\Delta_0}}\right] \left(e^{-\lambda_q(z-z_q)} + R_{q(+)}^{(M)} e^{\lambda_q(z-z_q)}\right)$$

$$\tilde{G}_{q0}^{(M)'}(z,z') = \left[\frac{e^{-\lambda_0(z_0-z')} - R_{0(-)}^{(M)} e^{-\lambda_0(z_0+z'-2z_{-1})}}{1 - R_{0(+)}^{(M)} R_{0(-)}^{(M)} e^{-2\lambda_0\Delta_0}}\right] \left(e^{-\lambda_q(z-z_q)} - R_{q(+)}^{(M)} e^{\lambda_q(z-z_q)}\right)$$

$$\tilde{G}_{-q0}^{(E)}(z,z') = \left[\frac{e^{-\lambda_0(z'-z_{-1})} + R_{0(+)}^{(E)} e^{-\lambda_0(2z_0-z'-z_{-1})}}{1 - R_{0(+)}^{(E)} R_{0(-)}^{(E)} e^{-2\lambda_0\Delta_0}}\right]$$
$$\left(e^{\lambda_{-q}(z-z_{-(q+1)})} + R_{-q(-)}^{(E)} e^{-\lambda_{-q}(z-z_{-(q+1)})}\right)$$

$$\tilde{G}_{-q0}^{(M)}(z,z') = \left[\frac{e^{-\lambda_0(z'-z_{-1})} + R_{0(+)}^{(M)} e^{-\lambda_0(2z_0-z'-z_{-1})}}{1 - R_{0(+)}^{(M)} R_{0(-)}^{(M)} e^{-2\lambda_0\Delta_0}}\right]$$
$$\left(e^{\lambda_{-q}(z-z_{-(q+1)})} + R_{-q(-)}^{(M)} e^{-\lambda_{-q}(z-z_{-(q+1)})}\right)$$

$$\tilde{G}_{-q0}^{(M)'}(z,z') = \left[\frac{e^{-\lambda_0(z'-z_{-1})} - R_{0(+)}^{(M)} e^{-\lambda_0(2z_0-z'-z_{-1})}}{1 - R_{0(+)}^{(M)} R_{0(-)}^{(M)} e^{-2\lambda_0\Delta_0}}\right]$$
$$\left(e^{\lambda_{-q}(z-z_{-(q+1)})} - R_{-q(-)}^{(M)} e^{-\lambda_{-q}(z-z_{-(q+1)})}\right). \qquad (10.3)$$

The reflection coefficients, $R_{0(+)}^{(E)}$, $R_{0(-)}^{(E)}$, $R_{0(+)}^{(M)}$, $R_{0(-)}^{(M)}$, correspond, respectively, to $R_1^{(E)}$, $R_{-1}^{(E)}$, $R_1^{(M)}$, $R_{-1}^{(M)}$, of Chap. 2, and Δ_0 corresponds to T, of Chap. 2. The reflection coefficients, $R_{(\pm)q(\pm)}^{(E,M)}$ are defined in Fig. 2.5 of Chap. 2, where we also derived a recursion relation for computing these coefficients. The transfer functions, $\tilde{A}_{-q0}$, $\tilde{B}_{q0}$, $\tilde{C}_{-q0}$, and $\tilde{D}_{q0}$, are defined by recursion relations that follow from the four-vector approach, but will be omitted here.

We discretize the functional equations of (10.1) in the usual way, by first substituting the electric- and magnetic-current expansions, (4.4) and (4.8) of

Chap. 4, into (10.1), and then testing with the electric facet elements, (4.2) of Chap. 4. The result is:

$$
\begin{aligned}
E_{klm}^{(x)} &= \sum_{KLM} T_{klm,KLM}^{(ee)(xx)(q0)} J_{KLM}^{(x)(e)} + \sum_{KLM} T_{klm,KLM}^{(ee)(xy)(q0)} J_{KLM}^{(y)(e)} + \sum_{KLM} T_{klm,KLM}^{(ee)(xz)(q0)} J_{KLM}^{(z)(e)} \\
&\quad + \sum_{KLM} T_{klm,KLM}^{(em)(xx)(q0)} M_{KLM}^{(x)} + \sum_{KLM} T_{klm,KLM}^{(em)(xy)(q0)} M_{KLM}^{(y)} + \sum_{KLM} T_{klm,KLM}^{(em)(xz)(q0)} M_{KLM}^{(z)} \\
E_{klm}^{(y)} &= \sum_{KLM} T_{klm,KLM}^{(ee)(yx)(q0)} J_{KLM}^{(x)(e)} + \sum_{KLM} T_{klm,KLM}^{(ee)(yy)(q0)} J_{KLM}^{(y)(e)} + \sum_{KLM} T_{klm,KLM}^{(ee)(yz)(q0)} J_{KLM}^{(z)(e)} \\
&\quad + \sum_{KLM} T_{klm,KLM}^{(em)(yx)(q0)} M_{KLM}^{(x)} + \sum_{KLM} T_{klm,KLM}^{(em)(yy)(q0)} M_{KLM}^{(y)} + \sum_{KLM} T_{klm,KLM}^{(em)(yz)(q0)} M_{KLM}^{(z)} \\
E_{klm}^{(z)} &= \sum_{KLM} T_{klm,KLM}^{(ee)(zx)(q0)} J_{KLM}^{(x)(e)} + \sum_{KLM} T_{klm,KLM}^{(ee)(zy)(q0)} J_{KLM}^{(y)(e)} + \sum_{KLM} T_{klm,KLM}^{(ee)(zz)(q0)} J_{KLM}^{(z)(e)} \\
&\quad + \sum_{KLM} T_{klm,KLM}^{(em)(zx)(q0)} M_{KLM}^{(x)} + \sum_{KLM} T_{klm,KLM}^{(em)(zy)(q0)} M_{KLM}^{(y)} + \sum_{KLM} T_{klm,KLM}^{(em)(zz)(q0)} M_{KLM}^{(z)} \\
B_{klm}^{(x)} &= \sum_{KLM} T_{klm,KLM}^{(me)(xx)(q0)} J_{KLM}^{(x)(e)} + \sum_{KLM} T_{klm,KLM}^{(me)(xy)(q0)} J_{KLM}^{(y)(e)} + \sum_{KLM} T_{klm,KLM}^{(me)(xz)(q0)} J_{KLM}^{(z)(e)} \\
&\quad + \sum_{KLM} T_{klm,KLM}^{(mm)(xx)(q0)} M_{KLM}^{(x)} + \sum_{KLM} T_{klm,KLM}^{(mm)(xy)(q0)} M_{KLM}^{(y)} + \sum_{KLM} T_{klm,KLM}^{(mm)(xz)(q0)} M_{KLM}^{(z)} \\
B_{klm}^{(y)} &= \sum_{KLM} T_{klm,KLM}^{(me)(yx)(q0)} J_{KLM}^{(x)(e)} + \sum_{KLM} T_{klm,KLM}^{(me)(yy)(q0)} J_{KLM}^{(y)(e)} + \sum_{KLM} T_{klm,KLM}^{(me)(yz)(q0)} J_{KLM}^{(z)(e)} \\
&\quad + \sum_{KLM} T_{klm,KLM}^{(mm)(yx)(q0)} M_{KLM}^{(x)} + \sum_{KLM} T_{klm,KLM}^{(mm)(yy)(q0)} M_{KLM}^{(y)} + \sum_{KLM} T_{klm,KLM}^{(mm)(yz)(q0)} M_{KLM}^{(z)} \\
B_{klm}^{(z)} &= \sum_{KLM} T_{klm,KLM}^{(me)(zx)(q0)} J_{KLM}^{(x)(e)} + \sum_{KLM} T_{klm,KLM}^{(me)(zy)(q0)} J_{KLM}^{(y)(e)} + \sum_{KLM} T_{klm,KLM}^{(me)(zz)(q0)} J_{KLM}^{(z)(e)} \\
&\quad + \sum_{KLM} T_{klm,KLM}^{(mm)(zx)(q0)} M_{KLM}^{(x)} + \sum_{KLM} T_{klm,KLM}^{(mm)(zy)(q0)} M_{KLM}^{(y)} + \sum_{KLM} T_{klm,KLM}^{(mm)(zz)(q0)} M_{KLM}^{(z)},
\end{aligned}
\tag{10.4}
$$

where the transfer matrices satisfy $T_{klm,KLM}^{(q0)} = T_{k-K,l-L;m,M}^{(q0)}$; i.e., they are Töplitz (2D-convolution) in (X,Y). Thus, one can still use two-dimensional FFTs to efficiently execute the computations. The superscript (ee) on the transfer matrices denotes "electric–electric," which means that these matrices "transfer" an electric current in region 0 into an electric field in region q. Similar interpretations attach to "electric–magnetic," "magnetic–electric," and "magnetic–magnetic" transfer matrices; all transfer a source in region 0 into a field in region q. The other transfer matrices can be deduced from the electric–electric ones by virtue of Maxwell's equations.

For example, the magnetic-field moments shown on the left-hand side of the bottom three equations in (10.4) can be easily computed in terms of the electric-field moments:

$$
\begin{aligned}
B_{klm}^{(i)(x)} &= \int \mathbf{B}^{(i)}(\mathbf{r}) \cdot T_{klm}^{(x)(m)}(\mathbf{r})\mathbf{a}_x d\mathbf{r} \\
&= -\frac{1}{j\omega} \int \nabla \times \mathbf{E}^{(i)}(\mathbf{r}) \cdot T_{klm}^{(x)(m)}(\mathbf{r})\mathbf{a}_x d\mathbf{r} \\
&= -\frac{1}{j\omega} \int \mathbf{E}^{(i)}(\mathbf{r}) \cdot \nabla \times T_{klm}^{(x)(m)}(\mathbf{r})\mathbf{a}_x d\mathbf{r} \\
&= -\frac{1}{j\omega} \int \mathbf{E}^{(i)}(\mathbf{r}) \cdot \left[\pi_{1k}(x)\pi_{2l}(y)\pi_{2m}'(z)\mathbf{a}_y \right. \\
&\quad \left. -\pi_{1k}(x)\pi_{2l}'(y)\pi_{2m}(z)\mathbf{a}_z\right] d\mathbf{r} \\
&= -\frac{1}{j\omega}\left[\frac{E_{klm}^{(y)} - E_{klm+1}^{(y)}}{\delta z} - \frac{E_{klm}^{(z)} - E_{kl+1m}^{(z)}}{\delta y}\right] \\
B_{klm}^{(y)} &= -\frac{1}{j\omega}\left[\frac{E_{klm}^{(z)} - E_{k+1lm}^{(z)}}{\delta x} - \frac{E_{klm}^{(x)} - E_{klm+1}^{(x)}}{\delta z}\right] \\
B_{klm}^{(z)} &= -\frac{1}{j\omega}\left[\frac{E_{klm}^{(x)} - E_{kl+1m}^{(x)}}{\delta y} - \frac{E_{klm}^{(y)} - E_{k+1lm}^{(y)}}{\delta x}\right].
\end{aligned}
\tag{10.5}
$$

This is simply $-(1/j\omega) \times$ the discrete curl operator, as we expected.

10.1.1 Summary of Discrete Equations With Transfer Matrices

The master discrete equations, (4.9) and (4.10), are augmented by the transfer matrices when there is a second grid containing anomalous currents, yielding the following equations for a coupled problem between grids q and 0:

$$
\begin{bmatrix} \mathbf{E}^{(i)(q)} \\ \mathbf{B}^{(i)(q)} \\ \mathbf{E}^{(i)(0)} \\ \mathbf{B}^{(i)(0)} \end{bmatrix} =
\begin{bmatrix}
\mathbf{Q}^{(ee)(qq)} - \mathbf{G}^{(ee)(qq)} & -\mathbf{G}^{(em)(qq)} \\
\mathbf{G}^{(me)(qq)} & \mathbf{Q}^{(mm)(qq)} + \mathbf{G}^{(mm)(qq)} \\
-\mathbf{T}(ee)(0q) & -\mathbf{T}^{(em)(0q)} \\
-\mathbf{T}^{(me)(0q)} & -\mathbf{T}^{(mm)(0q)}
\end{bmatrix.
$$

$$
\begin{matrix}
-\mathbf{T}^{(ee)(q0)} & -\mathbf{T}^{(em)(q0)} \\
-\mathbf{T}^{(me)(q0)} & -\mathbf{T}^{(mm)(q0)} \\
\mathbf{Q}^{(ee)(q0)} - \mathbf{G}^{(ee)(00)} & -\mathbf{G}^{(em)(00)} \\
\mathbf{G}^{(me)(00)} & \mathbf{Q}^{(mm)(00)} + \mathbf{G}^{(mm)(00)}
\end{matrix} \Bigg]
\begin{bmatrix} \mathbf{J}^{(e)(q)} \\ \mathbf{M}^{(q)} \\ \mathbf{J}^{(e)(0)} \\ \mathbf{M}^{(0)} \end{bmatrix}. \tag{10.6}
$$

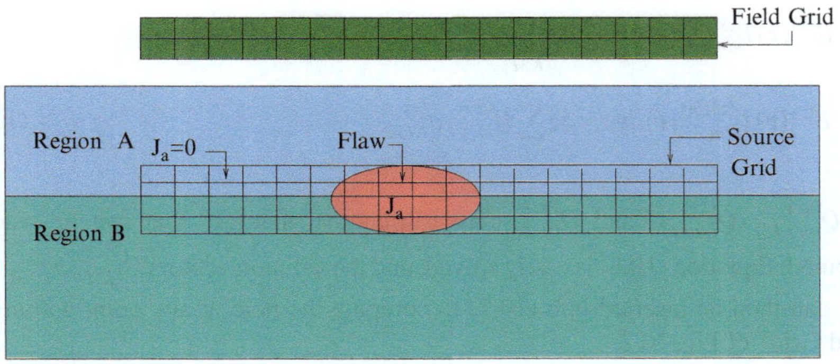

Fig. 10.2 Showing a flaw that spans two regions and a grid in the probe region

10.1.2 Application: Computing the Scattered Fields in the Vicinity of a Probe.

Consider Fig. 10.2, which shows a flaw spanning two regions, together with its "source grid," and a "field grid" in the upper-region where the probe is located. The objective is to calculate the scattered magnetic field in the probe region due to the anomalous current, $\mathbf{J}_a$, of the flaw, because this is the variable that a magnetic sensor, such as a Hall-effect probe or a giant magnetoresistive (GMR) sensor, would measure.

We assume that we have used (10.4) to compute the magnetic-field moments on the "Field Grid" of Fig. 10.2, so now we want to compute the **B**-field, itself. To do this, we expand the **B**-field in terms of the magnetic edge elements:

$$\begin{aligned}B^{(x)}(\mathbf{r}) &= \sum_{KLM} b^{(x)}_{KLM} T^{(x)(m)}_{KLM}(\mathbf{r}) \\ B^{(y)}(\mathbf{r}) &= \sum_{KLM} b^{(y)}_{KLM} T^{(y)(m)}_{KLM}(\mathbf{r}) \\ B^{(z)}(\mathbf{r}) &= \sum_{KLM} b^{(z)}_{KLM} T^{(z)(m)}_{KLM}(\mathbf{r}),\end{aligned} \tag{10.7}$$

where the expansion coefficients, $b^{(x)}_{KLM}$, $b^{(y)}_{KLM}$, $b^{(z)}_{KLM}$, are to be determined from knowledge of the field moments. This is done as follows:

$$\begin{aligned}\int B^{(x)}(\mathbf{r}) T^{(x)(m)}_{klm}(\mathbf{r}) d\mathbf{r} &= \sum_{KLM} b^{(x)}_{KLM} \int T^{(x)(m)}_{KLM}(\mathbf{r}) T^{(x)(m)}_{klm}(\mathbf{r}) d\mathbf{r} \\ &= \delta x \sum_{KLM} b^{(x)}_{KLM} \delta_{kK} \int\int \pi_{2L}(y)\pi_{2M}(z)\pi_{2l}(y)\pi_{2m}(z) dy dz \\ &= \delta x \sum_{LM} Q^{(yz)}_{lm,LM} b^{(x)}_{kLM}\end{aligned}$$

$$\int B^{(y)}(\mathbf{r})T_{klm}^{(y)(m)}(\mathbf{r})d\mathbf{r} = \delta y \sum_{KM} Q_{km,KM}^{(xz)} b_{KlM}^{(y)}$$

$$\int B^{(z)}(\mathbf{r})T_{klm}^{(z)(m)}(\mathbf{r})d\mathbf{r} = \delta z \sum_{KL} Q_{kl,KL}^{(xy)} b_{KLm}^{(z)}. \tag{10.8}$$

The $Q_{lm,LM}^{(yz)}$, $Q_{lm,LM}^{(xz)}$, and $Q_{lm,LM}^{(xy)}$ are tri-diagonal matrices that can be readily computed. Equation (10.8) is easily solved, and the solution vectors $b_{KLM}^{(x)}$, $b_{KLM}^{(y)}$ and $b_{KLM}^{(z)}$ can then be inserted into (10.7) to compute the field at any point within the "Field Grid" of Fig. 10.2.

We can ask "In what sense does (10.7) satisfy the condition, $\nabla \cdot \mathbf{B} = 0$?" To answer this question, we observe from the preceding equations that the **b**'s satisfy

$$\begin{aligned} &\frac{B_{k+1lm}^{(x)} - B_{klm}^{(x)}}{\delta x} + \frac{B_{kl+1m}^{(y)} - B_{klm}^{(y)}}{\delta y} + \frac{B_{klm+1}^{(z)} - B_{klm}^{(z)}}{\delta z} = 0 \\ &= \sum_{LM} Q_{lm,LM}^{(yz)} \left(b_{k+1LM}^{(x)} - b_{kLM}^{(x)} \right) + \sum_{KM} Q_{km,KM}^{(xz)} \left(b_{Kl+1M}^{(y)} - b_{KlM}^{(y)} \right) \\ &+ \sum_{KL} Q_{kl,KL}^{(xy)} \left(b_{KLm+1}^{(z)} - b_{KLm}^{(z)} \right), \end{aligned} \tag{10.9}$$

where the first equality comes from (10.5). Hence, the divergence condition is met in the discrete sense involving moments.

10.1.3 Test Results of Field Calculation in Probe Region

In a test problem, in which a probe is centered on a cubical flaw within a workpiece, we calculate the coefficients, $b_{KLM}^{(x)}$, $b_{KLM}^{(y)}$, $b_{KLM}^{(z)}$, of the magnetic edge elements of the scattered field in the probe region (recall (10.7)). The results, which were computed at 100 Hz, are shown in Fig. 10.3. Note that $B^{(x)}$ is antisymmetric in x and $B^{(y)}$ is antisymmetric in y.

We validate the computation of the z-component by applying Faraday's law to compute the impedance of the probe coil, and get a number very close to that calculated using the incident electric field and anomalous currents within the flaw. Figure 10.4 shows the z-component of the **B**-field. The four dimples occur at the corners of the flaw.

The details of the test problem are as follows: the host is a half space with $\sigma = 10^6$ S/m, $\mu = \mu_0$, $\varepsilon = \varepsilon_0$, and the flaw is a surface-breaking 1-mm cube. The probe is a conventional 100-turn pancake coil, with inner radius 1 mm, outer radius 1.1 mm, and height of 1 mm. The probe is centered over the flaw, with a lift-off of zero. The flaw grid is $16 \times 16 \times 8$, and the field grid, whose bottom is at $z = 0$, spans 2 mm $\times$ 2 mm $\times$ 1 mm with a $32 \times 32 \times 4$ grid.

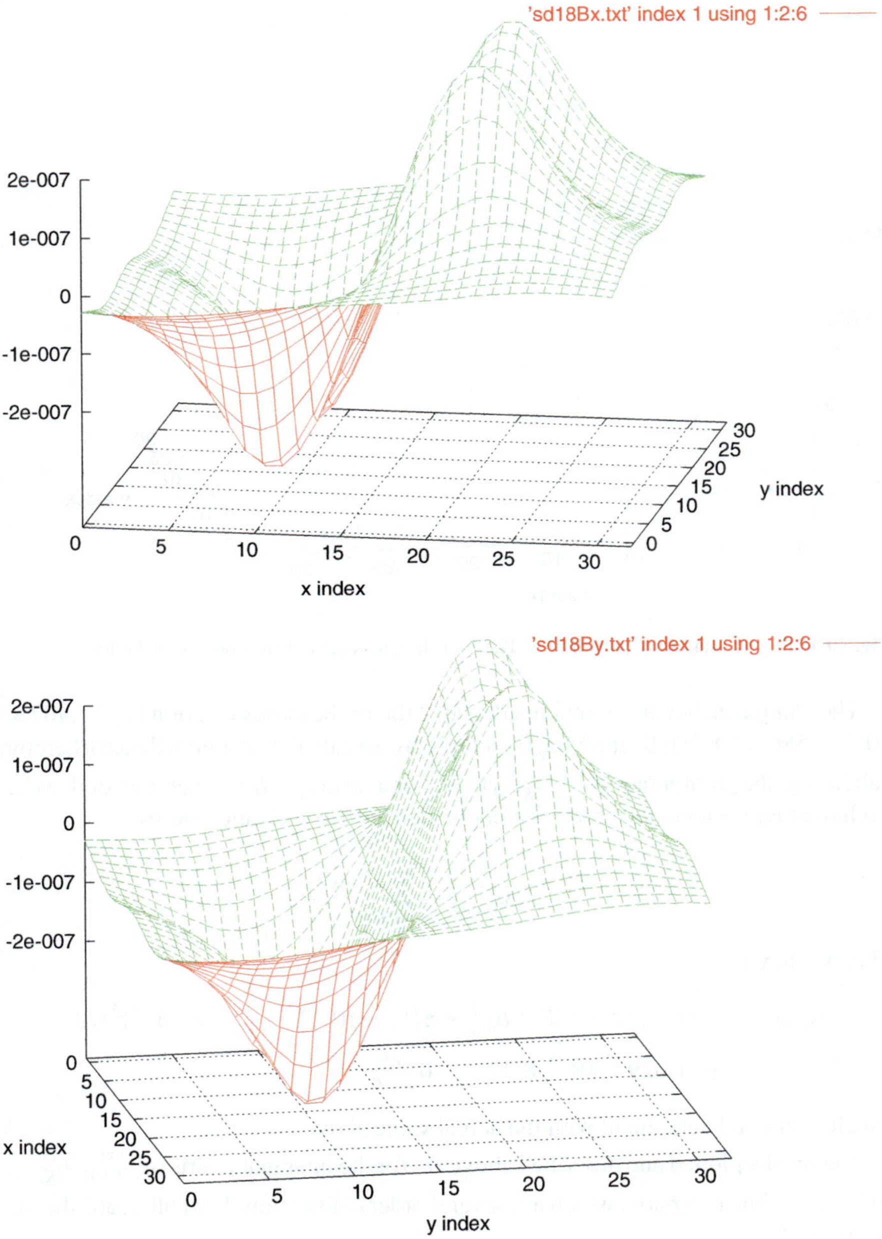

Fig. 10.3 Plot of imaginary part of $B^{(x)}$ (*top*) and imaginary part of $B^{(y)}$ (*bottom*)

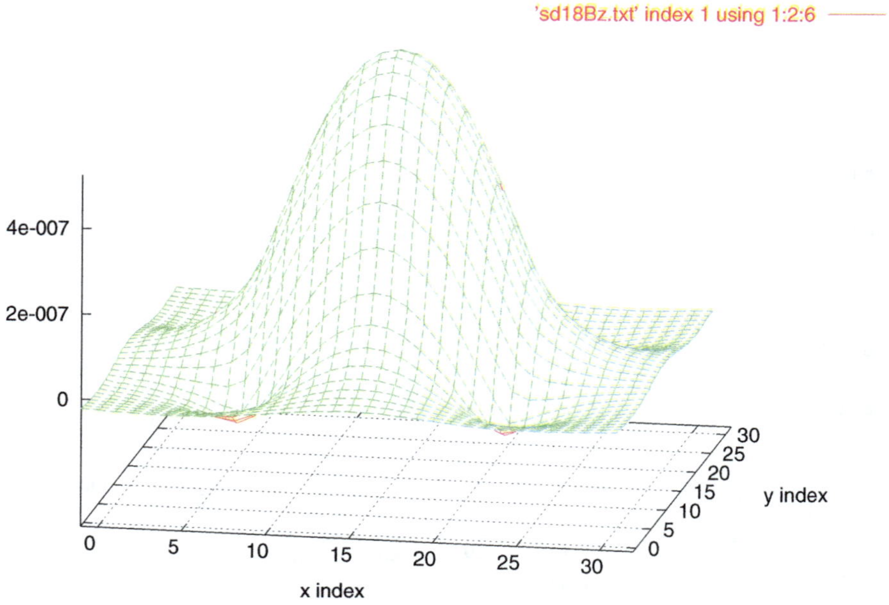

Fig. 10.4 Plot of imaginary part of $B^{(z)}$. The four dimples occur at the corners of the flaw

The change in driving-point impedance of the probe in this location is $(-1.407\times 10^{-8}, 8.586\times 10^{-12})$. In applying Faraday's law to calculate the impedance (thereby validating the computation of $b^{(z)}_{KLM}$), we first average $B^{(z)}$ over the coil axis, excluding edge elements whose centers lie outside the coil outer radius:

$$B^{(z)}_{\text{avg}} = \frac{1}{N_{\text{inside}}} \sum_{\text{coilinterior}} b^{(z)}_{klm}.$$

Then we have

$$\begin{aligned} Z_{\text{faraday}} &= 100\times j2\pi \times 100\times B^{(z)}_{\text{avg}} \times \pi((1\times 10^{-3})^2 + (1.1\times 10^{-3})^2)/2 \\ &= (-1.329\times 10^{-8}, 8.272\times 10^{-12}), \end{aligned}$$

which is in good agreement with the actual value above.

The results plotted and described above are for the imaginary parts of $b^{(x)}_{KLM}, b^{(y)}_{KLM}$, and, $b^{(z)}_{KLM}$. The real parts, which are several orders of magnitude smaller, are shown in Fig. 10.5.

As a final test, we compute the divergence expressions of (10.9). In Fig. 10.6, we plot the sum of the three partial derivatives (i.e., the divergence) divided by the square root of the sum of the squares of their absolute values. These are computed from the expansion coefficients written to an external file and, therefore, are limited by having only six digits. When double precision values are used, both expressions of (10.9) give values several orders of magnitude smaller, 10^{-10} to 10^{-13}. Clearly, the zero-divergence condition holds!

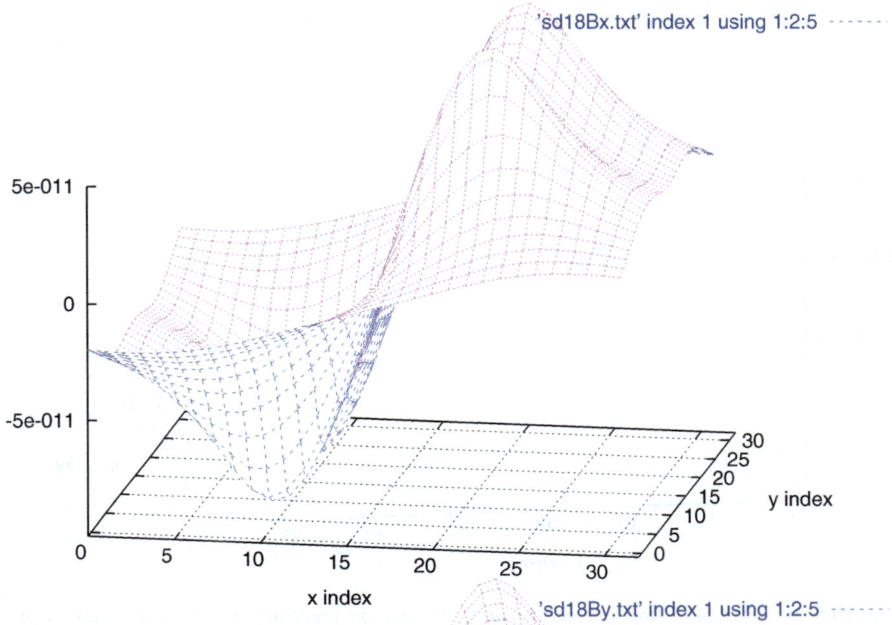

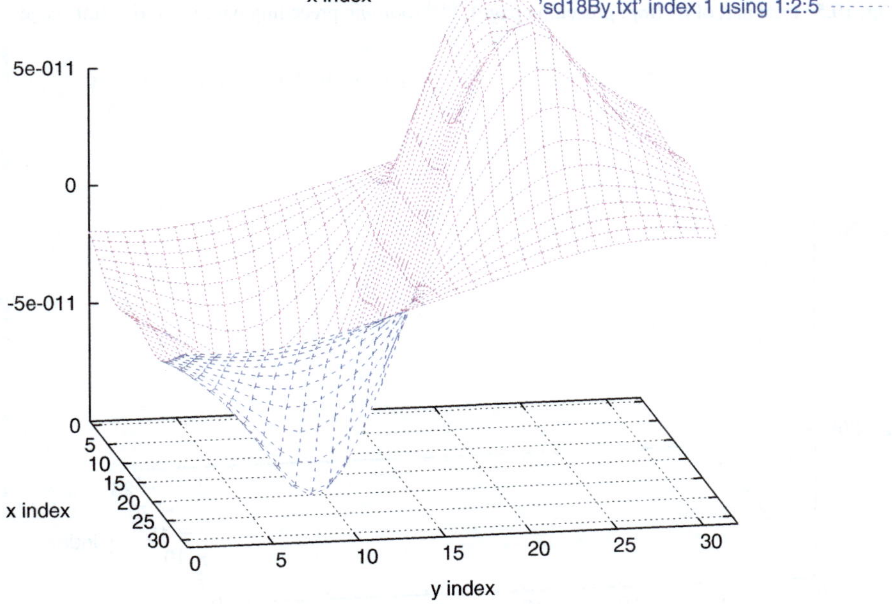

It may be thought that the expansion coefficients would satisfy a zero-divergence expression of the type $0 = (b_{k+1lm} - b_{klm})/dx + (b_{kl+1m} - b_{klm})/dy + (b_{klm+1} - b_{klm})/dz$, but that clearly doesn't hold, as Fig. 10.7 illustrates. It is necessary to use the Q-matrices in (10.9) in order to change bases.

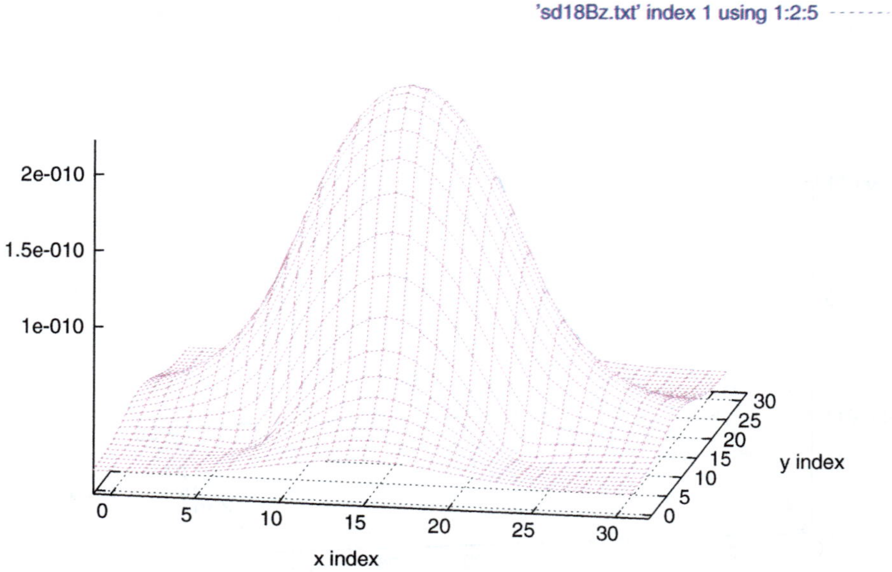

Fig. 10.5 Real part of B^x (*top*: preceding page), $B^{(y)}$ (*bottom*: preceding page), and $B^{(z)}$ (this page)

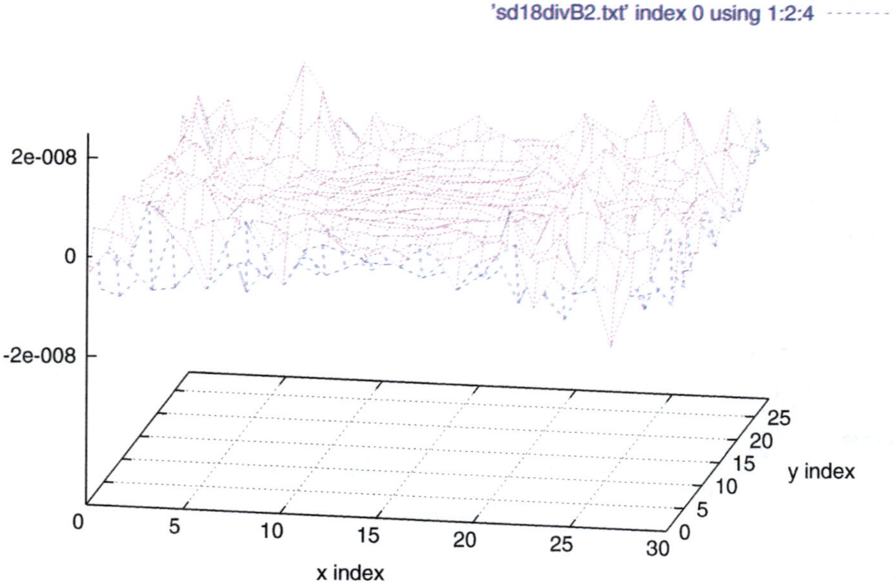

Fig. 10.6 Plot of the "normalized divergence" expression on the left of (10.9). *Top*: real part; *bottom*: imaginary part

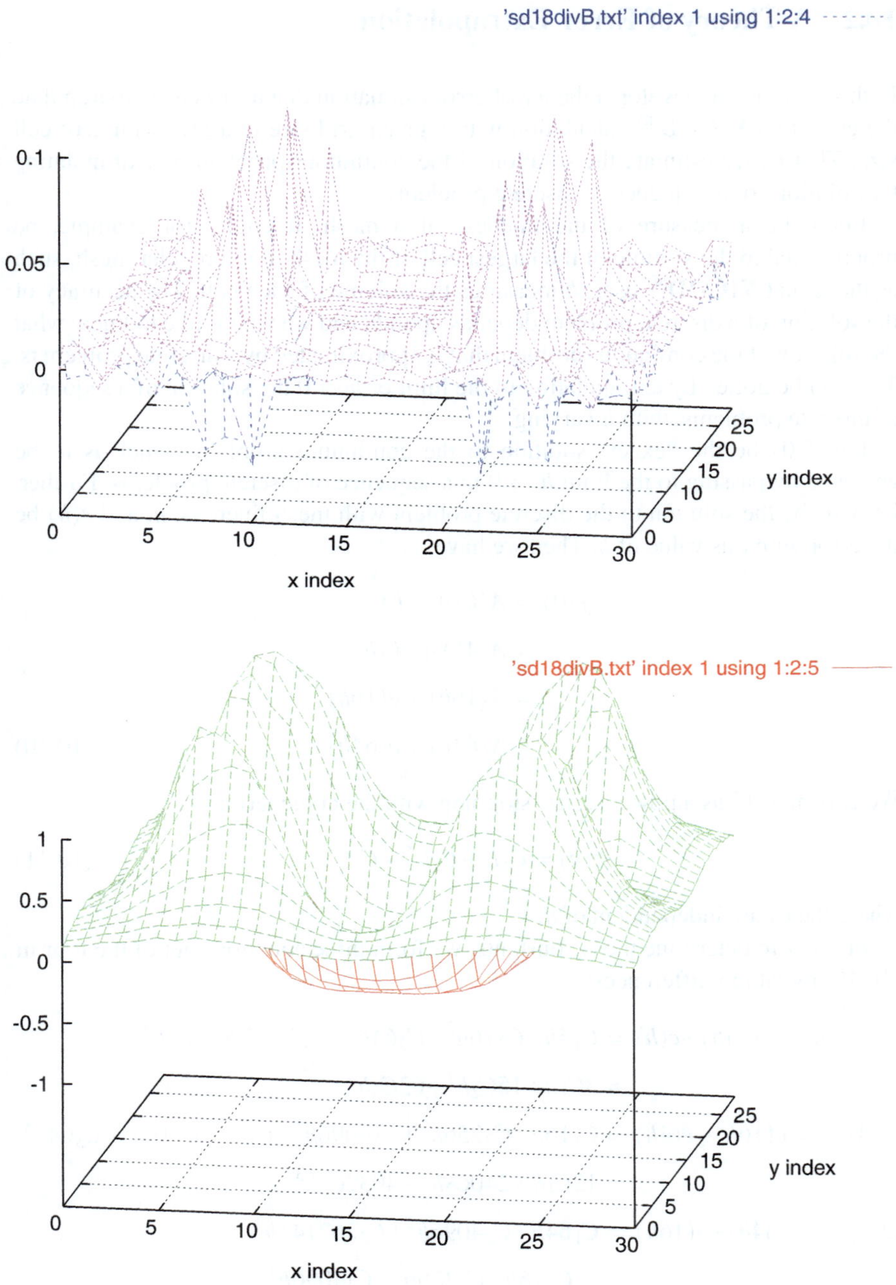

Fig. 10.7 A "pseudo expression" for the divergence in terms of the expansion coefficients, only

10.2 A Theory of Error Extrapolation

In this section, we develop a theory of error estimation that allows us to extrapolate the error in a **VIC-3D®** calculation with a given grid size to a grid with zero cell size. That is, we estimate the solution of the continuous problem by extrapolating the solutions of a sequence of discrete problems.

Let h be a measure of the fineness of a mesh. It could, for example, be proportional to the volume of the largest cell of the mesh. For a regular mesh, such as those that **VIC-3D®** uses, h measures the volume of each cell. The accuracy of the solution of a discrete problem depends upon h, and we would like to know what the solution of the continuous problem that is approximated by a discrete problem is. This can be gotten by extrapolating to the limit of $h \to 0$ the solution of a sequence of discrete problems, with h varying.

Let $f(0)$ be the "exact" solution to the continuum problem, which is to be achieved by passing to the limit $h \to 0$ in a sequence of discrete problems. Further, let $A(h)$ be the solution to the discrete problem with mesh fineness, h, and $e(h)$ be the error with this value of h. Then we have

$$\begin{aligned} f(0) &= A(h) + e(h) \\ &= A(4h) + e(4h) \\ &= A(16h) + e(16h) \\ &= A(64h) + e(64h). \end{aligned} \tag{10.10}$$

We expand $e(h)$ as a power series, stopping with the cubic term:

$$e(h) = C_1 h + C_2 h^2 + C_3 h^3 , \tag{10.11}$$

where the C_i are independent of h.

In order to determine these constants, we form three equations out of the four in (10.10) by taking differences;

$$\begin{aligned} \Delta_h A = e(4h) - e(h) &= C_1 4h + C_2 16h^2 + C_3 64h^3 - C_1 h - C_2 h^2 - C_3 h^3 \\ &= 3C_1 h + 15C_2 h^2 + 63C_3 h^3 \\ \Delta_{4h} A = e(16h) - e(4h) &= C_1 16h + C_2 256h^2 + C_3 4096h^3 - C_1 4h - C_2 16h^2 - C_3 64h^3 \\ &= 12C_1 h + 240C_2 h^2 + 4032C_3 h^3 \\ \Delta_{16h} A = e(64h) - e(16h) &= C_1 64h + C_2 4096h^2 + C_3 262144h^3 \\ &\quad - C_1 16h - C_2 256h^2 - C_3 4096h^3 \\ &= 48C_1 h + 3840C_2 h^2 + 258048C_3 h^3, \end{aligned} \tag{10.12}$$

where $\Delta_h A = A(h) - A(4h)$, $\Delta_{4h} A = A(4h) - A(16h)$, and $\Delta_{16h} A = A(16h) - A(64h)$.

The solution of (10.12) is easily obtained:

$$C_1 h = \frac{1024\Delta_h A - 80\Delta_{4h} A + \Delta_{16h} A}{2160}$$
$$C_2 h^2 = \frac{-256\Delta_h A + 68\Delta_{4h} A - \Delta_{16h} A}{8640}$$
$$C_3 h^3 = \frac{64\Delta_h A - 20\Delta_{4h} A + \Delta_{16h} A}{181,440}. \tag{10.13}$$

Upon substituting this result into (10.11), we get an expression for the error, $e(h)$, which, when substituted into (10.10) gives the final result

$$f(0) = A(h) + \Delta_h A \frac{1261}{2835} - \Delta_{4h} A \frac{83}{2835} + \Delta_{16h} A \frac{1}{2835}$$

$$= \frac{4096A(h) - 1344A(4h) + 84A(16h) - A(64h)}{2835}. \tag{10.14}$$

We apply this theory to model-data that correspond to an experiment in which a bobbin coil is scanned past a through-wall notch drilled into an aluminum tube.[1] The notch is 1.420 inches long and 0.017 inches wide. The wall thickness is 0.220 inch. The conductivity of the tube is 2.577×10^7 S/m, and the frequency of operation is 500 Hz. The grids have cell counts of $2 \times 16 \times 4$, $2 \times 32 \times 8$, $2 \times 64 \times 16$, and $2 \times 128 \times 32$, where the width of the flaw contains two cells in all the grids. The results are shown in Fig. 10.8.

Next, we do a linear interpolation on the two coarsest grids, using the two-point extrapolation formula, $f(0) = \big(4A(h) - A(4h)\big)/3$, which can be easily deduced using the ideas set out above, and get the results shown in Fig. 10.9. In this example it is clear that a simple linear extrapolation based on two relatively coarse grids will suffice to give a very good response. This could represent a significant cost savings in computer time.

The large discrepancy in resistance in both figures is due to the manner in which the data were taken. Four slots were cut into the aluminum cylinder, equally spaced in azimuth by 90°, and data were recorded for this composite anomaly. This was done to get a sufficiently large signal for the sensitivity of the instrument. We then divided this response by four and compared that with **VIC-3D®**'s response to a single slot. Because of the size of the slots, compared to the inner radius of the aluminum tube, 1.53 inch, there is considerable interaction between the slots, which means that simple division of the measurements by four will produce an error. This error is manifested mostly in the resistance values.

Finally, we apply extrapolation theory to the proximity sensor problem that is shown in Fig. 10.10. The sensor problem is characterized by, among other things,

[1]Caius V. Dodd, Oak Ridge National Laboratory, private communication.

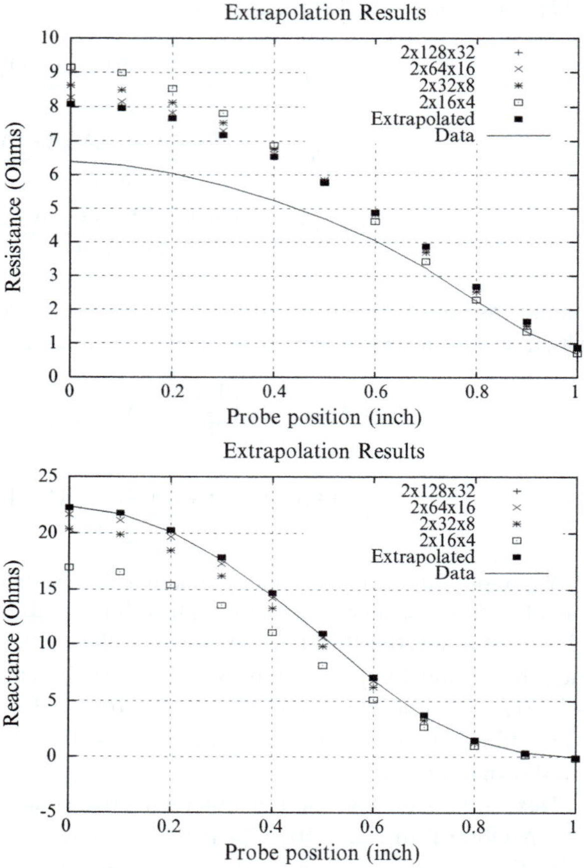

Fig. 10.8 Comparing the convergence of model impedances with mesh refinement to measured data. The extrapolated results, based on equation (10.14), are also shown. *Top*: resistance; *bottom*: reactance

the need to apply the magnetic-dipole model of nonstandard probes to the racetrack transmit and receive coils, as well as model the nickel mesa anomalous region. This can result in a problem with a large number of unknowns. To help alleviate this problem, we use extrapolation techniques to generate increasingly accurate solutions, while simultaneously conserving computer resources and time. We present the results of some numerical experiments that demonstrate the application of extrapolation techniques to this particular sensor problem.

Table 10.1 lists the convergence results for the racetrack T/R coils in freespace. Note that the results converge nicely with N_z, for a fixed $N_x \times N_y$. In each of the first three cases, the change in going from $N_z = 8$ to $N_z = 16$ is 0.51 %, so this suggests that $N_z = 16$ is sufficient to model this sensor.

As for the extrapolation analysis, if we use the three-point formula,

$$f(0) = \frac{768 \times A(h) - 240 \times A(4h) + 12 \times A(16h)}{540}, \tag{10.15}$$

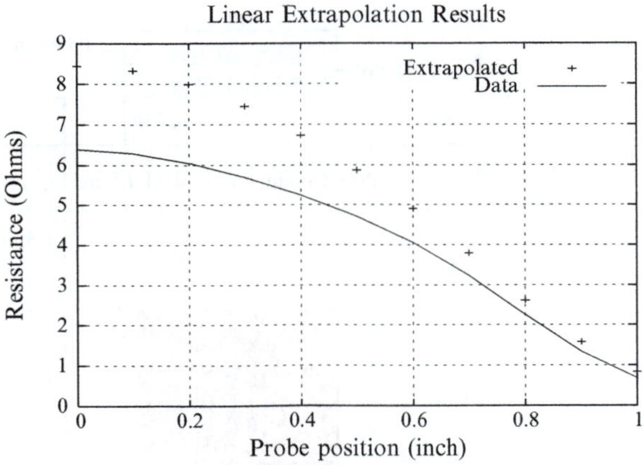

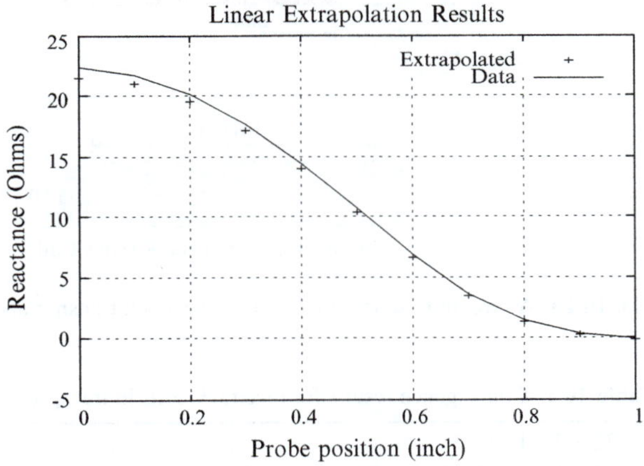

Fig. 10.9 Comparing the convergence of model impedances with mesh refinement to measured data. The linearly extrapolated results, based on the two coarsest grids, $2 \times 16 \times 4$ and $2 \times 32 \times 8$, are also shown. *Top*: resistance; *bottom*: reactance

on the first three bottom entries (with $N_z = 16$), we get an "ideal" solution of 0.107308, which is higher by less than 0.2 % than the computed solution for $32 \times 256 \times 16$. If we apply the four-point formula, (10.14), to the bottom four entries, we get 0.108 for the new "ideal" solution. This result differs from 0.10711 by 0.8 %.

In order to determine the convergence profile for modeling the anomalous region within the host material, we ran a number of different models, whose results are shown in Table 10.2. The first four entries demonstrate convergence under the fixed conditions of frequency and host conductivity, while varying the cell-density count, $N_y \times N_z$, keeping N_x, the density across the width of the mesa, fixed. If we use the three-point extrapolation over the first three entries, we get an ideal result of $f(0) = (-8.0617 \times 10^{-7},\ 3.3687 \times 10^{-6})$, which should be compared to $(-8.197 \times 10^{-7},\ 3.3913 \times 10^{-6})$, the value corresponding to the finest grid with $8 \times 64 \times 32$ cells. The error in the real part is 1.68 %, and in the imaginary 0.67 %. If we use the four-point formula over the first four entries, we get $f(0) = (-8.3576 \times 10^{-7},\ 3.4295 \times 10^{-6})$.

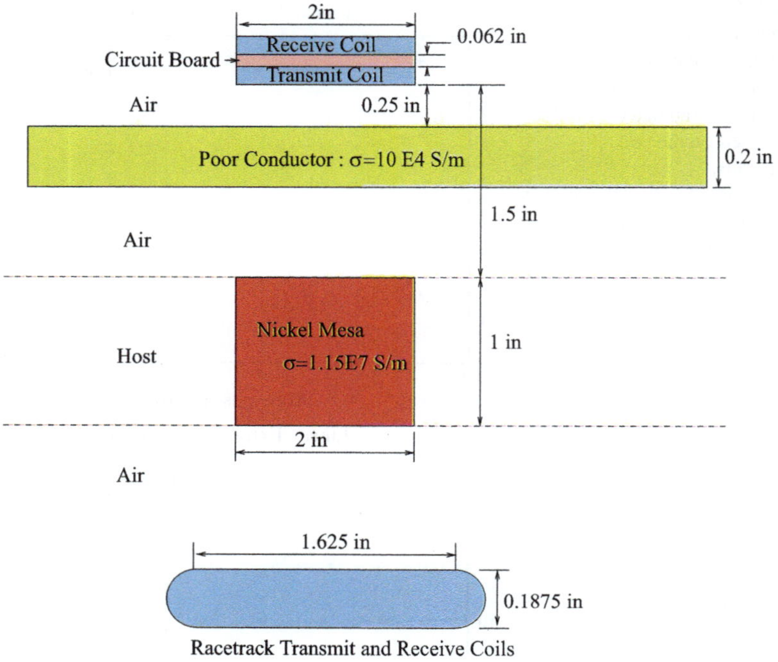

Fig. 10.10 A proximity sensor that utilizes racetrack transmit and receive coils

Table 10.1 Convergence results for racetrack coils in freespace

$4 \times 32 \times 2$	0.10376						
$4 \times 32 \times 4$	0.077742	$8 \times 64 \times 4$	0.10123				
$4 \times 32 \times 8$	0.074941	$8 \times 64 \times 8$	0.097598	$16 \times 128 \times 8$	0.10517		
$4 \times 32 \times 16$	0.074559	$8 \times 64 \times 16$	0.0971	$16 \times 128 \times 16$	0.10463	$32 \times 256 \times 16$	0.10711
Cells	Reactance	Cells	Reactance	Cells	Reactance	Cells	Reactance

Table 10.2 Convergence profile (iterations/% error in residuals) for iterative solver

Cells	No. unknowns	Frequency	σ_{host}	Profile	Solution
$8 \times 8 \times 4$	768	1 MHz	10	1389/1.00 %	(−6.0793E−7,2.8248E−6)
$8 \times 16 \times 8$	3072	1 MHz	10	1643/1.00 %	(−7.0964E−7,3.0987E−6)
$8 \times 32 \times 16$	12288	1 MHz	10	2016/1.00 %	(−7.791E−7,3.2928E−6)
$8 \times 64 \times 32$	49152	1 MHz	10	3756/1.00 %	(−8.197E−7,3.3913E−6)
$8 \times 64 \times 32$	49152	100 kHz	10	500/8.62 %	(−1.6599E−6,−1.9207E−6)
$8 \times 64 \times 32$	49152	100 kHz	100	500/5.96 %	(−1.7655E−6,−1.9889E−6)
$8 \times 64 \times 32$	49152	1 MHz	10	500/6.41 %	(−7.1739E−7,3.0734E−6)
$8 \times 64 \times 32$	49152	1 MHz	100	500/4.36 %	(−3.7532E−7,3.1366E−6)

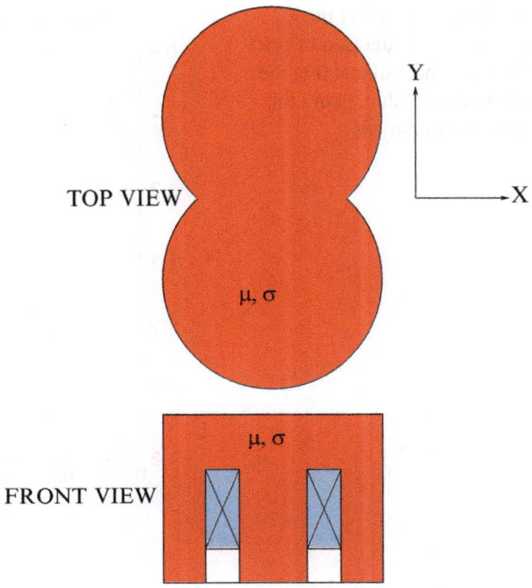

Fig. 10.11 A ferrite core probe

Table 10.3 Results, using a cubic error term, for grids of increasing refinement in a nonconducting ferrite core

Grid($N_x \times N_y \times N_z$)	Z	Grid($N_x \times N_y \times N_z$)	Z
$4 \times 4 \times 8$	$-j1.5078$	$4 \times 4 \times 16$	$-j1.7702$
$8 \times 8 \times 8$	$-j1.6049$	$8 \times 8 \times 16$	$-j1.9335$
$16 \times 16 \times 8$	$-j0.92719$	$16 \times 16 \times 16$	$-j0.98303$
$32 \times 32 \times 8$	$-j1.1119$	$32 \times 32 \times 16$	$-j1.0733$
Extrapolated Value	$-j1.2139$	Extrapolated Value	$-j1.1413$

The last four entries in the table indicate convergence profiles as a function of either frequency or host conductivity. Clearly, for a given frequency, convergence is better at the higher host conductivity, whereas for a given host conductivity, convergence is better at the higher frequency.

10.2.1 Application to a Ferrite Core Probe

A ferrite core probe is shown in Fig. 10.11. Consider the ferrite core problem with $\sigma = 0$, $\mu = 4300$, $\varepsilon = 1$, and grids of increasing refinement, with a cubic error term as shown in Table 10.3. Since the core is nonconducting, the impedances are purely imaginary. The extrapolated values differ by 6.4 %, which is reasonable for this computation. The value for the $32 \times 32 \times 8$ grid differs by 9.2 % from its extrapolated value, whereas that for $32 \times 32 \times 16$ differs by 6.3 %, which makes sense since the finer mesh should produce a better estimate.

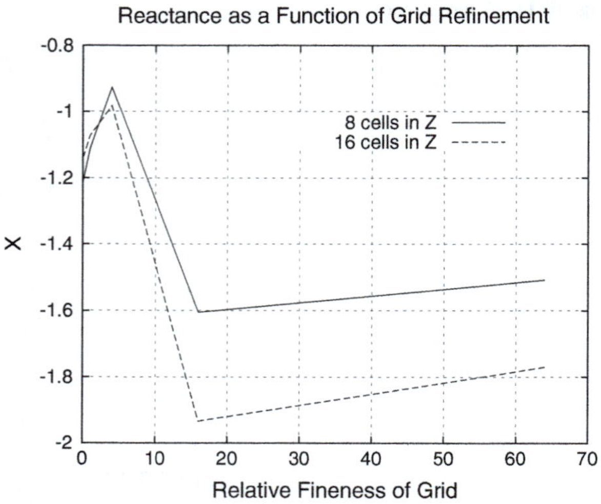

Fig. 10.12 Showing the reactance as a function of grid fineness. The value at 0 is the extrapolated value, using the cubic error analysis

It is clear that the complex shape of the probe produces an oscillating result for the impedance as the grid is refined (see Fig. 10.12), contrary to what we saw in Fig. 10.8 for a regular slot, for which the convergence was monotonic. In the case of the slot, the finer mesh produces a more accurate estimate of the anomalous currents, while doing nothing to improve the accuracy of the representation of the geometry of the anomaly. In the case of the probe, however, we need a refined grid to accurately represent both the anomaly and the anomalous currents.

It may be argued that the data for the four coarsest grids in Table 10.3 unduly skews the extrapolation, or, to put it another way, these data may be outside the range of a cubic error analysis. This argument may justify a simple linear extrapolation over the results for grids $32 \times 32 \times 16$ and $16 \times 16 \times 8$:

$$\begin{aligned} f(0) &= A(h) + \frac{A(h) - A(8h)}{7} \\ &= \frac{8A(h) - A(8h)}{7} \\ &= -j\frac{8(1.0733) - 0.92719}{7} \\ &= -j1.0942. \end{aligned} \tag{10.16}$$

The result for the finer grid differs by less than 2 % from the linearly extrapolated value. Clearly, there is a numerical advantage in using finer grids, but the disadvantage is in computational time. In any case, this analysis gives us an estimate of the discretization error when using any of these grids.

These results hold for computations in freespace and suggest that using a relatively coarse grid in X, Y may not be serious from a numerical viewpoint. When the probe is placed over a host that includes a flaw, it is important to have enough

Fig. 10.13 Schematic configuration for the tilted notch benchmark problem with a 30° EDM notch

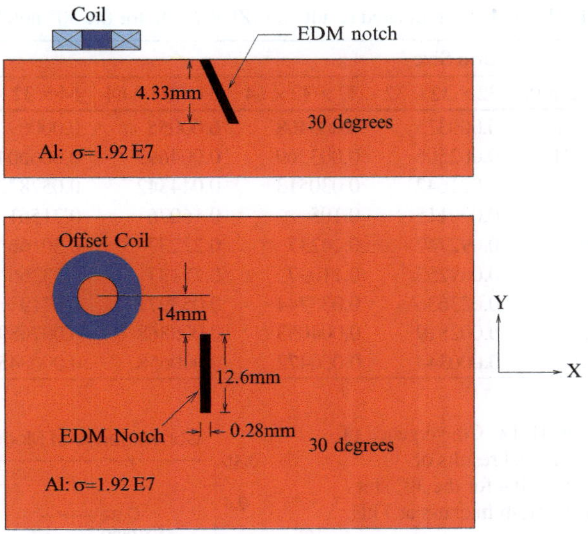

cells in the transverse plane to produce an accurate incident field on the flaw, since it is this field that will not only couple the probe to the host, but will produce the flaw "signature" that will be used in an inversion process.

10.2.2 Application to Tilted Notch Benchmark Problems

Figure 10.13 illustrates the configuration for the tilted notch benchmark problem with a 30° EDM notch. Data are taken at 7 kHz, which, with a conductivity of 1.92×10^7 S/m for the host, yields a skin depth of 1.38 mm. This implies that the problem is quite large electrically, given the dimensions and extent of the EDM notch anomaly in Fig. 10.13. Thus, we will apply extrapolation theory with cell sizes of $32 \times 32 \times 32$, $32 \times 32 \times 64$, $64 \times 32 \times 64$, and $64 \times 32 \times 128$. The model results, using the previous notation, are, therefore, $A(h)$, $A(2h)$, $A(4h)$, and $A(8h)$, and the extrapolation formula, using a cubic error term, becomes

$$f(0) = \frac{64A(h) - 56A(2h) + 14A(4h) - 1A(8h)}{21}. \quad (10.17)$$

We tabulate in Table 10.4 the results for $|Z|$ at 7 kHz for scans over the 30° notch with these four grids.

These results are plotted in Figs. 10.14 and 10.15. The convergence is monotonic from below, in contrast to the previous results shown in Fig. 10.12, confirming, once again, that the geometry is well defined with a rather coarse grid and that the refinement of the grid goes into creating a better approximation of the anomalous currents. The asymmetry in the extrapolated model response is clearly evident, with

Table 10.4 Extrapolated results for $|Z|$ at 7 kHz for the 30° notch using four grids

	Grid Size					
Scan Pt	32 × 32 × 32	32 × 32 × 64	64 × 32 × 64	64 × 32 × 128	Extrapolated	Data
−28	0.000315	0.000408	0.00055	0.00069	0.000893	0.013287
−21	0.002506	0.003369	0.004699	0.006004	0.007894	0.016114
−14	0.021642	0.030513	0.044342	0.057872	0.077437	0.069345
−7	0.070115	0.105	0.16076	0.21561	0.29506	0.312066
0	0.092742	0.14237	0.22237	0.30166	0.41685	0.426732
7	0.069253	0.11037	0.17197	0.23366	0.32380	0.316074
14	0.023636	0.035744	0.053971	0.0719	0.09790	0.080692
21	0.002819	0.004053	0.005868	0.007689	0.01035	0.008785
28	0.000344	0.000472	0.000658	0.000845	0.001119	0.008045

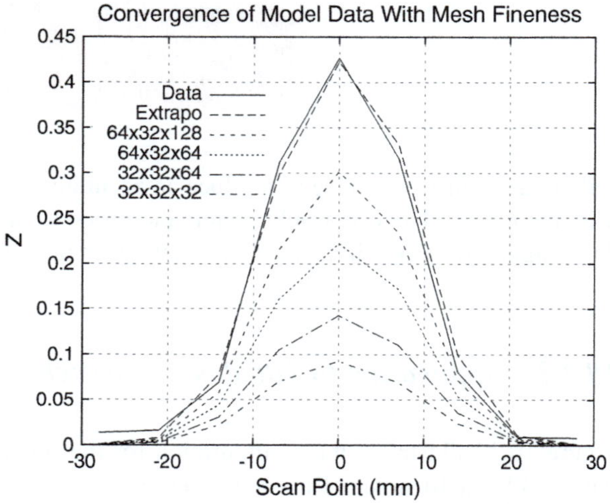

Fig. 10.14 Convergence of the model results of Table 10.4 for the 30° slot with mesh fineness at 7 kHz

the larger signal occurring for positive values of x (the slot is located at $x = 0$), where the slot is tilted towards the surface. The symmetry is less noticeable in the actual data, probably because of a slightly non-flat surface. At lower frequencies, this effect probably will be less of a problem.

Figure 10.16 illustrates the measured and extrapolated-model responses in the impedance plane of the measured data for the 30° notch. The enclosed areas are roughly equal, which is a measure of the asymmetry associated with the 30° tilt.

10.2.2.1 Notch With a 10° Tilt

We continue with the problem defined in Fig. 10.13, except that the 5-mm long slot is tilted 10° from the vertical. We use two grids, $16 \times 32 \times 32$ and $32 \times 32 \times 64$, and extrapolate with a linear error term over these two grids. The formula is

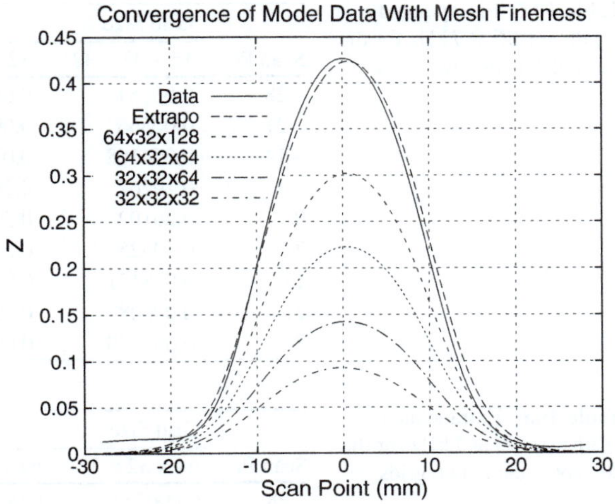

Fig. 10.15 Convergence of the model results for the 30° slot with mesh fineness at 7 kHz. The data of Table 10.4 have been smoothed with cubic splines

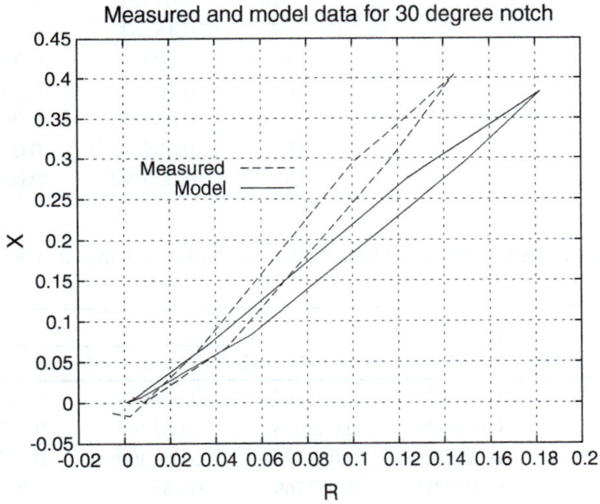

Fig. 10.16 The measured and extrapolated-model responses in the impedance plane for the 30° notch

$$f(0) = \frac{4A(h) - A(4h)}{3}, \tag{10.18}$$

and the corresponding data are shown in Table 10.5. It is clear, when comparing the extrapolated data of Table 10.4 with the corresponding data of Table 10.5, that the asymmetry associated with the 30° slot is considerably greater than that with the 10° slot, as we would expect intuitively.

Table 10.5 Extrapolated results for $|Z|$ at 7 kHz for the 10° notch using two grids

	Grid Size		
Scan Pt	16 × 32 × 32	32 × 32 × 64	Extrapolated Value
−28	0.00064	0.00089	0.000973
−21	0.00566	0.008013	0.0088
−14	0.054168	0.078135	0.086124
−7	0.19262	0.28445	0.31506
0	0.26102	0.39007	0.433
7	0.19925	0.29269	0.32384
14	0.058121	0.084437	0.09321
21	0.006087	0.008690	0.00956
28	0.000680	0.000951	0.00104

Table 10.6 Extrapolated results for $|Z|$ at 7 kHz for the 20° notch using two grids

	Grid Size		
Scan Pt	64 × 32 × 32	64 × 32 × 128	Extrapolated Value
−28	0.000531	0.000845	0.00095
−21	0.004583	0.007518	0.0085
−14	0.043252	0.073263	0.08242
−7	0.15363	0.27134	0.31058
0	0.20956	0.3763	0.432
7	0.1608	0.28571	0.32735
14	0.049172	0.084589	0.09639
21	0.005265	0.008869	0.01007
28	0.000594	0.000968	0.001093

Table 10.7 Extrapolated results for $|Z|$ at 7 kHz for the 20° notch using four grids and algorithm (7.5)

	Grid Size				
Scan Pt	8 × 32 × 16	16 × 32 × 32	32 × 32 × 64	64 × 32 × 128	Extrapolated Value
−28	0.000157	0.000263	0.000614	0.000845	0.000938
−21	0.001046	0.002053	0.005365	0.007518	0.008379
−14	0.006663	0.017137	0.051309	0.073263	0.082031
−7	0.013207	0.052769	0.18542	0.27134	0.305687
0	0.013525	0.068195	0.25435	0.3763	0.425
7	0.013263	0.054705	0.19471	0.28571	0.322102
14	0.007017	0.018769	0.058905	0.084589	0.094842
21	0.001132	0.002282	0.006216	0.008869	0.009934
28	0.000167	0.000286	0.000692	0.000968	0.001079

10.2.2.2 Notch With a 20° Tilt

We'll use the same linear extrapolation formula, (10.18), as for the 10° notch, but with grids of 64 × 32 × 32 and 64 × 32 × 128 cells. Table 10.6 gives the results. Table 10.4 showed the results for a cubic interpolation using four rather fine grids. If we use the cubic interpolation algorithm of (7.5), we can get by with coarser grids, as shown in Table 10.7 for the 20° notch.

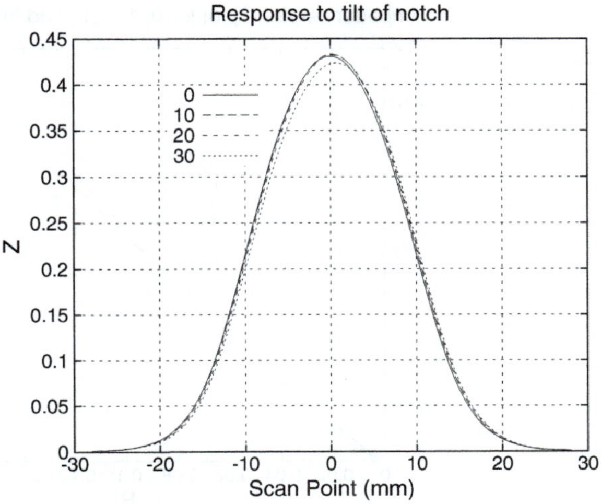

Fig. 10.17 Comparing $|Z|$ for notches of 0°, 10°, 20°, and 30° tilts

Table 10.8 Asymmetry ratio for notch tilt

Tilt	(Asymmetry ratio − 1) (%)
0°	0
10°	8.2
20°	17
30°	26.5

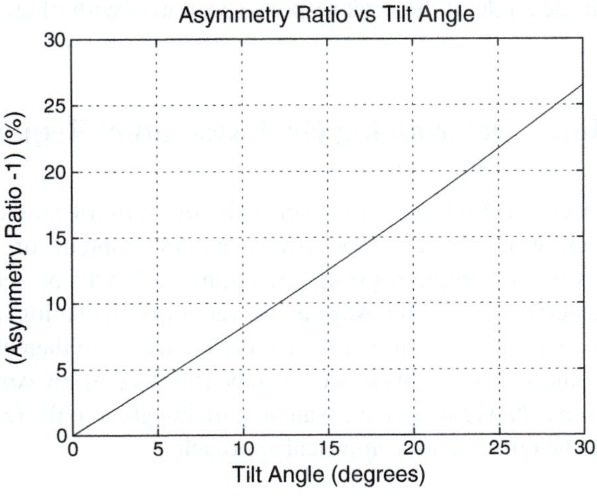

Fig. 10.18 Asymmetry ratio vs. tilt angle

Figure 10.17 compares the $|Z|$ response to EDM notches at 0°, 10°, 20°, and 30° tilts. It is difficult to eyeball these results and draw conclusions as to the slant of each notch, because each notch has the same volume. If, however, we refer to the raw data in Tables 10.4–10.6, and consider the "asymmetry ratio" $|Z(14)|/|Z(-14)|$, then the results tabulated in Table 10.8 give a clear estimate of the tilt. Indeed, it is a virtually linear relation, with a slightly increasing slope with angle (see Fig. 10.18).

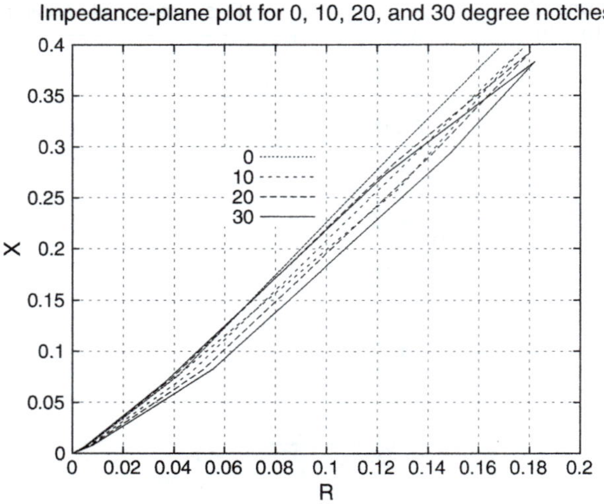

Fig. 10.19 Illustrating the increase in area with tilt angle enclosed by the impedance-plane plots of each of the notches

The impedance-plane plots of the responses for the 0°, 10°, 20°, and 30° notches are shown in Fig. 10.19. As we would expect, the area enclosed by each curve increases with the tilt angle, indicating a greater asymmetry. Note, further, that the phase angle of the peak response decreases with tilt angle.

10.3 Determining Grid Resolution Requirements

In this section, we will address the question of how to estimate the required grid resolution to accurately solve a model problem, or, to put it another way, how do we determine a priori how many cells will be required to solve the problem accurately. We can always do the recommended thing, and that is to test several grid resolutions, chosen in a systematic manner, and then check the rate of convergence of the solution. This process can be followed by an extrapolation, as we have shown above. Nevertheless, we want to gain insight into the requirements without resorting at the outset to this empirical approach.

10.3.1 The Volume-Integral Equation

Our theoretical consideration starts with an analysis of the volume-integral equation that is the basis of **VIC-3D®**:

$$\mathbf{E}_0(\mathbf{r}) = \frac{\mathbf{J}_a(\mathbf{r})}{\sigma_a(\mathbf{r})} - \int \mathbf{G}(\mathbf{r},\mathbf{r}') \cdot \mathbf{J}_a(\mathbf{r}'), \tag{10.19}$$

which can be rearranged to read

$$\mathbf{J}_a(\mathbf{r}) = \sigma_a(\mathbf{r}) \left[\mathbf{E}_0(\mathbf{r}) + \int \mathbf{G}(\mathbf{r},\mathbf{r}') \cdot \mathbf{J}_a(\mathbf{r}') \right]. \tag{10.20}$$

$\mathbf{E}_0(\mathbf{r})$ is the incident electric field, which is known, as is the anomalous conductivity, $\sigma_a(\mathbf{r})$ (for a forward problem). $\mathbf{G}(\mathbf{r},\mathbf{r}')$ is the dyadic Green's function, and $\mathbf{J}_a(\mathbf{r})$ is the unknown anomalous current.

The gridding question can be answered by analyzing the spatial-frequency content of $\mathbf{J}_a(\mathbf{r})$. This requires taking the spatial Fourier transform of (10.20), which involves the product of two spatially-varying functions, $\sigma_a(\mathbf{r})$, and the total electric field within the brackets. Once we get an estimate of the highest spatial frequencies that must be present in $\mathbf{J}_a(\mathbf{r})$, we can estimate the size of the cells that are required to produce these frequencies.

10.3.2 Review of Fourier Analysis

We are going to apply Fourier analysis to study the question, using the notion of spatial frequency in the same manner that temporal frequency is used to answer resolution questions in system theory. Equation (10.20) shows that we need to determine the spatial Fourier transform of the product of two functions. This is given by the convolution in the spatial-frequency domain of the individual transforms, as we'll now prove.

$$\begin{aligned} \int e^{-j\mathbf{k}\cdot\mathbf{r}} f(\mathbf{r})g(\mathbf{r})d\mathbf{r} &= \int e^{-j\mathbf{k}\cdot\mathbf{r}} f(\mathbf{r}) \left\{ \frac{1}{(2\pi)^3} \int e^{j\mathbf{k}'\cdot\mathbf{r}} \tilde{G}(\mathbf{k}')d\mathbf{k}' \right\} d\mathbf{r} \\ &= \frac{1}{(2\pi)^3} \int \tilde{G}(\mathbf{k}') \left\{ \int e^{-j(\mathbf{k}-\mathbf{k}')\cdot\mathbf{r}} f(\mathbf{r})d\mathbf{r} \right\} d\mathbf{k}' \\ &= \frac{1}{(2\pi)^3} \int \tilde{G}(\mathbf{k}')\tilde{F}(\mathbf{k}-\mathbf{k}')d\mathbf{k}'. \end{aligned} \tag{10.21}$$

Consider, for example, that $g(\mathbf{r})$ is uniform in space, with a value of g. Then $\tilde{G}(\mathbf{k}) = (2\pi)^3 g\delta(\mathbf{k})$, so that the right-hand side of (10.21) is equal to $g\tilde{F}(\mathbf{k})$, as expected.

To simplify our lives a bit, we'll ignore the integral term in (10.20), and just work with the approximation, $\mathbf{j}_a(\mathbf{r}) \approx \mathbf{E}_0(\mathbf{r})\sigma_a(\mathbf{r})$, in determining the frequency content of $\mathbf{j}_a(\mathbf{r})$:

$$\tilde{\mathbf{J}}_a(\mathbf{k}) \approx \frac{1}{(2\pi)^3} \int \widetilde{\mathbf{E}_0}(\mathbf{k}')\tilde{\Sigma}_a(\mathbf{k}-\mathbf{k}')d\mathbf{k}'. \tag{10.22}$$

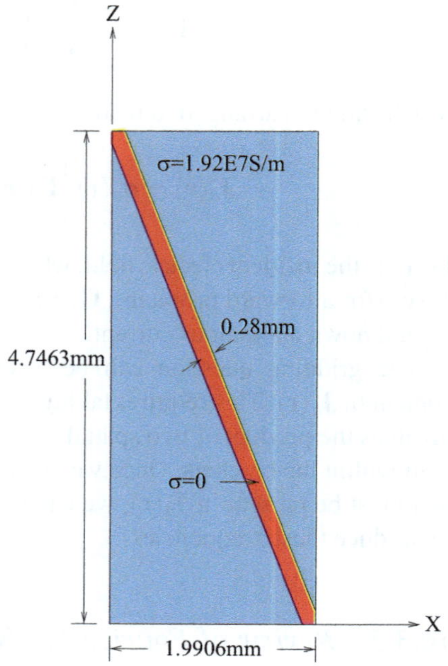

Fig. 10.20 Illustrating the bounding box for the 20° notch problem of Sect. 10.2.2.2

We will call $\tilde{\Sigma}_a(\mathbf{k})$ the spectrum of the conductivity scene, $\tilde{\mathbf{E}}_0(\mathbf{k})$ the spectrum of the electromagnetic scene, and the convolution in (10.22) the composite spectral scene. As a simple example of what to expect, let the conductivity and electromagnetic spectral scenes in (10.22) each be a "baseband" unit pulse extending from 0 to $K_{\max}$, then the composite spectrum will be a unit tent function extending from 0 to $2K_{\max}$. Clearly, the spatial-frequency content has been doubled for $\tilde{\mathbf{J}}_a$, when compared to the original spectra.

10.3.3 Application to the 20° Notch Problem

The **VIC-3D**®-generated bounding box for the 20° notch problem of Sect. 10.2.2.2 is shown in Fig. 10.20. We'll start by computing the spectrum of the conductivity scene, $\tilde{\Sigma}_a(k_x,k_z)$. Since our "computational universe" is bounded by the rectangle $X_0 = 1.9906$, $Z_0 = 4.7463$, we will use a Fourier series representation with discrete spatial frequencies for all spectra. For the conductivity spectrum we write

$$\tilde{\Sigma}_a(k_x,k_z) = \frac{1}{X_0 Z_0}\int_0^{X_0}\int_0^{Z_0} \sigma_a(x,z)e^{-j2\pi(k_x x+k_z z)}dxdz \qquad (10.23)$$

where $k_x = n_x/X_0$, $k_z = n_z/Z_0$, $n_x, n_z = 0, \pm 1, \pm 2, \ldots$. The inverse relation is

$$\sigma_a(x,z) = \sum_{k_x=-\infty}^{\infty} \sum_{k_z=-\infty}^{\infty} \tilde{\Sigma}_a(k_x,k_z) e^{j2\pi(k_x x + k_z z)}, \quad (10.24)$$

as is well known from the theory of Fourier series.

The equation of the line passing through (1.9906, 0) and (0, 4.7463) is $x/1.9906 + z/4.7463 - 1 = 0$, or $z = 4.7463 - 2.384x$. The unit normal to this line is $\mathbf{a}_n = 0.922\mathbf{a}_x + 0.387\mathbf{a}_z$, so the equation of the line parallel to this line, and whose distance from the origin is d, is given by the inner-product of $\mathbf{a}_n$ with the arbitrary vector, $x\mathbf{a}_x + z\mathbf{a}_z$: $0.922x + 0.387z = d$. Now, the boundaries of the notch are separated by 0.28 mm, which means that the top boundary of the notch is given by $0.922x + 0.387z = d + 0.14$, and the bottom boundary by $0.922x + 0.387z = d - 0.14$. Thus, the length of the notch in the z direction, at a fixed x coordinate, is $z_1 - z_2 = 0.28/0.387 = 0.724$ mm.

We'll evaluate (10.23) as an iterated integral, starting with z:

$$\begin{aligned}
\frac{1}{Z_0}\int_0^{Z_0} \sigma_a(x,z) e^{-j2\pi k_z z} dz &= \frac{-1.92 \times 10^7}{Z_0} \int_{z(x)-0.362}^{z(x)+0.362} e^{-j2\pi k_z z} dz \\
&= \frac{-1.92 \times 10^7}{Z_0} e^{-j2\pi k_z z(x)} \frac{\sin(2\pi k_z \times 0.362)}{\pi k_z} \\
&= -1.92 \times 10^7 e^{-j2\pi n_z(1-0.502x)} \frac{\sin(0.153\pi n_z)}{\pi n_z} \\
&= -2.94 \times 10^6 e^{-j2\pi n_z} \frac{\sin(0.153\pi n_z)}{0.153\pi n_z} e^{j2\pi n_z(0.502x)}.
\end{aligned} \quad (10.25)$$

The number 0.153 is the ratio of the area of the notch in the (X,Z) plane to the total area of the bounding box.

We complete the calculation of the iterated integral by substituting (10.25) into (10.23):

$$\begin{aligned}
\tilde{\Sigma}_a(k_x,k_z) &= -2.94 \times 10^6 e^{-j2\pi n_z} \frac{\sin(0.153\pi n_z)}{0.153\pi n_z} \frac{1}{X_0} \int_0^{X_0} e^{j2\pi n_z(0.502x)} e^{-j2\pi k_x x} dx \\
&= -2.94 \times 10^6 e^{-j2\pi n_z} \frac{\sin(0.153\pi n_z)}{0.153\pi n_z} \frac{1}{X_0} \int_0^{X_0} e^{-j2\pi x(k_x - 0.502 n_z)} dx \\
&= -2.94 \times 10^6 e^{-j2\pi n_z} \frac{\sin(0.153\pi n_z)}{0.153\pi n_z} \left[\frac{e^{-j2\pi(k_x - 0.502 n_z)X_0} - 1}{-j2\pi(n_x - n_z)}\right] \\
&= -2.94 \times 10^6 e^{-j2\pi n_z} \frac{\sin(0.153\pi n_z)}{0.153\pi n_z} \left[\frac{e^{-j2\pi(n_x - n_z)} - 1}{-j2\pi(n_x - n_z)}\right]
\end{aligned}$$

Table 10.9 Comparing spatial frequency responses

n_x	$\sin(0.153\pi n_x)/0.153\pi n_x$	$\sin(0.612\pi n_x)/0.612\pi n_x$
0	1	1
1	0.962	0.488
2	0.853	−0.168
10	−0.207	0.019
100	−0.017	−0.00306
200	0.0096	0.00247
1,000	−0.00069	0

$$= 0, \text{ if } n_x \neq n_z$$

$$= -2.94 \times 10^6 e^{-j2\pi n_x} \frac{\sin(0.153\pi n_x)}{0.153\pi n_x}, \text{ if } n_x = n_z. \tag{10.26}$$

The fact that $n_x = n_z$ means, from (10.24), that the sinusoidal expansion functions for $\sigma_a(x,z)$ have precisely the same spatial frequencies in the x and z directions. This means that in order to reconstruct the conductivity, we must sample equally finely in the x and z directions. (This holds only for characterizing the conductivity scene.) To get an idea of what the sampling requirements are, consider Table 10.9. At a spatial frequency of 100, the magnitude of the spectrum in (10.26) is reduced to slightly less than 2 %. Hence, if we are to reconstruct $\sigma_a(x,z)$ to this precision we will need 200 samples in each direction to avoid aliasing. For a flaw with four times the area, however, we can get by with one-tenth the number of samples for the same precision.

10.3.4 The Electromagnetic Scene

Now, we'll turn our attention to the electromagnetic scene for the 20° notch problem. Let the coil lie in region 1, above the workpiece, and let region 2 be the workpiece, which we assume is a homogeneous half-space. The appropriate axisymmetric Green's functions are [27]:

$$G_{11}(r,z;r',z') = \frac{1}{2\pi}\int_0^\infty \left[\frac{Re^{-\alpha_0(z+z')} + e^{-\alpha_0|z-z'|}}{2\alpha_0}\right] J_1(r'l)J_1(rl)l\,dl, \quad (r',z') \in 1,\ (r,z) \in 1$$

$$G_{21}(r,z;r',z') = \frac{1}{2\pi}\int_0^\infty \frac{T}{2\alpha_0} e^{(\alpha_1 z - \alpha_0 z')} J_1(r'l)J_1(rl)l\,dl, \quad (r',z') \in 1,\ (r,z) \in 2, \tag{10.27}$$

where

$$R = \frac{\alpha_0 - \alpha_1}{\alpha_0 + \alpha_1}, \quad T = \frac{2\alpha_0}{\alpha_0 + \alpha_1} \tag{10.28}$$

are the reflection and transmission coefficients, respectively, $\alpha_0 = (l^2 - \omega^2\mu_0\varepsilon_0)^{1/2}$, $\alpha_1 = (l^2 - \omega^2\mu_0\varepsilon_0 + j\omega\mu_0\sigma)^{1/2}$, $l^2 = k_x^2 + k_y^2$ is the spatial wave-number in Fourier space, σ is the conductivity of the host medium, $r^2 = x^2 + y^2$, and $J_1(rl)$ is the first-order Bessel function.

The ϕ-directed electric field within region 2, due to the axisymmetric coil, is given by

$$\begin{aligned}
E_2(r,z) &= -j\omega 2\pi\mu_0 \int\int_{\text{coil}} G_{21}(r,z;r',z')J_c(r',z')r'dr'dz' \\
&= -j\omega\mu_0 N_c \int_0^\infty \frac{T}{2\alpha_0} e^{\alpha_1 z} J_r(rl) \left\{ \int_L^{L+H} e^{-\alpha_0 z'}dz' \int_{R_1}^{R_2} J_1(r'l)r'dr' \right\} ldl \\
&= -j\omega\mu_0 N_c \int_0^\infty \left[\frac{T}{2\alpha_0} \frac{e^{-\alpha_0 L}(1-e^{-\alpha_0 H})}{\alpha_0} \frac{\mathcal{I}(R_1 l, R_2 l)}{l^2} e^{\alpha_1 z} \right] J_1(rl)ldl,
\end{aligned} \tag{10.29}$$

where $\mathcal{I}(a,b) = \int_a^b zJ_1(z)dz$. We assume that a current of one ampere is passing through coil, whose turns density is N_c. L is the lift-off parameter of the coil, H the height, and R_1 and R_2 the inner- and outer-radii of the coil, respectively.

The expression within the square brackets of (10.29) is the Bessel transform, $\tilde{E}_2(l,z)$, of the electric field. It is clearly an explicit function of the depth, $z(\leq 0)$. It is the relationship between the electric field and z that we need to sample in order to get an estimate of the number of cells in z that the EM scene will contribute. To do this, we start by computing the finite Fourier transform in z of $\tilde{E}_2(l,z)$:

$$\begin{aligned}
\tilde{E}_2(l,k_z) &= \frac{1}{Z_0}\int_0^{Z_0} \tilde{E}_2(l,z)e^{-j2\pi k_z z}dz \\
&= \frac{-j\omega\mu_0 N_c}{Z_0} \frac{T}{2\alpha_0} \frac{e^{-\alpha_0 L}(1-e^{-\alpha_0 H})}{\alpha_0} \frac{\mathcal{I}(R_1 l, R_2 l)}{l^2} \int_0^{Z_0} e^{-\alpha_1 z} e^{-j2\pi k_z z}dz \\
&= \frac{-j\omega\mu_0 N_c}{Z_0} \frac{T}{2\alpha_0} \frac{e^{-\alpha_0 L}(1-e^{-\alpha_0 H})}{\alpha_0} \frac{\mathcal{I}(R_1 l, R_2 l)}{l^2} \frac{1-e^{-(\alpha_1 + j2\pi k_z)Z_0}}{\alpha_1 + j2\pi k_z}.
\end{aligned} \tag{10.30}$$

Consider

$$\begin{aligned}
\alpha_1 + j2\pi k_z &= (l^2 - k_1^2)^{1/2} + j2\pi k_z \\
&= (l^2 - \omega^2\mu_0\varepsilon_0 + j\omega\mu_0\sigma)^{1/2} + j\frac{2\pi n_z}{Z_0} \\
&\approx (l^2 + j\omega\mu_0\sigma)^{1/2} + j\frac{2\pi n_z}{Z_0}, \quad n_z = 0, \pm 1, \pm 2, \ldots.
\end{aligned} \tag{10.31}$$

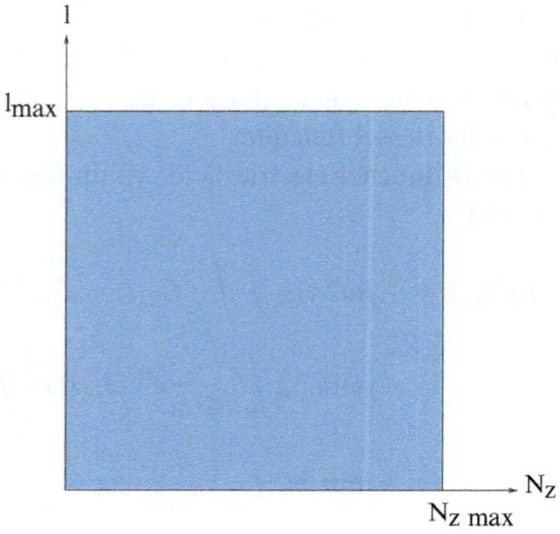

Fig. 10.21 Showing the bounding box for $\tilde{E}_2(l,k_z)$ in spatial-frequency space. $\tilde{E}_2(l,k_z) \approx 0$ for spatial frequencies greater than (l_{max}, N_{zmax})

Then

$$\tilde{E}_2(l,k_z) \approx \frac{-j\omega\mu_0 N_c}{Z_0}\frac{T}{2\alpha_0}\frac{e^{-\alpha_0 L}(1-e^{-\alpha_0 H})}{\alpha_0}\frac{\mathscr{I}(R_1 l, R_2 l)}{l^2}\frac{1-e^{-(l^2+j\omega\mu_0\sigma)^{1/2}Z_0+j2\pi n_z}}{(l^2+j\omega\mu_0\sigma)^{1/2}+j\dfrac{2\pi n_z}{Z_0}}. \tag{10.32}$$

In computing $\mathscr{I}(R_1 l, R_2 l)$ we can make use of

$$\begin{aligned}\mathscr{I}(a,b) &= \int_a^b zJ_1(z)dz \\ &= -\int_a^b z\frac{dJ_0}{dz}dz \\ &= \int_a^b J_0(z)dz - \int_a^b \frac{d}{dz}(zJ_0(z))dz \\ &= aJ_0(a) - bJ_0(b) + \int_a^b J_0(z)dz, \end{aligned} \tag{10.33}$$

where the final integral may be tabulated or easily computed.

The "bounding box" for $\tilde{E}_2(l,k_z)$ in spatial-frequency, (l,k_z), space is the smallest rectangle that "supports" $\tilde{E}_2(l,k_z)$, or, to put it another way, is the smallest rectangle beyond which $\tilde{E}_2(l,k_z) \approx 0$. Figure 10.21 shows such a bounding box with boundaries (l_{max}, N_{zmax}). We can get an estimate for N_{zmax}, by setting $l^2 = 0$ in the denominator of the final term of (10.32), and then arguing that $\dfrac{2\pi n_z}{Z_0}$ must be 50 (say) times as large as $(j\omega\mu_0\sigma)^{1/2}$ to ensure resolution in z. This yields $\dfrac{2\pi N_{zmax}}{Z_0} = 50 \times (\omega\mu_0\sigma)^{1/2}$. Using the values shown in Fig. 10.20, together with a frequency of 7 kHz, we have $(\omega f \mu_0 \sigma)^{1/2} = 1030$, and $\frac{2\pi}{Z_0} = 1319$, which yields $N_{z\max} = 39$. At this frequency, the bounding box is about 3.5 skin-depths deep, so we can argue that we need about 10 cells per skin depth just to account for the electromagnetic scene. We'll come back to this.

We can estimate the extent of the bounding box in (k_x, k_y, k_z)-space by simply postulating that $\frac{2\pi N_{x\max}}{X_0} = l_{\max}$, $\frac{2\pi N_{y\max}}{Y_0} = l_{\max}$, where, as before, X_0 and Y_0 are the lengths of the physical bounding box in the x and y directions, respectively. This gives the plausible results that $N_{x\max} = \frac{X_0 l_{\max}}{2\pi}$, $N_{y\max} = \frac{Y_0 l_{\max}}{2\pi}$, from which we also conclude that the cells should satisfy a "similarity relationship": $\frac{N_{x\max}}{X_0} = \frac{N_{y\max}}{Y_0}$.

Now we can put everything together, and say that, as far as gridding requirements for $\mathbf{J}_a(\mathbf{r})$ are concerned, this can be determined by convolving the spectrum of the conductivity scene with the spectrum of the EM scene, or $\tilde{\mathbf{J}}_a(\mathbf{k}) = \tilde{\Sigma}_a(\mathbf{k}) \otimes \tilde{E}_2(l, k_z)$. From our previous comments about the extension of the frequency response that comes from convolutions, we can estimate the composite frequency response to be $(N^c_{x\max} + N^e_{x\max}, N^c_{y\max} + N^e_{y\max}, N^c_{z\max} + N^e_{z\max})$, where the superscript "c" denotes the conductivity scene, and "e" denotes the electromagnetic scene. In other words, each scene contributes to the number of cells required for $\mathbf{J}_a(\mathbf{r})$.

The gridding requirement for an accurate impedance calculations requires an extra convolution, as the following argument shows. We have

$$\begin{aligned} \delta Z &= \int \mathbf{J}_a(\mathbf{r}) \cdot \mathbf{E}_0(\mathbf{r}) d\mathbf{r} \\ &\approx \int \sigma_a(\mathbf{r}) \mathbf{E}_0^2(\mathbf{r}) d\mathbf{r}, \end{aligned} \tag{10.34}$$

where we are using the same approximation as in (10.22). Thus, in order to get an accurate estimation of the integrand in (10.34), we should sample at the rate of $(N^c_{x\max} + 2N^e_{x\max}, N^c_{y\max} + 2N^e_{y\max}, N^c_{z\max} + 2N^e_{z\max})$, which will determine the number of cells in the grid.

10.4 Modeling Multiscale Problems

10.4.1 The Multiscale Algorithm

The use of integral equations and anomalous currents allows us to efficiently remove "background effects" in either forward or inverse modeling. Consider the anomalous region within a background host, as shown in Fig. 10.22. Let $\sigma_b(\mathbf{r})$ be the conductivity when the flaw is absent. Outside the background region, $\sigma_b(\mathbf{r})$ is equal to the host conductivity, σ_h. Inside the background regions it varies with position, $\mathbf{r}$.

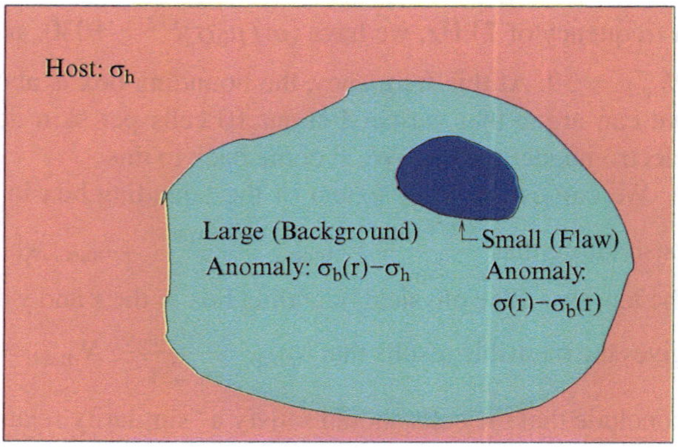

Fig. 10.22 Defining an anomalous region within a background host. $\sigma_b(\mathbf{r}) - \sigma_h$ is the anomalous conductivity associated with the "large" background and $\sigma(\mathbf{r}) - \sigma_b(\mathbf{r})$ is the anomalous conductivity associated with the "small" flaw

Let $\sigma(\mathbf{r})$ be the conductivity when the flaw is present. Outside the background region, $\sigma(\mathbf{r})$ is equal to the host conductivity. Inside the background region, but outside the flaw, $\sigma(\mathbf{r}) = \sigma_b(\mathbf{r})$. Inside the flaw, $\sigma(\mathbf{r})$ is not equal to $\sigma_b(\mathbf{r})$, but varies with position.

First, consider the unflawed background region. The anomalous current for this problem satisfies

$$\mathbf{J}_b(\mathbf{r}) = (\sigma_b(\mathbf{r}) - \sigma_h)(\mathbf{E}_{\text{in}}(\mathbf{r}) + \mathbf{E}(\mathbf{r})[\mathbf{J}_b]) \tag{10.35}$$

Next, consider the flawed background region and define anomalous currents

$$\mathbf{J}_f(\mathbf{r}) = (\sigma(\mathbf{r}) - \sigma_b(\mathbf{r}))(\mathbf{E}_{\text{in}}(\mathbf{r}) + \mathbf{E}(\mathbf{r})[\mathbf{J}_f]) \tag{10.36}$$

$$\begin{aligned}\mathbf{J}_d(\mathbf{r}) &= (\sigma_b(\mathbf{r}) - \sigma_h)(\mathbf{E}_{\text{in}}(\mathbf{r}) + \mathbf{E}(\mathbf{r})[\mathbf{J}_f] + \mathbf{E}(\mathbf{r})[\mathbf{J}_d]) \\ &\quad + (\sigma(\mathbf{r}) - \sigma_b(\mathbf{r}))\mathbf{E}(\mathbf{r})[\mathbf{J}_d].\end{aligned} \tag{10.37}$$

The anomalous current, $\mathbf{J}_a = \mathbf{J}_d + \mathbf{J}_f$, satisfies $\mathbf{J}_a = (\sigma(\mathbf{r}) - \sigma_h)(\mathbf{E}_{\text{in}} + \mathbf{E}(\mathbf{r})[\mathbf{J}_f] + \mathbf{E}(\mathbf{r})[\mathbf{J}_d])$.

The change in impedance due to the flaw is

$$\begin{aligned}\mathbf{E}_{\text{in}} \cdot \mathbf{J}_a - \mathbf{E}_{\text{in}} \cdot \mathbf{J}_b &= \mathbf{E}_{\text{in}} \cdot (\mathbf{J}_f + \mathbf{J}_d - \mathbf{J}_b) \\ &= \mathbf{E}_{\text{in}} \cdot \mathbf{J}_f + \mathbf{E}_{\text{in}} \cdot \mathbf{J}_{\text{int}},\end{aligned} \tag{10.38}$$

where we have defined $\mathbf{J}^{\text{int}} = \mathbf{J}_d - \mathbf{J}_b$.

Now, from (10.35) and (10.37), $\mathbf{J}^{\text{int}}$ satisfies the equation

$$\begin{aligned}\mathbf{J}^{\text{int}} &= (\sigma_b(\mathbf{r}) - \sigma_h)(\mathbf{E}(\mathbf{r})[\mathbf{J}_f] + \mathbf{E}(\mathbf{r})[\mathbf{J}^{\text{int}}]) + (\sigma(\mathbf{r}) - \sigma_b(\mathbf{r}))\mathbf{E}(\mathbf{r})[\mathbf{J}_d] \\ &= (\sigma_b(\mathbf{r}) - \sigma_h)(\mathbf{E}(\mathbf{r})[\mathbf{J}_f] + \mathbf{E}(\mathbf{r})[\mathbf{J}^{\text{int}}]) \\ &\quad + (\sigma(\mathbf{r}) - \sigma_b(\mathbf{r}))\mathbf{E}(\mathbf{r})[\mathbf{J}^{\text{int}}] + (\sigma(\mathbf{r}) - \sigma_b(\mathbf{r}))\mathbf{E}(\mathbf{r})[\mathbf{J}_b] \\ &= (\sigma_b(\mathbf{r}) - \sigma_h)\mathbf{E}(\mathbf{r})[\mathbf{J}_f] + (\sigma(\mathbf{r}) - \sigma_b(\mathbf{r}))\mathbf{E}(\mathbf{r})[\mathbf{J}_b] + (\sigma(\mathbf{r}) - \sigma_h)\mathbf{E}(\mathbf{r})[\mathbf{J}^{\text{int}}].\end{aligned} \tag{10.39}$$

From (10.35), (10.36), and (10.39) we obtain the uncoupled integral equations

$$\mathbf{E}_{\text{in}}(\mathbf{r}) = \frac{\mathbf{J}_b(\mathbf{r})}{\sigma_b(\mathbf{r}) - \sigma_h} - \mathbf{E}(\mathbf{r})\,[\mathbf{J}_b] \tag{10.40}$$

$$\mathbf{E}_{\text{in}}(\mathbf{r}) = \frac{\mathbf{J}_f(\mathbf{r})}{\sigma(\mathbf{r}) - \sigma_b(\mathbf{r})} - \mathbf{E}(\mathbf{r})\,[\mathbf{J}_f] \tag{10.41}$$

$$\mathbf{E}_{\text{ef}}(\mathbf{r}) = \frac{\mathbf{J}^{\text{int}}(\mathbf{r})}{\sigma(\mathbf{r}) - \sigma_h} - \mathbf{E}(\mathbf{r})\left[\mathbf{J}^{\text{int}}\right], \tag{10.42}$$

where the effective incident field, $\mathbf{E}_{\text{ef}}(\mathbf{r})$, is given by

$$\mathbf{E}_{\text{ef}}(\mathbf{r}) = \frac{\sigma_b(\mathbf{r}) - \sigma_h}{\sigma(\mathbf{r}) - \sigma_h}\mathbf{E}\left[\mathbf{J}_f\right] + \frac{\sigma(\mathbf{r}) - \sigma_b(\mathbf{r})}{\sigma(\mathbf{r}) - \sigma_h}\mathbf{E}\left[\mathbf{J}_b\right]. \tag{10.43}$$

We construct $\mathbf{J}^{\text{int}}(\mathbf{r})$ from two currents, $\mathbf{J}_f^{\text{int}}(\mathbf{r})$ and $\mathbf{J}_b^{\text{int}}(\mathbf{r})$, defined on the flaw and background grids, respectively. With $\mathbf{J}^{\text{int}}(\mathbf{r}) = \mathbf{J}_f^{\text{int}}(\mathbf{r}) + \mathbf{J}_b^{\text{int}}(\mathbf{r})$, we have

$$\mathbf{E}_{\text{ef}}(\mathbf{r}) = \frac{\mathbf{J}_f^{\text{int}}(\mathbf{r})}{\sigma(\mathbf{r}) - \sigma_h} + \frac{\mathbf{J}_b^{\text{int}}(\mathbf{r})}{\sigma(\mathbf{r}) - \sigma_h} - \mathbf{E}(\mathbf{r})\left[\mathbf{J}_f^{\text{int}}\right] - \mathbf{E}(\mathbf{r})\left[\mathbf{J}_b^{\text{int}}\right]. \tag{10.44}$$

We will expand $\mathbf{J}_b^{\text{int}}(\mathbf{r})$ and $\mathbf{J}_f^{\text{int}}(\mathbf{r})$ in terms of tent functions that are defined on the background and flaw grids, respectively.

Testing (10.44) with the expansion functions for $\mathbf{J}_b^{\text{int}}(\mathbf{r})$ gives

$$\mathbf{E}^{\text{ef},b} = [Q^b - G^b] \cdot \mathbf{J}^{\text{int}(b)} + [Q^{bf} - G^{bf}] \cdot \mathbf{J}^{\text{int}(f)}\ . \tag{10.45}$$

$\mathbf{E}^{\text{ef},b}$ is the vector of moments, $E_{klm}^{\text{ef},b}$, for the effective field, evaluated on the background grid. $\mathbf{J}^{\text{int}(b)}$ and $\mathbf{J}^{\text{int}(f)}$ are the vector expansion coefficients for $\mathbf{J}_b^{\text{int}}(\mathbf{r})$ and $\mathbf{J}_f^{\text{int}}(\mathbf{r})$, respectively. Testing (10.44) with the expansion functions for $\mathbf{J}_f^{\text{int}}(\mathbf{r})$ gives

$$\mathbf{E}^{\text{ef},f} = [Q^f - G^f] \cdot \mathbf{J}^{\text{int}(f)} + [Q^{fb} - G^{fb}] \cdot \mathbf{J}^{\text{int}(b)}, \tag{10.46}$$

where $\mathbf{E}^{\text{ef},f}$ is the vector of moments, $E_{klm}^{\text{ef},f}$, for the effective field, evaluated on the flaw grid.

The various matrices in (10.45) and (10.46) are defined here: (Q^b, G^b) and (Q^f, G^f) are the tridiagonal and Töplitz/Hankel matrices with which we are familiar, defined on the background and flaw grids, respectively. Q^{fb} is a sparse matrix involving integrals of $1/(\sigma(\mathbf{r}) - \sigma_h)$ multiplied by the product of an expansion function for $\mathbf{J}_b^{\text{int}}(\mathbf{r})$ and an expansion function for $\mathbf{J}_f^{\text{int}}(\mathbf{r})$. Q^{bf} is the transpose of Q^{fb}. G^{fb} (G^{bf}) is the transfer matrix that multiplies the expansion coefficients for $\mathbf{J}_b^{\text{int}}(\mathbf{r})$ ($\mathbf{J}_f^{\text{int}}(\mathbf{r})$) to give the moments of the fields that they produce, evaluated on the flaw (background) grid. G^{bf} is the transpose of G^{fb}.

We will use an approximation to simplify the evaluation of the moments of the effective field. When evaluating $E_{klm}^{(x)\mathrm{ef},f}$, we replace the values of the ratios, $(\sigma_b(\mathbf{r}) - \sigma_h)/(\sigma(\mathbf{r}) - \sigma_h)$ and $(\sigma(\mathbf{r}) - \sigma_b(\mathbf{r}))/(\sigma(\mathbf{r}) - \sigma_h)$, in flaw cells klm and $k+1lm$ with the averages, $V_{klm}^{x,ff}$ and $V_{klm}^{x,fb}$, over those two cells, defined by

$$V_{klm}^{x,ff} = \frac{1}{2}\left[\frac{\sigma_{b,klm} - \sigma_{h,klm}}{\sigma_{klm} - \sigma_{h,klm}} + \frac{\sigma_{b,k+1lm} - \sigma_{h,k+1lm}}{\sigma_{k+1lm} - \sigma_{h,k+1lm}}\right]$$

$$V_{klm}^{x,fb} = \frac{1}{2}\left[\frac{\sigma_{klm} - \sigma_{b,klm}}{\sigma_{klm} - \sigma_{h,klm}} + \frac{\sigma_{k+1lm} - \sigma_{b,k+1lm}}{\sigma_{k+1lm} - \sigma_{h,k+1lm}}\right]. \tag{10.47}$$

Treating the y and z directions similarly, we get

$$\mathbf{E}^{\mathrm{ef},f} = V^{fb} \cdot G^{fb} \cdot \mathbf{J}^{(b)} + V^{ff} \cdot G^{f} \cdot \mathbf{J}^{(f)}, \tag{10.48}$$

where the diagonal matrices, V^{fb} and V^{ff}, are given by

$$V^{fb} = \begin{bmatrix} V^{x,fb} & 0 & 0 \\ 0 & V^{y,fb} & 0 \\ 0 & 0 & V^{z,fb} \end{bmatrix}$$

$$V^{ff} = \begin{bmatrix} V^{x,ff} & 0 & 0 \\ 0 & V^{y,ff} & 0 \\ 0 & 0 & V^{z,ff} \end{bmatrix}. \tag{10.49}$$

The contribution of $\mathbf{J}_f(\mathbf{r})$ to $\mathbf{E}^{\mathrm{ef},b}$ is evaluated in a similar way and is given by

$$\mathbf{E}^{\mathrm{ef},b}[\mathbf{J}_f(\mathbf{r})] = V^{bf} \cdot G^{bf} \cdot \mathbf{J}^{(f)}, \tag{10.50}$$

where

$$V^{bf} = \begin{bmatrix} V^{x,bf} & 0 & 0 \\ 0 & V^{y,bf} & 0 \\ 0 & 0 & V^{z,bf} \end{bmatrix}. \tag{10.51}$$

V^{bf} is exactly analogous to V^{ff}, except that the ratios are averaged over the background cells instead of the flaw cells.

In evaluating the contribution of $\mathbf{J}_b(\mathbf{r})$ to $\mathbf{E}^{\mathrm{ef},b}$, we approximate the tent functions that are defined on the background grid in terms of the tent functions that are defined on the flaw grid. Then we have

$$\begin{aligned} \mathbf{E}^{\mathrm{ef},b}[\mathbf{J}_b(\mathbf{r})] &= A^{bf} \cdot \mathbf{E}^{\mathrm{ef},f}[\mathbf{J}_b(\mathbf{r})] \\ &= A^{bf} \cdot V^{fb} \cdot G^{fb} \cdot \mathbf{J}^{(b)}. \end{aligned} \tag{10.52}$$

The sparse matrix, A^{bf}, holds the expansion coefficients and has the structure

$$A^{bf} = \begin{bmatrix} A^{x,bf} & 0 & 0 \\ 0 & A^{y,bf} & 0 \\ 0 & 0 & A^{z,bf} \end{bmatrix}. \tag{10.53}$$

Combining the contributions of $\mathbf{J}_f(\mathbf{r})$ and $\mathbf{J}_b(\mathbf{r})$ to $\mathbf{E}^{\text{ef},b}$, we get

$$\mathbf{E}^{\text{ef},b} = A^{bf} \cdot V^{fb} \cdot G^{fb} \cdot \mathbf{J}^{(b)} + V^{bf} \cdot G^{bf} \cdot \mathbf{J}^{(f)}. \tag{10.54}$$

After computing $\mathbf{E}^{\text{ef},b}$ and $\mathbf{E}^{\text{ef},f}$, we solve the two coupled equations given in (10.45) and (10.46) using an iterative procedure. First, we rewrite these equations in the following form:

$$[Q^b - G^b] \cdot \mathbf{J}^{\text{int}(b)} = \mathbf{E}^{\text{ef},b} - [Q^{bf} - G^{bf}] \cdot \mathbf{J}^{\text{int}(f)} \tag{10.55}$$

$$[Q^f - G^f] \cdot \mathbf{J}^{\text{int}(f)} = \mathbf{E}^{\text{ef},f} - [Q^{fb} - G^{fb}] \cdot \mathbf{J}^{\text{int}(b)}. \tag{10.56}$$

We first solve (10.55), assuming that $\mathbf{J}^{\text{int}(f)} = 0$. Then we use the solution, $\mathbf{J}^{\text{int}(b)}$, to compute the right-hand side of (10.56), and solve this equation for $\mathbf{J}^{\text{int}(f)}$. Then we use this solution to update the right-hand side of (10.55) and re-solve. We continue this process until the norm of the residual vector

$$\begin{aligned} \mathbf{R} = {} & (\mathbf{E}^{\text{ef},b} - [Q^{bf} - G^{bf}] \cdot \mathbf{J}^{\text{int}(f)} - [Q^b - G^b] \cdot \mathbf{J}^{\text{int}(b)}) \\ & + (\mathbf{E}^{\text{ef},f} - [Q^{fb} - G^{fb}] \cdot \mathbf{J}^{\text{int}(b)} - [Q^f - G^f] \cdot \mathbf{J}^{\text{int}(f)}) \end{aligned} \tag{10.57}$$

is sufficiently small.

The algorithm for eliminating the background is:

1. Solve (10.40) for $\mathbf{J}_b$ using a coarse background grid, G_b
2. Solve (10.41) for $\mathbf{J}_f$ using a fine grid, G_f, covering only the flaw
3. Solve (10.42) for $\mathbf{J}^{\text{int}}$ using (10.55) and (10.56)

and then compute the change in the probe impedance, due to the flaw, as

$$\Delta Z_{\text{flaw}} = \mathbf{E}_{\text{in}} \cdot \mathbf{J}_f + \mathbf{E}_{\text{in}} \cdot \mathbf{J}^{\text{int}(b)} + \mathbf{E}_{\text{in}} \cdot \mathbf{J}^{\text{int}(f)} \tag{10.58}$$

where the dot products are the usual expressions for impedances. An attractive feature of this system is that the flaw and background may be gridded separately in a manner appropriate to their size and characteristic.

10.5 Notch at a Bolt Hole Benchmark Problems

The advantage of using dual integral equations to compute the change in impedance due to a small flaw in the presence of a larger background region is to avoid the more straightforward, but less accurate, method of simply computing Z with and without the flaw and subtracting the two nearly equal values to obtain the small difference.

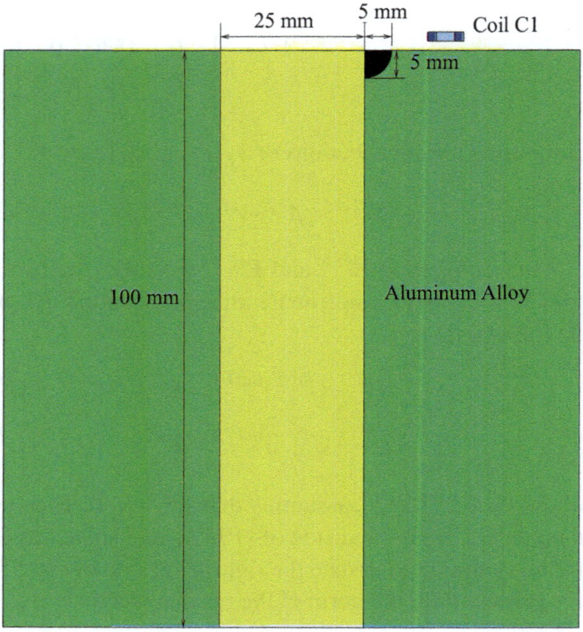

Fig. 10.23 Benchmark Problem 1: notch in a bolt hole through an aluminum block. The coil for the test, C1, is shown roughly to scale

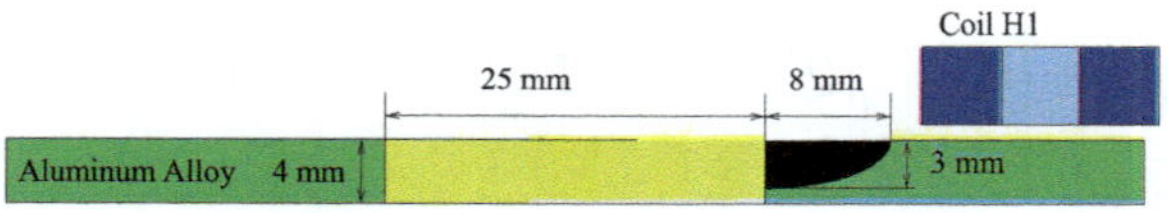

Fig. 10.24 Benchmark Problem 2: notch in a bolt hole through a 4 mm thick aluminum plate. The coil for the test, H1, is shown roughly to scale

Further, when it comes to inversion problems, we can subtract the background signal and concentrate on solving the much smaller flaw problem. In the following discussion, we apply a "graduated-grid" extension of the above algorithm to more accurately model two benchmark problems. The data for these two problems were supplied by Prof. J.R. Bowler and Yuan Ji of Iowa State University.

10.5.1 Sketch of Benchmark Problems 1 and 2

Benchmark Problem 1 is shown in Fig. 10.23, and Benchmark Problem 2 is shown in Fig. 10.24. The two coils that are used in the benchmark problems are described in Table 10.10. C1 will be used to collect data for Benchmark 1 and H1 for Benchmark 2.

Table 10.10 Data for Coils C1 and H1: dimensions in mm

Coil	Outer radius	Inner radius	Height	No. turns	C (pF)	L (H)	R (Ω)
C1	4	1.58	1.042	305	13.85	463×10^{-6}	19.08
H1	7.38	2.51	4.99	4000	72.8	0.10052	688.4

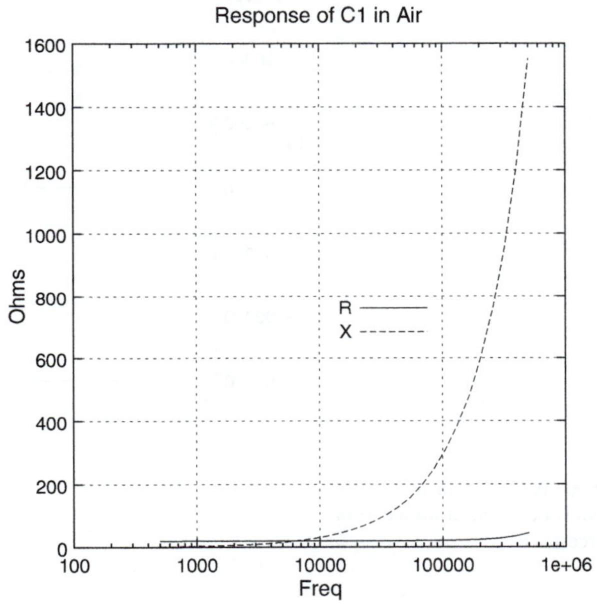

Fig. 10.25 Response of coil C1 in air

The response of C1 in air is shown in Fig. 10.25 and of H1 in air in Fig. 10.26. Though it can't be inferred from Fig. 10.25 because of the scale of the ordinate, the slope of the reactance curve at low frequencies yields an equivalent inductance of 0.463 μH for C1, and the same analysis of Fig. 10.26 yields a low-frequency inductance of 0.10052 H for H1, as shown in Table 10.10. The low-frequency resistance values for each coil are also given by the data plotted in Figs. 10.25 and 10.26 and are also shown in Table 10.10. Clearly, the small inductance of C1 has pushed its resonant frequency beyond the limits shown in Fig. 10.25, whereas the large inductance of H1 produces a relatively low-frequency resonance point.

10.5.2 One-Port Circuit Models

Consider the typical equivalent circuit for a coil in freespace shown in Fig. 10.27. Because there is only a single input port for the model, we refer to this as a "one-port circuit model." By applying the usual algorithm, we can determine the

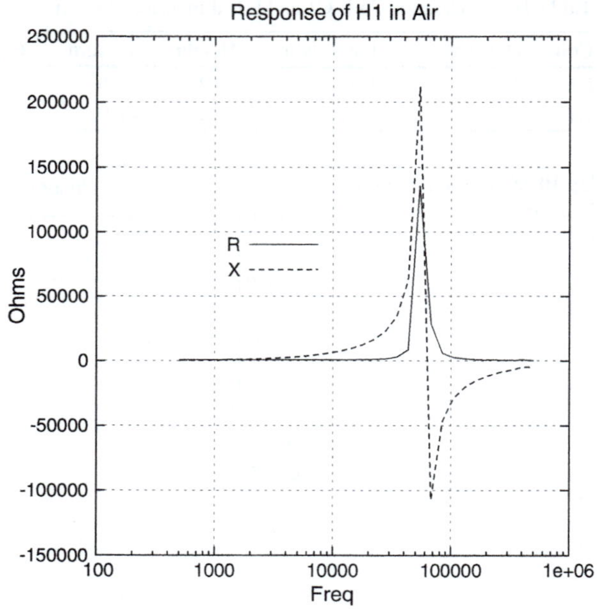

Fig. 10.26 Response of coil H1 in air

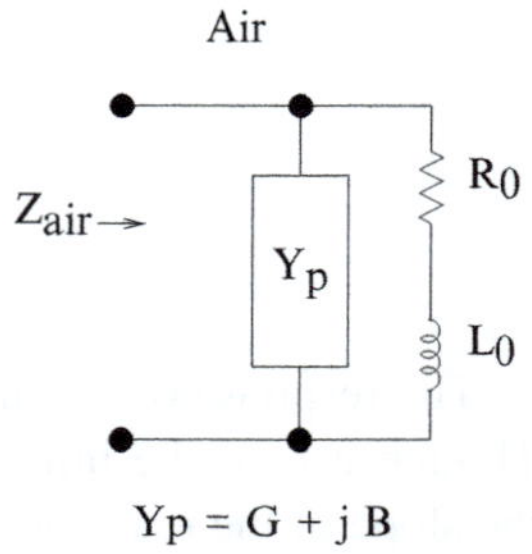

Fig. 10.27 A typical equivalent circuit for a coil in freespace

shunt admittance, $Y_p = G_p + jB_p$, and with this we can determine the equivalent series impedance of the series-parallel circuit:

$$
\begin{aligned}
Z_{\text{air}} &= \frac{1}{Y_p + \dfrac{1}{R_0 + j\omega L_0}} \\
&= \frac{1}{G_p + jB_p + \dfrac{1}{R_0 + j\omega L_0}} \\
&= \frac{R_0 + G_p(R_0^2 + \omega^2 L_0^2)}{(1 + R_0 G_p - \omega L_0 B_p)^2 + (\omega L_0 G_p + R_0 B_p)^2}
\end{aligned}
$$

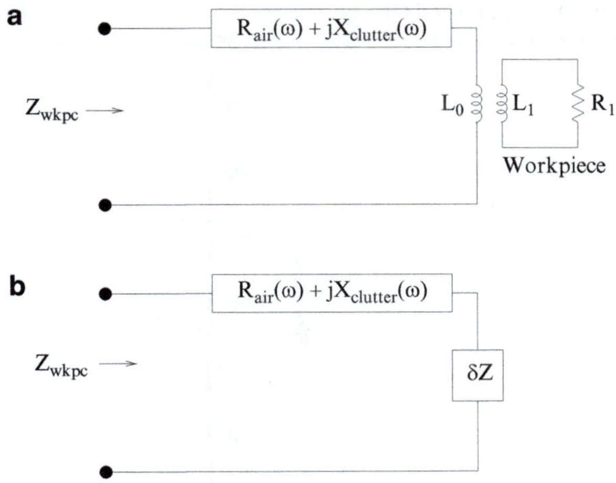

Fig. 10.28 Depicting the coil coupled to the workpiece. (**a**) Standard representation. (**b**) **VIC-3D**® representation in which δZ is "air balanced"

$$
\begin{aligned}
&+j\frac{\omega L_0(1-\omega L_0 B_p) - R_0^2 B_p}{(1+R_0G_p-\omega L_0 B_p)^2+(\omega L_0 G_p + R_0 B_p)^2} \\
&= R_{\text{air}}(\omega) + jX_{\text{air}}(\omega) \\
&= R_{\text{air}}(\omega) + j[X_{\text{air}}(\omega) - \omega L_0] + j\omega L_0 \\
&= R_{\text{air}}(\omega) + jX_{\text{clutter}}(\omega) + j\omega L_0,
\end{aligned} \tag{10.59}
$$

where we have singled out $j\omega L_0$ for purposes of introducing the coupled circuit representation in the next step.

When the coil is coupled to the workpiece, we have the situation depicted in Fig. 10.28. The top part of the figure is the standard coupled-circuit representation, whereas the bottom part is the **VIC-3D**® representation in which δZ is "air balanced." Thus, in order to get the circuit representation to agree with **VIC-3D**® we must subtract $R_{\text{air}}(\omega) + jX_{\text{clutter}}(\omega)$ from the measured impedance, Z_{wkpc}. We could attach another shunt element, Y_q, to the driving-point terminals of Fig. 10.28 to account for loading effects on the probe that are not accounted for by inductive coupling. For example, it is possible that the presence of the metallic host may cause a capacitive coupling that is not accounted for by Y_p in the freespace model of the probe. If that is the case, we can determine Y_q by repeating the same process that was used to determine Y_p, but this time with a known δZ in Fig. 10.28.

Figure 10.29 shows Y_p for coils C1 and H1. The slope of a line joining the origin to the extreme point for $B(f)$ in the top of Fig. 10.29 yields an equivalent capacitance of 13.85 pF for C1, which is a reasonable approximation for frequencies greater than 150 kHz, or so. The slope of $B(f)$ in the bottom of Fig. 10.29 for $0 \leq f \leq 320$ kHz yields an equivalent capacitance of 72.8 pF. These are the values listed in Table 10.10.

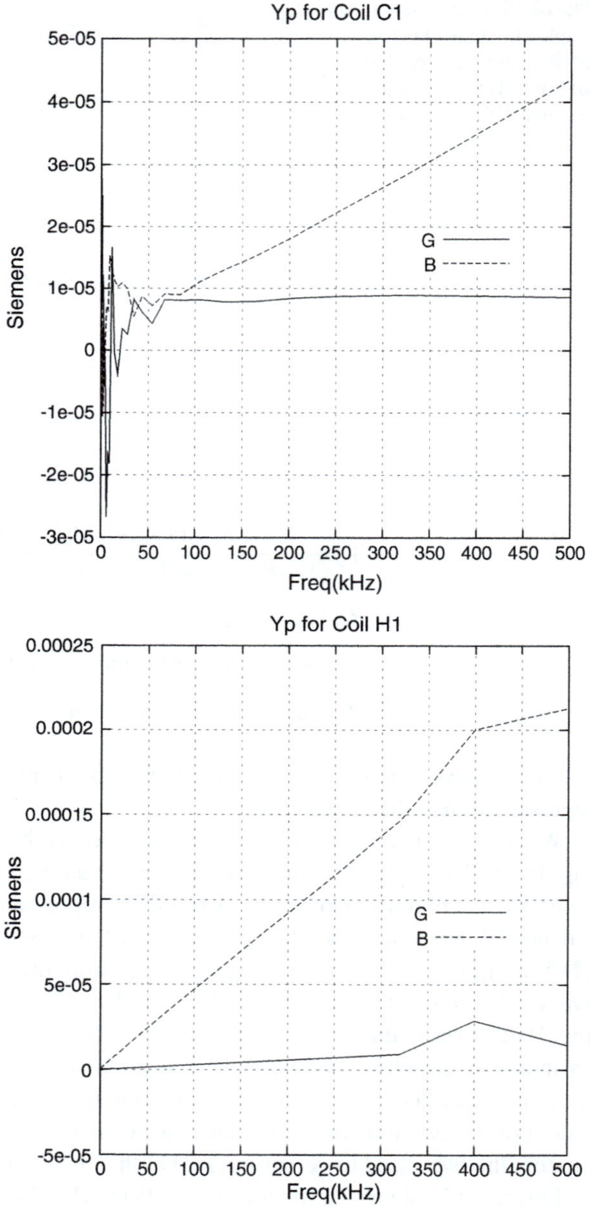

Fig. 10.29 Y_p for coil C1 (*top*) and H1 (*bottom*)

Table 10.11 NLSE inversion results for Benchmark 1 data sets

Data set	Φ	σ (S/m)/sensitivity	LO(mm)/sensitivity
M1	4.9166	1.8156(7)/2.73(−2)	0.1336/1.262(−2)
M2	4.8827	1.8151(7)/2.74(−2)	0.1364/1.255(−2)
M3	4.6704	1.8178(7)/2.63(−2)	0.1420/1.205(−2)
M4	4.8004	1.8175(7)/2.69(−2)	0.1389/1.236(−2)

Table 10.12 NLSE inversion results for Benchmark 2 data sets

Data set	Φ	σ(S/m)/sensitivity	LO(mm)/sensitivity
M1	0.6385	1.8340(7)/4.03(−3)	0.3480/1.542(−2)
M2	0.8585	1.8420(7)/5.46(−3)	0.3560/2.075(−2)
M3	0.7850	1.8390(7)/4.97(−3)	0.3550/1.899(−2)
M4	0.3887	1.8227(7)/2.41(−3)	0.3254/0.933(−2)

10.5.3 Model-Based Inversion with NLSE

After removing the shunt parasitic admittance element, Y_p (which includes the capacitance, C), of each coil in the usual manner, and then removing the freespace impedance, $R + j\omega L$, we are left with the "air-balanced" change in impedance, δZ, when the coil is placed over the workpiece. This is the input to NLSE[2] for inverting each of the data sets known as M1, M2, M3, and M4 for each of the two benchmark tests. (The data sets correspond to four different locations on the unflawed workpiece.) Even though each data set contained values of δZ at 32 frequencies, logarithmically spaced between 500 Hz and 500 kHz, we used only five frequencies for the inversions. For Benchmark 1, we used the five highest frequencies, 205.06, 256.24, 320.2, 400.13, and 500.0 kHz, and for Benchmark 2 the five lowest frequencies, 500, 624.8, 780.76, 975.65, and 1219.2 Hz. The reason for this separation is due to the fact that the coil for Benchmark 1, C1, has a very small inductance, as is seen from Table 10.10, whereas that for H1 is quite large. Therefore, we get more reliable data at high frequencies for C1 and at low frequencies for H1.

Following the customary procedure for model-based inversions with NLSE, we establish a 4×4 interpolation grid for host conductivity and coil lift-off for each of the benchmark tests. For Benchmark 1, the values of the parameters at the nodes of the grid are: $\sigma = 1 \times 10^7,\ 2 \times 10^7,\ 3 \times 10^7,\ 4 \times 10^7\,\mathrm{S/m}$; LO $= 0.1$, 0.2, 0.3, 0.4 mm. The conductivity nodal values are the same for Benchmark 2, whereas the lift-off values are 0.2, 0.4, 0.6, 0.8 mm.

The inversion results are shown in Table 10.11 for Benchmark 1, and in Table 10.12 for Benchmark 2.

[2]NLSE, a nonlinear least-squares estimator, will be defined and described in Chap. 12. The material in this section can be deferred until after reading that chapter, or can simply be accepted on faith, "in anticipation."

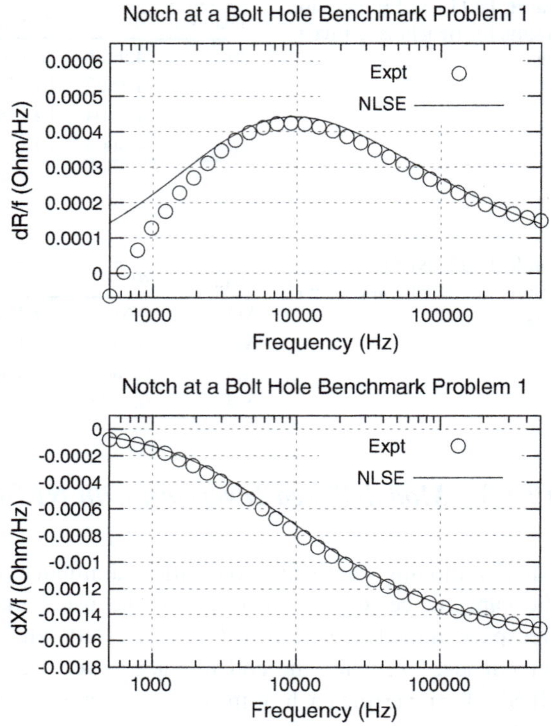

Fig. 10.30 Comparing the normalized model results and data sets for M1 of benchmark 1 in Table 10.11 over the entire frequency range

When we use the coil data of Table 10.10, and the results of data set M1 of Benchmark 1 in Table 10.11 to compute the forward model, and then plot the results over the entire frequency range of 500 Hz to 500 kHz, we obtain Fig. 10.30. (The data are normalized with respect to frequency.) That there is a breakdown in the measured resistance data at low frequencies (dR goes negative!) was anticipated because of the considerable uncertainty in Y_p for coil C1 at low frequencies in Fig. 10.29. The corresponding results for M1 of Benchmark 2 in Table 10.12 are shown in Fig. 10.31. The results for the other data sets in Tables 10.11 and 10.12 lie on top of the results shown in Figs. 10.30 and 10.31. The high-frequency breakdown in the resistance and reactance in Fig. 10.31 was anticipated because of the uncertainty in Y_p for H1 at high frequencies in Fig. 10.29. Note, however, that the normalized model reactance, dX/f, due to NLSE in Fig. 10.31 "saturates" at high frequencies, as expected by classical coupled-circuit theory and Förster plots. The asymptote is a measure of the coupling of the coil to the host. The asymptotic value of $dR/f = 0$ is reached at a higher frequency in Fig. 10.31, as is also predicted by coupled-circuit theory.

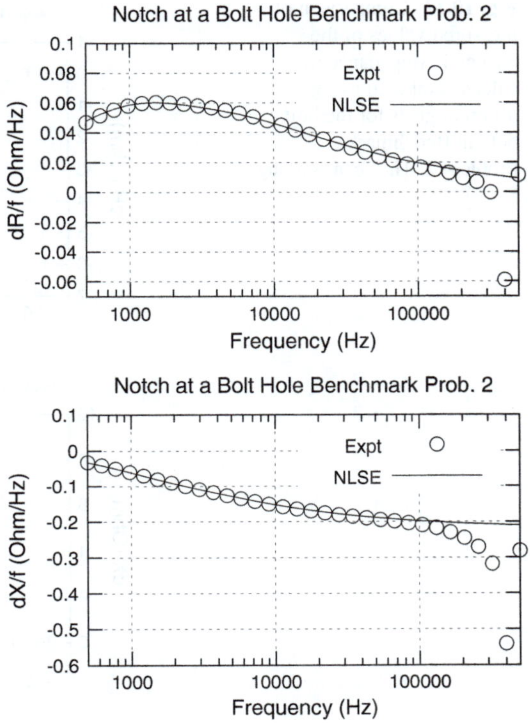

Fig. 10.31 Comparing the normalized model results and data sets for M1 of benchmark 2 in Table 10.12 over the entire frequency range

The final results of our validation studies for the multiscale algorithm are plotted in Fig. 10.32 for Benchmark Problem 2 and Fig. 10.33 for Benchmark Problem 1. Numerical experiments show that accurate results for Benchmark Problem 2 require a background grid that is much finer near the notch. We can obtain results that effectively use a graduated grid which becomes progressively finer in regions closer to the notch. It is clear, however, from varying the number of notch cells that the impedances we are trying to compute have not converged with respect to number of cells along the x, y, or z directions for a grid of $32 \times 2 \times 16$ cells.

It is also clear from varying the number of background (bolt-hole) cells that the impedances have not converged with respect to number of cells along the x, y, or z directions for a grid of $16 \times 16 \times 4$ cells. We can, however, improve these impedance values by combining them with values obtained from an $8 \times 8 \times 2$ cell graduated grid for the bolt hole and a $16 \times 2 \times 8$ cell notch grid. This allows us to perform a linear extrapolation to zero cell dimensions. That is the result labeled "Extrapol'd" in Fig. 10.32.

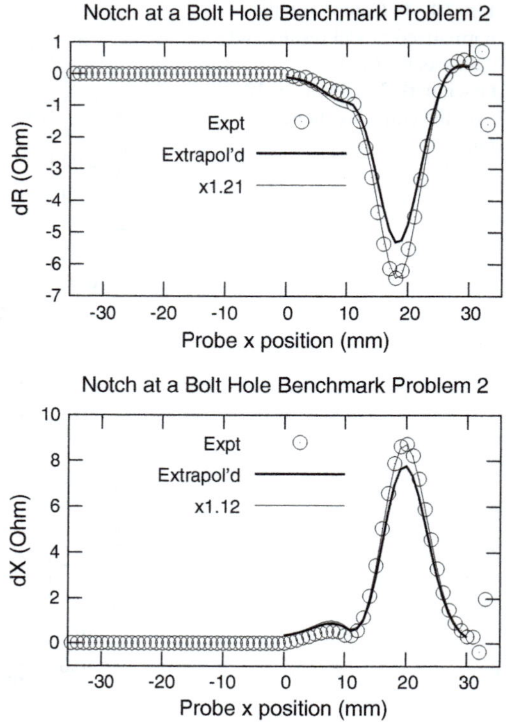

Fig. 10.32 Comparison of measured values of the change in impedance to values computed using different grids for the bolt hole in Benchmark Problem 2. The scan is along $Y = 0$

As with Benchmark Problem 2, accurate results for Benchmark Problem 1 require a background grid that is much finer near the notch, and so we again apply the graduated-grid scheme that was used for that problem. We also use the extrapolation procedure that was used in Problem 2, with the result labeled "Extrapol'd" in Fig. 10.33.

The response of the small notch is clearly evident in both benchmark problems, and its shape agrees with the measured data, which were the goals of the algorithm, but the absolute values differ by 12–39 %, with the smaller errors occurring in Benchmark Problem 2. The reason for the discrepancies in scale is due to the size of the problems. Benchmark Problem 1 contains a hole that is abnormally large compared to the notch, and at the excitation frequency of 5 kHz, this hole requires a huge number of cells. Benchmark Problem 2 is somewhat more realistic in its size, and the extrapolated results are closer in scale to the measured data. In any case, we should note that the raw data shown in Fig. 10.34 has a peak value of resistance $\sim$950 Ω and reactance $\sim$3,000 Ω, which means that the values shown in Figs. 10.32 and 10.33 are stable to within a very small fraction of the peak values. This attests to the accuracy of the algorithm and to the quality of the benchmark data. Had we attempted to compute the signal due to the small notch by subtracting two much larger signals, we would expect to get very noisy results.

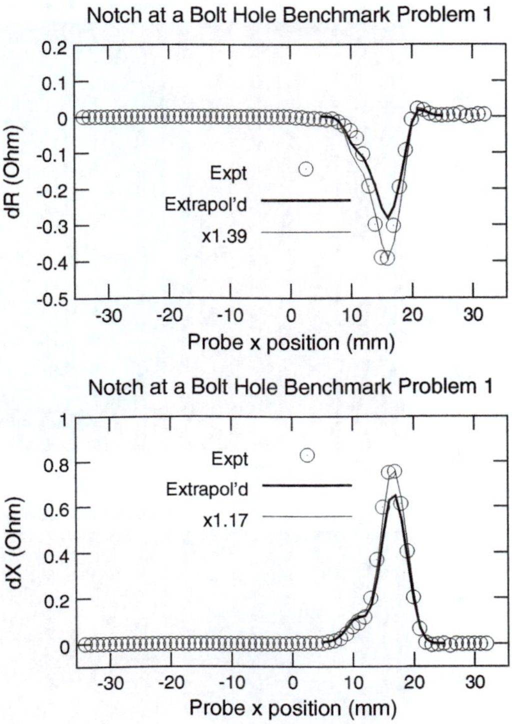

Fig. 10.33 Comparison of measured values of the change in impedance to values computed using different grids for the bolt hole in Benchmark Problem 1. The scan is along $Y = 0$

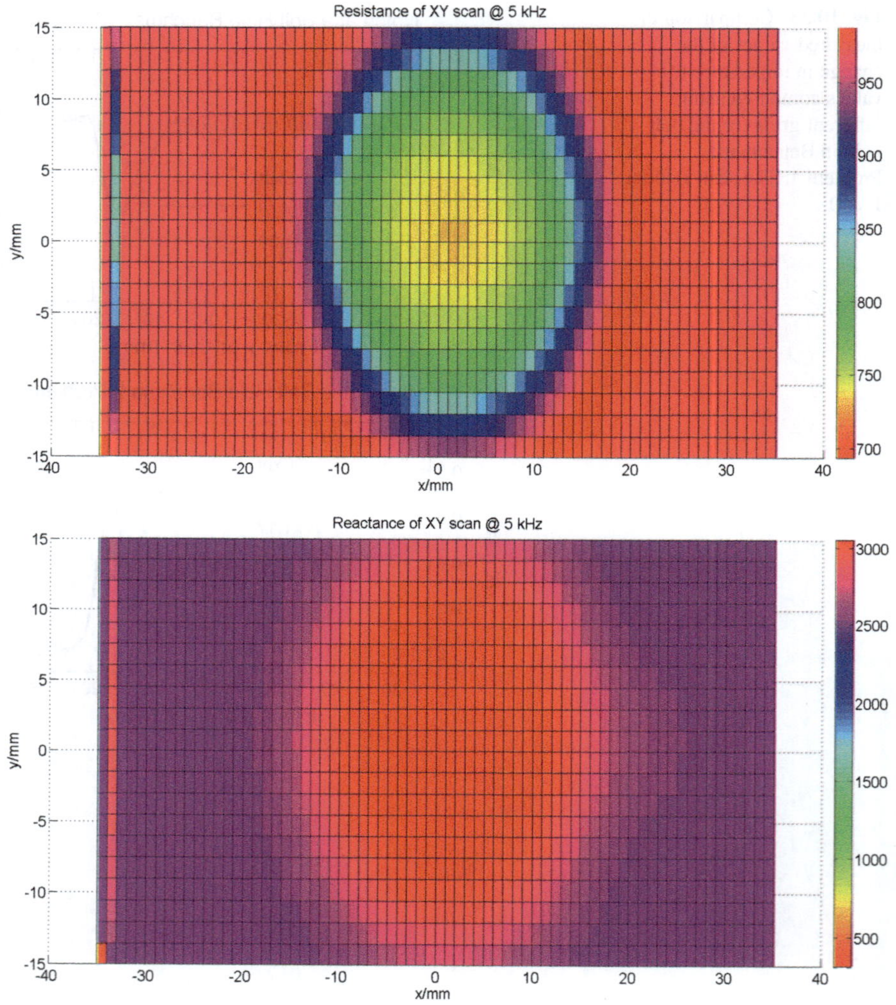

Fig. 10.34 Resistance (*top*) and reactance (*bottom*) of *XY* scan at 5 kHZ

Part II
Inversion Algorithms and Signal-Processing

Chapter 11
Examples of Basic Inverse Problems

11.1 Overview of the General Inversion Process for Parameter Estimation

Now that we have the electromagnetics down pat, let's look ahead to see where that leads us. In this part of the book, a general inverse method process is introduced that provides a framework for incorporating the numerical forward models discussed in Part I in an algorithm that estimates an unknown set of values of the test. A diagram of a general inverse method process is presented in Fig. 11.1. The heart of the inversion process is the inverse method step. Typical parameters to be evaluated include crack dimensions (e.g., surface length or depth), material properties (e.g., conductivity) or state(s) of the nondestructive evaluation system (e.g., probe lift-off or tilt.) Here, parameter values are estimated using an iterative scheme such as nonlinear least-squares estimation (NLSE) which equates the model results with experimental data through adjusting the model parameter values. Such general inverse methods will be presented in detail in Chap. 12. To ensure accurate estimation results, statistical estimators can be used to evaluate convergence to the solution and quantify the error bounds on the results. Noise and uncertainty analysis is first introduced in Sect. 11.2.2, with a more complete study developed in Sects. 11.6 and 11.7. The Cramer–Rao Lower Bound, another stochastic metric, is introduced in Sect. A.1 of Chap. 12. The theory and use of robust statistical estimation metrics will be presented in Chap. 13.

Although inverse methods have been studied for many years, to some degree there has been limited progress to transition them to practical applications. There are three primary reasons for the difficulty in transitioning inverse methods to NDE applications in particular: long solution times for accurate NDE measurement models, the inherent ill-posedness of certain inverse problems in NDE, and the lack of robustness of the inverse method schemes to the presence of noise and uncontrolled variation with in-field NDE measurements. To address these challenges, a comprehensive approach is proposed in Fig. 11.1 for model-based

H.A. Sabbagh et al., *Computational Electromagnetics and Model-Based Inversion*, Scientific Computation, DOI 10.1007/978-1-4419-8429-6_11,

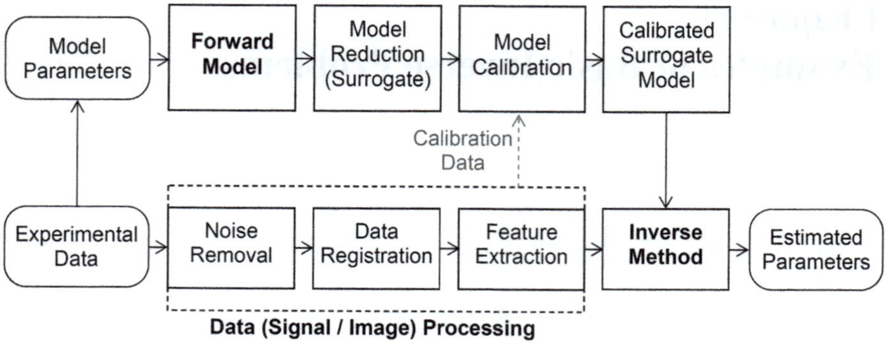

Fig. 11.1 General inversion process for parameter estimation

inversion design to ensure the reliability of the inverse method step through (a) key parameter assessment and model parameterization, (b) model reduction through surrogate models, (c) model calibration, (d) noise removal, (e) data registration, and (f) feature extraction techniques.

First, the inverse method development process must always begin with a full assessment of the key factors controlling the experimental data and driving inversion performance. All critical factors for the NDT technique, part material, part geometry, and discontinuity characteristics that control signal and noise must be identified and well understood. Clearly, the flaw characteristics such as the crack dimensions are the key factors of interest in the forward and inverse models. This step provides a critical opportunity to reduce the complexity of the inverse problem by parameterizing the flaw morphology, potentially addressing some levels of ill-posedness by simplifying the inverse problems. It is important to also consider all varying conditions in the NDE measurement that affect the measurement response, for example lift-off or scan resolution. If such key factors cannot be controlled during measurements, these parameters should also be incorporated in the inverse problem with the characteristics of the flaw. An excellent case study supporting this point is the inversion of inner diameter (ID) pits presented in Sect. 18.3.

Although the volume integral method introduced in Part I is quite efficient numerically for the simulation of eddy current NDE measurements, for inverse problem applications it can still be time consuming to perform such repeated numerical calculations within an inversion scheme. Surrogate models or structures can be created from the results of the numerical model to greatly improve the solution time and performance of inverse methods. Frequently, data or look-up tables are constructed and interpolation schemes are applied in order to provide a fast means of sampling the model over a wide range of conditions. The use of surrogate structures, first mentioned in Sects. 1.4 and 6.6, is formally introduced in Chap. 12. Emerging stochastic simulation methods, such as the probabilistic collocation method (PCM), are briefly introduced in Chap. 21 on model-assisted probably of detection (POD) analysis. These methods will be further developed and applied extensively in the second volume of this series.

Eddy-current simulations are designed to represent the change in impedance of an eddy-current probe interacting with a test specimen. However, conventional eddy-current measurement systems typically operate with the probe balanced over the test specimen and output voltage data. Thus, a model calibration step is a necessary step in practice to equate the measurement data with simulated results in a way that mimics the NDE technique procedure. In many cases, gains and thresholds are set based on the desired response to a known calibration standard. Parameters of the model can essentially be fit to obtain the best match with a limited set of empirical data acquired according to a calibration procedure. Chapter 15 introduces a method for transforming impedance calculations into voltage measurements using calibration data with known samples. At this stage in the general inversion process, a fast surrogate model calibrated according to the inspection procedure is available for inverse method applications.

There are several key steps for data pre-processing that can greatly improve the reliability of inverse methods with respect to all possible varying test conditions. Noise and clutter removal is an important first step in preparing data for inversion. Typical sources for clutter and noise found in eddy-current scans include probe lift-off variation, local changes in material properties, sensitivity to sub-surface structures, thermal variation of measurement components, and measurement (electrical) noise. For noise removal, a variety of conventional signal and image processing algorithms can be used to clean up the signals, for example: averaging, high-pass and low-pass filters, and median filters. However, clutter is defined here as coherent noise associated with departures in the impedance of the workpiece due to real non-flaw conditions such as surface conditions (probe lift-off), part (sub-surface) geometry, and material properties. Such variation cannot be removed through basic averaging or filtering techniques but typically requires a more sophisticated removal process. Several special clutter removal strategies are proposed in this part of the book. An iterative algorithm, reminiscent of the projection onto convex sets (POCS), is described in Chap. 14 to address coherent noise isolated to select data channels. In Chap. 15, several model-based clutter fitting and removal approaches are introduced to address more challenging background feature removal problems.

In addition to noise removal, data registration and feature extraction are two important steps to ensure the measurement data is of high quality necessary for the inversion process. Data registration is the process of aligning the raw data signals in time or images in space to facilitate proper data comparison and inversion with simulated data. Typically, cross-correlation routines are used to align signals to the necessary high degree for the inverse method to be successful. Lastly, feature extraction is a useful step in order to minimize the dimension and size of data being inverted. The effective use of feature extraction can reduce the likelihood an

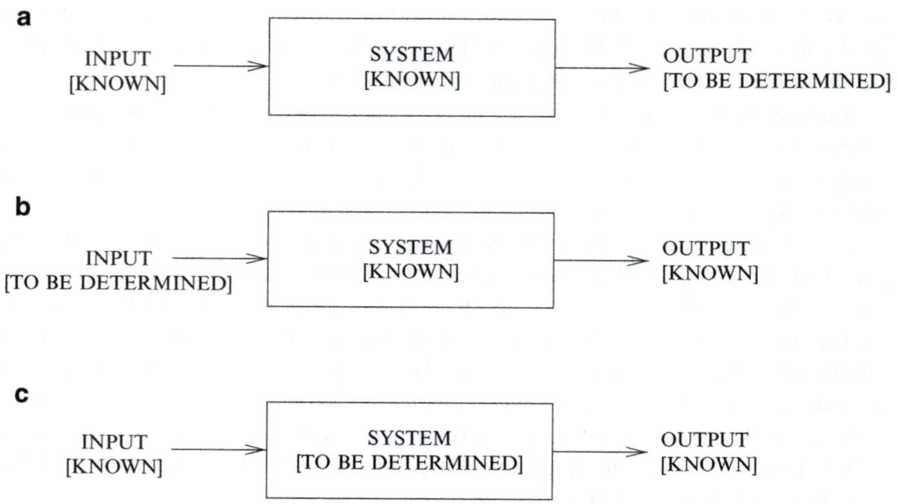

Fig. 11.2 System representation of direct and inverse problems: (**a**) the direct problem; (**b**) the signal-detection (communication) problem; (**c**) the inverse problem

inverse problem will be ill-posed, while reducing calculation time and improving the inverse solution performance. Several examples demonstrating the important of data registration and feature extraction steps are included in the remainder of the book.

Nondestructive evaluation (NDE) is to materials and structures what computer-aided tomography (CAT) scanning is to the human body—an attempt to look inside without opening the body. It is in nature an inverse problem. Figure 11.2 illustrates a system representation for three important problems: (a) a direct problem, in which the input and system are known and the output is to be determined; (b) a signal-detection (communication) problem, in which the system (a communication channel) and output are known, and the problem is to determine the input signal; and (c) the inverse problem, in which the input and outputs are known, and we must determine the system.

The results that we showed in Part I of this text were direct problems; we assumed knowledge of the probe and flaw and determined the response of the probe, namely the driving-point or transfer impedances. The second problem is dealt with in communication and information theory texts, such as [50], and has a close relation to inverse problems. In the typical inverse problem of eddy-current NDE, we assume that impedances (or instrument voltages) are measured, given the probe and measurement conditions, and the intent is to determine the anomaly that produced the output. By this we mean to quantify the size, shape, and location of a flaw, or other departure from an assumed-known condition.

The results of Part I have given us an engine, **VIC-3D®**, that is designed to solve the forward problem, and, by doing so, will provide the engine for solving the inverse problem. In this chapter, we will demonstrate the notion of eddy-current inverse problems by solving three basic problems: (a) a thickness measurement, (b) determining a conductivity profile, and (c) a combination of the two in the semiconductor chip industry. These problems are simple examples of the application of eddy-current NDE to process control and in-service inspection, and they also give us our first examples of model-based inversion, in which the physical problem can be modeled with just a few unknown parameters. We will defer to the next chapter a detailed discussion of the mathematical algorithms that will be used in this chapter and throughout the remainder of the book.

11.2 Thickness Measurements with Eddy-Current Probes

The problem is illustrated in Fig. 11.3 [52]. The thickness, t, of the plate is to be determined from measurements with an air-core probe. The probe lift-off, l, is often poorly known. This can be due to a nonconducting layer of unknown thickness on the surface, as illustrated in Fig. 11.3b, or simply because the probe geometry is not well known. Since the probe response can be very sensitive to this parameter, it is desirable to have an inversion scheme that does not require the lift-off to be known. We use a scheme that determines this parameter as well as the plate thickness.

The impedance of the probe is a complicated nonlinear function of several variables:

$$Z = g(\text{coil, workpiece, } t, l, f) \ , \tag{11.1}$$

where "coil" denotes coil parameters and "workpiece" denotes workpiece parameters. The plate thickness, probe lift-off and frequency are denoted by t, l, and f, respectively. The problem is to invert (11.1), obtaining t and l when Z, f, and the coil and plate parameters are known.

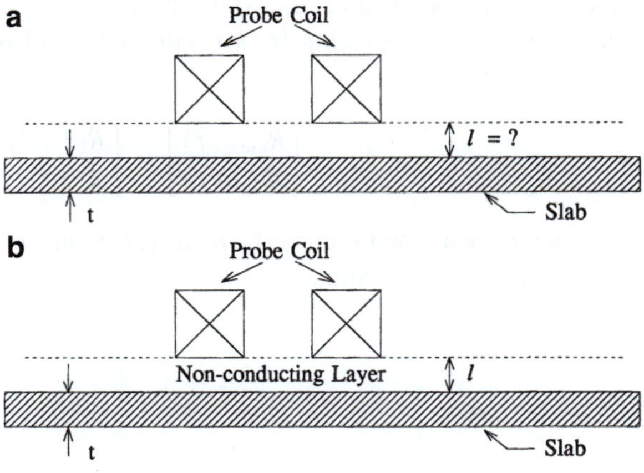

Fig. 11.3 Thickness measurement and lift-off problem. (**a**) Unknown lift-off. (**b**) Unknown nonconducting layer thickness

In principle t and l can be determined from a single measurement of impedance, since the resistance, R, and reactance, X, provide two independent equations. A more accurate determination can be made, however, when more than one measurement of Z is performed, thereby providing an overdetermined system of equations for t and l. Z is most often measured as a function of frequency, since this is usually the most easily varied parameter. In our analysis, we suppress the variation of Z which is only due to frequency by dividing all impedance values by the isolated probe reactance. We denote these normalized values simply by Z, R, and X. The statement of the problem then becomes: Given Z at several frequencies, and the coil and plate parameters, determine t and l.

Because the problem is nonlinear, we use a Gauss–Newton iteration scheme to perform the inversion. At a single frequency, f, the linear approximation to R and X can be written:

$$\begin{bmatrix} R(f,t,l) \\ X(f,t,l) \end{bmatrix} \approx \begin{bmatrix} \tilde{R}(f,t,l) \\ \tilde{X}(f,t,l) \end{bmatrix} = \begin{bmatrix} R(f,t_0,l_0) \\ X(f,t_0,l_0) \end{bmatrix} + \begin{bmatrix} \dfrac{\partial R}{\partial t} & \dfrac{\partial R}{\partial l} \\ \dfrac{\partial X}{\partial t} & \dfrac{\partial X}{\partial l} \end{bmatrix}_{(t_0,l_0)} \begin{bmatrix} t-t_0 \\ l-l_0 \end{bmatrix}. \tag{11.2}$$

The partial derivatives are computed numerically by

$$\begin{aligned} \frac{\partial Z}{\partial t} &\approx \frac{Z(t+\Delta t)-Z(t)}{\Delta t} \\ \frac{\partial Z}{\partial l} &\approx \frac{Z(l+\Delta l)-Z(l)}{\Delta l}. \end{aligned} \tag{11.3}$$

The left-hand side of (11.2) is taken to be the measured values of resistance and reactance. An initial value, (t_0,l_0), is chosen, and (11.2) is solved for (t,l). Iteration then proceeds by replacing (t_0,l_0) with (t,l), and re-solving (11.2) until t and l converge, and

$$\begin{bmatrix} R_{\text{meas}}(f) \\ X_{\text{meas}}(f) \end{bmatrix} = \begin{bmatrix} \tilde{R}(f,t,l) \\ \tilde{X}(f,t,l) \end{bmatrix}. \tag{11.4}$$

When measurements are performed at N frequencies, the overdetermined system of equations can be written:

$$\begin{bmatrix} \Delta R(f_1) \\ \Delta X(f_1) \\ \vdots \\ \Delta R(f_N) \\ \Delta X(f_N) \end{bmatrix} = \begin{bmatrix} \frac{\partial R}{\partial t}(f_1) & \frac{\partial R}{\partial l}(f_1) \\ \frac{\partial X}{\partial t}(f_1) & \frac{\partial X}{\partial l}(f_1) \\ \vdots & \vdots \\ \frac{\partial R}{\partial t}(f_N) & \frac{\partial R}{\partial l}(f_N) \\ \frac{\partial X}{\partial t}(f_N) & \frac{\partial X}{\partial l}(f_N) \end{bmatrix}_{(t_0,l_0)} \begin{bmatrix} t-t_0 \\ l-l_0 \end{bmatrix} \equiv A \begin{bmatrix} \Delta t \\ \Delta l \end{bmatrix}$$

$$\begin{aligned} \Delta R(f) &= R_{\text{meas}}(f) - R(f,t_0,l_0) \\ \Delta X(f) &= X_{\text{meas}}(f) - X(f,t_0,l_0). \end{aligned} \tag{11.5}$$

Iteration proceeds as before, except that now we solve the overdetermined system by finding the least-squares solution that minimizes the norm of the residuals (Gauss–Newton):

$$\|\text{Residual}\|^2 \equiv \sum_i \Delta R(f_i)^2 + \Delta X(f_i)^2. \tag{11.6}$$

The least-squares solution can be obtained by solving the normal equations:

$$A^T A \begin{bmatrix} \Delta t \\ \Delta l \end{bmatrix} = A^T \begin{bmatrix} \Delta R(f_1) \\ \Delta X(f_1) \\ \vdots \\ \Delta R(f_N) \\ \Delta X(f_N) \end{bmatrix}, \tag{11.7}$$

but improved numerical stability can be achieved by using the QR-decomposition [53], in which an orthogonal matrix, Q, is found that satisfies

$$QA = \begin{bmatrix} R \\ 0 \end{bmatrix}, \tag{11.8}$$

and R is nonsingular and upper-triangular. Then the least-squares solution is

$$\begin{bmatrix} \Delta t \\ \Delta l \end{bmatrix} = R^{-1} Q_1 \begin{bmatrix} \Delta R(f_1) \\ \Delta X(f_1) \\ \vdots \\ \Delta R(f_N) \\ \Delta X(f_N) \end{bmatrix}, \tag{11.9}$$

where Q_1 consists of the first two rows of Q.

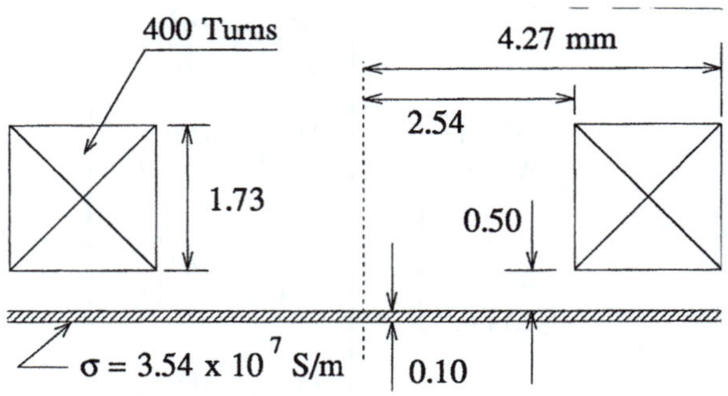

Fig. 11.4 Test Case 1: a model problem with an aluminum plate. The probe is cylindrical

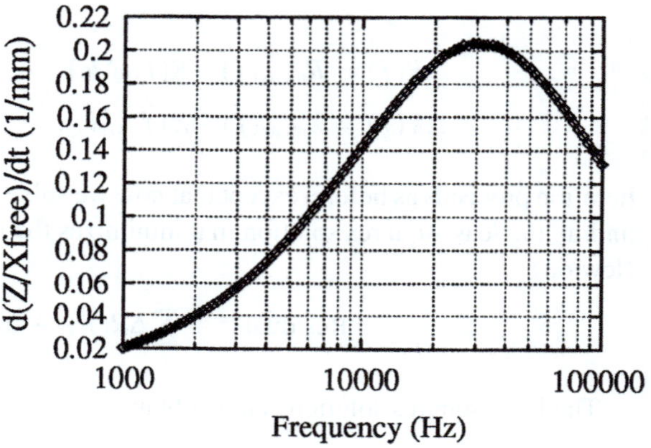

Fig. 11.5 Test Case 1: derivative of impedance with respect to thickness.

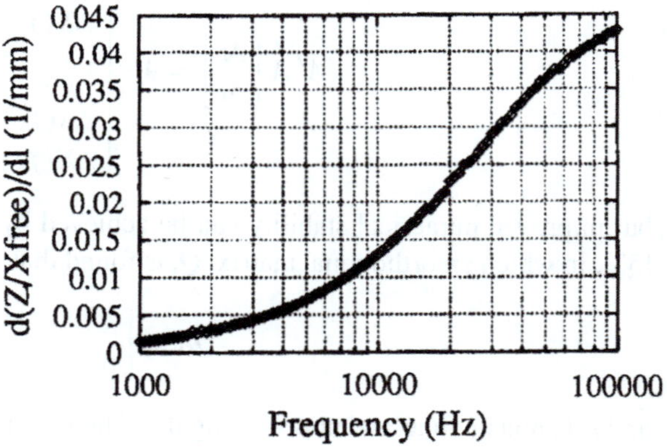

Fig. 11.6 Test Case 1: derivative of impedance with respect to lift-off

11.2.1 Test Case 1

Frequencies are selected to maximize the sensitivity to the plate thickness. We use**VIC-3D®** to obtain numerical derivatives of the impedance. For the test case of Fig. 11.4, these are shown in Figs. 11.5 and 11.6. From these derivative values,

Table 11.1 Deduced thickness and lift-off for Test Case 1. The plate is 0.1 mm thick, and the lift-off is 0.5 mm. Starting values: 0.2 mm thickness, 0.2 mm lift-off

Frequencies	Thickness	Lift-off	No. of iterations
20000.0	0.100003	0.499985	3
22963.1			
26365.1			
30271.2			
34756.0			
39905.2			

we choose frequencies in the range 20–40 kHz, where the variation with thickness is largest. We note that if the primary interest had been in determining the lift-off, then higher frequencies would be preferable. O. Baltzersen [55] gives an empirical formula, attributed to Förster and Libby [56], for the optimum frequency:

$$f_{\mathrm{opt}} = \frac{700}{\sigma t A_{\mathrm{eff}}}, \tag{11.10}$$

where f is in kHz, σ is the conductivity of the plate in mega/ohm-m, t is the plate thickness in mm, and A_{eff} is essentially the coil diameter in mm. In this case the formula gives 29 kHz, in good agreement with our choice of frequencies.

Six frequencies are chosen to cover the range of 20–40 kHz. Table 11.1 shows the thickness and lift-off that are obtained from our inversion procedure. The initial guess for both the thickness and lift-off is 0.2 mm. Convergence is achieved in three iterations.

11.2.2 Noise and Uncertainties in Δt

The uncertainties in the thickness that is obtained from the inversion procedure that we have described can be estimated from the uncertainties in the measured probe impedances. For small changes in resistance and reactance, the change in the deduced thickness is given by

$$\Delta t = \sum_{i=1}^{N} \frac{\partial t}{\partial R_i} \Delta R_i + \frac{\partial t}{\partial X_i} \Delta X_i. \tag{11.11}$$

If ΔR_i and ΔX_i have Gaussian distributions with half-width σ_{R_i} and σ_{X_i}, then Δt is Gaussian with width

$$\sigma_t = \left[\sum_i \left(\frac{\partial t}{\partial R_i} \right)^2 \sigma_{R_i}^2 + \left(\frac{\partial t}{\partial X_i} \right)^2 \sigma_{X_i}^2 \right]^{1/2}. \tag{11.12}$$

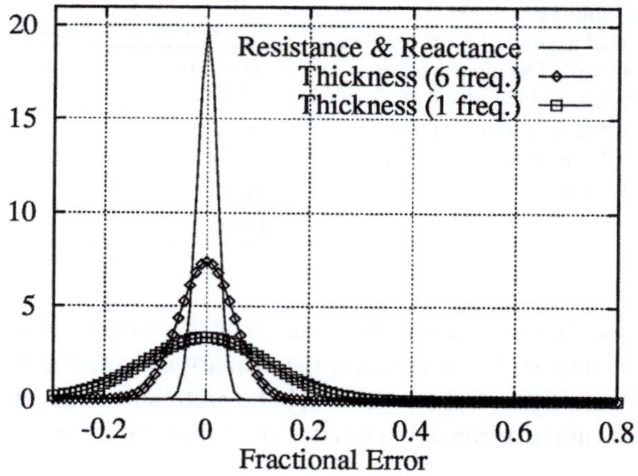

Fig. 11.7 Test Case 1: distribution of errors for impedance, thickness and lift-off

In the simple case that $\sigma_{R_i}/R_i = \sigma_{X_i}/X_i = \sigma_Z/Z$, the half-width for the thickness is given by

$$\frac{\sigma_t}{t} = \left[\sum_i \left(\frac{\partial t}{\partial R_i}\frac{R_i}{t}\right)^2 + \left(\frac{\partial t}{\partial X_i}\frac{X_i}{t}\right)^2\right]^{1/2} \frac{\sigma_Z}{Z}. \tag{11.13}$$

The partial derivatives in (11.13) can be evaluated by performing the inversion procedure for values of R and X near the measured values and using the deduced thicknesses to compute numerical derivatives. Using this procedure to evaluate the partial derivatives for Test Case 1, we obtain a half-width for the deduced thickness given by $\sigma_t/t = 0.0542$, assuming a half-width for the measured impedance given by $\sigma_z/Z = 0.02$. This is the half-width that is obtained with the inversion scheme using six frequencies. If only the single frequency, 30271.2 Hz, is used for the analysis, then the half-width is given by $\sigma_t/t = 0.120$. The corresponding probability distributions are illustrated in Fig. 11.7. We see that inclusion of multiple frequencies substantially reduces the spread in the deduced thickness. The uncertainties in the determination of the probe lift-off can be estimated in the same manner. The half-width for the deduced lift-off is given by $\sigma_l/l = 0.0267$ when six frequencies are used.

If ΔR_i and ΔX_i are due to systematic errors, so that $\Delta R_i/R_i = \Delta X_i/X_i = \Delta Z/Z$, then

$$\frac{\Delta t}{t} = \left[\sum_{i=1}^{N} \frac{\partial t}{\partial R_i}\frac{R_i}{t} + \frac{\partial t}{\partial X_i}\frac{X_i}{t}\right] \frac{\Delta Z}{Z}. \tag{11.14}$$

If the errors in Test Case 1 are systematic, with $\Delta R_i/R_i = \Delta X_i/X_i = \Delta Z/Z < 0.05$, then the errors in thickness and lift-off, using six frequencies, are given by

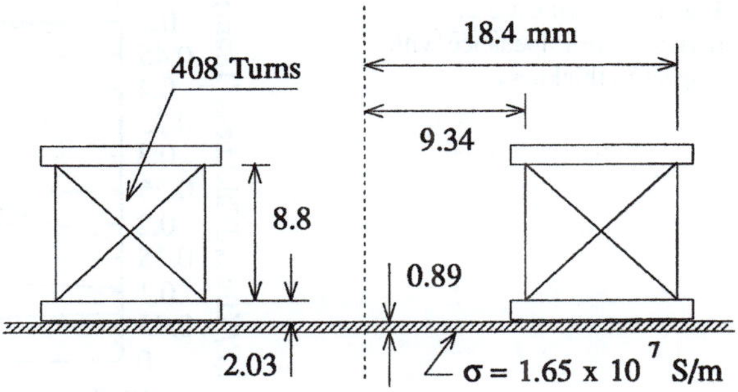

Fig. 11.8 Test Case 2: experimental arrangement with brass plate. The probe is cylindrical

$$\frac{\Delta t}{t} = -7.36\frac{\Delta Z}{Z}$$
$$\frac{\Delta l}{l} = -1.85\frac{\Delta Z}{Z}.$$

If the analysis includes only the single frequency, 30271.2 Hz, then $\Delta t/t$ is about the same, while $\Delta l/l$ is twice as large. Thus, if only one frequency is used in the analysis, then the systematic errors and statistical errors make comparable contributions to the thickness error, while the inclusion of six frequencies makes the statistical errors much less significant. In the six-frequency analysis, changes in reactance at high and low frequencies produce opposite effects on the deduced lift-off. In this case, systematic errors can be cancelled by an appropriate choice of frequencies.

11.2.3 Test Case 2

We now consider a second test case, for which we have experimental measurements. The problem is illustrated in Fig. 11.8. The measurements are those of Burke (1992, private communication) and cover the range 100–10,000 Hz. The computed derivatives of impedance with respect to thickness and lift-off are shown in Figs. 11.9 and 11.10. From these plots, the range 1–2 kHz is chosen for determining the thickness. The thickness and lift-off values obtained from a number of analyses, each using a different number of frequencies, are shown in Table 11.2.

In Fig. 11.11 the ratio of the deduced thickness and lift-off to the measured values are plotted versus the number of frequencies used in the analysis. The improvement observed in thickness determination with the addition of frequencies to the analysis suggests that there exists a statistical component of the errors in the measured impedance values. The remaining 1 % difference between the measured and deduced thickness is likely due to systematic errors in the measurements or calculations. The probe lift-off is not determined quite as well as the thickness. This could be anticipated from Figs. 11.9 and 11.10, which show the impedance to be less sensitive to the lift-off.

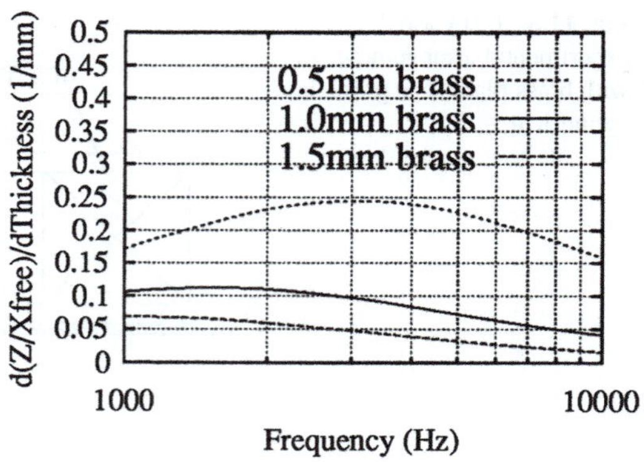

Fig. 11.9 Test Case 2: derivative of impedance with respect to thickness

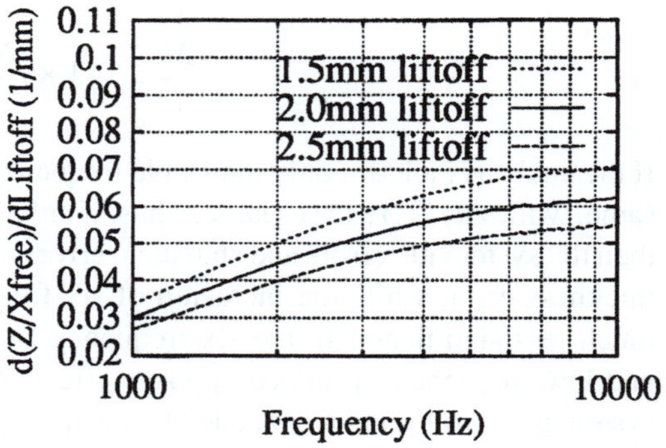

Fig. 11.10 Test Case 2: derivative of impedance with respect to lift-off

11.2.4 Conclusions

This first example illustrates several important features of model-based inversion that will become manifest throughout the remainder of the book. We have shown how uncertain parameters, such as lift-off, can be accounted for when determining a "primary parameter," such as plate thickness. We simply add such parameters to the bag of unknowns and solve all simultaneously. There is a limit to this, of course, but in many practical cases that will be illustrated later, this approach succeeds well.

A second point to be made is that with modern computational tools, such as **VIC-3D®**, one can dispense with "rules of thumb" and other empirical criteria that saturate the current practice of eddy-current NDE. Such tools provide a means of quickly and efficiently optimizing probe design and measurements to maximize the accuracy needed in modern engineering practice.

Table 11.2 Deduced thickness and lift-off for Test Case 2. The brass plate is 0.89 mm thick and the lift-off is 2.03 mm. Starting values: 2 mm thickness, 0 mm lift-off

Frequencies	Thickness	Lift-off	No. of iterations
1584.89	0.8779	2.063	3
1412.54 1584.89	0.8779	2.062	3
1412.54 1584.89 1778.28	0.8788	2.063	3
1258.93 1412.54 1584.89 1778.28	0.8793	2.062	3
1258.93 1412.54 1584.89 1778.28 1995.26	0.8801	2.065	3
1122.02 1258.93 1412.54 1584.89 1778.28 1995.26	0.8803	2.065	3
1000.00 1122.02 1258.93 1412.54 1584.89 1778.28 1995.26	0.8801	2.064	3

11.3 Conductivity Profile Measurements

It is often necessary to inspect materials for hardness, residual stress, fiber-volume content, etc. These properties may be the results of a manufacturing process, or the results of exposure to a harsh environment. Whatever the cause, when these physical attributes can be correlated with electrical conductivity or permeability, then eddy-current inspection can be used to monitor them. We do not address the question of correlating conductivity to the material properties, leaving that to materials scientists. What we will do is show how modeling can (1) tell what is required of the eddy-current instrument in order to achieve satisfactory performance and (2) help us understand the measurements we obtain from our instruments.

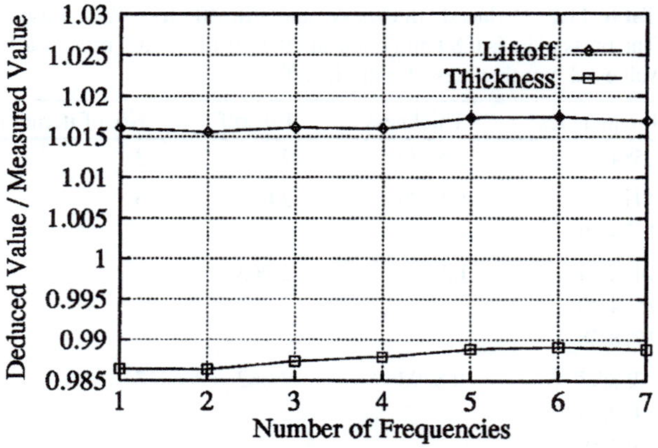

Fig. 11.11 Test Case 2: thickness and lift-off versus the number of frequencies used

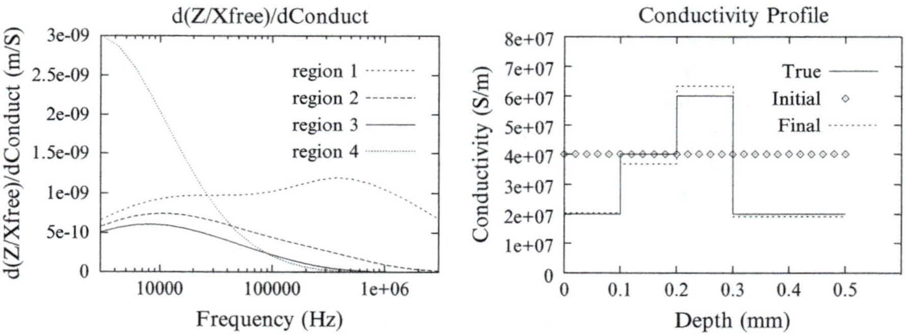

Fig. 11.12 A model problem with three layers over a half-space. Derivative of normalized impedance with respect to conductivity, versus frequency (*left*), and conductivity profile. Region 1 lies between 0 and 0.1 mm; region 4 lies below 0.3 mm

Suppose a flat, layered structure has layers with the conductivities shown (solid) in Fig. 11.12 [54]. We can reconstruct this profile using impedance measurements from a simple air-core probe. To model this problem, we use **VIC-3D®** to compute the impedance seen by an air-core probe resting on our layered structure. Our probe has 400 turns, is 1.73 mm in height, and has inner and outer radii of 2.54 and 4.27 mm. To determine the best frequencies for our probe, we look at the derivative of the normalized impedance with respect to the conductivities of the four regions of our structure. For all regions, this derivative peaks between 3 kHz and 3 MHz, as shown in Fig. 11.12. We use this frequency range.

We use the same nonlinear least-squares algorithm (Gauss–Newton) that was used in the thickness measurements to recover the conductivities from the impedances. The starting point for our iterative algorithm is a uniform conductivity

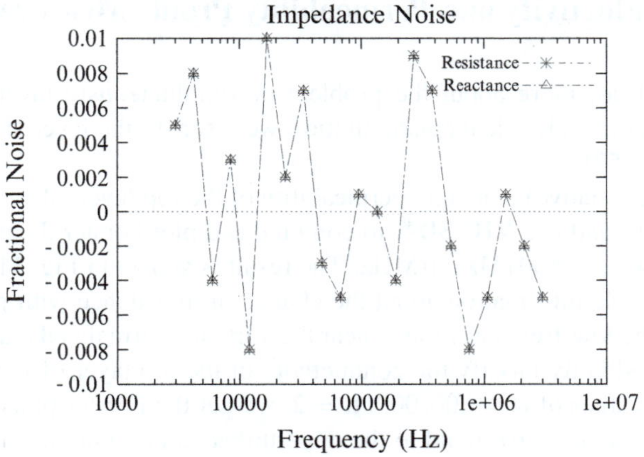

Fig. 11.13 Perturbations in the computed impedances at 21 frequencies. This is the input data that represents the effects of "noise"

of 4×10^7 S/m. Three to six frequencies, in a fairly narrow band, are sufficient to exactly reconstruct the conductivities, if our computed "noiseless" impedances are used. To investigate the effect of "noise" in the data, we perturb the computed impedances, which are our input data, by the fractional amounts shown in Fig. 11.13. This is not representative of any particular noise model, but simply illustrates the effects that perturbations have on the reconstructions. Additional frequencies are now required to produce a satisfactory inversion. The conductivities obtained using 21 frequencies are shown (dashed) in Fig. 11.12. The frequencies chosen are simply those that form a geometric progression between 3 kHz, and 3 MHz. Ten iterations were required.

We can better understand the quality of the reconstruction we can expect from our measurements by referring to the derivatives shown in Fig. 11.12. Clearly, at low frequencies, the impedance is much more sensitive to the conductivity of the half-space, while at high frequencies it is much more sensitive to that of the top layer. Hence, our recovered conductivity profile is most accurate in these regions.

Our modeling allows us to test our inversion algorithm, determine the proper frequencies to measure, establish noise requirements for our measurements, and understand the quality of reconstruction we can expect.

We believe that this is an important point to stress. Eddy-currents have long been applied to the measurement of thickness and conductivity, but the techniques have generally been quite empirical, and used discrete standards. The results of blindly using the nearest available instrument have often been disappointing. Modeling must be performed before equipment is purchased, or expensive experiments undertaken.

11.4 Conductivity and Permeability Profile Measurements

In order to learn more about the problem of simultaneously inverting magnetic permeabilities and electrical conductivities, we resort to the layered system shown in Fig. 11.14 [57].

We let the (relative) magnetic permeability of the top layer take on the values 1, 2, 4, 8, and 16 and use **VIC-3D®** to compute and plot Förster diagrams[1] over the frequency range of 500 kHz–10 MHz. The result is shown in Fig. 11.15. It is clear that considerable information about the change in impedance with permeability is revealed in the low-frequency range near the vertical (normalized reactance) axis.

When we slightly modify the conductivity of the top layer of Fig. 11.14, using the nominal values of $\sigma = 800{,}000$, $\mu = 2$, we get the Förster plots of Fig. 11.16, in which the circles correspond to the unperturbed configuration, and the triangles to the configuration in which the conductivity is slightly perturbed. Clearly, the conductivity changes are along the Förster plot, in the direction of frequency.

Next, we repeat the experiment, changing the relative permeability of the top layer slightly from its nominal value of two, and plot the results in Fig. 11.17. It appears, now, that the change is roughly normal to the original curve.

Indeed, this is borne out in Table 11.3, which shows the cosine of the angle between the vectors representing conductivity and permeability changes in Figs. 11.16 and 11.17. The frequency range in Table 11.3 extends from 50 kHz to 1 MHz, and the nominal value of the relative permeability of the top layer is 16.

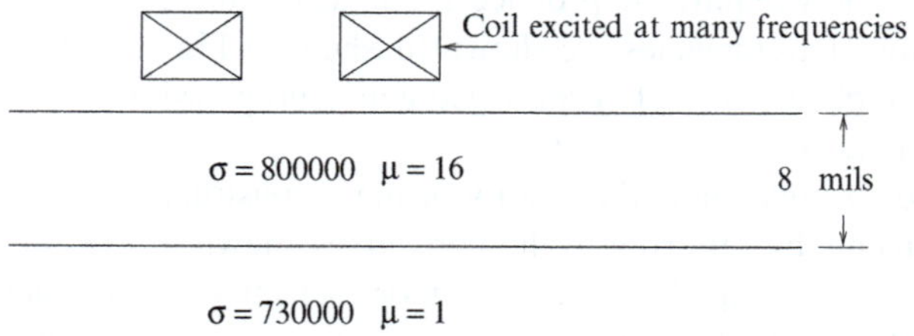

Fig. 11.14 A model configuration to study the problem of simultaneously reconstructing magnetic permeabilities and electrical conductivities. For this study, the relative permeability of the top layer varies from 1 to 16

[1]Recall that a Förster diagram is a plot of normalized reactance to normalized resistance, with the normalizing factor being the freespace reactance of the probe.

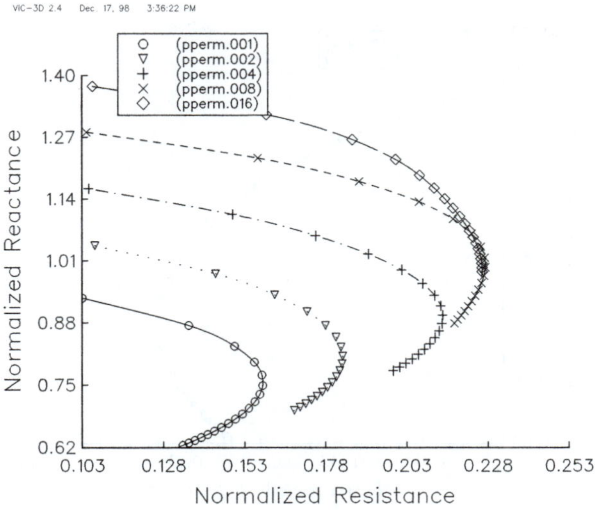

Fig. 11.15 Förster plots, as computed by **VIC-3D®** for $\mu = 1, 2, 4, 8, 16$, (*bottom* to *top*), over the frequency range of 500 kHz to 10 MHz. The curves evolve in the clockwise direction with frequency

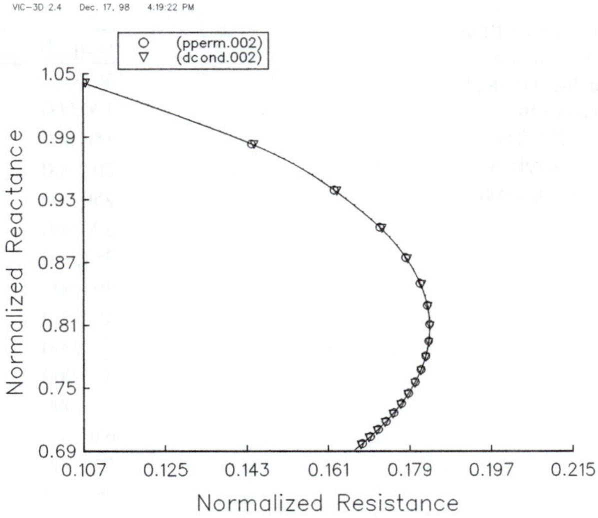

Fig. 11.16 Förster plots of the unperturbed configuration of Fig. 11.14 (*circles*) and the configuration in which the conductivity of the top layer is slightly perturbed (*triangles*). The frequency range is the same as in Fig. 11.15

Clearly, there is an advantage to working in the lower-frequency range when trying to discriminate between conductivity and permeability changes, because the correlation between these two parameters is smallest there.

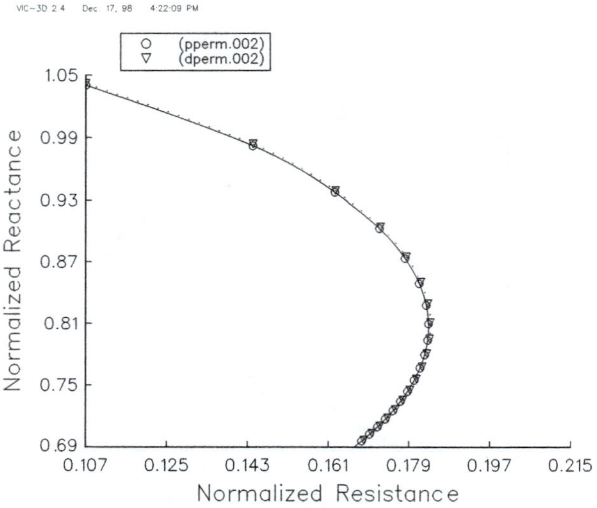

Fig. 11.17 Förster plots of the unperturbed configuration of Fig. 11.14 (*circles*) and the configuration in which the permeability of the top layer is slightly perturbed from its nominal value of two (*triangles*). The frequency range is the same as in Fig. 11.15

Table 11.3 The cosine of the angle between the vectors representing conductivity and permeability changes in Figs. 11.16 and 11.17. The nominal value of the relative permeability of the top layer is 16

Freq. (Hz)	Cosine
50,000	−0.130716
100,000	−0.206825
150,000	−0.270171
200,000	−0.326754
250,000	−0.375093
300,000	−0.420795
350,000	−0.463919
400,000	−0.503752
450,000	−0.542168
500,000	−0.579675
550,000	−0.615551
600,000	−0.650175
650,000	−0.683748
700,000	−0.716262
750,000	−0.747107
800,000	−0.775634
850,000	−0.803411
900,000	−0.828341
950,000	−0.851567
1,000,000	−0.872808

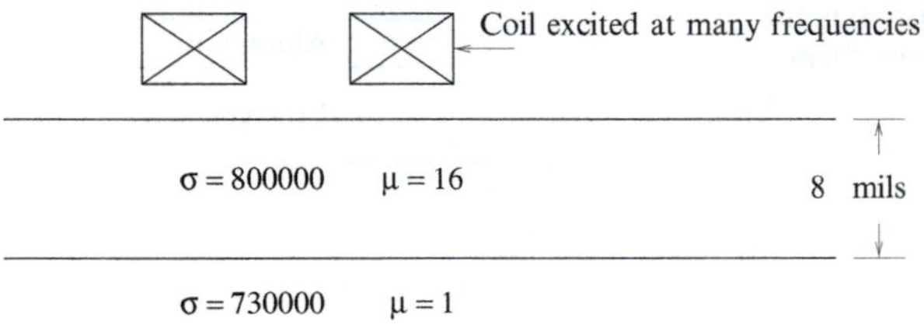

Fig. 11.18 A two-layer ferromagnetic system, whose conductivities and permeabilities are to be reconstructed using multifrequencies. The frequency range extends from 50 kHz to 1 MHz in 19 equal steps

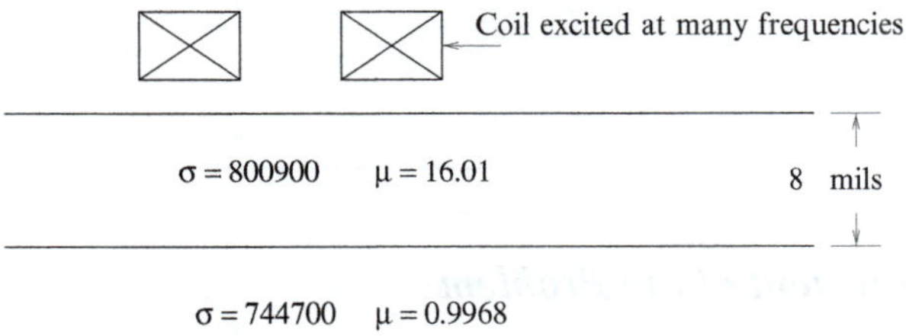

Fig. 11.19 Reconstruction of the system shown in Fig. 11.18, when 20 frequencies are used from 50 kHz to 1 MHz. The impedance data are noiseless

11.4.1 Multifrequency Reconstruction of a Two-Layered Ferromagnetic System

This problem, which is shown in Fig. 11.18, addresses the issue of reconstructing magnetic permeabilities, as well as electrical conductivities.

The impedance data are taken over the frequency range of 50 kHz to 1 MHz, in 19 equal intervals. These data, without the addition of noise, are used to reconstruct both permeability and conductivity, with the results shown in Fig. 11.19. The starting points for the iterations were $\sigma_L = \sigma_S = 750,000$, $\mu_L = \mu_S = 2$, where the subscript "*L*" denotes the layer and "*S*" denotes the substrate.

11.5 A Problem in the Semiconductor Chip Industry

The geometry of the problem is illustrated in Fig. 11.20 (TENCOR, Inc., 1996, private communication):

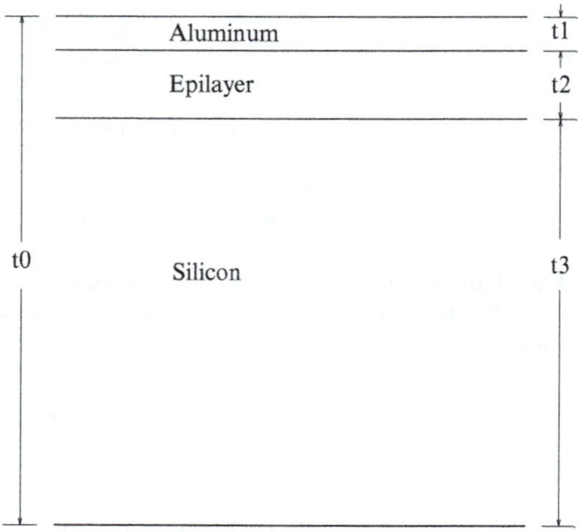

Fig. 11.20 Illustrating the geometry of the problem.

11.5.1 Statement of the Problem

Statement No. 1

Given:

(a) t_0, the total thickness of the package
(b) $\sigma_{\text{Al}} = 3.57 \times 10^7$ S/m, the conductivity of aluminum
(c) $\sigma_{\text{Epi}} = 10$ S/m, $t_2 = 10 \times 10^{-6}$ m

Determine:

(a) t_1, the thickness of the aluminum layer
(b) σ_{Si}, the conductivity of the silicon substrate

Statement No. 2

Given:

(a) $t_3 = 550 \times 10^{-6}$ m, the thickness of the silicon substrate
(b) $\sigma_{\text{Al}} = 3.57 \times 10^7$, the conductivity of aluminum
(c) $\sigma_{\text{Epi}} = 10$ S/m, $t_2 = 10 \times 10^{-6}$ m

Determine:

(a) t_1, the thickness of the aluminum layer
(b) σ_{Si}, the conductivity of the silicon substrate

The only difference between the two statements lies in the information that is given about the thickness. Mathematically, the two problems are identical, because one piece of thickness information is given in each of the statements.

11.5.2 Method of Approach

We will follow the approach described in Sect. 11.2, but now there are two approaches that we can take to gather the data. As before, we can use multifrequency excitation of a coil that is placed at a known lift-off above the aluminum layer and invert the impedance data at these frequencies to determine t_1 and σ_{Si}, jointly. The second possibility is to vary the lift-off of the coil that is excited at a fixed frequency, measure the impedance at each lift-off value, and invert these data to determine t_1 and σ_{Si}, jointly. The latter is a realistic possibility, because the manufacturing and measurement processes for semiconductor chips are performed in clean rooms, with well-defined conditions and instruments. The lift-off can be controlled and measured accurately by ultrasonic means.

We'll briefly examine the multifrequency case and discuss more thoroughly numerical experiments with the multi lift-off case. It is convenient for analysis purposes to normalize impedances with respect to the free-space reactance of the coil, when data are taken over a broad frequency range. Figure 11.21 shows derivatives of the normalized resistance and reactance with respect to t_1 and σ_{Si}, as a function of frequency over the range 10^6–10^9 Hz. The nominal values of the unknowns are $t_1 = 560\,\text{Å} = 5.6 \times 10^{-5}$ mm and $\sigma_{\mathrm{Si}} = 2 \times 10^4$ S/m, with a (known) lift-off of 0.5 mm.

The results indicate that care must be taken in choosing an operating frequency. The thin aluminum layer shields the probe coil from the silicon substrate at frequencies greater than 100 MHz, which is why we see a significant decrease in the conductivity sensitivity coefficients (the derivatives with respect to σ_{Si}) over this frequency range in the bottom part of Fig. 11.21.

We'll turn now to the second possibility, in which lift-off is varied over 16 values, starting at 0.0762 mm and ending at 1.01614 mm in equal logarithmic increments. We use three different frequencies, 10, 25 and 100 MHz. The sensitivity coefficients (derivatives) for each of these frequencies are shown in Figs. 11.22–11.24. It is interesting to note that $dR/d\sigma_{\mathrm{Si}}$, as a function of lift-off, is positive for 10 MHz and "pivots" around the zero-value at a lift-off of 1 mm and is negative at 100 MHz. It appears that at a frequency somewhat higher than 25 MHz, this derivative is actually zero for all lift-off values, which suggests that at this frequency the resistance is insensitive to changes in conductivity.

The input data for the multi lift-off inversion problem are obtained by scanning a coil of inner radius 1 mm, outer radius 1.2 mm, height 0.2 mm, with 18 turns above the system shown in Fig. 11.20 at 16 lift-off values, and then recording the change in impedance from freespace conditions. The freespace resistance is zero, of course, and the freespace reactance is 73.0313 Ω at 10 MHz, 182.578 Ω at 25 MHz, and 730.314 Ω at 100 MHz, as computed by **VIC-3D®**. The model data are shown in Table 11.4.

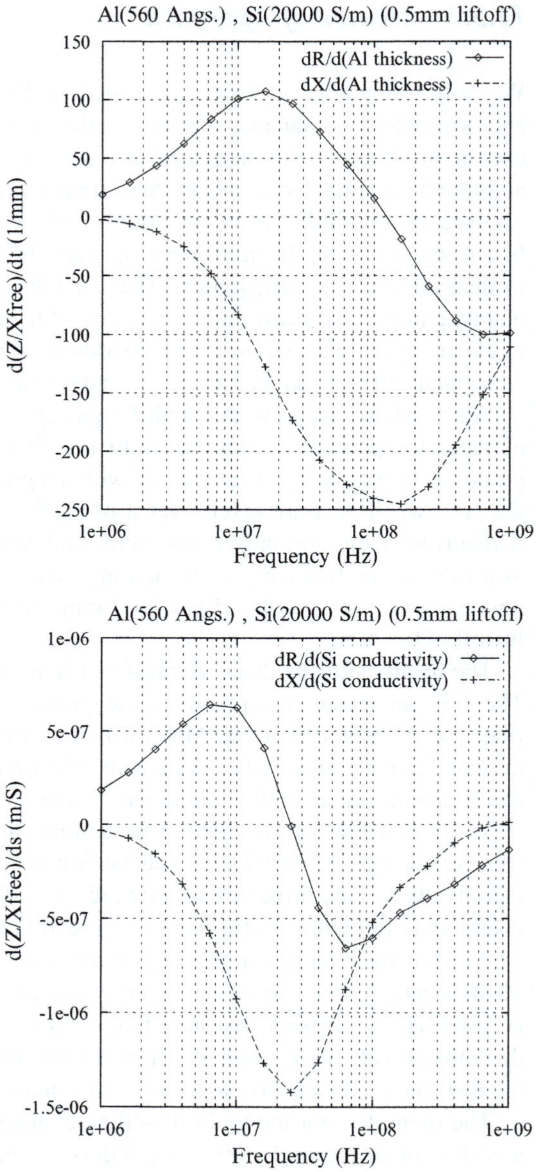

Fig. 11.21 Derivatives of normalized resistance and reactance with respect to t_1 and σ_{Si}

11.6 Noise Analysis: The Covariance Matrix

The matrix, A, of (11.5) is called the Jacobian of the system, and the derivatives shown in Figs. 11.22–11.24 form the columns of A:

$$A = [\mathbf{t},\ \sigma], \tag{11.15}$$

Fig. 11.22 Derivatives of normalized resistance and reactance with respect to t_1 and σ_{Si}. Lift-off is varied at $f = 10\,\text{MHz}$

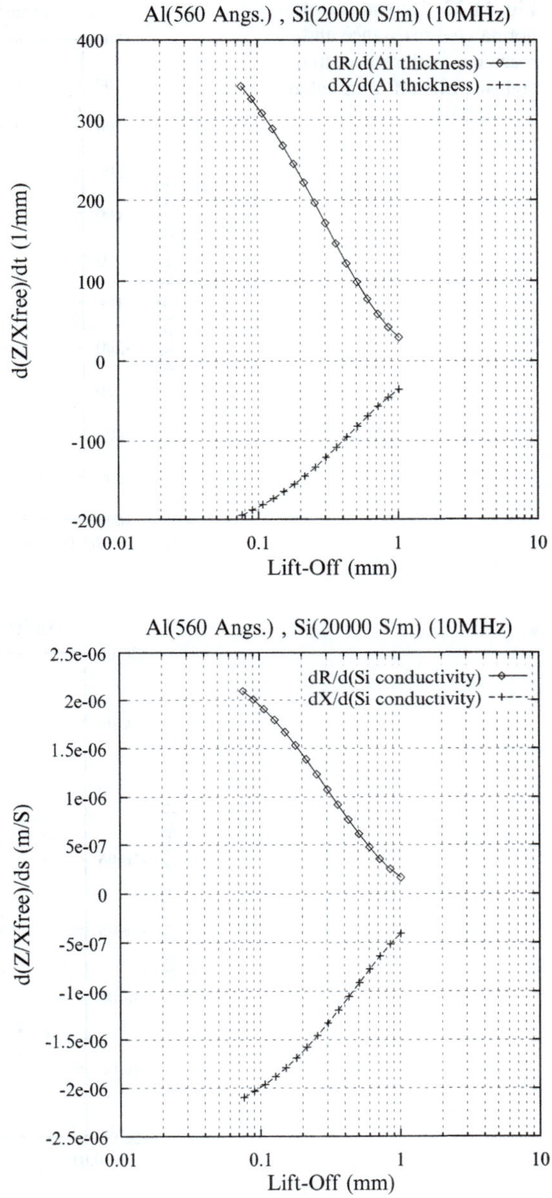

where **t** denotes the column-vector containing derivatives with respect to t_1 and σ denotes the column-vector containing derivatives with respect to σ_{Si}. Each of these vectors is of length 32.

The 2×2 matrix of the dot-products of the columns of A is called the Fisher Information Matrix:

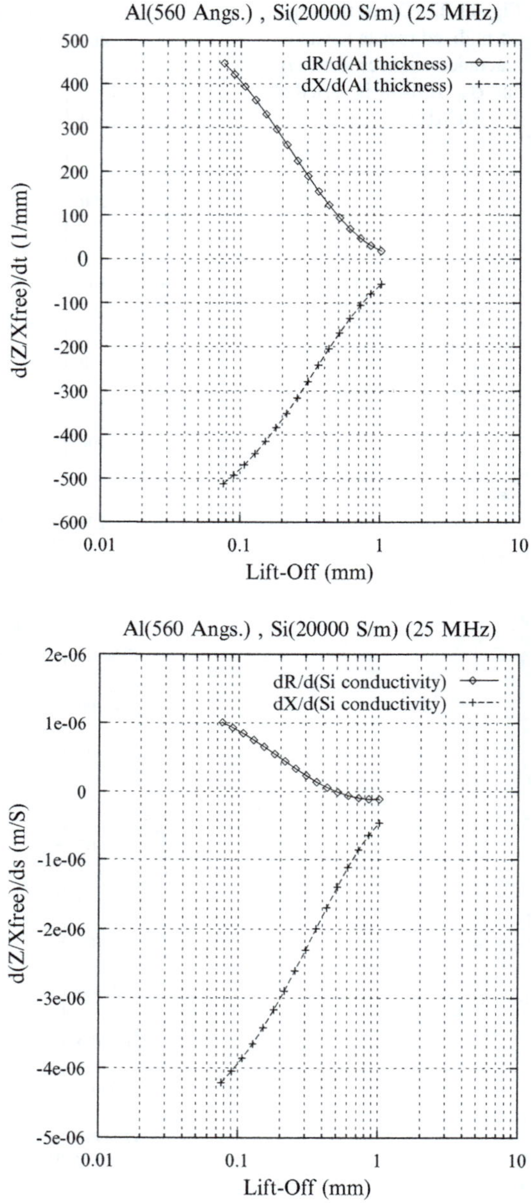

Fig. 11.23 Derivatives of normalized resistance and reactance with respect to t_1 and σ_{Si}. Lift-off is varied at $f = 25\,\mathrm{MHz}$

$$F = \begin{bmatrix} \mathbf{t}\cdot\mathbf{t} & \mathbf{t}\cdot\sigma \\ \sigma\cdot\mathbf{t} & \sigma\cdot\sigma \end{bmatrix}. \tag{11.16}$$

If our data have a uniform variance, σ^2_{meas}, then the covariance matrix, C, for our computed parameters is given by

Fig. 11.24 Derivatives of normalized resistance and reactance with respect to t_1 and σ_{Si}. Lift-off is varied at $f = 100\,\text{MHz}$

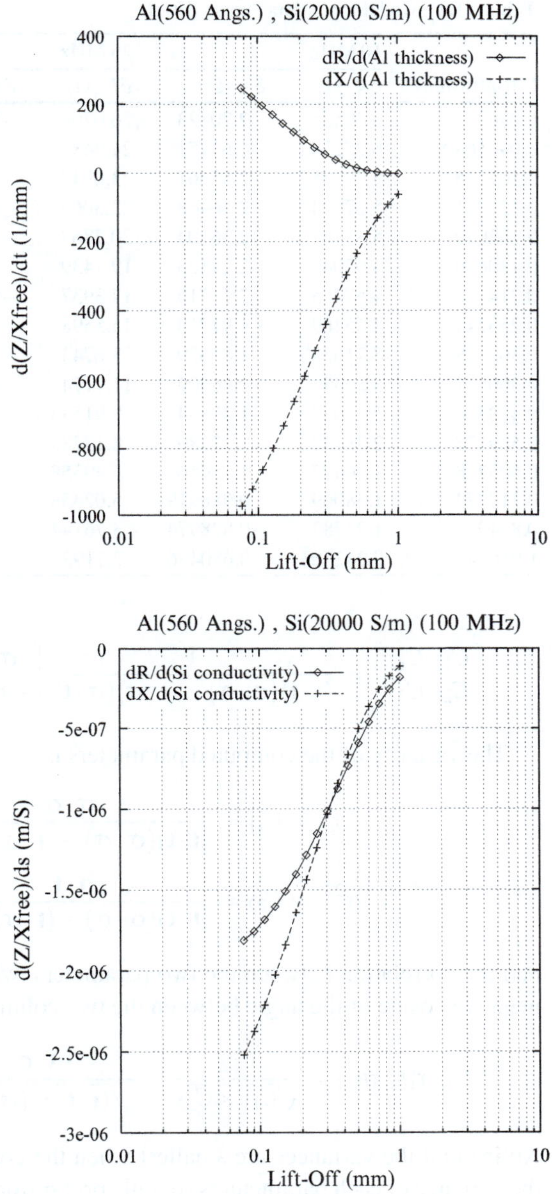

$$C = F^{-1}\sigma^2_{\text{meas}}. \tag{11.17}$$

where σ_{meas} is the variance in the measurement noise. There should be no confusion between σ implying a variance and σ representing electrical conductivity. From (11.16) and (11.17) we get

Table 11.4 Model input data versus lift-off

	10 MHz		25 MHz		100 MHz	
Lift-off (mm)	δR (Ω)	δX (Ω)	δR (Ω)	δX (Ω)	δR (Ω)	δX (Ω)
0.0762	6.49585	−2.70823	26.0706	−20.723	109.204	−204.725
0.0905639	6.25278	−2.63751	24.9487	−20.0579	103.151	−196.281
0.107635	5.97935	−2.55661	23.6947	−19.3018	96.5301	−186.805
0.127925	5.67444	−2.46468	22.3065	−18.4487	89.3745	−176.274
0.152039	5.33781	−2.36104	20.7867	−17.4951	81.7439	−164.697
0.180699	4.97041	−2.24523	19.1439	−16.4401	73.7279	−152.135
0.214761	4.57465	−2.11715	17.3937	−15.287	65.4471	−138.7
0.255244	4.15469	−1.97718	15.5598	−14.0442	57.0521	−124.571
0.303358	3.71658	−1.82629	13.6743	−12.726	48.7193	−109.995
0.360541	3.2683	−1.66609	11.7774	−11.3531	40.6436	−95.2812
0.428504	2.81956	−1.49894	9.91533	−9.95243	33.0258	−80.7884
0.509278	2.38128	−1.32789	8.13789	−8.5563	26.057	−66.9023
0.605278	1.96477	−1.15656	6.49358	−7.20021	19.8993	−53.9995
0.719374	1.58069	−0.988919	5.02458	−5.91994	14.6663	−42.4069
0.854978	1.23787	−0.828974	3.76144	−4.74799	10.4069	−32.3626
1.01614	0.942303	−0.680426	2.71921	−3.71	7.09821	−23.9863

$$\begin{bmatrix} C_{11} & C_{12} \\ C_{21} & C_{22} \end{bmatrix} = \frac{1}{(\mathbf{t}\cdot\mathbf{t})(\sigma\cdot\sigma)-(\mathbf{t}\cdot\sigma)(\sigma\cdot\mathbf{t})}\begin{bmatrix} \sigma\cdot\sigma & -\mathbf{t}\cdot\sigma \\ -\sigma\cdot\mathbf{t} & \mathbf{t}\cdot\mathbf{t} \end{bmatrix} \times \sigma_{\text{meas}}^2. \tag{11.18}$$

The variance of the computed parameters is

$$\sigma_{t_1}^2 = C_{11} = \frac{\sigma\cdot\sigma}{(\mathbf{t}\cdot\mathbf{t})(\sigma\cdot\sigma)-(\mathbf{t}\cdot\sigma)(\sigma\cdot\mathbf{t})} \times \sigma_{\text{meas}}^2$$

$$\sigma_{\sigma_{\text{Si}}}^2 = C_{22} = \frac{\mathbf{t}\cdot\mathbf{t}}{(\mathbf{t}\cdot\mathbf{t})(\sigma\cdot\sigma)-(\mathbf{t}\cdot\sigma)(\sigma\cdot\mathbf{t})} \times \sigma_{\text{meas}}^2. \tag{11.19}$$

and the correlation between the two parameters, which lies between 0 and 1, is the negative cosine of the angle between the two column-vectors of A:

$$\text{cor}(t_1, \sigma_{\text{Si}}) = \frac{C_{12}}{\sqrt{C_{11}\times C_{22}}} = \frac{-\mathbf{t}\cdot\sigma}{\sqrt{(\mathbf{t}\cdot\mathbf{t})\times(\sigma\cdot\sigma)}} = -\cos(\theta) \tag{11.20}$$

Notice that the variances are smallest when the correlation is small, in which case the variance of each parameter is roughly proportional to the inverse of the derivative of the impedance with respect to that parameter.

Consider, for example, the conditions at the nominal values of the parameters: $t_1 = 5.6\times 10^{-5}$ mm and $\sigma_{\text{Si}} = 2\times 10^4$ S/m. From Figs. 11.22–11.24 we compute the covariance matrix at $f = 10$, 25, and 100 MHz to be:

$$C[10] = \begin{bmatrix} 0.1392\times 10^{-4}(\text{mm})^2 & -0.1716\times 10^{4}(\text{S/m})(\text{mm}) \\ -0.1716\times 10^{4}(\text{mm})(\text{S/m}) & 0.2280\times 10^{12}(\text{S/m})^2 \end{bmatrix} \times \sigma_{\text{meas}}^2$$

$$C[25] = \begin{bmatrix} 0.1629 \times 10^{-5}(\mathrm{mm})^2 & -0.2185 \times 10^{3}(\mathrm{S/m})(\mathrm{mm}) \\ -0.2185 \times 10^{3}(\mathrm{mm})(\mathrm{S/m}) & 0.3764 \times 10^{11}(\mathrm{S/m})^2 \end{bmatrix} \times \sigma^2_{\mathrm{meas}}$$

$$C[100] = \begin{bmatrix} 0.3028 \times 10^{-6}(\mathrm{mm})^2 & -0.6073 \times 10^{2}(\mathrm{S/m})(\mathrm{mm}) \\ -0.6073 \times 10^{2}(\mathrm{mm})(\mathrm{S/m}) & 0.3019 \times 10^{11}(\mathrm{S/m})^2 \end{bmatrix} \times \sigma^2_{\mathrm{meas}} \tag{11.21}$$

If we assume that the measurements have a noise variance of 0.1 % of the freespace reactance, then we compute the results for 10 MHz:

$$\begin{aligned}
\sigma^2_{t_1} &= 0.1392 \times 10^{-4}\,\mathrm{mm}^2 \sigma^2_{\mathrm{meas}} \\
&= 0.1392 \times 10^{-4}\,\mathrm{mm}^2 \times (0.1 \times 10^{-2})^2 \\
&= 0.1392 \times 10^{-10}\,\mathrm{mm}^2 \\
\sigma_{t_1} &= 0.373 \times 10^{-5}\,\mathrm{mm} \\
\%\ \mathrm{Error} &= \frac{37.3}{5.6} \\
&= 6.66 \\
\sigma^2_{\sigma_{\mathrm{Si}}} &= 0.2280 \times 10^{12}(\mathrm{S/m})^2 \sigma^2_{\mathrm{meas}} \\
&= 0.2280 \times 10^{12}(\mathrm{S/m})^2 \times (0.1 \times 10^{-2})^2 \\
&= 0.2280 \times 10^{6}(\mathrm{S/m})^2 \\
\sigma_{\sigma_{\mathrm{Si}}} &= 477.49\,\mathrm{S/m} \\
\%\ \mathrm{Error} &= \frac{47,749}{20,000} \\
&= 2.39 \\
\mathrm{cor}(t_1, \sigma_{\mathrm{Si}}) &= \frac{-0.1716 \times 10^4}{\sqrt{0.1392 \times 10^{-4} \times 0.2280 \times 10^{12}}} \\
&= -0.963\,,
\end{aligned} \tag{11.22}$$

which means that t_1 and σ_{Si} are highly anti-correlated.

In a similar manner we compute results for 25 MHz

$$\begin{aligned}
\sigma_{t_1} &= 0.1276 \times 10^{-5}\,\mathrm{mm} \\
\%\ \mathrm{Error} &= 2.28 \\
\sigma_{\sigma_{\mathrm{Si}}} &= 194.01\,\mathrm{S/m} \\
\%\ \mathrm{Error} &= 0.97 \\
\mathrm{cor}(t_1, \sigma_{\mathrm{Si}}) &= -0.882,
\end{aligned} \tag{11.23}$$

Table 11.5 Results of inversions with four different input noise sets at 10 MHz

Noise set	t_1 (mm)	σ_{Si} (S/m)	C	
1	5.72505×10^{-5}	19794.9	0.1403×10^{-4} -0.1727×10^{4}	-0.1727×10^{4} 0.2292×10^{12}
2	5.99395×10^{-5}	19543.5	0.1423×10^{-4} -0.1751×10^{4}	-0.1751×10^{4} 0.2320×10^{12}
3	5.80033×10^{-5}	19671.8	0.1412×10^{-4} -0.1739×10^{4}	-0.1739×10^{4} 0.2305×10^{12}
4	5.22920×10^{-5}	20441.1	0.1360×10^{-4} -0.1678×10^{4}	-0.1678×10^{4} 0.2236×10^{12}

and

$$\begin{aligned} \sigma_{t_1} &= 0.055 \times 10^{-5}\,\text{mm} \\ \%\ \text{Error} &= 0.982 \\ \sigma_{\sigma_{Si}} &= 173.75\,\text{S/m} \\ \%\ \text{Error} &= 0.869 \\ \text{cor}(t_1, \sigma_{Si}) &= -0.635, \end{aligned} \tag{11.24}$$

for 100 MHz. It is clear that there is significant improvement in noise reduction at the higher frequencies. Furthermore, t_1 and σ_{Si} become less correlated at higher frequencies, which leads to a more reliable inversion.

11.7 Noise Analysis: Monte Carlo Method

Monte Carlo methods imply performing a number of numerical experiments with random input data. In the experiments reported here, we add random noise whose variance is 0.1 % of the freespace reactance to the input data of Table 11.4. Thus, the input impedance at a lift-off of 0.0762 mm is:

$$\begin{aligned} Z &= 6.4958 \pm 0.073 + j(73.0313 - 2.70823) \pm 0.073\,\Omega \text{ at } 10\,\text{MHz} \\ Z &= 26.070 \pm 0.183 + j(182.578 - 20.723) \pm 0.183\,\Omega \text{ at } 25\,\text{MHz} \\ Z &= 109.204 \pm 0.730 + j(730.314 - 204.725) \pm 0.73\,\Omega \text{ at } 100\,\text{MHz} \end{aligned}$$

The noise is simulated by using a FORTRAN random number generator.

Our first experiment is conducted at 10 MHz using four different noise sets, all with the same variance. The results of the inversions with these four input data vectors are listed in Table 11.5.

Upon computing the mean and standard deviation of this data set (including the covariance matrix, C), and then performing the usual variance analysis, as in

Table 11.6 Results of Monte Carlo runs at 10 MHz

Variable	Average	Error	Standard deviation	Standard deviation	Variance from covariance matrix
t_1	5.68713×10^{-5} mm	1.6 %	3.25623×10^{-6} mm	5.7 %	6.7 %
σ_{Si}	19862.8 S/m	0.69 %	398.947 S/m	2.0 %	2.4 %

(11.20)–(11.22), we obtain the results shown in Table 11.6. The closeness of the values in the last two columns of Table 11.6 indicates that the Monte Carlo process was reasonably well represented with these four data sets.

Appendix

A.1 A Generalized Error Analysis

Given a set of measurements, b_i, we want to estimate the errors in the parameters, x_i, which are the solution to the equation

$$Ax = b, \tag{11.25}$$

from the error estimates for the b_i. We will evaluate the covariance matrix, C, for the x_i, which will give us estimates of their errors as well as their correlations.

We assume that the matrix A is $m \times n$ with $m > n$. The least squares solution can be obtained be solving the normal equations:

$$A^T Ax = A^T b. \tag{11.26}$$

Solving for x gives

$$x = Db, \tag{11.27}$$

where the matrix D is defined as

$$D = [A^T A]^{-1} A^T. \tag{11.28}$$

The errors, Δx_i, in x are then given by

$$\begin{aligned} \Delta x_i &= \sum_{k=1}^{m} \frac{\partial x_i}{\partial b_k} \Delta b_k \\ &= \sum_{k=1}^{m} D_{ik} \Delta b_k. \end{aligned} \tag{11.29}$$

Thus

$$\Delta x_i \Delta x_j = \sum_{k=1}^{m} \sum_{l=1}^{m} D_{ik} D_{jl} \Delta b_k \Delta b_l. \tag{11.30}$$

Assuming that the errors Δb_k are distributed so as to give cancellation of the terms with $k \neq l$ (e.g., they have a normal distribution), the expected values of $\Delta x_i \Delta x_j$ are then given by

$$\begin{aligned} E(\Delta x_i \Delta x_j) &= \sum_{k=1}^{m} D_{ik} D_{jk} E(\Delta b_k \Delta b_k) \\ &= \sum_{k=1}^{m} D_{ik} D_{jk} \sigma_{b_k}^2 \\ &= \sum_{k=1}^{m} [DS]_{ik} [DS]_{kj}^T \\ &= C_{ij} , \end{aligned} \tag{11.31}$$

where S is a diagonal matrix with $S_{kk} = \sigma_{b_k}$, the standard deviation for b_k, and C is the covariance matrix

$$C = [DS][DS]^T = [(A^T A)^{-1} A^T S][(A^T A)^{-1} A^T S]^T. \tag{11.32}$$

When $S = \sigma_{\text{meas}} I$, i.e., the standard deviation for all measured values is the same, then

$$C = [A^T A]^{-1} A^T A [A^T A]^{-1} \sigma_{\text{meas}}^2 = [A^T A]^{-1} \sigma_{\text{meas}}^2. \tag{11.33}$$

The expected value of the error for x_i is now given by

$$\sigma_{x_i} = \sqrt{C_{ii}} \tag{11.34}$$

and the correlation between x_i and x_j, which lies between 0 and 1, is given by

$$\text{cor}(x_i, x_j) = \frac{C_{ij}}{\sqrt{C_{ii} C_{jj}}}. \tag{11.35}$$

Chapter 12
NLSE: Parameter-Based Inversion Algorithm

12.1 Introduction

Chapter 11 introduced us to the notion of an inverse problem and gave us some examples of the value of this idea to the solution of realistic industrial problems. The basic inversion algorithm described in Chap. 11 was based upon the Gauss–Newton theory of nonlinear least-squares estimation and is called NLSE in this book. In this chapter we will develop the mathematical background of this theory more fully, because this algorithm will be the foundation of inverse methods and their applications during the remainder of this book. We hope, thereby, to introduce the reader to the application of sophisticated mathematical concepts to engineering practice without introducing excessive mathematical sophistication.

We separate the discussion of inversion algorithms into two categories: (a) parameter-based, in which a few parameters are used to define the problem or anomalous region (as in Chap. 11) and (b) voxel-based, in which the anomalous region is constructed voxel-by-voxel on a grid. NLSE is the prototype of the former algorithm; the next volume in this series will introduce us to voxel-based algorithms.

12.2 NLSE: Nonlinear Least-Squares Parameter Estimation

12.2.1 Overview of the Algorithm: Nonlinear Least-Squares

Let

$$Z = g(p_1, \ldots, p_N, f) , \tag{12.1}$$

where $p_1, \ldots, p_N$ are the N parameters of interest, and f is a control parameter at which the impedance, Z, is measured. f can be frequency, scan-position, lift-off, etc. It is, of course, known; it is not one of the parameters to be determined. To be explicit during our initial discussion of the theory, we will call f "frequency."

H.A. Sabbagh et al., *Computational Electromagnetics and Model-Based Inversion*, Scientific Computation, DOI 10.1007/978-1-4419-8429-6_12,

In order to determine $p_1,\ldots,p_N$, we measure Z at M frequencies, $f_1,\ldots,f_M$, where $M > N$:

$$
\begin{aligned}
Z_1 &= g(p_1,\ldots,p_N,f_1) \\
&\vdots \\
Z_M &= g(p_1,\ldots,p_N,f_M).
\end{aligned}
\tag{12.2}
$$

The right-hand side of (12.2) is computed by applying the volume-integral code to a model of the problem, usually at a discrete number of values of the vector, **p**, forming a multidimensional interpolation grid.

Because the problem is nonlinear, we use a Gauss–Newton iteration scheme to perform the inversion [97]. First, we decompose (12.2) into its real and imaginary parts, thereby doubling the number of equations (we assume the $p_1,\ldots,p_N$ are real). Then we use the linear approximation to the resistance, R_i, and reactance, X_i, at the ith frequency:

$$
\begin{bmatrix} R_1 \\ X_1 \\ \vdots \\ R_M \\ X_M \end{bmatrix} \approx
\begin{bmatrix} R_1(p_1^{(q)},\ldots,p_N^{(q)}) \\ X_1(p_1^{(q)},\ldots,p_N^{(q)}) \\ \vdots \\ R_M(p_1^{(q)},\ldots,p_N^{(q)}) \\ X_M(p_1^{(q)},\ldots,p_N^{(q)}) \end{bmatrix} +
\begin{bmatrix} \frac{\partial R_1}{\partial p_1} & \cdots & \frac{\partial R_1}{\partial p_N} \\ \frac{\partial X_1}{\partial p_1} & \cdots & \frac{\partial X_1}{\partial p_N} \\ & \vdots & \\ \frac{\partial R_M}{\partial p_1} & \cdots & \frac{\partial R_M}{\partial p_N} \\ \frac{\partial X_M}{\partial p_1} & \cdots & \frac{\partial X_M}{\partial p_N} \end{bmatrix}_{(p_1^{(q)},\ldots,p_N^{(q)})}
\begin{bmatrix} p_1 - p_1^{(q)} \\ \vdots \\ p_N - p_N^{(q)} \end{bmatrix},
\tag{12.3}
$$

where the superscript (q) denotes the qth iteration and the partial derivatives are computed numerically by the software. The left side of (12.3) is taken to be the measured values of resistance and reactance. We rewrite (12.3) as

$$
0 \approx r + Jp\,,
\tag{12.4}
$$

where r is the $2M$-vector of residuals, J is the $2M \times N$ Jacobian matrix of derivatives, and p is the N-dimensional correction vector. Equation (12.4) is solved in a least-squares manner starting with an initial value, $(x_1^{(0)},\ldots,x_N^{(0)})$, for the vector of unknowns, and then continuing by replacing the initial vector with the updated vector $(x_1^{(q)},\ldots,x_N^{(q)})$ that is obtained from (12.3), until convergence occurs.

We are interested in determining a bound for the sensitivity of the residual norm to changes in some linear combination of the parameters. Given an $\varepsilon > 0$ and a unit vector, v, the problem is to determine a sensitivity (upper) bound, σ, such that

$$
\|r(x^* + \sigma v)\| \le (1+\varepsilon)\|r(x^*)\|\,.
\tag{12.5}
$$

We will derive a first-order estimate of σ. Equation (12.5) is equivalent to

$$\|r(x^* + \sigma v)\| - \|r(x^*)\| \leq \varepsilon \|r(x^*)\| \,. \tag{12.6}$$

The left-hand side of (12.6) can be approximated to the first order in σ by the first-order Taylor expansion:

$$\|r(x^* + \sigma v)\| - \|r(x^*)\| \approx \sigma v \cdot \nabla \|r(x^*)\| \,, \tag{12.7}$$

where ∇ is the gradient operator in N-dimensional space. We compute the gradient of the norm of the residual vector:

$$\nabla \|r(x)\| = \nabla \left[f_1^2(x) + f_2^2(x) + \cdots + f_{2M}^2(x) \right]^{1/2}$$

$$= \frac{1}{\|r(x)\|} \begin{bmatrix} f_1 \dfrac{\partial f_1}{\partial x_1} + \cdots + f_{2M} \dfrac{\partial f_{2M}}{\partial x_1} \\ \vdots \\ f_1 \dfrac{\partial f_1}{\partial x_N} + \cdots + f_{2M} \dfrac{\partial f_{2M}}{\partial x_N} \end{bmatrix}^T$$

$$= \frac{r(x)^T}{\|r(x)\|} \begin{bmatrix} \dfrac{\partial f_1}{\partial x_1} & \cdots & \dfrac{\partial f_1}{\partial x_N} \\ & \vdots & \\ \dfrac{\partial f_{2M}}{\partial x_1} & \cdots & \dfrac{\partial f_{2M}}{\partial x_N} \end{bmatrix}$$

$$= e^T(x) \cdot J \,, \tag{12.8}$$

where the superscript T denotes the transpose of a matrix (or vector) and $e(x) = r(x)/\|r(x)\|$ is a unit vector. Thus, (12.7) becomes

$$\|r(x^* + \sigma v)\| - \|r(x^*)\| \approx \sigma e^T(x) \cdot J \cdot v \,. \tag{12.9}$$

The factor multiplying σ in (12.9) is the dot product of the two vectors, $e(x)$ and $J \cdot v$. Hence, its value is less than the product of the magnitude of each vector, which means that (12.9) becomes

$$\|r(x^* + \sigma v)\| - \|r(x^*)\| \leq \sigma \|J \cdot v\| \,, \tag{12.10}$$

because $e(x)$ has unit magnitude. Upon equating the right-hand side of (12.10) to the right-hand side of (12.6), we obtain the first-order estimate of σ:

$$\sigma_v = \varepsilon \left(\frac{\|r(x^*)\|}{\|J(x^*) \cdot v\|} \right) . \tag{12.11}$$

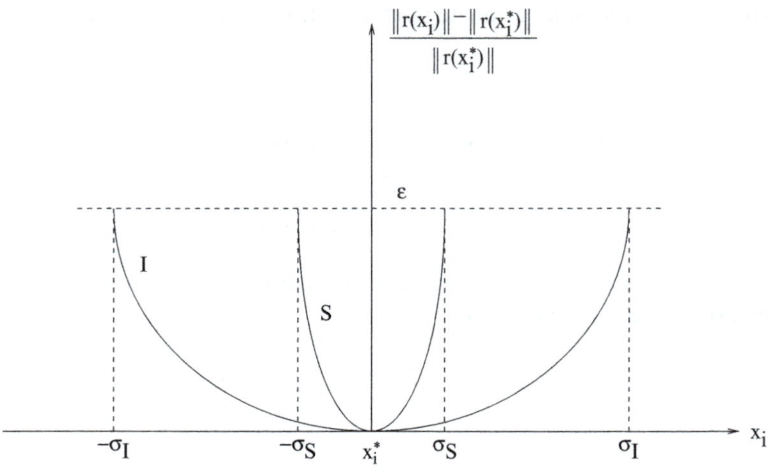

Fig. 12.1 Showing sensitivity parameters for two system responses to x_i. Response S is sensitive to x_i at x_i^*, whereas response I is not

Note that if $\|J(x^*)\cdot v\|$ is small compared to $\|r(x^*)\|$, then σ is large and the residual norm is insensitive to changes in the linear combination of the parameters specified by v. If $v = e_i$, the ith column of the $N\times N$ identity matrix, then (12.11) produces σ_i, the sensitivity bound for the ith parameter. Since σ_i will vary in size with the magnitude of x_i^*, it is better to compare the ratios σ_i/x_i^* for $i = 1,\cdots,N$ before drawing conclusions about the fitness of a solution.

The importance of these results is that we now have metrics for the inversion process: $\Phi = \|r(x^*)\|$, the norm of the residual vector at the solution, tells us how good the fit is between the model data and measured data. The smaller this number the better, of course, but the "smallness" depends upon the experimental setup and the accuracy of the model to fit the experiment. Heuristic judgement based on experience will help in determining the quality of the solution for a given Φ.

The sensitivity coefficient, σ, is more subtle, but just as important. It, too, should be small, but, again, the quality of the "smallness" will be determined by heuristics based upon the problem. If σ is large in some sense, it suggests that the solution is relatively independent of that parameter, so that we cannot reasonably accept the value assigned to that parameter as being meaningful, as suggested in Fig. 12.1, which shows a system, S, for which the system is sensitive to variable, x_i, at the solution point, x_i^*, and another system, I, for which the system is insensitive to x_i.

An example occurs when one uses a high-frequency excitation, with its attendant small skin depth, to interrogate a deep-seated flaw. The flaw will be relatively invisible to the probe at this frequency, and whatever value is given for its parameters will be highly suspect. When this occurs we will either choose a new parameter to characterize the flaw, or acquire data at a lower frequency.

These metrics are not available to us in the current inspection method, in which analog instruments acquire data that are then interpreted by humans using hardware standards. The opportunity to use these metrics is a significant advantage to the model-based inversion paradigm that we propose in this book.

If the residual norm is relatively insensitive to changes in some linear combination of the parameters, then the Jacobian matrix at the solution is nearly rank-deficient, and it may be useful to determine a set of linearly-independent parameters. The covariance matrix $(J^T J)^{-1}$ can be used for this purpose.

12.2.2 Stochastic Methods for Global Optimization

The problem defined above leads to the global optimization problem: finding the lowest minimizer of a nonlinear function of several variables that has multiple local minimizers. In the stochastic approach to global optimization, one applies a strictly descent local search procedure to a subset of a sample of starting points drawn from a uniform distribution over $\mathscr{R}$, so as to find all the local minima of $\|r\|$ that are potentially global [58–60]. One such stochastic approach, the *multilevel, single-linkage*, will, with probability one, find all relevant local minima of the objective function with the smallest possible number of local searches [59, 61]. We do not implement the multilevel, single-linkage approach in this book, but use a uniform distribution of starting points (in each coordinate direction), as in the multilevel approach. We do not reduce the sample size, however, but use the entire sample, comprising, perhaps, 500 points in each variable. We have found that, even with four variables, the procedure is so fast with modern machines, that it is quite efficient.

Hence, our global optimization algorithm starts by generating a uniform distribution of 500 points in each coordinate, and then immediately applying the Gauss–Newton iteration to each of these points. The result is 500 local minima, which are then ordered to give the smallest to largest values of the norm of the residual. The location of the smallest of these local minima is presumed to be the global minimum.

12.2.3 Computation of Function Values

NLSE is a post-processing feature of **VIC-3D®**. **VIC-3D®** is applied a priori to compute function values at certain equi-spaced values of the parameters, and these results are then stored in a table for interpolation. This speeds up the application of the algorithm, though it will require a large database for interpolation, the size, of course, depending upon the number of unknowns, and the number of precomputed function values for each unknown. The order of the interpolator is arbitrary, but we typically use first-order to fourth-order polynomial-splines for each variable, which means that we need to compute the norm of the residuals at two to five values of each variable. The derivatives in the Jacobian matrix are computed by interpolating in this table of precomputed function values.

12.2.4 Application of Statistical Communication Theory

Additional insight can be gained by appealing to the statistical theory of communication [50]. Pretend that the measured impedance vector, $\mathbf{Z} = [Z_1, \ldots, Z_M]$, of (12.2) is a random vector, along with the parameter vector, $\mathbf{p} = [p_1, \ldots, p_N]$. Think of $\mathbf{Z}$ as a received message over some communication channel, and $\mathbf{p}$ as a transmitted message. These two random vectors have a joint probability density, $\mathscr{P}_{\mathbf{Z},\mathbf{p}}(\mathbf{Z},\mathbf{p})$, which is to be maximized. That is to say, for a given $\mathbf{Z}$, we want to determine that $\mathbf{p}$ which maximizes $\mathscr{P}_{\mathbf{Z},\mathbf{p}}$.

From probability theory, we have Bayes rule:

$$\begin{aligned} \mathscr{P}_{\mathbf{Z},\mathbf{p}} &= \mathscr{P}_{\mathbf{Z}}(\mathbf{Z})\mathscr{P}(\mathbf{p}|\mathbf{Z}) \\ &= \mathscr{P}_{\mathbf{p}}(\mathbf{p})\mathscr{P}(\mathbf{Z}|\mathbf{p}) \\ &= \mathscr{P}_{\mathbf{p},\mathbf{Z}}(\mathbf{p},\mathbf{Z}) \, , \end{aligned} \tag{12.12}$$

where $\mathscr{P}_{\mathbf{Z}}(\mathbf{Z})$ and $\mathscr{P}_{\mathbf{p}}(\mathbf{p})$ are called a priori probability density functions, and $\mathscr{P}(\mathbf{p}|\mathbf{Z})$, $\mathscr{P}(\mathbf{Z}|\mathbf{p})$ are called conditional, or a posteriori, probability density functions. The variable to the right of the vertical bar in these latter functions is called the conditioning variable; for example, $\mathscr{P}(\mathbf{p}|\mathbf{Z})$ is called the a posteriori probability density for $\mathbf{p}$, conditioned on the fact that $\mathbf{Z}$ was received (or measured). This a posteriori probability density function is the object of our interest, as we shall now see.

From (12.12) we want to maximize

$$\mathscr{P}_{\mathbf{Z}}(\mathbf{Z})\mathscr{P}(\mathbf{p}|\mathbf{Z}) = \mathscr{P}_{\mathbf{p}}(\mathbf{p})\mathscr{P}(\mathbf{Z}|\mathbf{p}) \tag{12.13}$$

over $\mathbf{p}$. Because $\mathscr{P}_{\mathbf{Z}}(\mathbf{Z})$ is independent of $\mathbf{p}$, we can ignore it and maximize the a posteriori probability density

$$\mathscr{P}(\mathbf{p}|\mathbf{Z}) \propto \mathscr{P}_{\mathbf{p}}(\mathbf{p})\mathscr{P}(\mathbf{Z}|\mathbf{p}) \, . \tag{12.14}$$

That is, we want to choose that "message," $\mathbf{p}$, that is best associated with the "received signal," $\mathbf{Z}$.

If the measured $\mathbf{Z}$ of (12.2) is corrupted by "noise," of density $\mathscr{P}_{\mathbf{n}}$, then we can replace the conditional density $\mathscr{P}(\mathbf{Z}|\mathbf{p})$ by $\mathscr{P}_{\mathbf{n}}(Z_1 - g_1, Z_2 - g_2, \ldots, Z_M - g_M|\mathbf{p})$, where $g_1 = g(\mathbf{p}, f_1), \ldots, g_M = g(\mathbf{p}, f_M)$ in (12.2). Because the noise in the measurements is independent of the transmitted message, $\mathbf{p}$, we can ignore the conditioning variable, and replace (12.14) by

$$\mathscr{P}(\mathbf{p}|\mathbf{Z}) \propto \mathscr{P}_{\mathbf{p}}(\mathbf{p})\mathscr{P}_{\mathbf{n}}(\mathbf{Z} - \mathbf{g}(\mathbf{p})) \, . \tag{12.15}$$

If we have no prior knowledge of $\mathbf{p}$, or if all transmitted messages are a priori equally likely, then we can ignore $\mathscr{P}_{\mathbf{p}}(\mathbf{p})$ in (12.15) and work with the "likelihood function," $\mathscr{P}_{\mathbf{N}}(\mathbf{Z} - \mathbf{g}(\mathbf{p}))$. Maximizing the likelihood function over $\mathbf{p}$ is called "maximum likelihood estimation." We usually have some prior knowledge of $\mathbf{p}$, however, so we incorporate that knowledge in the a priori function, $\mathscr{P}_{\mathbf{p}}(\mathbf{p})$.

If **p** and **n** are jointly Gaussian processes, then we have the classical problem of communicating a Gaussian signal in Gaussian noise, which reduces to a classical least-squares problem. Typically, we work with the negative logarithm of the a posteriori density in (12.15), so we need to minimize

$$-\ln \mathscr{P}(\mathbf{p}|\mathbf{Z}) = -\ln \mathscr{P}_{\mathbf{p}}(\mathbf{p}) - \ln \mathscr{P}_{\mathbf{n}}(\mathbf{Z}-\mathbf{g}(\mathbf{p})) + f(\mathbf{Z}) . \qquad (12.16)$$

Now, we let the M components of the noise vector, **n**, be statistically independent, zero-mean, Gaussian random variables, each with variance, σ_n^2, and the N components of **p** be statistically independent Gaussian random variables, with mean values $\overline{\mathbf{p}}$, and each with variance, σ_p^2. Then, upon taking the negative logarithm of the appropriate density functions, and discarding unimportant factors, we replace (12.16) with the objective function

$$\begin{aligned}\Phi(\mathbf{p}|\mathbf{Z}) &= \frac{1}{2\sigma_n^2}\sum_{i=1}^{M}(Z_i - g_i(\mathbf{p}))^2 + \frac{1}{2\sigma_p^2}\sum_{i=1}^{N}(p_i - \overline{p}_i)^2 \\ &= \frac{|\mathbf{Z}-\mathbf{g}(\mathbf{p})|^2}{2\sigma_n^2} + \frac{|\mathbf{p}-\overline{\mathbf{p}}|^2}{2\sigma_p^2} , \qquad (12.17)\end{aligned}$$

which is to be minimized over **p**. Multiply (12.17) by $2\sigma_n^2$ to get the final expression for the objective function

$$\Phi(\mathbf{p}|\mathbf{Z}) = |\mathbf{Z}-\mathbf{g}(\mathbf{p})|^2 + \frac{\sigma_n^2}{\sigma_p^2}|\mathbf{p}-\overline{\mathbf{p}}|^2 , \qquad (12.18)$$

where we use the same notation for the objective function.

The ratio, σ_p^2/σ_n^2, of the variances is called the signal-to-noise ratio and is usually known. In the context of least-squares problems, this ratio is known as the Levenberg–Marquardt parameter, and in mathematical inverse theory it is called the Tichonov–Miller parameter.

Setting the value of the LM-parameter determines the certainty with which we choose to assert the a priori constraint on **p**. If σ_p is very small, then the LM-parameter is large, and we are more certain to impose the constraint. Unless we know the variances, we cannot determine the LM-parameter at the outset. There are numerical techniques, such as *ridge regression* [53] and *cross-validation* [62], that can be useful in selecting the LM-parameter.

Appendix

A.1 Cramer–Rao Lower Bound

The Cramer–Rao lower bound (CRLB), being the lower bound on the variance of any unbiased estimator, plays a role in statistical estimation theory [64, 65] that is similar to the sensitivity (upper) bound, σ, of (12.11). As with σ, the CRLB finds application in electromagnetic scattering and inverse problems [66].

A.1.1 Inverse Method Quality Metrics

Given the potential of inverse methods, it is important to develop a rigorous method for quantifying the performance and reliability of inversion schemes [105, 107]. Although empirical studies provide the means for evaluating the quality of NDE techniques incorporating inverse methods, opportunities also exist with inverse methods to use the model calculations with quantitative measures to evaluate key estimation performance metrics without considerable experimental burden.

In estimation theory, the Cramer–Rao Lower Bound (CRLB) provides the minimum variance that can be expected for an unbiased estimator of a set of unknown parameters. In other words, the CRLB provides a way of quantifying the inversion algorithm performance. For Gaussian noise, there is a simple inverse relationship between the CRLB and the Fisher information [65]:

$$\operatorname{var}\left(\hat{\theta}_i\right) = \left[C_{\hat{\theta}}\right]_{ii} \geq \left[I^{-1}(\theta\right]_{ii} , \tag{12.19}$$

where **C** is the covariance matrix, the Fisher information is defined as

$$I(\theta)_{ii} = -E\left[\frac{\partial^2 \ln f(Z;\theta)}{\partial \theta_i \partial \theta_j}\right] , \tag{12.20}$$

θ is the parameter being estimated, and Z is the measurement vector. Fisher information represents the amount of information contained in a measurement and depends on the derivatives of the likelihood function which is based on the forward model and the noise parameters. The variance in a measurement is inversely related to the amount of information contained in the measurement, so it is not a surprise that (12.19) shows that the variance in the measurement is greater than or equal to the inverse of the Fisher information matrix. In eddy-current NDE, the measurement is often the real and imaginary component of the impedance, $Z = [R,X]$, and the Fisher information becomes a square matrix with dimensions equal to the number of parameters being estimated.

The covariance matrix can be evaluated as a performance metric for inverse methods. First, the diagonal terms of the covariance matrix (the CRLB variances) provide a metric of sensitivity of a parameter estimated using inverse methods to measurement variation. Second, the off-diagonal terms represent the interdependence between select parameters being estimated to measurement variation. The corresponding metric is the correlation coefficient given by

$$\rho_{i,j} = \frac{C_{ij}}{\sqrt{C_{ii}C_{jj}}} . \tag{12.21}$$

These metrics can be used with parametric studies involving frequency or other probe parameters to optimize the NDE system design. As a general design rule for inverse methods, it is desirable to minimize the sensitivity to variation (the CRLB variances) and to have the correlation coefficient between the parameters being estimated approach zero.

Another tool used in numerical linear algebra for sensitivity analysis is singular value decomposition (SVD). SVD essentially provides a measure of sensitivity of measurements to perturbations in the unknown parameters [106]. To evaluate the sensitivity of an inverse problem for a set of measurements to changes in fit parameters, SVD can be applied to the Jacobian matrix where

$$\mathbf{J} = \begin{bmatrix} \frac{\partial Z_1}{\partial \theta_1} & \cdots & \frac{\partial Z_1}{\partial \theta_n} \\ \vdots & \ddots & \vdots \\ \frac{\partial Z_m}{\partial \theta_1} & \cdots & \frac{\partial Z_m}{\partial \theta_n} \end{bmatrix} = \mathbf{U\Sigma V}' \,. \tag{12.22}$$

The condition number (CN) of the matrix is defined as the ratio of the largest and smallest singular values resulting from SVD. For inversion, CN has been used to quantify the well-posedness of the inverse problem for select parameters. The ability to estimate parameters independently increases as the condition number approaches unity. It should be noted that SVD does not incorporate noise; it depends only on the noiseless relationship between the measurement output and the parameter changes.

A.1.2 Optimizing Layer Estimation Using Metrics

An inversion experiment is revisited [55] for the purpose of demonstrating estimation theory metrics [107]. In this experiment, the thickness of an AISI-304 stainless steel plate and probe lift-off were estimated. A thickness and lift-off model, similar to the one shown in Chap. 11, was used to solve the forward problem. The estimation procedure is represented in (12.23). The left side is the measured impedance, the Jacobian is simply the derivative information from the forward model, and the thickness and lift-off parameters are updated until this equation converges,

$$\begin{bmatrix} R(f,t,l) \\ X(f,t,l) \end{bmatrix} \approx \begin{bmatrix} R(f,t_0,l_0) \\ X(f,t_0,l_0) \end{bmatrix} + \begin{bmatrix} \frac{\partial R}{\partial t} & \frac{\partial R}{\partial l} \\ \frac{\partial X}{\partial t} & \frac{\partial X}{\partial l} \end{bmatrix}_{t_0,l_0} \begin{bmatrix} t-t_0 \\ l-l_0 \end{bmatrix} . \tag{12.23}$$

Four scenarios in particular are investigated. Impedance values were generated for combinations of lift-off values of 0.75 and 1.5 mm and a plate thickness values of 1.0 mm and 2.0 mm with Gaussian noise of 1 % of the impedance value added as shown in Fig. 12.2a. For each of these measurements, the NLSE algorithm is applied to estimate the thickness and lift-off simultaneously. Figure 12.2b shows the inversion results in the parameter space. Note that for high lift-off, visual inspection indicates the variance in the estimation is much greater for lift-off and likewise for the thicker plate, the variance of the estimation of thickness is greater.

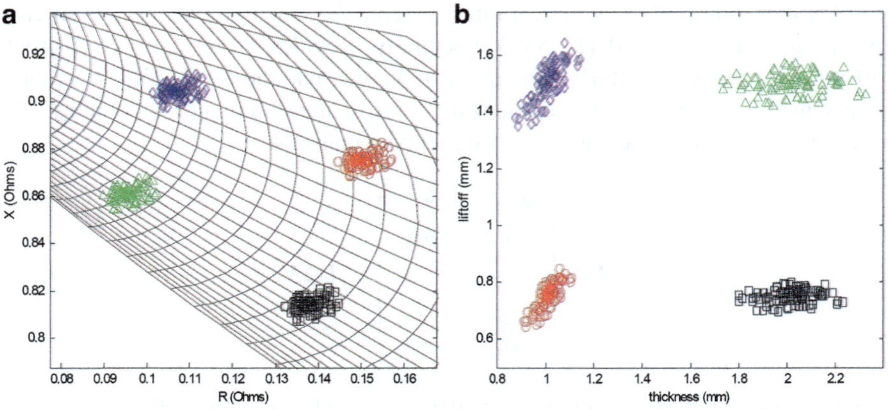

Fig. 12.2 (**a**) Distribution of source data in impedance plane, and (**b**) corresponding estimated values in lift-off-thickness parameter space

The calculations required for the CRLB involve taking numerical derivatives of the impedance changes with respect to the parameter changes from the forward model. These calculations thus require far less computational expense with respect to Monte-Carlo simulation. Following (12.20), the Fisher information for this particular case is given by:

$$\mathbf{I} = \begin{bmatrix} J_{11}^2 + J_{21}^2 & J_{12}J_{11} + J_{22}J_{21} \\ J_{11}J_{12} + J_{21}J_{22} & J_{12}^2 + J_{22}^2 \end{bmatrix} . \quad (12.24)$$

The covariance matrix is then calculated from the Fisher information (by (12.19)):

$$\mathbf{C} = \sigma^2 \mathbf{I}^{-1} . \quad (12.25)$$

The Jacobian is also decomposed into its singular values and singular vectors in the form of the right-hand side of (12.22). The ratio of the smallest to largest singular values provides the condition number.

Figure 12.3 shows the CRLB of the estimation of the thickness and lift-off of a 1 mm thick plate and 1 mm lift-off for multiple frequencies. The agreement between the CRLB and the Monte-Carlo approach is quite good. This analysis demonstrates that there is an optimal frequency to achieve highest accuracy in the estimation of thickness. Estimating conductivity and thickness simultaneously is typically more ill-conditioned than estimating thickness and lift-off simultaneously. The CRLB for conductivity and thickness estimation along with the condition number and correlation number as a function of frequency are all displayed in Fig. 12.4. The behavior of the CRLB as a function of frequency for estimating conductivity and thickness simultaneously follows a similar trend and this is expected since the impedance changes due to conductivity and thickness are similar. The condition number reaches a maximum around 95 kHz which implies that selectivity is good and the correlation is zero at this frequency which further confirms that point.

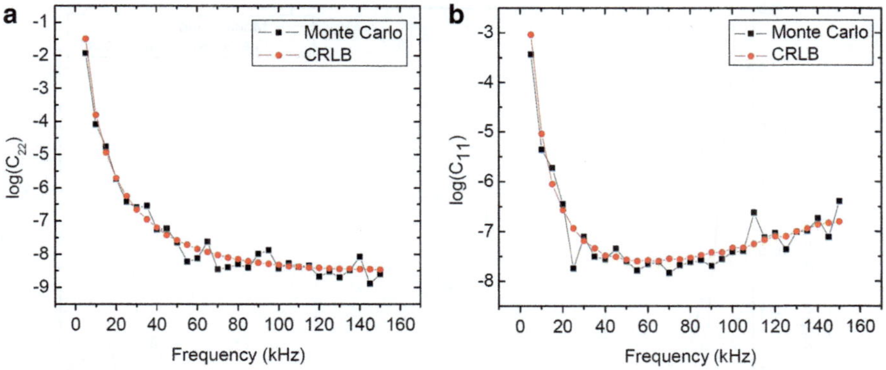

Fig. 12.3 Comparison of variance with varying frequency using CRLB and Monte Carlo methods for estimating (**a**) lift-off and (**b**) thickness, respectively

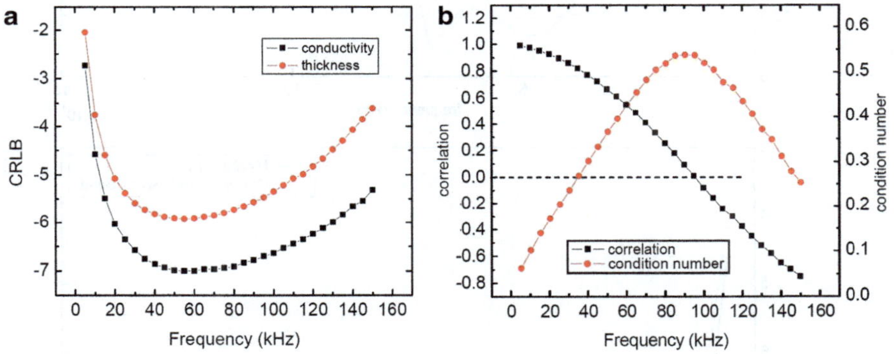

Fig. 12.4 Comparison of inversion metrics with varying frequency: (**a**) CRLB variance for thickness and conductivity estimation and (**b**) correlation and condition number

A.1.3 Two Examples from Chap. 11

Figure 11.5 of Chap. 11 shows the variation with frequency of the derivative of impedance with respect to thickness for Test Case 1 (shown in Fig. 11.4). There is a relatively broad peak in this derivative extending from roughly 20 to 50 kHz. Thus, we would expect that the optimum range of frequencies to be used for determining the thickness should lie in this range and that was essentially confirmed when frequencies in the range of 20–40 kHz were found to give good results.

Figure 12.5(top) illustrates the CRLB thickness response for this same test case, and we see that there is a relatively broad minimum centered near 50 kHz, but generally including the same frequency range chosen in executing the test case. This confirms the consistency between the CRLB result and the information contained in the first derivative.

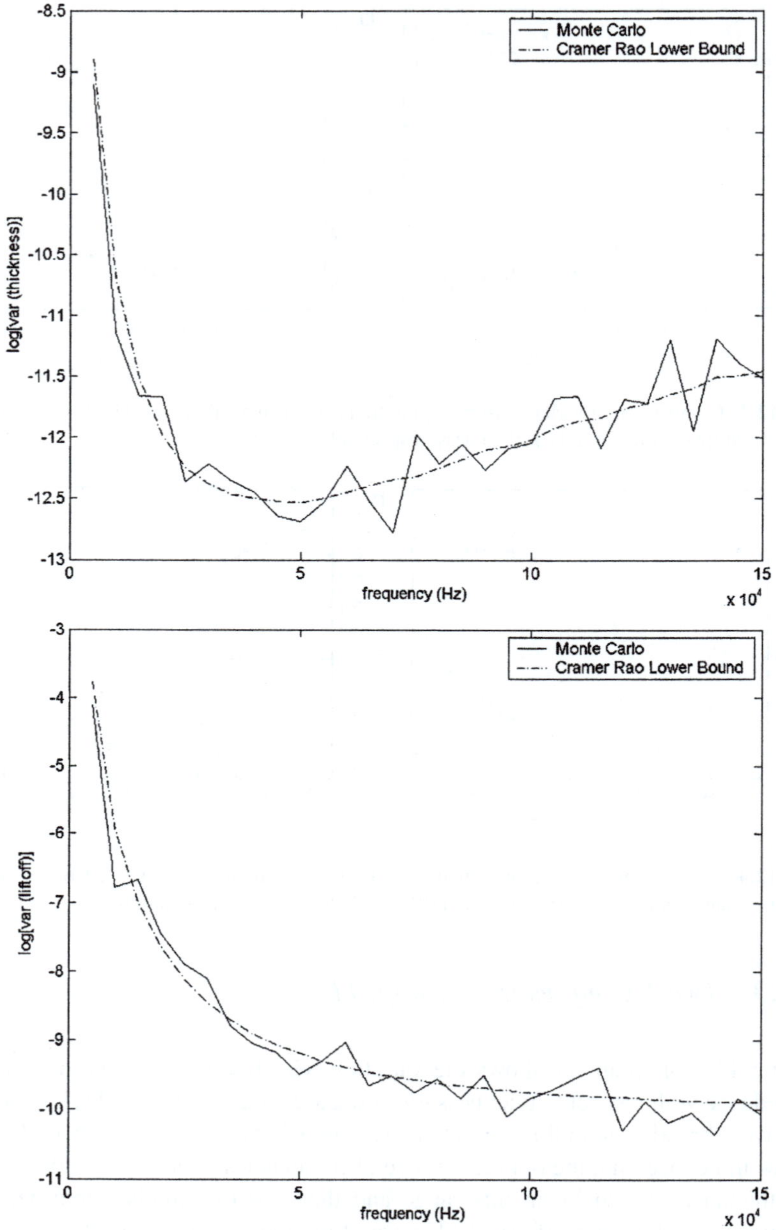

Fig. 12.5 Comparison of variance with varying frequency for Test Case 1 in Chap. 11, using CRLB and Monte Carlo methods for estimating (*top*) thickness and (*bottom*) lift-off, respectively

As for the lift-off problem in Test Case 1, we see from Fig. 11.6 that the derivative with respect to lift-off continues to increase with frequency even beyond 100 kHz, which corresponds to what we might expect for the CRLB associated with lift-off,

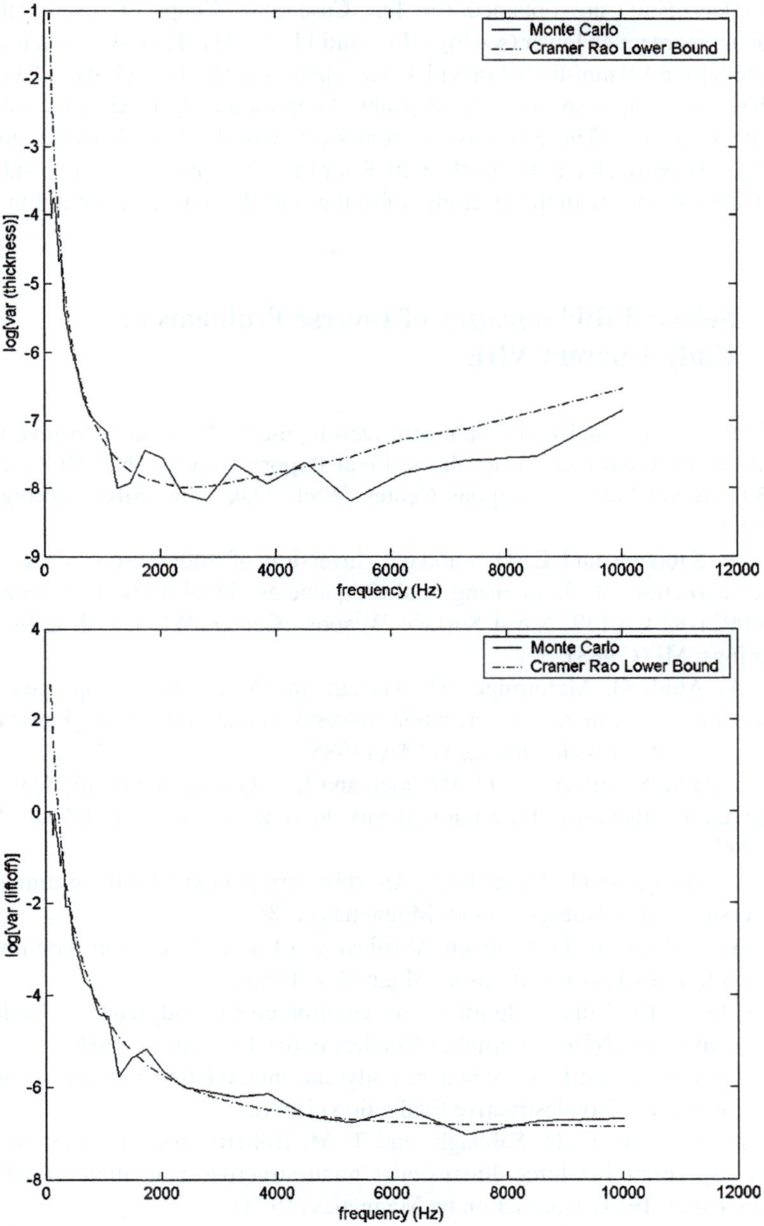

Fig. 12.6 Comparison of variance with varying frequency for Test Case 2 in Chap. 11, using CRLB and Monte Carlo methods for estimating (*top*) thickness and (*bottom*) lift-off, respectively

as shown in the bottom part of Fig. 12.5. In the latter figure, we see that the CRLB never does achieve a true minimum, even out to 150 kHz, but the decrease is leveling off at this frequency.

Similar results are obtained for Test Case 2 of Chap. 11, except that the frequencies are much lower (see Figs. 11.9 and 11.10). The derivative with respect to thickness for a 1.0 mm-thick brass plate has a maximum in the vicinity of 2 kHz, and the derivative with respect to lift-off peaks at approximately 10 kHz, for a nominal lift-off of 2.0 mm. These results are consistent with the CRLB results shown in Fig. 12.6. In particular, note that the CRLB for lift-off is virtually flat beyond 8 kHz, which corresponds with the levelling off of the peak derivative shown in Fig. 11.10.

A.2 Selected Bibliography of Inverse Problems in Eddy-Current NDE

- H. A. Sabbagh and L. D. Sabbagh, Development of a system to invert eddy-current data and reconstruct flaws, Final Report: Contract No. N60921-81-C-0302, Naval Surface Weapons Center, White Oak Labs, Silver Springs, MD (1982).
- H. A. Sabbagh and L. D. Sabbagh, Inversion of eddy-current data and the reconstruction of flaws using multifrequencies, Final Report: Contract No. N60921-82-C-0139, Naval Surface Weapons Center, White Oak Labs, Silver Springs, MD (1983).
- B. A. Auld, G. Mcfetridge, M. Riaziat, and S. Jefferies, Improved probe-flaw interaction modeling, inversion processing, and surface roughness clutter, Review of Progress in QNDE, Vol 4A (1985).
- B. A. Auld, S. Jefferies, J. C. Moulder, and J. C. Gerlitz, Semi-elliptical surface flaw EC interaction and inversion: theory, Review of Progress in QNDE, Vol 5A (1986).
- H. A. Sabbagh and L. D. Sabbagh, An eddy-current model for three-dimensional inversion, IEEE Transactions on Magnetics (1986).
- H. A. Sabbagh, L. D. Sabbagh, Verification of an eddy-current flaw inversion algorithm, IEEE Transactions on Magnetics (1986).
- L. Udpa and S. Udpa, Solution of inverse problems in eddy-current nondestructive evaluation (NDE), Journal of Nondestructive Evaluation (1988).
- L. D. Sabbagh, and H. A. Sabbagh, Eddy-current modeling and flaw reconstruction, Journal of Nondestructive Evaluation (1988).
- H. A. Sabbagh, L. D. Sabbagh, and T. M. Roberts, An eddy-current model and algorithm for three-dimensional nondestructive evaluation of advanced composites, IEEE Transaction on Magnetics (1988).
- S. M. Nair and J. H. Rose, Reconstruction of three-dimensional conductivity variations from eddy current (electromagnetic induction) data, Inverse Problems (1990).
- J. C. Moulder, E. Uzal, and J. H. Rose, Thickness and conductivity of metallic layers from eddy current measurements, Rev. Sci. Instrum (1992).

- S. J. Norton and J. R. Bowler, Theory of eddy current inversion, J. Appl. Phys (1993).
- O. Baltzersen, Model-based inversion of plate thickness and lift-off from eddy current probe coil measurments, Materials Evaluation (1993).
- N. J. Goldfine, Magnetometers for improved materials characterization in aerospace applications, Materials evaluation (1993).
- H. A. Sabbagh and R. G. Lautzenheiser, Inverse problems in electromagnetic nondestructive evaluation, International Journal of Applied Electromagnetics in Materials (1993).
- J. R. Bowler, S. Norton, D. J. Harrison, Eddy-current interaction with an ideal crack. II. The inverse problem, J. Appl. Phys. (1994).
- S. K. Burke, Eddy-current inversion in the thin-skin limit: Determination of depth and opening for a long crack, J. Appl. Phys. (1994).
- C. Tai, J. H. Rose, and J. C. Moulder, Thickness and conductivity of metallic layers from pulsed eddy-current measurments, Rev. Sci. Instrum. (1996).
- L. Udpa and S. Udpa, Application of signal processing and pattern recognition techniques to inverse problems in NDE, Int J. of Applied Electromagnetics and Mechanics (1997).
- Z. Chen and K. Miya, ECT inversion using a knowledge-based forward solver, Journal of Nondestructive Evaluation (1998).
- H. T. Banks et al, Nondestructive evaluation using a reduced-order computational methodology, Inverse Problems (2000).
- H. T. Banks et al, Real time computational algorithms for eddy-current-based damage detection, Inverse Problems (2002).
- W. Yin, S. J. Dickinson, and A. J. Peyton, Imaging the continuous conductivity profile within layered metal structures using inductance spectroscopy, IEEE Sensors Journal (2005).
- W. Yin and A. J. Peyton, Thickness measurement of non-magnetic plates using multi-frequency eddy current sensors, NDT&E International (2007).

Chapter 13
Robust Statistical Estimators

13.1 Introduction

"Robust" estimators are resistant to outliers in data, contrary to the usual classical least-squares estimator such as NLSE. We will describe two robust estimators in this chapter and give an example of the application of them to pre-processing some experimental data. Both of the robust estimators are taken from [63], and we closely follow those authors in our presentation.

13.2 Robust Estimators

The mathematical regression problem that we are addressing is based on the linear model given by $y_i = \theta x_i + e_i$ for $i = 1, \ldots, N$. y_i are the data points corresponding to the dependent variable, x_i the data points corresponding to the independent variable, e_i the error term, and θ the parameter to be determined.

Central to the notion of robust estimation is the concept of the "breakdown point," which we define following [63]: Take any sample X of N data points (x_i, y_i), and any estimator T of the parameter θ. Let $\beta(M,T,X)$ be the supremum of $\|T(X') - T(X)\|$ for all corrupted samples X', where any M of the original points of X are replaced by arbitrary values. Then the breakdown point of T at X is defined as

$$\varepsilon_N^*(T,X) = \min\left[\frac{M}{N}; \beta(M,T,X) \text{ is infinite}\right]. \tag{13.1}$$

In words, ε_N^* is the smallest fraction of contaminated data that can cause the estimator to take on values arbitrarily far from $T(X)$. For least-squares, only one bad observation is needed to cause breakdown, so that $\varepsilon_N^*(T,X) = 1/N$, which tends to 0 % when the sample size N becomes large.

H.A. Sabbagh et al., *Computational Electromagnetics and Model-Based Inversion*, Scientific Computation, DOI 10.1007/978-1-4419-8429-6_13,

The best breakdown point that we can expect is 50 %; otherwise it becomes impossible to distinguish the "good" data from the "bad." During Phase I we studied two estimators that possess a 50 % breakdown point: the least median of squares (LMS) and a scale ("S-estimator") estimator [63].

13.3 Least Median of Squares Estimator

Definition 13.1. *Let $(x_1, y_1), \ldots, (x_n, y_n)$ be a sample of regression data. For each slope, θ, we obtain residuals $r_i(\theta) = y_i - \theta x_i$. Then the LMS-estimator $\hat{\theta}$ that is defined by*

$$\min_{\theta} \operatorname{median}(r_i^2) , \tag{13.2}$$

possesses a 50 % breakdown point, is affine equivariant (that is, it is independent of the choice of the coordinate axes of the x_i) and has an asymptotic convergence rate of $n^{-1/3}$.

We will define the asymptotic convergence rate shortly.

The LMS estimator is easy to code, and runs quickly. Nevertheless, we investigated the following scale-estimator that has a 50 % breakdown point, is affine equivariant, and has a better asymptotic convergence rate of $n^{-1/2}$.

13.4 Scale (S) Estimator

Our one-dimensional estimator of scale ("S-estimator") is defined by a function ρ satisfying [63]:

(R1) ρ is symmetric, continuously differentiable and $\rho(0) = 0$.
(R2) there exists $c > 0$ such that ρ is strictly increasing on $[0, c]$ and is constant on $[c, \infty)$.

Definition 13.2. *For any sample $\{r_1, \ldots, r_n\}$ of real numbers, we define the scale estimate $s(r_1, \ldots, r_n)$ as the solution of*

$$\frac{1}{n} \sum_{i=1}^{n} \rho(r_i/s) = K , \tag{13.3}$$

where K is taken to be $E_{\Phi}[\rho]$, and Φ is the standard normal distribution. If there is more than one solution to (13.3), then we put $s(r_1, \ldots, r_n)$ equal to the supremum of the set of solutions; if there is no solution to (13.3), then we put $s(r_1, \ldots, r_n) = 0$.

Definition 13.3. *Let $(x_1, y_1), \ldots, (x_n, y_n)$ be a sample of regression data. For each slope, θ, we obtain residuals $r_i(\theta) = y_i - \theta x_i$, from which we can calculate the dispersion $s(r_i(\theta), \ldots, r_n(\theta))$ by (13.3), where ρ satisfies (R1) and (R2). Then the S-estimator $\hat{\theta}$ is defined by*

$$\min_{\theta} s(r_1(\theta), \ldots, r_n(\theta)) \,, \tag{13.4}$$

and the final scale parameter is

$$\hat{\sigma} = s(r_1(\hat{\theta}), \ldots, r_n(\hat{\theta})). \tag{13.5}$$

An example of a ρ-function for (13.3) is [63]

$$\begin{aligned} \rho(x) &= \frac{x^2}{2} - \frac{x^4}{2c^2} + \frac{x^6}{6c^4} \quad \text{for } |x| \le c \\ &= \frac{c^2}{6} \qquad\qquad\qquad\; \text{for } |x| \ge c. \end{aligned} \tag{13.6}$$

In order to achieve a 50 % breakdown point using the ρ-function of (13.6), we require that

$$\frac{E_{\Phi}[\rho]}{\rho(c)} = \frac{1}{2} \,, \tag{13.7}$$

which can be achieved by letting $c = 1.547$. In this case, $K = c^2/12 = 0.1994341$, in (13.3).

The function defined in (13.6) should be compared with Huber's classic function

$$\begin{aligned} \rho_H(x) &= \frac{x^2}{2} \quad \text{for } |x| \le k \\ &= |x| \quad \text{for } |x| \ge k. \end{aligned} \tag{13.8}$$

The Huber function produces an L_2 norm (least-squares) for $x \le k$, and an L_1 norm otherwise. Though the Huber function gives rise to an estimator that is robust against outliers in y, it still possesses a breakdown point of 0 % in the presence of outliers in x, which are called "leverage points," and have a significant effect on the fit [63].

The following is an example of the resistance of this S-estimator to "bad data": suppose we have 100 data points, (x_i, y_i), and 51 of them lie exactly on a straight line. Then the first 51 residuals vanish in (13.3), which means that $\rho(r_i/s) = 0$, $i = 1, \ldots, 51$. Hence, the largest that the left-hand side of (13.3) can be $(49/100)(c^2/6)$, which is less than K. Hence, there is no solution to (13.3), and we set $s = 0$. This means that the estimator chooses the slope of the regression line to be equal to the slope of the "exact data line." The estimator is impervious to the remaining 49 points.

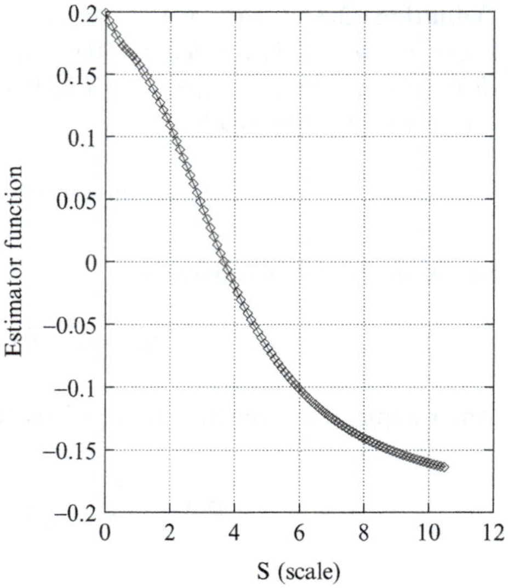

Fig. 13.1 The function (13.3) produced by a typical data set from an inversion problem

We sketch, in Fig. 13.1, the estimator function, $\frac{1}{n}\sum_{i=1}^{n}\rho(r_i/s) - K$, for a typical data set.

Finally, let us say something about the asymptotic behavior of S-estimators at the central Gaussian model, where (x_i, y_i) are independent, identically distributed random variables satisfying

$$y_i = x_i\theta_0 + e_i. \tag{13.9}$$

x_i follows some distribution H, and e_i is independent of x_i and distributed according to $\Phi(e/\sigma_0)$, for some $\sigma_0 > 0$, where Φ is the standard normal cumulative distribution function. Then Rousseeuw and Yohai [63] prove the following two theorems:

Theorem 13.1 (Consistency). *Let ρ be a function satisfying (R1) and (R2), with the derivative $\rho' = \psi$. Assume that*

(i) $\psi(u)/u$ *is nonincreasing for* $u > 0$*;*
(ii) $E_H[|x|] < \infty$*, and H has a density;*
(iii) Let (x_i, y_i) *be i.i.d. according to the model (13.9);*
(iv) Let $\hat{\theta}_n$ *be a solution of (13.4) for the first n points, and;*
(v) Let $\hat{\sigma}_n = s(r_1(\hat{\theta}_n), \ldots, r_n(\hat{\theta}_n))$.

Then

$$\hat{\theta}_n \to \theta_0 \text{ a.s.}$$
$$\hat{\sigma}_n \to \sigma_0 \text{ a.s.} \tag{13.10}$$

Theorem 13.2 (Asymptotic Normality). *Let $\theta_0 = 0$ and $\sigma_0 = 1$ for simplicity. If the conditions of Theorem 13.1 hold, and also*

(vi) *ψ is differentiable in all but a finite number of points, $|\psi'|$ is bounded, and $\int \psi' d\Phi > 0$;*
(vii) *$E_H[x^2] \neq 0$ and $E_H[|x|^3] \leq \infty$.*

Then

$$n^{1/2}(\hat{\theta}_n - \theta_0) \to N\left(0, E_H[x^2]^{-1} \int \psi^2 d\Phi / (\int \psi' d\Phi)^2\right)$$

$$n^{1/2}(\hat{\sigma}_n - \sigma_0) \to N\left(0, \int (\rho(y) - K)^2 d\Phi(y) / (\int y\psi(y) d\Phi(y))^2\right). \quad (13.11)$$

These theorems guarantee that, while the S-estimators do not break down easily when the data are contaminated, they continue to behave well when the data are not contaminated, that is, when they satisfy the classical assumptions [63]. The Asymptotic Normality Theorem defines the asymptotic convergence rate that was alluded to earlier.

13.5 An Application of Robust Estimation

Though our main interest in robust estimators is in their application to the voxel-based inversion algorithm that will be treated in the second volume of this series, we will here apply the LMS-estimator to the problem of determining a "baseline" for experimental data that will be the input to the actual inversion algorithm.

The problem is to determine a constant (the "baseline") from a set of data that are virtually constant, except for isolated signals due to flaws, tube supports, and removal of the probe from the end of the tube. Our interest is in the application to eddy-current NDE of steam generator tubes in nuclear power plants. The isolated signals constitute "impulsive noise" and may be considered to be the outliers of the data set. Clearly, if we averaged the data, the resulting baseline would be strongly affected by these signals, which would hardly result in a baseline. The current approach in the industry is simply to "eyeball" the data to produce the baseline, but this is hardly efficient, and is still subject to errors. What is needed is a reliable mathematical approach that will automatically produce the answer with no human intervention or preprocessing.

The baseline estimator was exercised on a set of experimental data, supplied by the Electric Power Research Institute (EPRI). The data consisted of a set of approximately 12,000 measurements, taken at regular intervals along the axis of a tube, using eddy-current instrumentation. The measurements were made at four different frequencies, and at each frequency, the "baseline" had a different value due to the way the instrumentation behaved. One of the post-processing procedures for these data required that the data at the different frequencies have the same baseline (the data at two or more different frequencies were scaled and mixed to produce a composite signal). We used the baseline estimator to "normalize" all frequencies to the same baseline. The figures below show the experimental data, before and after using the baseline estimator. The results are shown for the full scan, as well as for smaller portions of the full scan (Figs. 13.2–13.4).

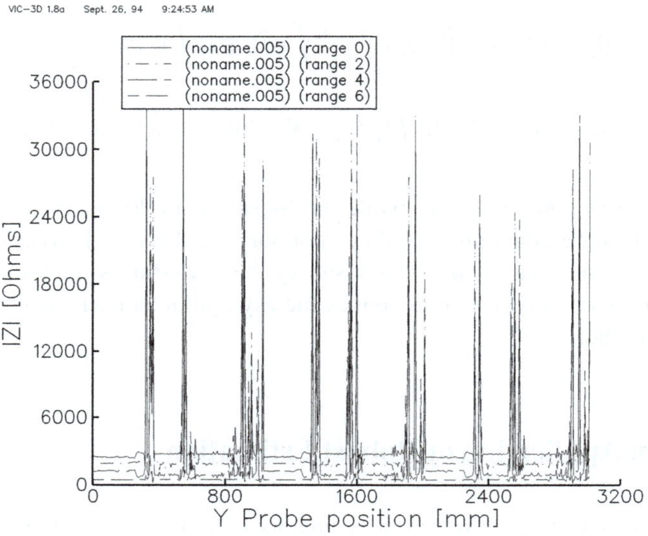

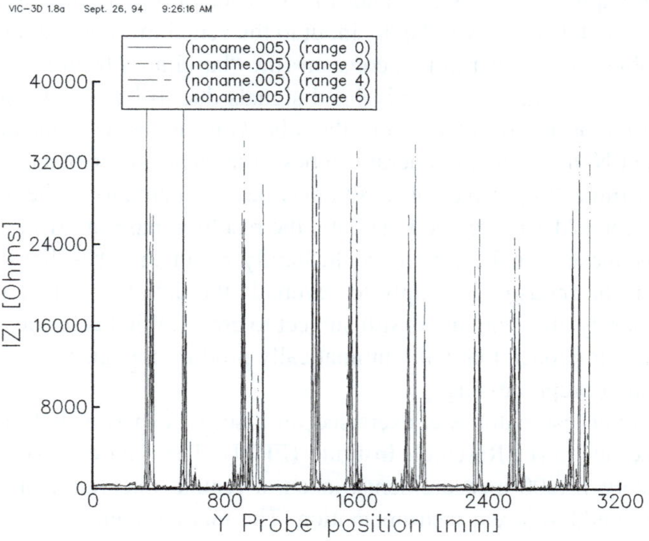

Fig. 13.2 Full scan of experimental data, at four different frequencies (approximately 12,000 points), before (*top*) and after (*bottom*) any "balancing" using the baseline estimator

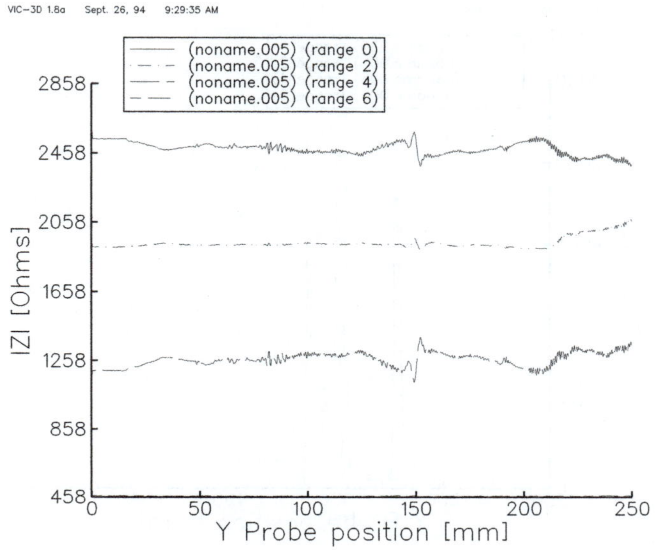

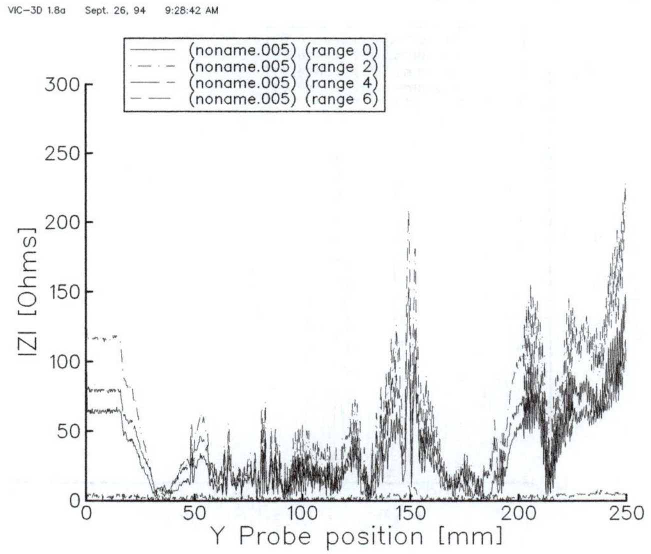

Fig. 13.3 Portion of the full scan, data points 0–1,000, of experimental data, before (*top*) and after (*bottom*) any "balancing" using the baseline estimator

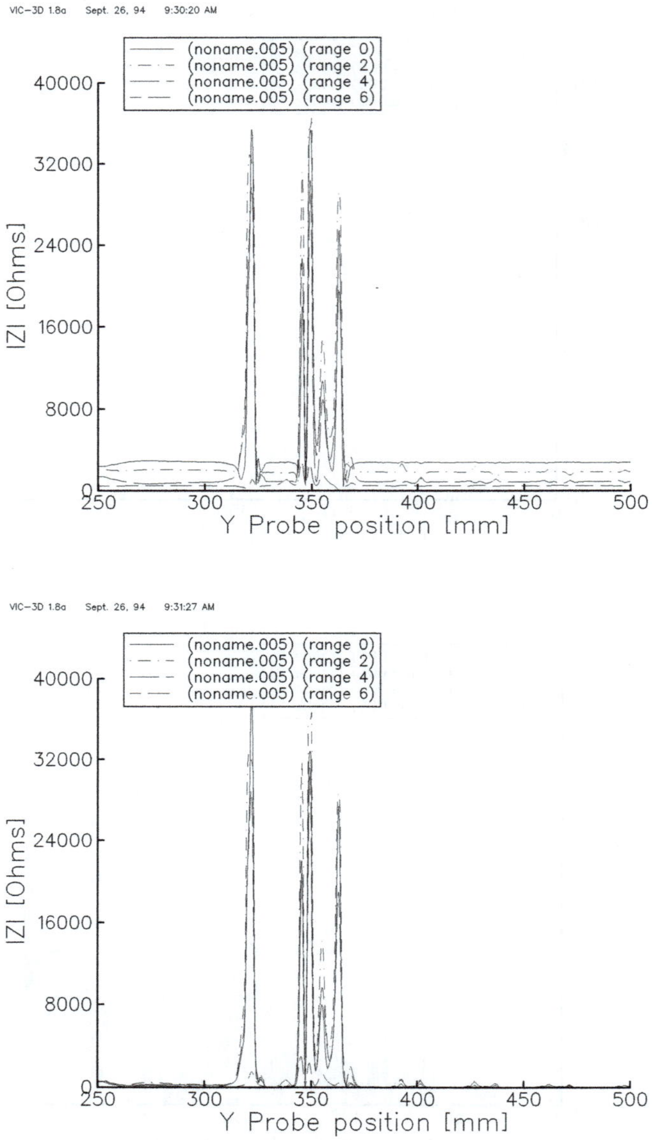

Fig. 13.4 Portion of the full scan, data points 1,000–2,000, of experimental data, before (*top*) and after (*bottom*) any "balancing" using the baseline estimator

Chapter 14
Some Special Signal-Processing Algorithms

In this chapter we sketch some rather elegant mathematical theorems that have had a significant impact on computational aspects of electrical engineering. We have used them over the years in performing eddy-current inversions and believe that they will have an expanded role to play in the future development of eddy-current NDE.

14.1 Projection Onto Convex Sets

The mathematical foundation of the theorems is the notion of projection onto convex sets (POCS) in a Hilbert space, which is proving to be quite valuable in engineering analysis and design [71–78]. We recall that a Hilbert space is simply a vector space in which an inner-product is defined.[1] Our entire development of volume-integral equations is done in a Hilbert space, though we don't emphasize that fact. A set, $\mathscr{S}$, is called convex if for every pair of vectors, $\mathbf{V}$ and $\mathbf{U}$ that lie in $\mathscr{S}$, the straight line joining these two vectors, $\mathbf{V}+\lambda(\mathbf{U}-\mathbf{V})$, $0\leq\lambda\leq 1$, also lies in $\mathscr{S}$. Hyperplanes, spheres, and cubes are simple examples of convex sets. The projection of a vector, $\mathbf{V}$, onto a convex set, $\mathscr{S}$, is that point, $\mathbf{U}$, on $\mathscr{S}$ that is closest to $\mathbf{V}$. We write $\mathbf{U}=P_{\mathscr{S}}\mathbf{V}$, where $P_{\mathscr{S}}$ is the projection operator.

Projection operators offer a computationally convenient algorithm for determining the intersections of sets, which can be translated into solving systems of linear and nonlinear equations. Consider Fig. 14.1, which shows two convex sets intersecting at the single point, $\mathbf{v}^*$. If we define the composition operator[2]

$$T=P_2P_1\ , \tag{14.1}$$

then the alternating projection scheme shown will converge to $\mathbf{v}^*$, for an arbitrary starting point, $\mathbf{v}$. Because of the tangency of the two boundaries, however, the convergence will not be geometric in the vicinity of $\mathbf{v}^*$, so we use over-relaxation

[1] The discussion in this section follows [73], which should be consulted for the details.

[2] The order of the P_i's is not important.

H.A. Sabbagh et al., *Computational Electromagnetics and Model-Based Inversion*, Scientific Computation, DOI 10.1007/978-1-4419-8429-6_14,

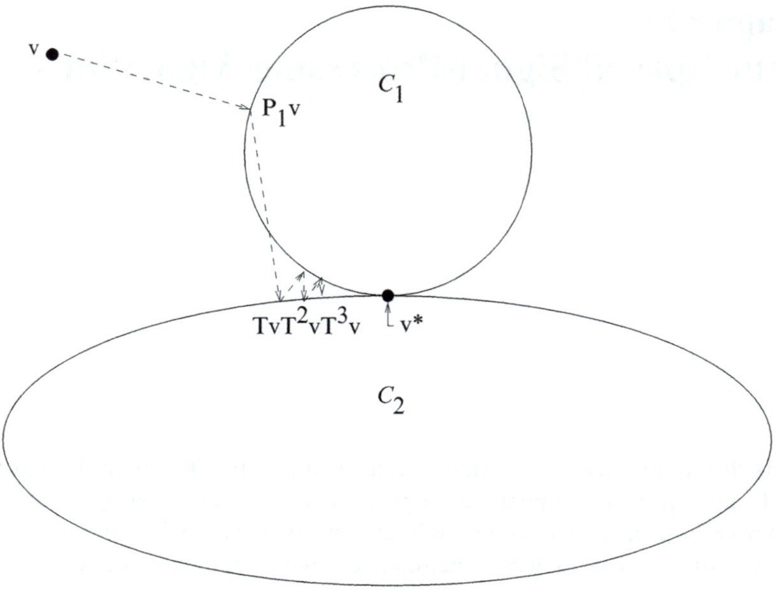

Fig. 14.1 Illustrating projections onto two convex sets, and the convergence of the iterates to $\mathbf{v}^*$, the intersection of the sets. The starting point, $\mathbf{v}$, is arbitrary

by extending the projections beyond the boundaries of the sets. This is done by selecting appropriate "relaxation" parameters, r_1, r_2 within the interval $0 < r < 2$, replacing P_1 and P_2 by

$$\begin{aligned} T_1 &= 1 + r_1(P_1 - 1) \\ T_2 &= 1 + r_2(P_2 - 1)\,, \end{aligned} \tag{14.2}$$

and then defining

$$T = T_2 T_1\,, \tag{14.3}$$

before proceeding with the scheme shown in Fig. 14.1. We will give in the next section an example of this over-relaxation procedure in the context of Kaczmarz' algorithm for solving large systems of linear equations. Following that, we will give examples of the use of POCS for signal processing and inversions with incomplete or noisy data.

14.2 Kaczmarz' Algorithm and the Algebraic Reconstruction Technique

Suppose that we want to find the minimum-norm solution of

$$||Ax - b||^2 + \lambda^2 ||x||^2\,, \tag{14.4}$$

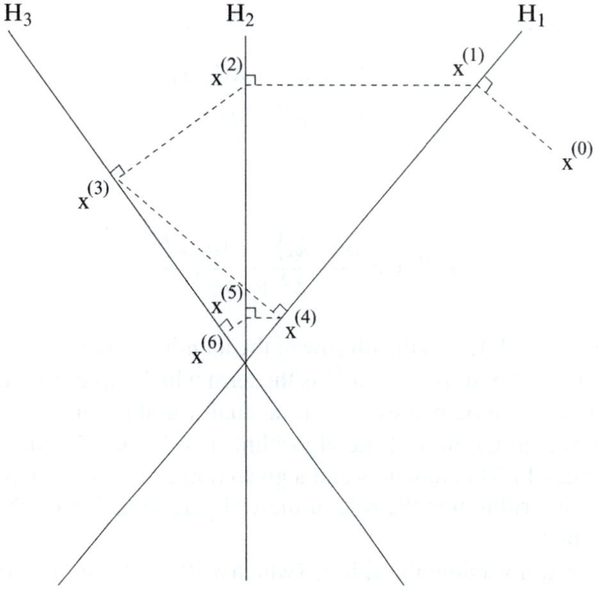

Fig. 14.2 Geometric interpretation of ART for three equations and two cycles of iterations. Each equation defines a hyperplane. At each stage, the current solution is projected onto the next hyperplane. Results after two cycles of iterations with unity relaxations are shown

where A is $m \times n$ and λ is the Levenberg–Marquardt (LM) parameter. The Kaczmarz algorithm is an iterative method of computing this solution, in which the new point in the sequence of iterates depends only on the most recent point and the next equation to be satisfied. The basic idea is to satisfy each equation, or hyperplane, by projecting the previous solution onto the next hyperplane. Figure 14.2 is an illustration of this geometric interpretation where three equations (hyperplanes) exist and the process has gone through two iterations. By an iteration we mean a complete cycle through the algorithm. In this example, $r_k = 1$, which is the case of unity relaxation. If $r_k > 1$, then we have the case of overrelaxation, and the projections "overshoot" the hyperplanes; if $r_k < 1$, then we have the case of underrelaxation, and the projections "undershoot" the hyperplanes. We note that the relaxation parameter can be changed at each iteration. This may help the convergence of this algorithm when used on certain problems.

The theoretical basis for this method is that if $x^{(k)}$ and $z^{(k)}$ are sequences generated by the following algorithm, then $x^{(k)}$ converges to the minimum of (14.4). We now state the algorithm. Initialization: $z^{(0)}$ is arbitrary, and

$$\lambda x^{(0)} = A^T z^{(0)} ,$$

step:

$$z^{(k+1)} = z^{(k)} + \lambda d^{(k)} e_i$$
$$x^{(k+1)} = x^{(k)} + d^{(k)} a_i ,$$

where

$$d^{(k)} = r_k \frac{b_i - \lambda z_i^{(k)} - (a_i, x^{(k)})}{\lambda^2 + \|a_i\|^2} ,$$

and a_i is the ith row of A, e_i is the ith row of the identity matrix I, $i = (k \bmod m) + 1$, and $0 \le \liminf r_k \le \limsup r_k \le 2$. $d^{(k)}$ is the term which represents a generalization of projections onto hyperplanes, $z^{(k)}$ is a dual variable, and r_k is a relaxation parameter. For each iteration of the algorithm, $1 \le i \le m$. The method converges for a given range of LM parameters and a given range of relaxation parameters. It is clear from the algorithm that the role of the LM parameter is to stabilize $d^{(k)}$ when $\|a_i\|^2$ is very small.

In voxel-based inversion algorithms (which will be the subject of volume 2 of this series) we acquire data via an array of sensors, each of which can be excited to serve as a transmitter as well as a receiver. This yields the system (14.5)

$$\begin{aligned} Z_1(\nu) &= \sum_{\mathrm{LMJ}} \mathbf{E}^{(1)}_{\mathrm{LMJ}}(\nu) \cdot \mathbf{J}_{\mathrm{LMJ}}(\nu) \\ Z_2(\nu) &= \sum_{\mathrm{LMJ}} \mathbf{E}^{(2)}_{\mathrm{LMJ}}(\nu) \cdot \mathbf{J}_{\mathrm{LMJ}}(\nu) \\ &\vdots \\ Z_{N_s}(\nu) &= \sum_{\mathrm{LMJ}} \mathbf{E}^{(N_s)}_{\mathrm{LMJ}}(\nu) \cdot \mathbf{J}_{\mathrm{LMJ}}(\nu) , \end{aligned} \tag{14.5}$$

where $\mathbf{E}^{(n)}_{\mathrm{LMJ}}(\nu)$ is the "incident" field produced by the sensor when it is in its nth scan position, during the νth view, $Z_n(\nu)$ is the measured impedance data, and $\mathbf{J}_{\mathrm{LMJ}}(\nu)$ is the unknown anomalous current within the flaw that is to be determined.

The minimum-norm solution of (14.5) is given by

$$\tilde{\mathbf{J}}_{\mathrm{LMJ}}(\nu) = \mathscr{M}^{\dagger}(\nu) \begin{bmatrix} Z_1(\nu) \\ \vdots \\ Z_{N_s}(\nu) \end{bmatrix} , \tag{14.6}$$

where $\mathscr{M}^{\dagger}(\nu)$ is the pseudoinverse of the matrix in (14.5). One can use the QR-decomposition of linear algebra to determine $\mathscr{M}^{\dagger}(\nu)$. This works well for fairly small problems, but for large, indeed huge, problems Kaczmarz' algorithm would be the choice [67–69].

We have used ART successfully in [70], in which it reconstructed a number of flaws in a heat-exchanger tube using a linear model consisting of 1,024 equations in 1,100 unknowns. In those numerical experiments $0.01 \le \lambda \le 0.02$, and the relaxation parameter was fixed to be 0.25 throughout the iterations, during which the equations were accessed in the order in which they appear in the given system. In [69] it is claimed that by a careful adjustment of the order in which the equations are accessed during the reconstruction procedure, and by adjusting the relaxation parameter at each iteration, ART can produce high-quality reconstructions with excellent computational efficiency.

The algorithm is robust, and the additional features of the Levenberg–Marquardt parameter and relaxation parameter give it added strength as an inversion algorithm. Its computational advantages derive from the facts that no changes are made to the original matrix, no operations are performed on the entire matrix, only one row of the matrix is required in a single step, and only $x^{(k)}$ is required for the calculation of $x^{(k+1)}$ during that step.

Kaczmarz proposed his algorithm in 1937 [79], and it was further developed by Tanabe in 1971 [80]. G.N. Hounsfield, in his 1972 patent disclosing the invention of the X-ray computed tomographic scanner [81], used this algorithm, which in this context is called the "algebraic reconstruction technique (ART)" to distinguish it from Fourier, Hilbert and Radon transform reconstruction algorithms, or from the convolution-backprojection algorithm [82].

14.3 Analytic Continuation with Constraints

A significant advantage to using POCS for solving inverse problems or performing signal-processing is that it easily accommodates the imposition of constraints on the solution, as long as those constraints can be expressed geometrically as convex sets. This often is the only way to get a unique solution to an inverse problem.

A typical problem is to reconstruct a function, such as the conductivity of a flawed region, $g(x,y) = \sigma_f(x,y)/\sigma_h - 1$, given its Fourier spectrum, $G(k_x,k_y)$, restricted to low spatial frequencies. The physical process of field-evanescence acts as a low-pass filter, thereby restricting the sensed magnetic field at high-spatial frequencies to lie below the system noise level. High-frequency components, however, are necessary for high-resolution reconstructions. Therefore, it is necessary to try to regain these missing frequencies in order to achieve the desired resolution. That this attempt at "super-resolution" is feasible and practical rests upon the notion of "analytic continuation" [83, 84]. We will state two theorems that are relevant (see [83, p. 133]):

Theorem 14.1. *The two-dimensional Fourier transform of a spatially bounded function is an analytic function in the* (k_x,k_y)*-plane.*

Theorem 14.2. *If any analytic function in the (k_x, k_y)-plane is known exactly in an arbitrarily small (nonzero) region of that plane, then the entire function can be found (uniquely) by means of analytic continuation.*

In order to use these theorems we recall that the support of $g(x,y)$ is bounded. Thus, by Theorem 14.1 $G(k_x, k_y)$ is analytic in the (k_x, k_y)-plane. If we have only limited information about G, say only its values at low spatial-frequencies, Theorem 14.2 tells us that we can uniquely extend G to the entire (k_x, k_y)-plane. Once we have continued G to the entire (k_x, k_y)-plane, we can take the inverse Fourier transform to recover g. The following algorithm is an application of POCS that performs analytic continuation under the conditions just stated:

ALGORITHM

Let the function $G(k_x, k_y)$ be given over a prescribed region $\mathscr{L}$, and let $\mathscr{F}$ be the Fourier transform. Then starting with:

$$
\begin{aligned}
f_0(x,y) &= \mathscr{F}^{-1}[G(k_x,k_y)];\\
r &= 0;\\
\text{REPEAT}&\\
f_r^{(1)} &= P_1 f_r;\\
f_r^{(2)} &= P_3 f_r^{(1)};\\
F_{r+1}^{(1)} &= \mathscr{F}[f_r^{(2)}];\\
F_{r+1} &= P_2 F_{r+1}^{(1)};\\
f_{r+1} &= \mathscr{F}^{-1}[F_{r+1}];\\
r &= r+1;\\
\text{UNTIL CONVERGENCE OCCURS.}&
\end{aligned}
$$

The important operations in the Algorithm, in addition to the Fourier and inverse Fourier transforms, are the various projection operators, P_i:

$$P_1 f = \begin{cases} f, \ (x,y) \in \mathscr{S}, \ \mathscr{S} = \text{support of } f, \\ 0, \ \text{otherwise.} \end{cases}$$

$$P_2 F = \begin{cases} G(k_x,k_y), \ (k_x,k_y) \in \mathscr{L}, \\ F(k_x,k_y), \ (k_x,k_y) \notin \mathscr{L}, \ \text{where } F(k_x,k_y) = \mathscr{F}[f(x,y)]. \end{cases}$$

$$P_3 f = \begin{cases} -1, & \text{if } f(x,y) < -1 \\ f(x,y), & \text{if } -1 \le f(x,y) \le 0 \\ 0, & \text{if } f(x,y) > 0. \end{cases} \tag{14.7}$$

From its definition, $-1 \le g(x,y) \le 0$, which is the reason for the definition of P_3. Hence, the algorithm successively applies the known properties of the function, $g(x,y)$. The analyses of [72,73,78] prove that the algorithm converges. In Volume 2 we will apply this algorithm to a problem of flaw reconstruction in graphite-epoxy advanced composites. Other applications of the algorithm can be found in [85–91].

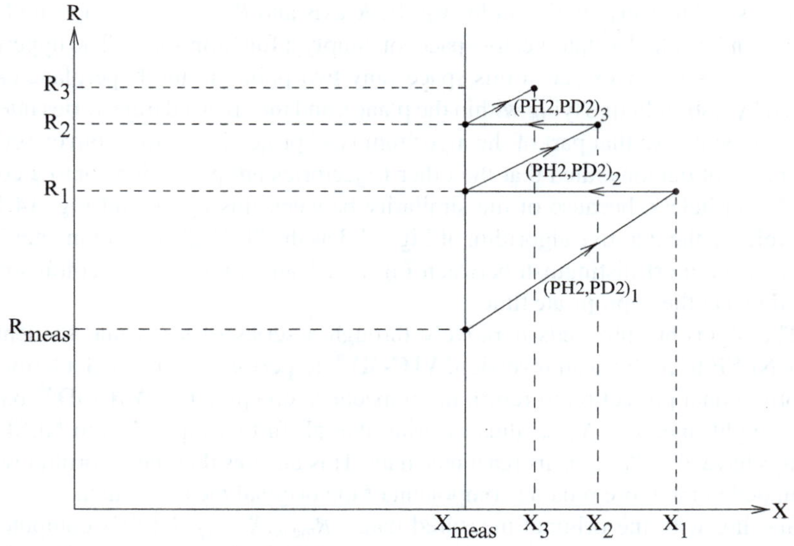

Fig. 14.3 Illustrating an iterative method, based on the idea of projections onto convex sets (POCS), for improving resistance data during the inversion process. It is likely that the points (R_1,X_1), (R_2,X_2), (R_3,X_3) lie on another convex set (or simply a locally-convex set) that approaches $X = X_{meas}$, but this has not been proved

14.4 Reconstructing Network Functions

In electric circuit theory a network function is the ratio of the Fourier or Laplace transform of a response to that of an excitation, when the network is initially relaxed. Typical network functions are driving-point and transfer impedances or admittances. The inverse problems described in this text use impedances as input data, so it is important to know that the quality of these impedance data is sufficient to allow a good reconstruction. It is abundantly clear throughout this text that measured impedance data usually have a reasonably good reactance, but the resistance is much noisier, generally because it is much smaller. Even though the resistance may be small compared to the reactance, it still plays a crucial role in determining the shape and location of a flaw. Thus, it is important to develop a means of "reconstructing" the resistance from the known reactance data in order to complete the inversion process. The POCS concept, together with NLSE, allows us to do that, as illustrated in Fig. 14.3.

The nomenclature in Fig. 14.3 refers to a corrosion topology problem that will be described in Chaps. 16 and 17. PH2 and PD2 refer, respectively, to the height and diameter of a "pillbox" model of a pit in layer 2 of a double-layer system. The input data for this problem are the impedances obtained from a 245-point scan of a probe past the pit. Each point on the X-axis, therefore, corresponds to a 245-vector of reactance data, with X_{meas} corresponding to the original measured data. A similar interpretation holds for the R-axis and R_{meas}. The composite (X,R) space can be called a state-vector space, or simply a function space. The hyperplane $X = X_{meas}$ is a convex set in this space (any two points in the hyperplane can be joined by a straight line lying within the plane), and the arrowed trajectories intersect it orthogonally, so that part of the algorithm is a "projection onto a convex set," but we have not demonstrated that the other trajectories are projections onto a convex set. Nevertheless, because of the similarity between this figure and Fig. 14.1, we will refer to the iterative algorithm of Fig. 14.3 as the "POCS" algorithm later in the book. If we must distinguish between Figs. 14.1 and 14.3 to avoid confusion, we will do so at the appropriate time.

The algorithm proceeds iteratively through a series of steps that sequentially calls NLSE to perform an inversion, **VIC-3D®** to perform a direct calculation, and an orthogonal projection to return the impedance computed by **VIC-3D®** back to the straight line, $X = X_{meas}$, thus assuring that all further input data to NLSE will always have $X = X_{meas}$ as its reactance data. This ensures that the algorithm will be grounded in the more reliable component of the original measured data.

Starting with the original measured data, (R_{meas}, X_{meas}), NLSE_0 computes the first approximation to the pit dimensions, $(\mathrm{PH2}, \mathrm{PD2})_1$, which is then fed to **VIC-3D®** to produce the first update to the impedance data, (R_1, X_1). X_1, however, is projected (orthogonally) back onto the hyperplane, $X = X_{meas}$, so that the new impedance data for NLSE_1 are (R_1, X_{meas}). The output of NLSE_1 is $(\mathrm{PH2}, \mathrm{PD2})_2$, which is then fed into **VIC-3D®** to produce (R_2, X_2), and then (R_2, X_{meas}) after projection. The process is continued until some stopping point. We have chosen the stopping point to be when the β-scale factor in $\beta \times Z_{k+1} - Z_k = 0$ is unity, or very close. This criterion states that there is not much difference (in the least-squares sense) between $Z_{k+1} = (R_{k+1}, X_{k+1})$ and $Z_k = (R_k, X_k)$, so there is no sense in running another NLSE step. A by-product of this algorithm for reconstructing the measured resistance is the reconstruction of the pit, which is what we wanted in the first place, and for that reason the algorithm will play an important role in Part III of the text.

Chapter 15
Preprocessing Data and Transformation of Signal Vectors

15.1 Clutter Modeling and Rejection

By "clutter" we mean a large background signal of known origin, such as probe lift-off, or a large background signal due to systematic errors of unknown origin, such as a variation in the material properties of the sample, or variations in the subsurface structure. In any case, clutter is usually more pernicious to the reliable detection and reconstruction of a flaw than random (electronic) noise that can be often eliminated by averaging. Hence, the ability to model clutter and reject it is paramount to the successful application of inverse methods and is the subject of the first part of this chapter.

15.1.1 A Benchmark Validation Experiment

Data were taken on an Inconel 600 plate, which is nonmagnetic (relative permeability, $\mu = 1$), and has a conductivity of $\sigma = 9.86 \times 10^5$ S/m. A notch is introduced into the "outer diameter" of the plate, which means the surface opposite to the probe. A coil is excited at 100 kHz and is scanned transversely to the slot, such that impedances are measured at 250 points, each separated by 4 mils (see Fig. 15.1). The OD of the coil is 0.112 inch, the ID is 0.034 inch, the height is 0.048 inch, and there are 131 turns in the coil. The experiment is run with a coil lift-off of 0.015 inch above the plate.

The impedance data were measured by Warren Junker of the Westinghouse Research Labs, using a Hewlett-Packard 4194A impedance analyzer, and recorded to six significant digits, which implies a dynamic range of 120 dB. The "host-only" impedance was measured to be $Z_{\text{host}} = 4.05115 + j10.8770\,\Omega$. When Z_{host} is subtracted from all 250 data points, the result is the "anomaly signal" shown in Fig. 15.2. Note the significant "clutter" in the resistance data, produced by a systematic error.

H.A. Sabbagh et al., *Computational Electromagnetics and Model-Based Inversion*, Scientific Computation, DOI 10.1007/978-1-4419-8429-6_15,

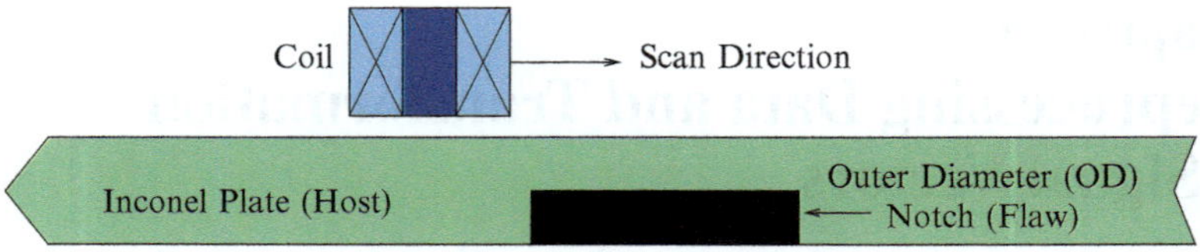

Fig. 15.1 Illustrating the setup for the experiment and model calculations

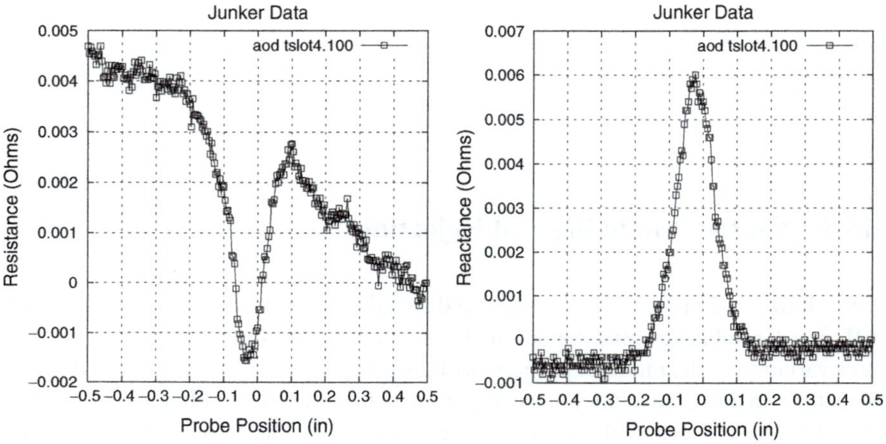

Fig. 15.2 Original data supplied by Warren Junker of the Westinghouse Research Labs. Note the significant "clutter" in the resistance data, produced by a systematic error

In order to determine the "flaw signal," we subtract the clutter from the anomaly signal. This is done by first modeling the clutter with a simple mathematical expression, in this case by the piecewise linear functions:

$$R_{\text{clutter}} = \begin{array}{ll} 0.0045 - 0.00225 \times (\text{pos} + 0.5) & \text{if pos} \leq -0.1 \\ 0.0036 + 0.0066 \times (\text{pos} + 0.1) & \text{otherwise} \end{array}$$

$$X_{\text{clutter}} = -0.0005 + 0.0005 \times (\text{pos} + 0.5), \tag{15.1}$$

where "pos" designates the probe position. When the clutter signal of (15.1) is subtracted from the data of Fig. 15.2, we get the "processed" data of Fig. 15.3. It is now clear that the original data resulted from a scan over a notch-type flaw.

We assume that the flaw is a rectangular paralleliped notch that breaks the back surface of the workpiece, i.e., the surface away from the probe. The problem then is to determine the size of the notch, namely its length, width, and height. In order to do that, we apply the processed data of Fig. 15.3 as the input to NLSE (recall Chap. 12). The three parameters are, of course, the length, width, and height of the notch. The results are: length = 0.2277 inch, width = 0.0049 inch, and height = 0.025 inch. When we repeat the inversion experiment, this time holding the height of the flaw to 0.024 inch, we get length = 0.2424, and width = 0.0050. In the first case (height = 0.025) the final norm of the residuals is 0.8112711×10^{-2}, whereas

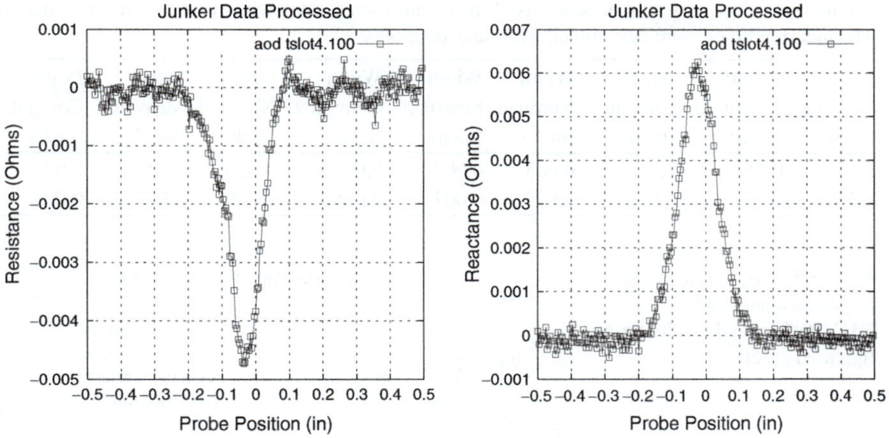

Fig. 15.3 Showing the original data with the clutter removed

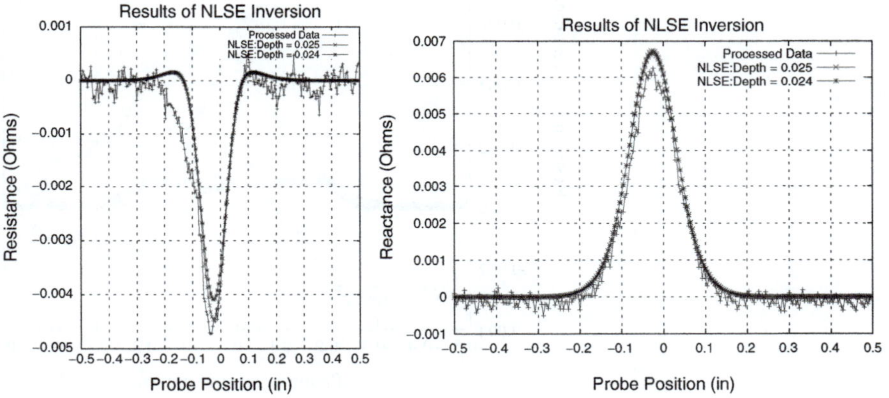

Fig. 15.4 Showing the effects of inverting the processed original data to reconstruct the flaw

with the height $= 0.024$ the norm is equal to 0.8522527×10^{-2}, which is 5 % larger. The effect of this is discussed next.

When we use **VIC-3D®** in the "forward mode" to compute the scanned impedances, given notches of the size just determined, and compare these results with the input data of Fig. 15.3, we get the results shown in Fig. 15.4. Clearly, the clutter-removal algorithm, together with **VIC-3D®**, has been effective. The resistance curve for the case of the notch with height $= 0.025$ falls closer to the negative peak of the input data than does the curve for the other notch. The reactance curves of the two notches, however, are virtually indistinguishable. This suggests that it is the better fit of the resistance data that gives the 0.025-inch notch the smaller residual-norm compared to the 0.024-inch notch.

Table 15.1 compares the nominal, measured, and computed data for these notches.

Table 15.1 Comparison of dimensions of the reconstructed A4 notches with two different depths. The measured data were determined from salastic molds

Length Nominal (inch)	Length Measured (inch)	Length Computed (inch)	Width Nominal (inch)	Width Measured (inch)	Width Computed (inch)	Depth Nominal (inch)	Depth Measured (inch)	Depth Computed (inch)
0.25	0.252	0.2277	0.005	0.007	0.0049	0.020	0.0238	0.025
0.25	0.252	0.2424	0.005	0.007	0.0050	0.020	0.0238	0.024

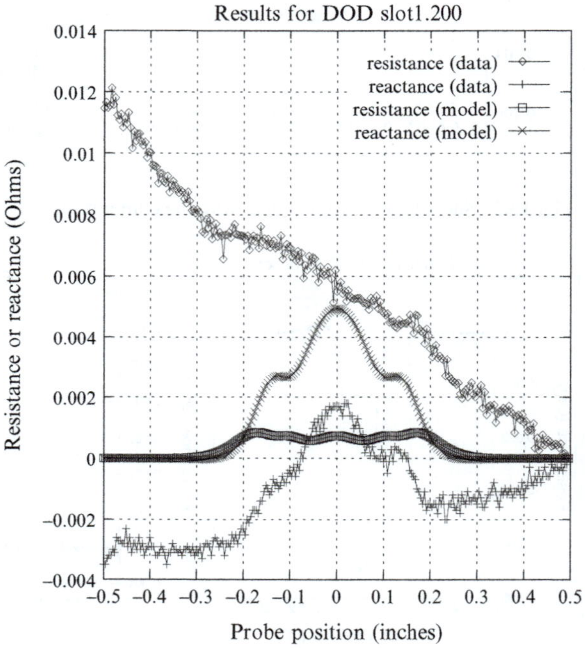

Fig. 15.5 Original measured impedance data and **VIC-3D®** model results for a ligatured crack

15.1.2 A Ligatured Outer-Diameter Slot at 200 kHz

We continue with another example from the data provided by the Westinghouse Research Labs. In this example we consider an "outer-diameter" flaw comprising 5 notches and 4 gaps, which could model a crack with periodic contact points between the surfaces, i.e., a partially closed crack.

The notches are each 0.041 inches long, 0.0065 inches wide, and 0.0138 inches deep. The plate is 0.048 inches thick, so this would be an outer-diameter flaw that extends 28.75 % into the host. The gaps (ligatures) are 0.0066 inches long. Hence, the crack is 0.2314 inches long and is 11.4 % closed.

The original measured impedance data and **VIC-3D®** model results at 200 kHz are shown in Fig. 15.5. We model the clutter signal by the piecewise linear functions:

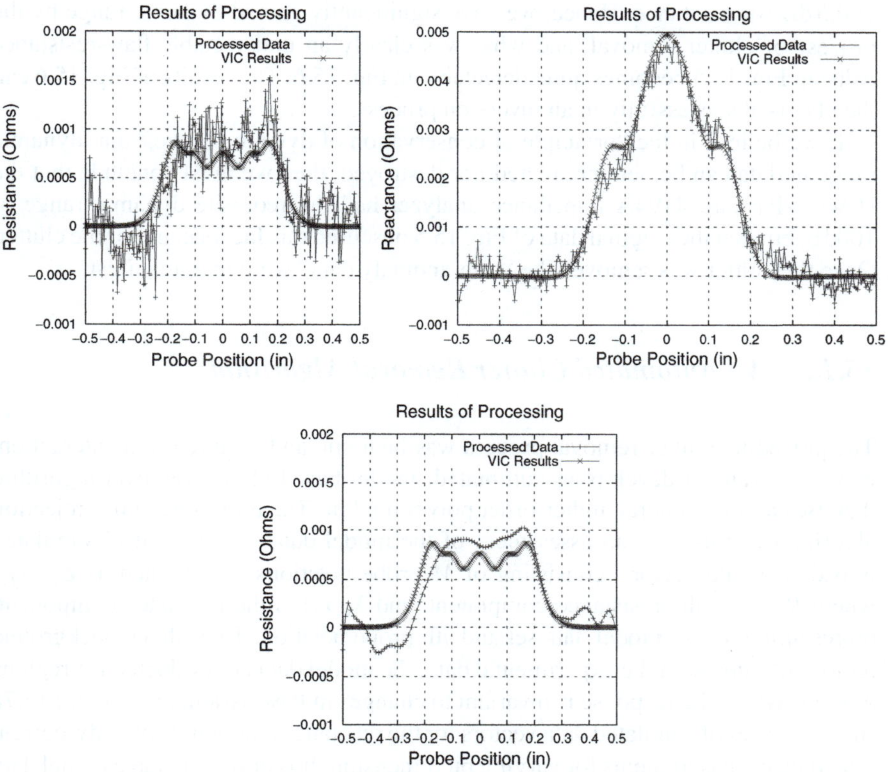

Fig. 15.6 Measured data with the clutter removed, together with the same **VIC-3D®** model results. *Top left*: resistance, *top right*: reactance. *Bottom*: Measured resistance data with the clutter removed and then passed through a Bezier filter

$$
\begin{aligned}
R_{\text{clutter}} &= 0.0117 - 0.01719 \times (\text{pos} + 0.5) && \text{if pos} \leq -0.244 \\
&= 0.0073 - 0.00986 \times (\text{pos} + 0.244) && \text{otherwise} \\
X_{\text{clutter}} &= -0.003 && \text{if pos} \leq 0.1 \\
&= -0.003 + 0.0075 \times (\text{pos} - 0.1) && \text{otherwise}
\end{aligned}
\tag{15.2}
$$

When this signal is subtracted from the measured data of Fig. 15.5, we get the results shown in the top portion of Fig. 15.6. The Bezier-filtered and processed resistance data are shown in the bottom portion of Fig. 15.6. The ripples that appear in the **VIC-3D®**-generated model results and in the original data over the peak range are a manifestation of the existence of the periodic contact points or "ligands."

At 200 kHz, the free-space impedance of the coil that was used to take these data is $4.02 + j21.91\,\Omega$, and the on-plate, host-only, impedance is $4.78 + j21.29\,\Omega$. Figure 15.6 shows that we can easily resolve a value of $\delta R = 0.0001\,\Omega$ (using the Bezier filter) and $\delta X = 0.0002\,\Omega$. This means that we have an effective dynamic range of $20\log(4.78/0.0001) = 93.6$ dB for resistance, and $20\log(21.29/0.0002) =$

100.5 dB for reactance. Hence, we have significantly gained dynamic range by the process of clutter removal, and what was clearly an undetectable flaw-resistance value in Fig. 15.5 becomes quite detectable in Fig. 15.6. The results of Fig. 15.6 can then be used successfully in an inversion process.

If we believe in the "principle of conservation of dynamic range," that dynamic range in data can be neither created nor destroyed, then we must conclude that the Hewlett-Packard 4194A impedance analyzer had the requisite dynamic range of 100dB, but that the original data of Fig. 15.5 obscured this fact because of the clutter. Once the clutter was removed the "true-anomaly data" were then manifest.

15.1.3 An Automated Clutter Removal Algorithm

The preceding clutter removal process was heuristic and required user interaction. Now, we want to develop an automated, model-based clutter removal algorithm that uses a more general higher-order polynomial fit. The proposed clutter rejection algorithm begins with an assessment of the model data. A series of N simulated impedance data vectors, consisting of M probe positions, are defined as $Z_{k,i}(x_j)$, where $R_i(x_j)$ is the resistance component, and $X_i(x_j)$ is the reactance component, representing the ith model data set and jth probe position. To evaluate background regions of interest in the experimental data, the model data are evaluated for regions of scans where the response is invariant to changes in flaw parameters. Figure 15.7a shows a series of simulated data vectors for the reactance component of eddy-current impedance measurements for varying pit dimension. It is desired to have model data with pertinent flaw parameters ranging over all expected levels in application.

A model-based variation measure, which is used to estimate regions of invariance, is given by:

$$G_k(x_j) = 1 - \frac{\sigma_k(x_j)}{\max(\sigma_k)}\,, \tag{15.3}$$

where

$$\sigma_k(x_j) = \sqrt{\frac{1}{N-1}\sum_{i=1}^{N}\left[Z_{k,i}(x_j) - \overline{Z}_k(x_j)\right]^2}\;; \tag{15.4}$$

$k = 1$ corresponds to $Z_1(x) = R(x)$, $k = 2$ corresponds to $Z_2(x) = X(x)$, and $\overline{Z}_k(x_j)$ is the mean of the appropriate impedance component. Thus, for scan regions with invariance to flaw size, G_k approaches 1, and for locations with greatest variation to flaw size, G_k approaches 0. Figure 15.7b shows an example of the model-based variation measure for the model data shown in Fig. 15.7a. For practical implementation, a threshold function, $G'(x_j)$, can be defined using a minimum variation criterion, γ, resulting in regions associated with the background (equal to 1) and flaw (equal to 0). Figure 15.7c displays the threshold function, $G'(x_j)$, defining a select scan region for background clutter fit based on a minimum variation criterion of $\gamma = 0.995$.

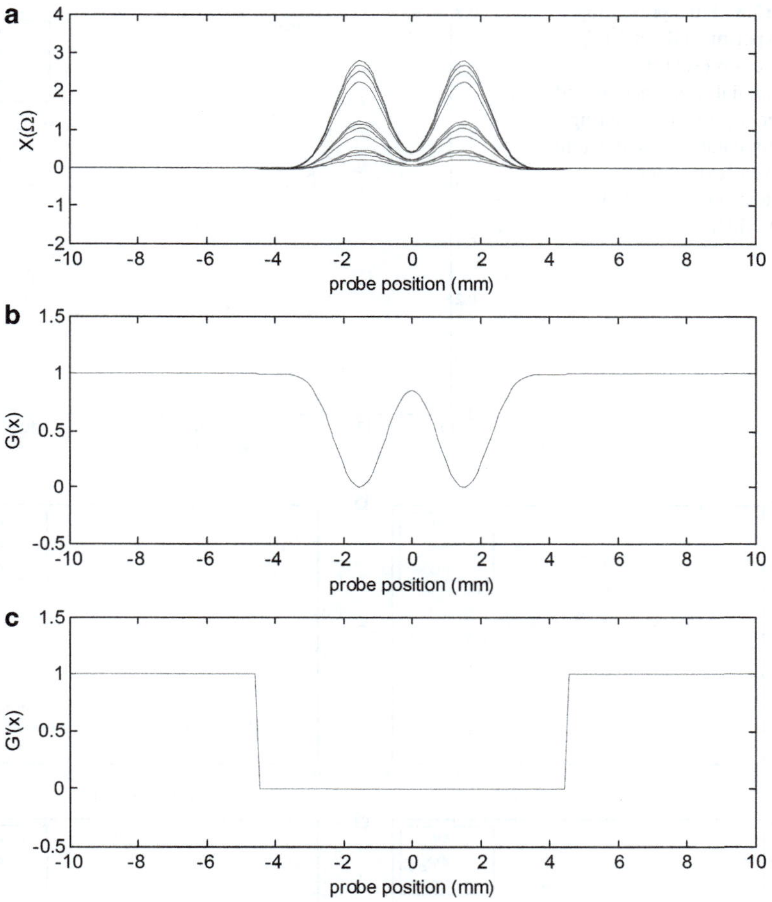

Fig. 15.7 Plots of (**a**) simulated data for the reactance component of impedance measurements for varying pit dimensions, (**b**) a model-based variation measure of the reactance components as a function of position, and (**c**) a threshold function defining a select region for background fit

Using the result of the model-based background region assessment, a fit of the background clutter can now be performed. To properly represent the background clutter found in experimental data, a 3rd-order polynomial representation,

$$f_{\mathrm{cr}}(x) = a_3x^3 + a_2x^2 + a_1x + a_0 \, , \tag{15.5}$$

is used. The coefficients of the polynomial are evaluated in a least-squares sense with the experimental data fit over the background data point regions. This function can then be applied to the experimental data to subtract the estimated error due to background clutter.

Figure 15.8 presents a plot of the resistance component at $f = 1.3\,\mathrm{kHz}$ for the transformed experimental data ($\exp_0$) for the case of a subsurface artificial pit (0.062″ diameter, 0.062″ height) on the back side of the first layer of a two-layer stack-up of aluminum panels of 0.125″ in thickness. Figure 15.8 also displays

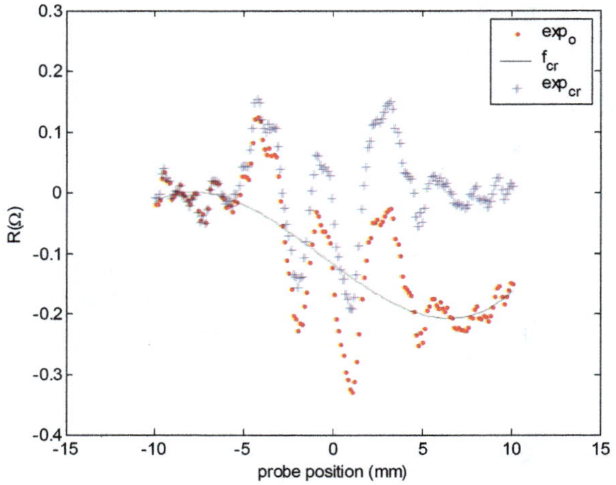

Fig. 15.8 Plots presenting the experimental data $(\exp_0)$ a curve representing a polynomial fit to background clutter, $f_{cr}(x)$, and resulting experimental data with clutter removed $(\exp_{cr})$ for resistance component at $f = 1.3\,\text{kHz}$

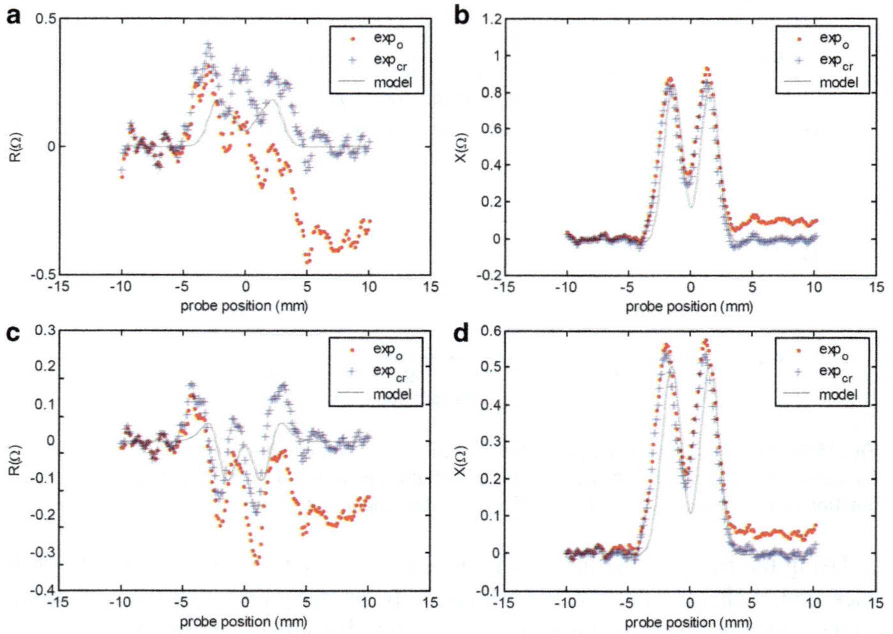

Fig. 15.9 Illustrating the results of applying the automated clutter removal algorithm to various measured data sets

the curve, $f_{cr}(x)$, representing a polynomial fit to background clutter, and the resulting experimental data with clutter removal $(\exp_{cr})$. Clearly, the application of the automated clutter removal algorithm is beneficial in eliminating the severe background variance across both the background and flaw regions. Figure 15.9 shows some results of applying the algorithm to various measured data.

15.2 Transformations of Signal Vectors

Suppose that we run two N-point scans with a probe, one at frequency f_1 and the other at frequency f_2. We arrange the output of each scan as the two complex N-vectors, Z_1 and Z_2, respectively. That is, these vectors represent the complex impedance associated with the two scans. We wish to determine a coupling coefficient, or scale factor, β, such that

$$\beta Z_1 + Z_2 = 0 . \tag{15.6}$$

If we form the dot-product of (15.6) with Z_1, we easily solve for β:

$$\beta = -\frac{Z_1 \cdot Z_2}{|Z_1|^2} , \tag{15.7}$$

where

$$Z_1 \cdot Z_2 = \sum_{n=1}^{N} Z_1^*(n) Z_2(n) , \tag{15.8}$$

and the asterisk denotes the complex-conjugate. This transformation can be used in a number of ways. The use of multifrequencies was introduced as a method of eliminating background signals due to tube-support plates in the inspection of heat-exchanger tubes in nuclear power plants. We use it extensively at a single frequency to transform instrument voltage readings into an equivalent impedance, thereby effectively transforming the instrument into an impedance analyzer.

If we wish to mix four frequencies, we first collect the impedance vectors, Z_1, Z_2, Z_3, Z_4, at the four frequencies, and then determine the three parameters, β_1, β_2, β_3, such that

$$\beta_1 Z_1 + \beta_2 Z_2 + \beta_3 Z_3 + Z_4 = 0 . \tag{15.9}$$

Upon taking dot-products of (15.9) sequentially with respect to Z_1, Z_2, and then Z_3, we form the system of three equations in the three unknowns, β_1, β_2, β_3:

$$\begin{aligned} c_{11}\beta_1 + c_{12}\beta_2 + c_{13}\beta_3 &= d_1 \\ c_{12}^*\beta_1 + c_{22}\beta_2 + c_{23}\beta_3 &= d_2 \\ c_{13}^*\beta_1 + c_{23}^*\beta_2 + c_{33}\beta_3 &= d_3 , \end{aligned} \tag{15.10}$$

where

$$\begin{aligned} c_{11} &= Z_1 \cdot Z_1 \quad c_{12} = Z_1 \cdot Z_2 \quad c_{13} = Z_1 \cdot Z_3 \\ c_{22} &= Z_2 \cdot Z_2 \quad c_{23} = Z_2 \cdot Z_3 \quad c_{33} = Z_3 \cdot Z_3 \\ d_1 &= -Z_1 \cdot Z_4 \quad d_2 = -Z_2 \cdot Z_4 \quad d_3 = -Z_3 \cdot Z_4 . \end{aligned} \tag{15.11}$$

The solution of (15.10) is

$$\begin{aligned}
\beta_1 &= \frac{d_1(c_{22}c_{33}-|c_{23}|^2)+d_2(c_{13}c_{23}^*-c_{12}c_{33})+d_3(c_{12}c_{23}-c_{13}c_{22})}{\Delta} \\
\beta_2 &= \frac{d_1(c_{13}^*c_{23}-c_{12}^*c_{33})+d_2(c_{11}c_{33}-|c_{13}|^2)+d_3(c_{12}^*c_{13}-c_{11}c_{23})}{\Delta} \\
\beta_3 &= \frac{d_1(c_{12}^*c_{23}^*-c_{13}^*c_{22})+d_2(c_{12}c_{13}^*-c_{11}c_{23}^*)+d_3(c_{11}c_{22}-|c_{12}|^2)}{\Delta}, \qquad (15.12)
\end{aligned}$$

where $\Delta = c_{11}(c_{22}c_{33}-|c_{23}|^2)+c_{12}^*(c_{13}c_{23}^*-c_{12}c_{33})+c_{13}^*(c_{12}c_{23}-c_{13}c_{22})$.

15.3 Further Developments and Applications of Scaling and Transformations

15.3.1 Modeling Differential-Bobbin Probes

The differential bobbin probe is often used to inspect heat-exchanger tubes in nuclear power plants. The probe comprises two identical bobbin coils that are coaxial and parallel to each other, as shown in Fig. 18.19. Each coil is connected to one leg of a bridge circuit, which produces the difference signal.

In modeling such a probe, we simply take the response of a single bobbin coil, and then interpolate within that response to compute the response of the second coil (assuming that the scan increment is different from the spacing of the two coils). The interpolated value is then subtracted from the primary response to model the response of the differential bobbin. In effect, we have transformed a single "true" response to create a "virtual" differential-bobbin probe response. As we will show in Chap. 18, this virtual probe yields excellent results when modeling the real probe response. This suggests, as an aside, that it may not be necessary to use complicated probes, together with analog circuitry (such as a bridge circuit), to create a desired response. It may be much easier (and cheaper) to use a simple probe and accomplish the desired response numerically.

15.3.2 A General Virtual Probe Model

We can generalize the development of a virtual probe beyond the differential-bobbin example. Suppose we are trying to invert data taken by a "large" probe, but find that the resolution is insufficient to reconstruct a "small" flaw. What is needed, we decide, is a "small" probe. It would be nice if we could use the original probe to take the data, but develop a virtual probe to transform the measured data into data that would allow the reconstruction of the small flaw. This would obviate the need to use a different piece of hardware, while still retaining the ability to satisfactorily invert the original measured data. That this concept is mathematically feasible will be demonstrated in this section.

Assume that we know the "bounding-box" of the flaw, which will allow us to establish a flaw grid, but we don't know what is in the box; i.e., we don't know what the flaw is, but only its position and maximum extent. With this assumption, we can compute the incident-field moments, $\mathbf{E}_0^{(L)}$, due to the large coil. Similarly, we can compute $\mathbf{E}_0^{(s)}$, the incident-field moments produced by the small coil, once we decide what the parameters of the small coil should be. With these results, we transform the large-coil incident-field moments into the small-coil moments via our usual least-squares β-relationship:

$$\beta \mathbf{E}_0^{(L)} = \mathbf{E}_0^{(s)} , \tag{15.13}$$

which has for its solution

$$\beta = \frac{\mathbf{E}_0^{(L)\mathrm{H}} \cdot \mathbf{E}_0^{(s)}}{||\mathbf{E}_0^{(L)}||^2} , \tag{15.14}$$

with the symbol, H, denoting the Hermitian (complex-conjugate) transpose of a vector or matrix.

If the anomalous-current response of the volume-integral equation to a flaw with $\mathbf{E}_0^{(L)}$ as the driver is $\mathbf{J}_a^{(L)}$, then the response to the same flaw with $\mathbf{E}_0^{(s)}$ as the driver is $\mathbf{J}_a^{(s)} = \beta \mathbf{J}_a^{(L)}$. This holds without knowing the nature of the flaw and is a property of the linearity of the volume-integral equation. Thus, the impedance measured by the virtual (small) probe is

$$\begin{aligned} Z_a^{(s)} &= \mathbf{J}_a^{(s)} \cdot \mathbf{E}_0^{(s)} \\ &= \beta^2 \mathbf{J}_a^{(L)} \cdot \mathbf{E}_0^{(L)} \\ &= \beta^2 Z_a^{(L)} ; \end{aligned} \tag{15.15}$$

$Z_a^{(L)}$ is the known (measured) impedance using the large probe. This relationship must be computed at each scan position, so we assume that the small probe follows the same scan as the large probe.

15.3.3 Another Virtual Probe Model

We'll apply the ABCD-matrix theory of Chap. 8 to the situation of a single probe, rather than a T/R system. A single probe is usually represented as a one-port, but in order to use ABCD-matrix theory, we must transform it into a two-port. If the driving-point impedance of the probe is Z_2, then an appropriate two-port is shown

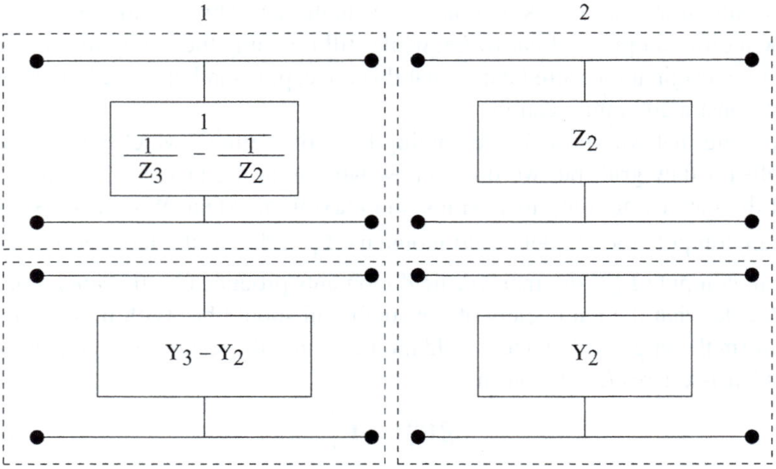

Fig. 15.10 Showing a network representation of one-ports. *Top*: Representation in terms of impedances. *Bottom*: Representation in terms of admittances

as system 2 in Fig. 15.10. The open-circuit driving-point/transfer impedance matrix of such a one-port is

$$\mathbf{Z}_2 = \begin{bmatrix} Z_2 & Z_2 \\ Z_2 & Z_2 \end{bmatrix}. \tag{15.16}$$

We want to transform this probe into one whose driving-point impedance is Z_3, and the transformation equations in terms of the ABCD matrices, M_1, M_2, and M_3 is $M_3 = M_1 \times M_2$, where

$$\mathbf{M}_2 = \begin{bmatrix} 1 & 0 \\ 1/Z_2 & 1 \end{bmatrix}$$

$$\mathbf{M}_3 = \begin{bmatrix} 1 & 0 \\ 1/Z_3 & 1 \end{bmatrix}. \tag{15.17}$$

$\mathbf{M}_1$ is, therefore, given by

$$\begin{aligned} \mathbf{M}_1 &= \mathbf{M}_2^{-1} \times \mathbf{M}_3 \\ &= \begin{bmatrix} 1 & 0 \\ -1/Z_2 & 1 \end{bmatrix} \times \begin{bmatrix} 1 & 0 \\ 1/Z_3 & 1 \end{bmatrix} \\ &= \begin{bmatrix} 1 & 0 \\ -1/Z_2 + 1/Z_3 & 1 \end{bmatrix}, \end{aligned} \tag{15.18}$$

from which we conclude that

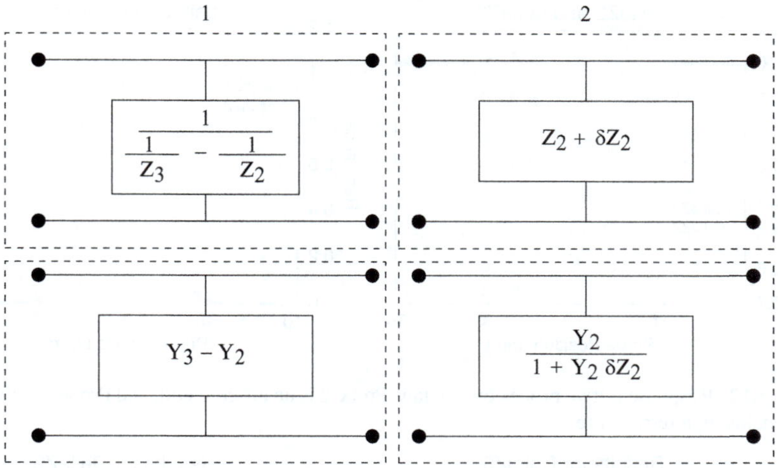

Fig. 15.11 Showing a network representation of one-ports when the virtual probe is loaded, producing a change in the driving-point impedance of δZ_2. *Top*: representation in terms of impedances. *Bottom*: representation in terms of admittances

$$\mathbf{Z}_1 = \begin{bmatrix} \dfrac{1}{1/Z_3 - 1/Z_2} & \dfrac{1}{1/Z_3 - 1/Z_2} \\ \dfrac{1}{1/Z_3 - 1/Z_2} & \dfrac{1}{1/Z_3 - 1/Z_2} \end{bmatrix} . \tag{15.19}$$

This result yields the network representation labeled "1" in the top part of Fig. 15.10.

Under load, the virtual probe has a driving-point impedance of $Z_2^L = Z_2 + \delta Z_2$, where δZ_2 is the change in impedance induced by the load. This situation is shown in Fig. 15.11. Under this condition, the "real" probe has a loaded admittance given by

$$\begin{aligned} Y_3^{(L)} &= Y_3 - Y_2 + \frac{Y_2}{1 + Y_2 \delta Z_2} \\ &= \frac{Y_3 + (Y_2 Y_3 - Y_2^2)\delta Z_2}{1 + Y_2 \delta Z_2} , \end{aligned} \tag{15.20}$$

from which we get for the change in driving-point impedance of the real probe

$$\begin{aligned} \delta Z_3 &= \frac{1 + Y_2 \delta Z_2}{Y_3 + (Y_2 Y_3 - Y_2^2)\delta Z_2} - \frac{1}{Y_3} \\ &= \frac{Y_2^2 \delta Z_2}{Y_3[Y_3 + (Y_2 Y_3 - Y_2^2)\delta Z_2]} . \end{aligned} \tag{15.21}$$

We can use this in several ways: given a measured δZ_3 due to a complex probe, perhaps with a conducting core, and a model result, $Y_2 \delta Z_2$, for the same situation that produced δZ_3, we have a free parameter, Y_3, that allows us to use (15.21) to replace the actual probe with the simpler virtual probe. Note that if we ignore the parenthetical term in the denominator of (15.21), then we have a linear relation between δZ_3 and δZ_2, with Y_2^2/Y_3^2 playing the role of β in (15.6) or (15.13).

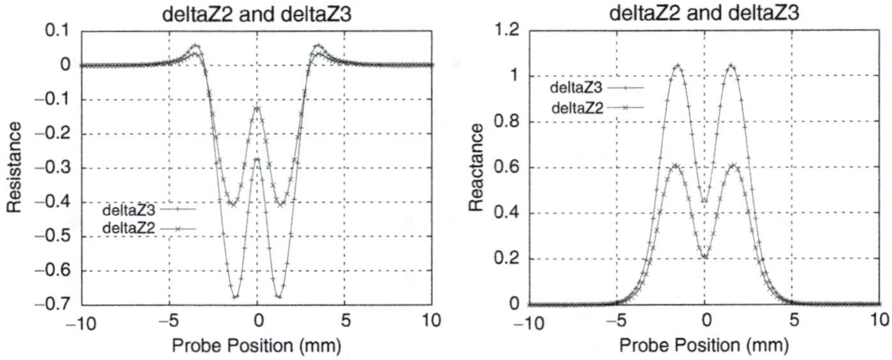

Fig. 15.12 Response of two probes over a flaw. Probe 2 is an air-core coil, and Probe 3 is the same coil enclosing a ferrite core

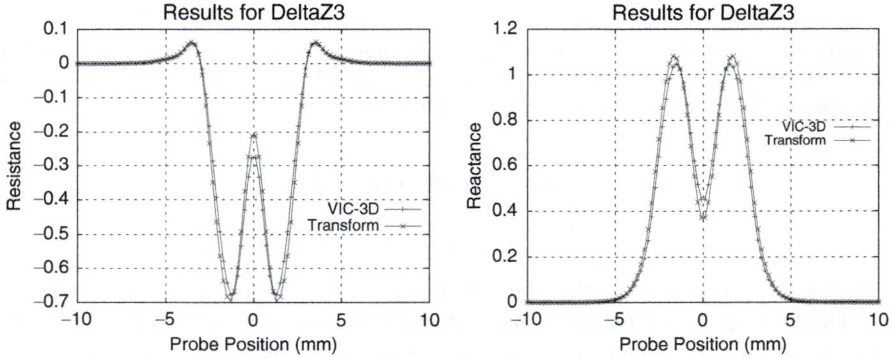

Fig. 15.13 Comparing δZ_3 when computed by **VIC-3D®** and when computed by the transformation equation (15.21)

Remember, we are treating Y_2 and Y_3 as complex scalars, whereas δZ_2 and δZ_3 are complex vectors, because they are the results of scans over a flaw. If we take Y_2 and Y_3 as conditions in freespace, and δZ_2 and δZ_3 as changes due to loading by an unflawed workpiece, then these two deltas are scalars, also.

15.3.4 An Example

Consider a small air-core coil whose impedance over an unflawed workpiece is $Z_2 = 2.860 + j43.9184\,\Omega$, and the same coil enclosing a nonconducting ferrite core with relative permeability of 2,000, whose impedance over the same workpiece at the same frequency is $Z_3 = 4.4197 + j57.982\,\Omega$. When each of these probes is scanned past a pit in the workpiece measuring $2 \times 2 \times 2$ mm, the resulting flaw responses are shown in Fig. 15.12. Figure 15.13 compares the results of the **VIC-3D®** ferrite-core calculation of Fig. 15.12 with the application of the transformation equation, (15.21), to the air-core calculation shown in Fig. 15.12.

Given the values of Y_3 and Y_2 in this example, together with the fact that $|\delta Z_2|$ is small, we can ignore the parenthetical expression in the denominator of (15.21), so that the ratio $\delta Z_3/\delta Z_2 \approx [(4.4197+j57.982)/(2.8603+j43.9184)]^2 \approx 1.746$. For this same problem, we get $\beta = 1.7053 - j0.0620$, which differs from 1.746 by less than 3 %.

15.4 A General Transformation Matrix

In developing the idea of the virtual probe, we came across a transformation matrix that is more general than the simple product of a vector with the scalar, β, yet should be very easy to implement numerically. The mathematical tool that is basic to our idea is the Householder transformation matrix that finds significant use in computational linear algebra.[1] Our presentation follows the footnoted reference closely.

The Householder matrix, $\mathbf{P}$, has the form

$$\mathbf{P} = \mathbf{I} - 2\mathbf{w}\cdot\mathbf{w}^H , \tag{15.22}$$

where $\mathbf{w}$ is a complex unit vector. As written, the second term in (15.22) is the *outer* product of the two vectors.[2] The *inner* or *scalar* product of two vectors, $\mathbf{a}$ and $\mathbf{b}$ is written $\mathbf{a}^H\cdot\mathbf{b}$. The matrix $\mathbf{P}$ is clearly Hermitian and orthogonal, because

$$\begin{aligned}\mathbf{P}^2 &= (\mathbf{I}-2\mathbf{w}\cdot\mathbf{w}^H)\cdot(\mathbf{I}-2\mathbf{w}\cdot\mathbf{w}^H)\\ &= \mathbf{I}-4\mathbf{w}\cdot\mathbf{w}^H+4\mathbf{w}\cdot(\mathbf{w}^H\cdot\mathbf{w})\cdot\mathbf{w}^H\\ &= \mathbf{I}\,.\end{aligned} \tag{15.23}$$

Therefore $\mathbf{P}=\mathbf{P}^{-1}=\mathbf{P}^H$, which establishes orthogonality. Consider the subspace of vectors that are orthogonal to $\mathbf{w}$; any vector in this subspace remains invariant under $\mathbf{P}$, but any vector, $\mathbf{v}=|\mathbf{v}|\mathbf{w}$, that is oriented in the direction of $\mathbf{w}$ is inverted, $\mathbf{P}\cdot\mathbf{v} = -\mathbf{v}$. As a final geometric interpretation of (15.22), consider $\mathbf{P}\cdot\mathbf{x}=\mathbf{x}-2\mathbf{w}\cdot\mathbf{w}^H\cdot\mathbf{x}$, where $\mathbf{x}$ is arbitrary. Rearranging, and taking the norms of each side, we have

$$\begin{aligned}4(\mathbf{w}\cdot\mathbf{w}^H\cdot\mathbf{x})^H\cdot(\mathbf{w}\cdot\mathbf{w}^H\cdot\mathbf{x}) &= 4|\mathbf{w}^H\cdot\mathbf{x}|^2\\ &= (\mathbf{x}-\mathbf{P}\cdot\mathbf{x})^H\cdot(\mathbf{x}-\mathbf{P}\cdot\mathbf{x})\\ &= |\mathbf{x}|^2-2\mathbf{x}^H\cdot\mathbf{P}^H\cdot\mathbf{x}+|\mathbf{P}\cdot\mathbf{x}|^2\\ &= 2|\mathbf{x}|^2-2|\mathbf{x}|\mathbf{x}^H\cdot\mathbf{a}\\ &= 2|\mathbf{x}|^2(1-\cos\phi)\,,\end{aligned} \tag{15.24}$$

[1] See W. H. Press, B. P. Flannery, S. A. Teukolsky, W. T. Vetterling, *Numerical Recipes: The Art of Scientific Computing*, Cambridge University Press, 1986.

[2] This is also called the *direct* or *tensor* or *Cartesian* product.

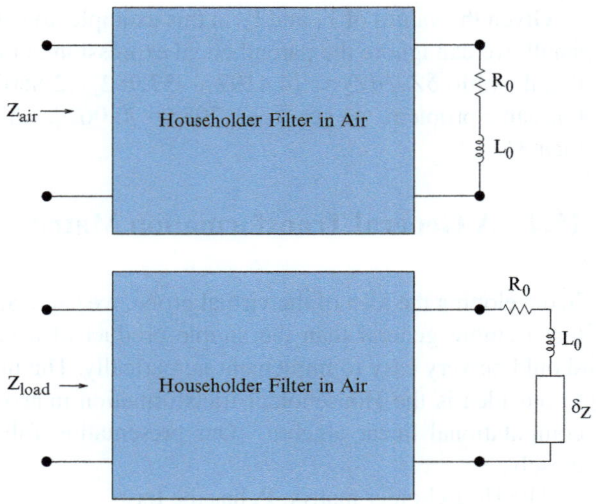

Fig. 15.14 Illustrating the use of a "Householder Filter" to simulate the coupling between the driving-point terminals of a probe and the **VIC-3D®** model. The low-frequency parameters, R_0 and L_0, have been independently measured

where **a** is a unit vector that points in the direction of the image vector, $\mathbf{P}\cdot\mathbf{x}$, and we have used the fact that $\mathbf{P}^H = \mathbf{P} = \mathbf{P}^{-1}$, which implies that $|\mathbf{P}\cdot\mathbf{x}| = |\mathbf{x}|$. ϕ is the angle between **x** and its image and is a measure of the amount that **x** has been rotated by the orthogonal transformation. Equation (15.24) is the law of cosines in hyperspace.

Now, let **u** be any vector, and rewrite **P** as

$$\mathbf{P} = \mathbf{I} - \frac{\mathbf{u}\cdot\mathbf{u}^H}{N}\,, \tag{15.25}$$

where $N = \dfrac{1}{2}|\mathbf{u}|^2$. We'll use this operator to rotate a vector, **x**, into another vector oriented in the direction of the unit vector, **a**. To that end, let $\mathbf{u} = \mathbf{x} + |\mathbf{x}|\mathbf{a}$. Then

$$\begin{aligned}
\mathbf{P}\cdot\mathbf{x} &= \mathbf{x} - \frac{\mathbf{u}}{N}\cdot(\mathbf{x}+|\mathbf{x}|\mathbf{a})^H\cdot\mathbf{x} \\
&= \mathbf{x} - \frac{\mathbf{u}}{N}(\mathbf{x}^H\cdot\mathbf{x} + |\mathbf{x}|^2\cos\alpha) \\
&= \mathbf{x} - \frac{\mathbf{u}}{N}\mathbf{x}^H\cdot\mathbf{x}(1+\cos\alpha) \\
&= \mathbf{x} - \frac{2\mathbf{u}\mathbf{x}^H\cdot\mathbf{x}(1+\cos\alpha)}{(\mathbf{x}+(\mathbf{x}^H\cdot\mathbf{x})^{1/2}\mathbf{a})^H\cdot(\mathbf{x}+(\mathbf{x}^H\cdot\mathbf{x})^{1/2}\mathbf{a})} \\
&= \mathbf{x} - \frac{2\mathbf{u}\mathbf{x}^H\cdot\mathbf{x}(1+\cos\alpha)}{(\mathbf{x}^H\cdot\mathbf{x}+2(\mathbf{x}^H\cdot\mathbf{x})\cos\alpha+\mathbf{x}^H\cdot\mathbf{x})} \\
&= \mathbf{x} - \mathbf{u} \\
&= -|\mathbf{x}|\mathbf{a}\,.
\end{aligned} \tag{15.26}$$

Thus, if our target vector is $\mathbf{A} = |\mathbf{A}|\mathbf{a}$, then we have

$$\mathbf{A} = -\frac{|\mathbf{A}|}{|\mathbf{x}|}\mathbf{P}\cdot\mathbf{x}\,, \tag{15.27}$$

as the required transformation of $\mathbf{x}$ into $\mathbf{A}$ that replaces the simpler scaling by β. The advantage of introducing $\mathbf{P}$, besides the fact that it is orthogonal, is that it is easily implemented as a stable computational algorithm.

We propose to use this transformation in the same way that we use the β transformation; we first determine the parameters of the orthogonal transformation by selecting a standard "source" vector, $\mathbf{x}_s$, with its corresponding standard "target" vector, $\mathbf{A}_s$, and establish the parameters, $\frac{|\mathbf{A}_s|}{|\mathbf{x}_s|}$, $\mathbf{u}_s = \mathbf{x}_s + |\mathbf{x}_s|\mathbf{a}_s$ and N_s which determine the standard $\mathbf{P}_s$, according to (15.25). We then use $\mathbf{P}_s$, together with an arbitrary source vector, $\mathbf{x}$, to determine the corresponding target vector, $\mathbf{A}$, by applying (15.27).

15.4.1 Application to Probe Characterization

We will apply the theory of the Householder transformation matrix to create a filter to transform measured impedance data into model data for the purpose of characterizing probes over a broad frequency range. Consider Fig. 15.14, which shows a situation in which a probe driving-point impedance is measured over a broad frequency range, and the intention is to isolate, and eventually remove, the coupling filter between the input terminals and the output circuit that represents the **VIC-3D®** model. It is assumed that the low-frequency parameters, R_0 and L_0, have been independently measured.

In this case, the standard source vector, $\mathbf{x}_s = \mathbf{Z}_{\text{air}}(\omega_i)$, $i = 1,\cdots,N$, and the standard target is $\mathbf{A}_s = R_0 + j\omega_i L_0$, $i = 1,\cdots,N$. Once the standard $\mathbf{P}_s$ is created, we use (15.27) with $\mathbf{x} = \mathbf{Z}_{\text{load}}(\omega_i)$, $i = 1,\cdots,N$, and compute $\mathbf{A} = R_0 + j\omega_i L_0 + \delta Z(\omega_i)$, from which $\delta Z(\omega_i)$ is determined by subtracting $R_0 + j\omega_i L_0$ from $\mathbf{A}$. It is implicitly assumed that the presence of the workpiece does not significantly change the coupling network that is represented by the Householder filter in air.

Part III
Applications

Chapter 16
Modeling Corrosion and Pitting Problems

16.1 Introduction and Overview of Approach

Given the material properties of typical aircraft structures and environmental conditions experienced over a service life, corrosion in aircraft structures is a major issue. For the United States Air Force alone, the total annual cost for corrosion management is on the order of a billion dollars. Avoidance of even a small portion of this annual expenditure would result in significant savings. To improve the management of corrosion, limits of existing inspection tools must be addressed. In particular, the ability to characterize the micro-topography of corrosion that forms fatigue and stress corrosion cracking is critical [114]. With this capability, repairs could be better directed to corrosion damage that only exceeds defect criteria with minimal aircraft disassembly. This chapter explores the problem of characterizing the surface topology of corrosion at faying surfaces in multilayered aircraft structures, with the focus on quantifying the size of corrosion pits in both first and second layers.

Although many corrosion techniques for aircraft structures only use a single NDE method, such as eddy-current, improvements may be achieved through the use of measurement data from multiple methods. Research has investigated data fusion of ultrasonic and eddy-current (EC) methods [115] and conventional and pulsed eddy-current techniques [116]. In particular, ultrasonic and eddy-current methods are quite complementary for multilayer corrosion problems, where ultrasonic methods provide excellent resolution of corrosion in the first layer, while eddy-current measurements are sensitive to the presence of corrosion in both the first and second layers. The approach described in this chapter uses ultrasonic data to characterize the first layer corrosion and thus simplify the second layer characterization problem using eddy-current.

In particular, a pragmatic approach is presented concerning the fusion of ultrasonic and eddy-current data, where ultrasonic measurements are used to accurately characterize first layer corrosion topology and thus minimize the number of unknowns in the second layer eddy-current inversion problem. This approach

H.A. Sabbagh et al., *Computational Electromagnetics and Model-Based Inversion*, Scientific Computation, DOI 10.1007/978-1-4419-8429-6_16,

contrasts with traditional data fusion where pixel-by-pixel fusion of measurement data is used. In reality, subsurface features such as pits are spread across image data and are dependent upon probe parameters. Thus, models in conjunction with inverse methods are needed to best extract features associated with the dimensions of the pit spread across the image data.

For the experiments reported in this chapter, artificial corrosion samples using drilled holes with conical tips to represent real pits were constructed to investigate optimizing the model-based inverse method approach of this book. Simultaneous acquisition of both ultrasonic and eddy-current data was performed using an eddy-current probe integrated with ultrasonic transducers.

16.2 Modeling the Corrosion Topology Problem

The problem that we are addressing is shown in Fig. 16.1. A probe, comprising a coil with a rather complex ferrite core, is scanned past a truncated cylindrical pit of diameter, *PD*, embedded in a quarter-inch aluminum slab. The pit extends only into the upper-half of the slab and has a height of $PH1$. For this analysis we will assume that the frequency of excitation is 2,200 Hz.

16.2.1 A Simplified Probe Model

It will facilitate running the many forward models that will be required for inverting data if we can replace the ferrite core in the probe model with a simpler probe that does not have a ferrite core. We have run a number of NLSE calculations to determine that coil whose impedance most closely approximates the measured results of a pit that is 75 % through-wall (PH1 = 2.38125 mm in Fig. 16.1), with a diameter of 0.125 in (PD = 0.125 in. in Fig. 16.1). The results, which are shown in Fig. 16.2, were obtained with a coil whose inner radius is 0 mm, outer radius is 2.75 mm, height of 2.54 mm, containing 2,500 turns, and tilted at an angle of 2°

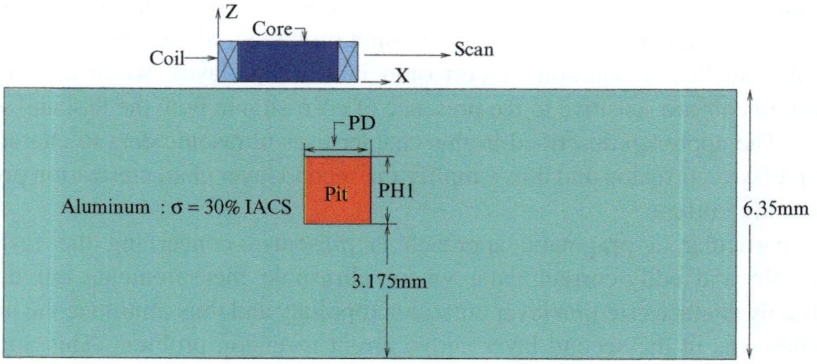

Fig. 16.1 The basic corrosion topology problem

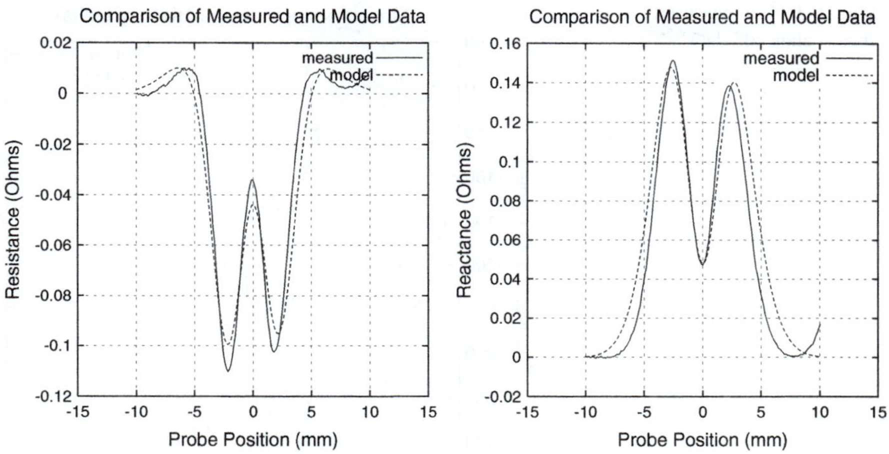

Fig. 16.2 Comparison of measured and model data for a probe that is tilted at an angle of 2°

about the Y-axis of Fig. 16.1. This probe model captures the peaks and valleys of the measured data reasonably well and will be used when creating the interpolation table for inverting the data for the class of problems defined in Fig. 16.1.

16.2.2 Data Transformation

The experimental data for this setup are recorded as A/D counts (or "voltage") in two channels that are 90° out of phase with each other. We have done some initial processing on this data by first defining the "baseline value" of each channel and then subtracting this baseline from the scanned data of each channel. The baseline value is taken to be the left-hand value of the data, because there are no other pits or anomalies to interact with this value (we are well away from any edges of the host plate). We call these data the "host-balanced" data, and they are shown in Fig. 16.3.

The measured data are clearly asymmetrical, even though the pit and probe are symmetrical. As we discussed above, we have created an accurate, though simplified, model of the probe, and have shown that if the probe is tilted about 2° about the Y-axis of Fig. 16.1, then the model also produces asymmetric lobes similar to those of Fig. 16.3. It remains, now, to transform data, measured in A/D counts or "voltage" into impedances, so that they can be further processed using **VIC-3D®**. We use the transformation of signal vectors described in Chap. 15 to accomplish this. The actual data used to determine β were for the same flaw (PD $=$ 0.125 in, PH1 $=$ 2.38125 mm) as above, and the value obtained was $\beta = 6.5276 \times 10^{-6} + j8.2605 \times 10^{-6}$. When the measured data of Z_1 (see Fig. 16.3) are multiplied by β, we get the "β-normalized" measured data shown in Fig. 16.4.

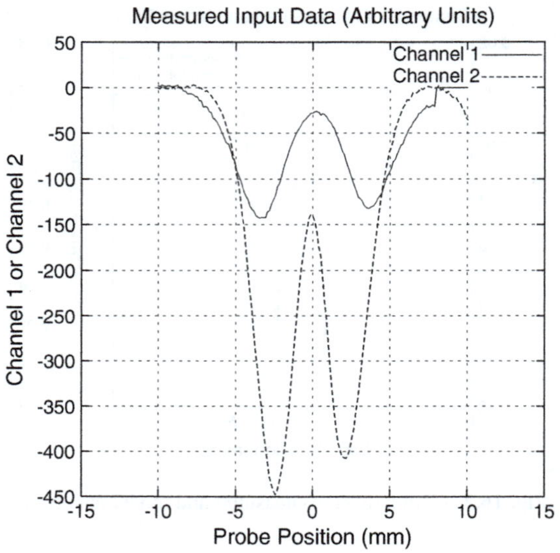

Fig. 16.3 The measured "host-balanced" data

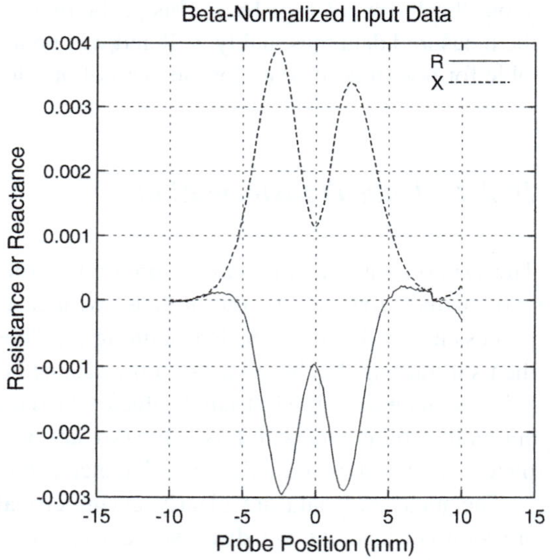

Fig. 16.4 The β-normalized measured "host-balanced" data

16.3 The Inverse Problem

The inverse problem can be addressed in terms of Fig. 16.1, in which two 1/8-inch panels are placed back-to-back, with corrosion pits emanating from the bottom of the first panel upward, and from the top of the second panel downward. The first pit is called pit 1, and the second, pit 2. The problem is to determine the height of each pit (PH1 and PH2), as well as the diameter (PD1 and PD2). We confine ourselves here to pits that emanate upward only; hence, $PH2 = PD2 = 0$.

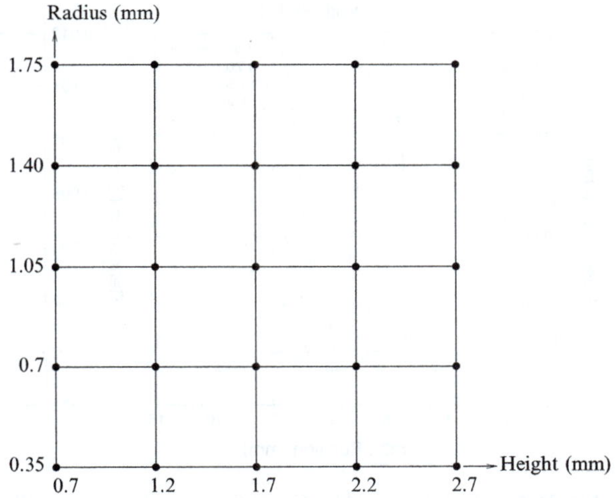

Fig. 16.5 A 5 × 5-grid that is suitable for interpolation with polynomials of order 1–4 in each variable

16.3.1 The Interpolator for NLSE

From here on we work with only the radius and height of the pit. The interpolation table is defined with respect to the grid shown in Fig. 16.5. Because this is a 5 × 5 grid, we can use polynomial splines of order 1–4 for interpolation in each variable.

The interpolating functions (sometimes called "blending" functions) associated with each of the 25 nodes of Fig. 16.5 are shown in Figs. 16.6–16.10. These functions were computed using the same probe-flaw model that produced the model data of Fig. 16.2.

It is clear that the challenge in the inversion process comes not only from the nonlinear relationship between the impedances and the two parameters, PD and PH1, but also from the significant change in appearance of the resistance scan-data with these two parameters. It appears that the inversion might be less confounding if we restricted ourselves to only reactance data; this will be discussed shortly.

16.3.2 Results at 2,200 Hz

We have run a number of inversions, and we show six that are typical of the results. In each of the following tabulated results, we show the norm of the residuals, Φ, the computed values of pit radius, R, and pit height, H, and the sensitivity of the solution, Σ_R, Σ_H, to each of these variables. These data are tabulated as a function of the order of the interpolating polynomial. Because there are five nodes in each variable, the interpolating polynomial in each variable can range in order from 1 to 4.

Table 16.1 shows the results of the inversion using the higher-order spline interpolator when applied to the model input data of Fig. 16.2.

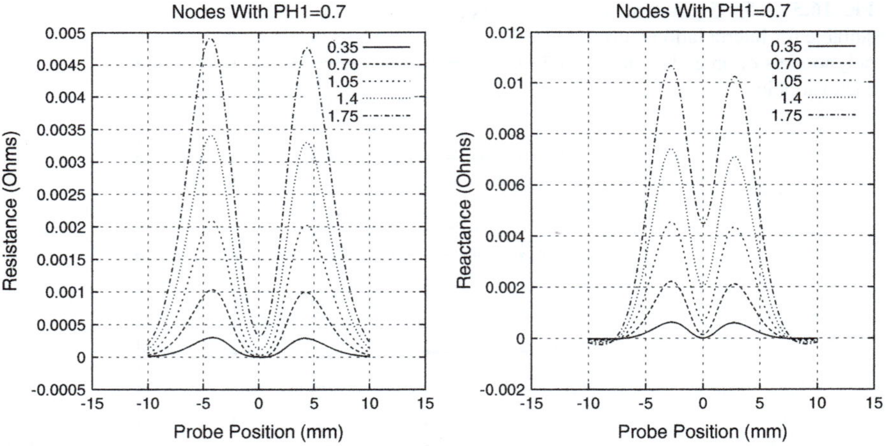

Fig. 16.6 Interpolating ("blending") functions for the case PH1 = 0.7. The numbers indicate the radius of the corresponding node in Fig. 16.5. *Left*: resistance; *Right*: reactance

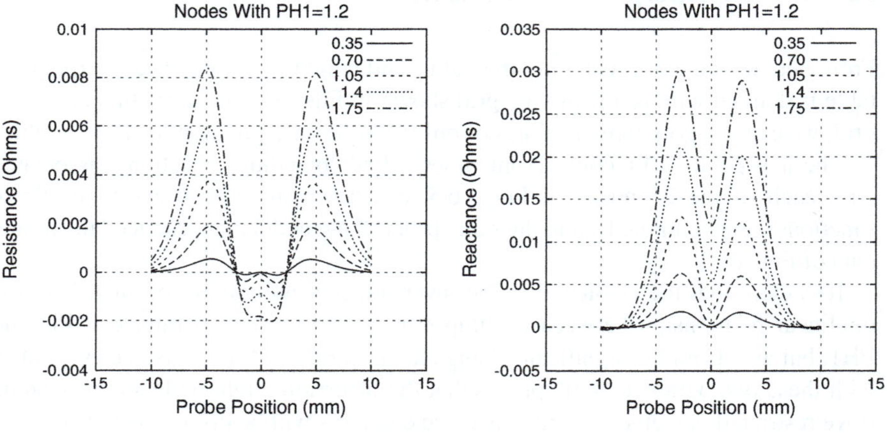

Fig. 16.7 Interpolating ("blending") functions for the case PH1 = 1.2. The numbers indicate the radius of the corresponding node in Fig. 16.5. *Left*: resistance; *Right*: reactance

The original data of Fig. 16.11, **pd-031_ph1-25_ph2-0_f1**, show a clear systematic error that we call generically "clutter" (see Chap. 15). The inverted results when using these cluttered data as input to NLSE are shown in Table 16.2. In order to determine the "flaw signal," we must remove the clutter from the original measured data. This is done, as described in Chap. 15, by modeling the clutter as a simple mathematical expression:

$$R_{\text{clutter}} = -0.001 + (0.0085/17.907) \times (\text{pos} + 10.033)$$

$$X_{\text{clutter}} = 0.00026 - (0.00246/17.907) \times (\text{pos} + 10.033)\,, \qquad (16.1)$$

where "pos" denotes the probe position.

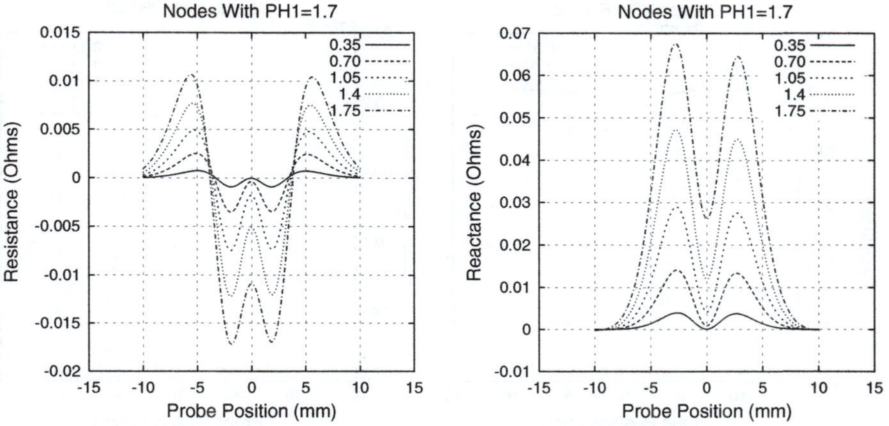

Fig. 16.8 Interpolating ("blending") functions for the case PH1 = 1.7. The numbers indicate the radius of the corresponding node in Fig. 16.5. *Left*: resistance; *Right*: reactance

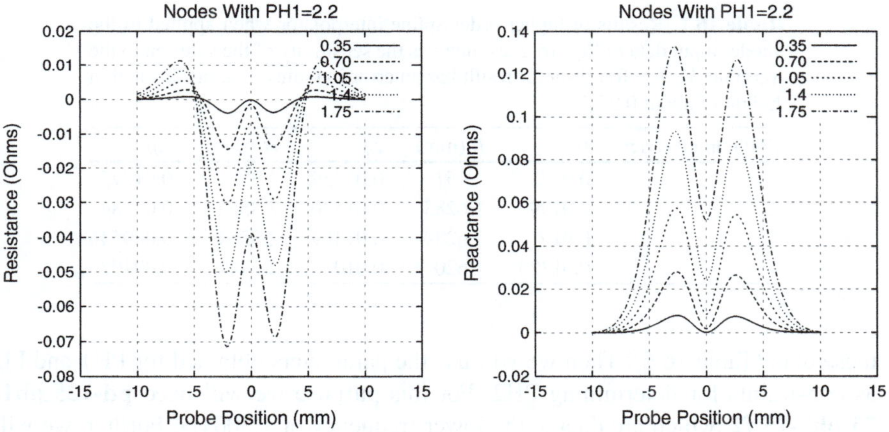

Fig. 16.9 Interpolating ("blending") functions for the case PH1 = 2.2. The numbers indicate the radius of the corresponding node in Fig. 16.5. *Left*: resistance; *Right*: reactance

After removing the clutter by subtracting (16.1) from the measured data, we get the results shown in Fig. 16.12, which are a clear improvement. The inverted results for the new data are shown in Tables 16.3 and 16.4.

In Fig. 16.13, we compare the original **pd-031_ph1-50** data with the reconstructed data computed from the fourth-order result of Table 16.5. Clearly, the reactance data, which are much cleaner, drive the inversion.

Now, we will test **pd-125_ph1-75_ph2-75_f1** as the first part of a two-step strategy to size the corrosion in layer 2. We will treat this input, since it is at the high frequency of 2,200 Hz, as containing only a pit in layer 1, so we will use the machinery already set up for a two-parameter problem, PH1 and PD, to invert

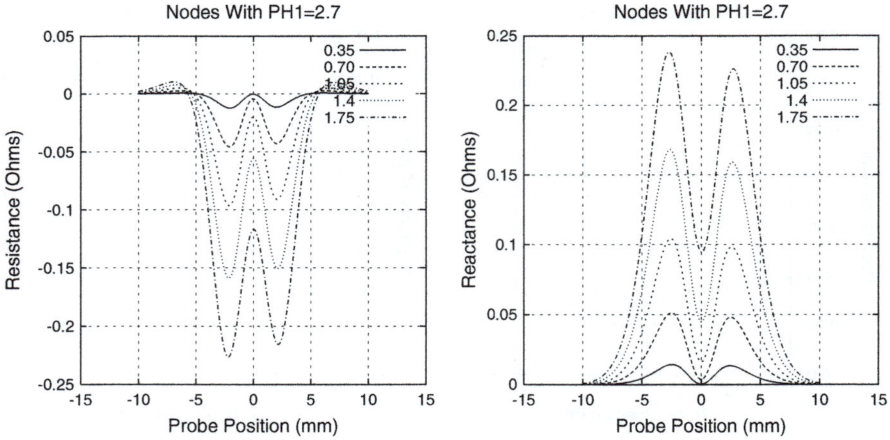

Fig. 16.10 Interpolating ("blending") functions for the case PH1 = 2.7. The numbers indicate the radius of the corresponding node in Fig. 16.5. *Left*: resistance; *Right*: reactance

Table 16.1 Results of higher-order spline interpolator when applied to the model input data of Fig. 16.2. Σ_V denotes the sensitivity of the solution to the variable, V. The fourth-order result has an error in radius of about 2% and in height of about 0.27%

Polynomial order	Φ	R (mm)	Σ_R	H (mm)	Σ_H
1	0.0278	1.651	0.02259	2.317	0.01275
2	0.0199	1.6283	0.01530	2.3554	0.00984
3	0.0191	1.6289	0.01469	2.3554	0.009546
4	0.01323	1.620	0.010	2.3747	0.00702

these data (Table 16.6). Then we will use the parameters obtained for PH1 and PD as constraints for determining PH2. For this purpose we will need **pd-125_ph1-75_ph2-75_f2**, which are data at the lower frequency of 1,300 Hz. Further, we will develop an entirely new interpolation table for operation at this frequency, as will be described in the next section.

The results for **pd-125_ph1-75_ph2-75_f1** are shown in Table 16.7 and confirm the reasonableness of our strategy of using high frequencies to reconstruct flaws in layer 1.

16.3.3 Results at 1,300 Hz

Our intention is to use the lower frequency to reconstruct flaws in layer 2. To this end we introduce a three-parameter interpolation grid and table for PH1, PH2, and PD. The grid is a three-dimensional version of Fig. 16.5, in which the third variable,

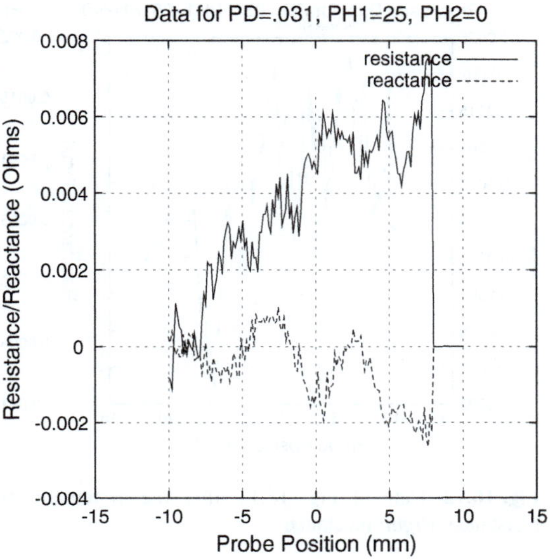

Fig. 16.11 Original data for PD = 0.03125 in., PH1 = 25% through-wall = 0.79375 mm

Table 16.2 Results of higher-order spline interpolator when applied to the original input data of Fig. 16.11. Σ_V denotes the sensitivity of the solution to the variable, V. The fourth-order result has an error in radius of about 75% and in height of about 9.7%

Polynomial order	Φ	R (mm)	Σ_R	H (mm)	Σ_H
1	0.0495	0.7005	1.0324	0.7000	0.9457
2	0.0495	0.7067	1.2621	0.7000	1.0554
3	0.0495	0.7087	1.2629	0.7000	1.1122
4	0.0494	0.6945	1.2043	0.7166	1.4225

Table 16.3 Results of higher-order spline interpolator when applied to the processed input data of Fig. 16.12. Σ_V denotes the sensitivity of the solution to the variable, V. The fourth-order result has an error in radius of about 15.5% and in height of about 0.5%

Polynomial order	Φ	R (mm)	Σ_R	H (mm)	Σ_H
1	0.01419	0.4896	0.4502	0.7000	0.4811
2	0.01419	0.5029	0.4874	0.7000	0.5406
3	0.01419	0.5050	0.4718	0.7000	0.5723
4	0.01416	0.4581	0.4030	0.7985	0.7539

PH2, is normal to the page. This variable will have the same five nodes as PH1, which means that the grid will have five copies of Fig. 16.5, giving a total of 125 nodes.

Our first result is for **pd-125_ph1-75_ph2-75_f2** and is shown in Table 16.8. The sensitivity parameter gives us insight into the quality of the results. The smallest

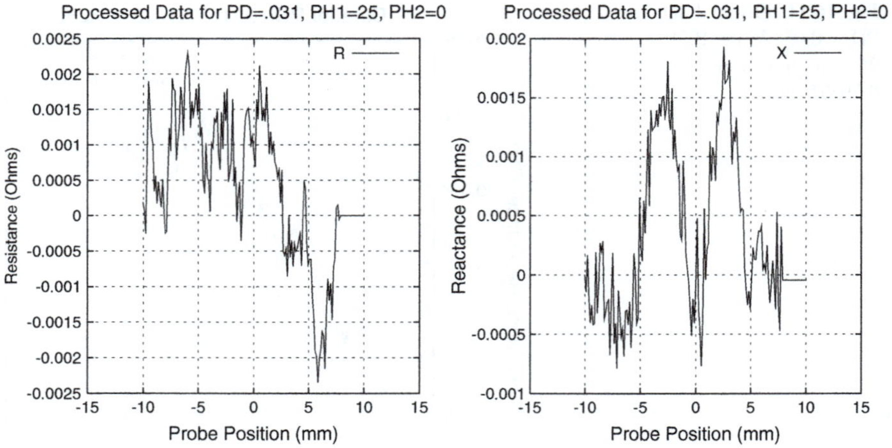

Fig. 16.12 Data of Fig. 16.11 after the systematic error "clutter" has been removed. *Left*: resistance; *Right*: reactance

Table 16.4 Results of higher-order spline interpolator when applied to the input data of pd-125_ph1-75_data2.in. Σ_V denotes the sensitivity of the solution to the variable, V. The fourth-order result has an error in radius of about 9.2% and in height of about 3.5%

Polynomial order	Φ	R (mm)	Σ_R	H (mm)	Σ_H
1	0.2022	1.4539	0.1407	2.4119	0.1175
2	0.1999	1.4552	0.1454	2.4364	0.1090
3	0.1998	1.4508	0.1445	2.4417	0.1141
4	0.1988	1.4422	0.1453	2.4640	0.1159

value of Φ occurs with the first-order polynomial, and the results for PD and PH1 are reasonable. The error in PD is 6.7%, and in PH1 6.9%. Note that the sensitivity of these two values at the solution is considerably smaller than that for PH2. This indicates that PD and PH1 are reasonably well determined, but PH2 is not. The solution is insensitive to PH2 because the frequency is too high for PH2 to be resolved. This is what we suspected at 2,200 Hz and holds true also for 1,300 Hz.

The result for **pd-031_ph1-25_ph2-25_f2**, however, is a little bit more encouraging, as shown in Table 16.9. The values given with the fourth-order polynomial for the three unknowns are in reasonable agreement with their nominal values, but the sensitivities for these parameters, though a little large due to the quality of the input data, are all comparable. This suggests that the value of PH2 can be resolved, even at 1,300 Hz, because PH1 is small.

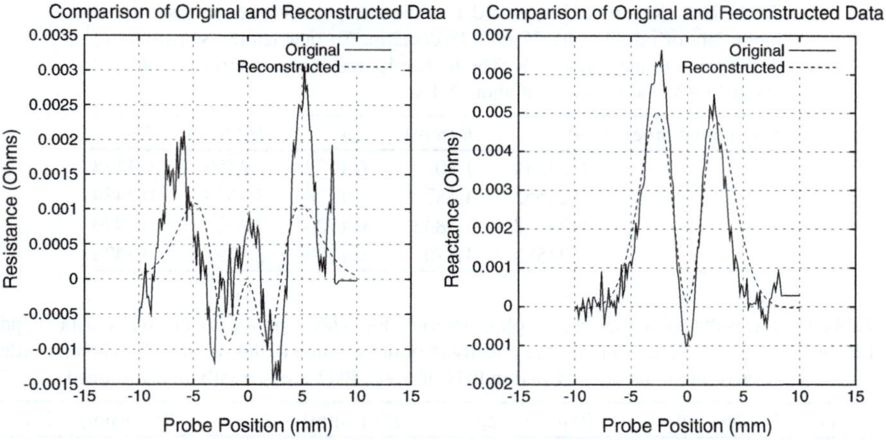

Fig. 16.13 Comparison of original pd-031_ph1-50 data with the reconstructed data computed from the fourth-order result of Table 16.5. *Left*: resistance; *Right*: reactance

Table 16.5 Results of higher-order spline interpolator when applied to the input data of pd-031_ph1-50_data2.in. Σ_V denotes the sensitivity of the solution to the variable, V. The fourth-order result has an error in radius of about 12.9% and in height of about 2.1%

Polynomial order	Φ	R (mm)	Σ_R	H (mm)	Σ_H
1	0.01404	0.4520	0.1041	1.4504	0.2823
2	0.01394	0.4394	0.1023	1.5452	0.2731
3	0.01392	0.4467	0.1053	1.5307	0.2726
4	0.01383	0.4480	0.1044	1.5534	0.2761

Table 16.6 Results of higher-order spline interpolator when applied to the input data of pd-031_ph1-75_data2.in. Σ_V denotes the sensitivity of the solution to the variable, V. The first-order result has an error in radius of about 0.9% and in height of about 13.4%

Polynomial order	Φ	R (mm)	Σ_R	H (mm)	Σ_H
1	0.0374	0.3932	0.0457	2.7	0.2441
2	0.0376	0.3968	0.0501	2.7	0.2264
3	0.0377	0.3986	0.0519	2.7	0.2185
4	0.0380	0.4060	0.0575	2.7	0.1896

16.3.4 Frequency Mixing

If we take the (first-order) solution for R in Table 16.8 and average it with the (fourth-order) solution in Table 16.4, we get 1.5719, which is in error by less than 1% from the nominal value of 1.5875 mm. Similarly, when we average the solutions for PH1 from the same tables, we get 2.3405, which is in error by 1.7% from the

Table 16.7 Results of higher-order spline interpolator when applied to the input data of pd-125_ph1-75_ph2-75_data2.in. Σ_V denotes the sensitivity of the solution to the variable, V. The first-order result has an error in radius of about 11.8% and in height of about 3.1%

Polynomial order	Φ	R (mm)	Σ_R	H (mm)	Σ_H
1	0.24544	1.40	0.1599	2.4559	0.1538
2	0.24553	1.387	0.1728	2.4858	0.1439
3	0.24558	1.3832	0.1720	2.4929	0.1479
4	0.24580	1.370	0.1723	2.5191	0.1474

Table 16.8 Results of higher-order spline interpolator when applied to the input data of pd-125_ph1-75_ph2-75_f2. Σ_V denotes the sensitivity of the solution to the variable, V. The first-order result has an error in radius of 6.7% and in PH1 of 6.9%. PH2 is essentially undetermined

Polynomial order	Φ	R (mm)	Σ_R	$PH1$ (mm)	Σ_{PH1}	$PH2$ (mm)	Σ_{PH2}
1	0.12293	1.7015	0.05962	2.2169	0.12605	0.7	3.5745
2	0.12344	1.7091	0.04978	2.2255	0.16515	0.7	3.5464
3	0.12346	1.7136	0.04671	2.2147	0.16735	0.7	3.5473
4	0.12358	1.7232	0.03667	2.2120	0.18108	0.7	3.5622

Table 16.9 Results of higher-order spline interpolator when applied to the input data of pd-031_ph1-25_ph2-25_f2. Σ_V denotes the sensitivity of the solution to the variable, V. The fourth-order result has an error in radius of 11.8%, in PH1 of 11.6% and 4.8% in PH2

Polynomial order	Φ	R (mm)	Σ_R	$PH1$ (mm)	Σ_{PH1}	$PH2$ (mm)	Σ_{PH2}
1	0.011785	0.35	0.8655	0.7019	0.8907	0.8817	4.9636
2	0.01180	0.35	0.9287	0.7095	1.0495	0.7434	5.1100
3	0.01182	0.3580	0.8385	0.7	1.1195	0.8822	5.01
4	0.011779	0.35	0.6277	0.7016	1.7667	0.8337	5.3144

nominal value of 2.38125 mm. This indicates that we can often use “frequency mixing” to get improved results when compared to the results at the individual frequencies.

16.3.5 Inverting Only Reactance Data

The resistance data for pd-031_ph1-25_ph2-00, shown in Fig. 16.11, are clearly quite noisy and play a significant role in producing the bad results shown in Table 16.2 for the estimate of the radius. This suggests that we remove the resistance data and invert only the reactance data. The result of this experiment is shown in Table 16.10. The estimates of the height and height-sensitivity remain about the same as before, but the estimates of the radius and radius-sensitivity are improved considerably.

We should not think, however, that the resistance should be discarded under all conditions. In several of our experiments, we found that discarding “good”

Table 16.10 Results of higher-order spline interpolator when applied to only the reactive part of the original input data of Fig. 16.11. Σ_V denotes the sensitivity of the solution to the variable, V. The fourth-order result has an error in radius of about 12% and in height of about 11.8%

Polynomial order	Φ	R (mm)	Σ_R	H (mm)	Σ_H
1	0.013566	0.3500	0.4804	0.7016	0.9694
2	0.013560	0.3503	0.5238	0.7000	1.1000
3	0.013561	0.3500	0.5438	0.7007	1.1724
4	0.013558	0.3501	0.6585	0.7000	1.5759

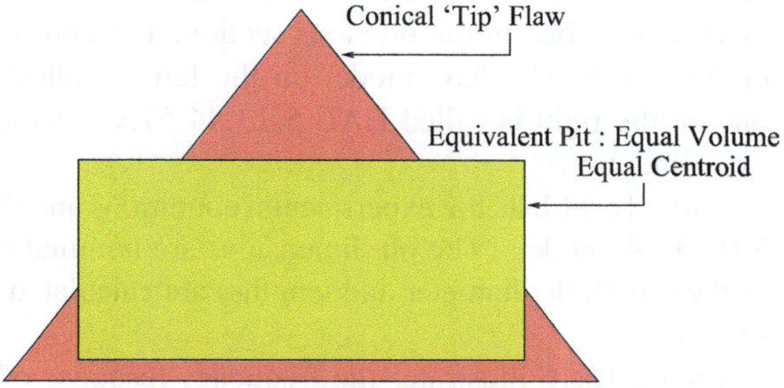

Fig. 16.14 A new flaw model that accounts for the conical tip produced by the drill bit. The usual truncated cylindrical pit model is shown in the case that the pit is not deeper than the tip. The pit model is assumed to have the same volume and centroid as the conical tip

resistance data degraded otherwise "good" solutions. It appears that resistance data stabilize the inversion process, especially in determining the height of a pit. Later, we will give an example of the use of the projection algorithm, POCS, to clean up the resistance and yield an improved inversion.

16.4 A New Flaw Model

The pits that have been used in obtaining data that are described in this chapter are the result of drilling into the aluminum hosts. The drill bits have pointed tips, so the result of the drilling is to produce a pit that will have a right-conical tip at its top. The resulting departure from the assumed truncated cylinder geometries is not significant in most cases, but will be quite significant if the desired "pit" is quite shallow. In this case we must modify our model of the pit to accommodate this geometry. We will call this a new flaw model.

Figure 16.14 illustrates the new flaw "tip" model together with an equivalent truncated cylinder "pit," in the case that the pit is not deeper than the conical tip. The truncated cylinder is assumed to have the same volume and centroid as the conical tip.

16.5 Low-Frequency Models and Experiments

In order to get good resolution for very deep-lying pits in the 0.125-inch panels, it is necessary to go to lower frequencies than we have used earlier. In this section we will describe the results of experiments that use larger-diameter probes of 0.44 and 0.57 inch, excited at 360 and 720 Hz for this purpose. We will carry out a few model calculations to determine such things as the dynamic range required of instruments in this frequency range.

Figure 16.15 shows two cases that we will consider. The dimensions of the pits (mm) are calculated using the "equivalent-pit" model of conical-tip flaw shapes, as was described in the previous section. The coil diameter, *D*, is either 0.44 inch or 0.57 inch. The flaw model on the left is called HAL_5_7_44_57.vic, and the one on the right is called HAL_5_13_44_57.vic, which follows the notation of the experiments.

Table 16.11 lists the experimental conditions and the names of the corresponding **VIC-3D®** models. The pit dimensions are nominal lengths; the actual dimensions of the pits (hole diameter and lengths) are calculated as explained in the preceding section.

Figure 16.16 illustrates the frequency response of the isolated pit in layer 1 as a function of the coil diameter. Note that we have extended the scan range from −19 mm to +19 mm in order to start and finish over unflawed host material.

We use the results of Fig. 16.16 to first determine the dynamic range required to detect the isolated pit. To do this we compare the peak-to-peak response of the probe as it scans the pit to the impedance of the probe over the unflawed host material. The base-10 logarithm of this number is roughly a measure of the number of significant digits that must be recorded by the test instrument, and is called the dynamic range.

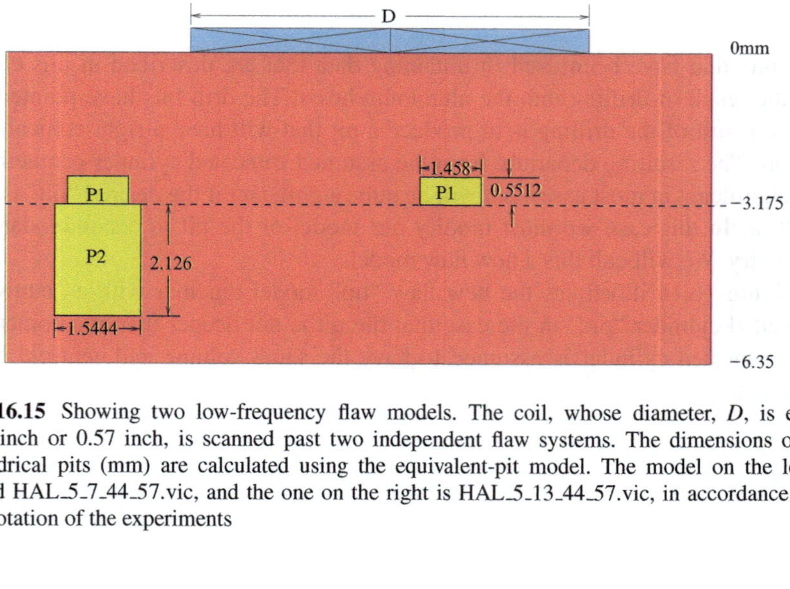

Fig. 16.15 Showing two low-frequency flaw models. The coil, whose diameter, *D*, is either 0.44 inch or 0.57 inch, is scanned past two independent flaw systems. The dimensions of the cylindrical pits (mm) are calculated using the equivalent-pit model. The model on the left is called HAL_5_7_44_57.vic, and the one on the right is HAL_5_13_44_57.vic, in accordance with the notation of the experiments

Table 16.11 Definition of experimental conditions for various **VIC-3D®** models. The nominal pit dimensions are lengths; the actual dimensions are calculated as explained in the preceding section

VIC-3D®-model	Panel thick (in)	Hole Diam (in-nom)	Pit 1 (in-nom)	Pit 2 (in-nom)
5_13_44_57	0.125	0.0625	0.03125	0
5_6_44_57	0.125	0.0625	0.03125	0.03125
5_8_44_57	0.125	0.0625	0.03125	0.0625
5_7_44_57	0.125	0.0625	0.03125	0.09375

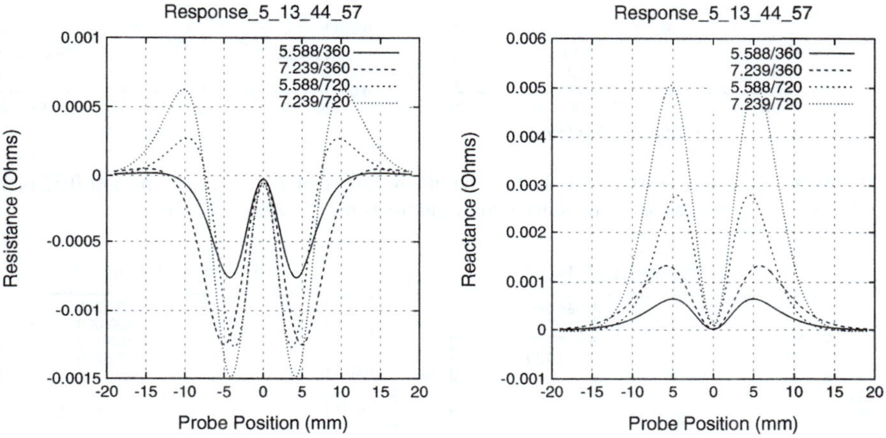

Fig. 16.16 Response of HAL_5_13_44_57. The title denotes coil diameter, 0.44 or 0.57 inch, whereas the legend denotes coil radius in mm, and frequency, 360 or 720 Hz

Table 16.12 Dynamic range requirements in dB for measuring response of the isolated pit in layer 1, as a function of coil diameter and frequency

Radius (mm)	Freq (Hz)	Z_{host}	$Z_{peak-peak}$	Dynamic range (dB)
5.588	360	$1.9617 + j37.691$	$0.0007 + j0.0007$	$69 + j94.7$
7.239	360	$4.1607 + j51.4588$	$0.00125 + j0.0013$	$70.4 + j92.0$
5.588	720	$5.7508 + j73.3232$	$0.0015 + j0.0028$	$71.7 + j88.4$
7.239	720	$11.445 + j98.2759$	$0.0021 + j0.005$	$74.7 + j85.9$

This number can also be expressed as A/D bits by dividing by 0.3. The dynamic range is sometimes expressed in decibels by $\mathrm{dB} = 20\log(\mathrm{ratio})$.

Table 16.12 tabulates the results of the calculations. The dynamic range requirements are not particularly severe, with the clear advantage going to the 0.57 inch coil excited at 720 Hz. The dynamic range in reactance is a rather modest 85.9 dB, which translates into 4.3 significant digits or an A/D converter with 14.3 bits of precision. Typical instruments start at 16 bits, which give about 4.8 significant digits, or 96 dB dynamic range.

Figures 16.17–16.20 illustrate the responses of the other model experiments. Figure 16.20 compares the responses at the single frequency of 720 Hz and coil

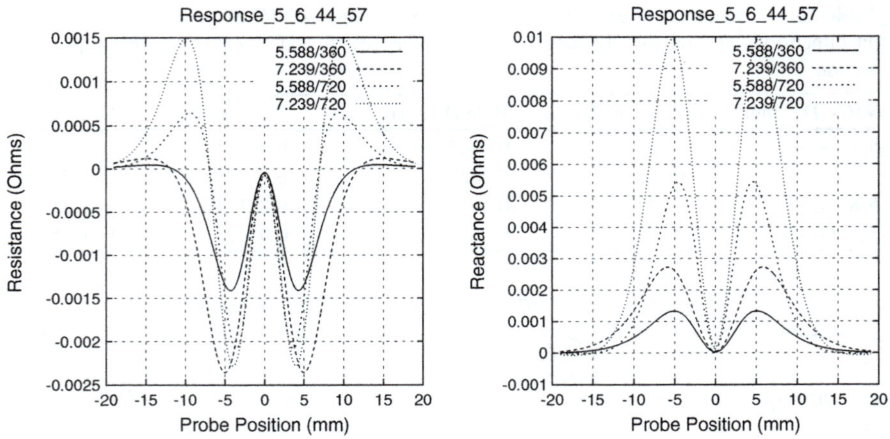

Fig. 16.17 Response of HAL_5_6_44_57. The title denotes coil diameter, 0.44 inch or 0.57 inch, whereas the legend denotes coil radius in mm, and frequency, 360 Hz or 720 Hz

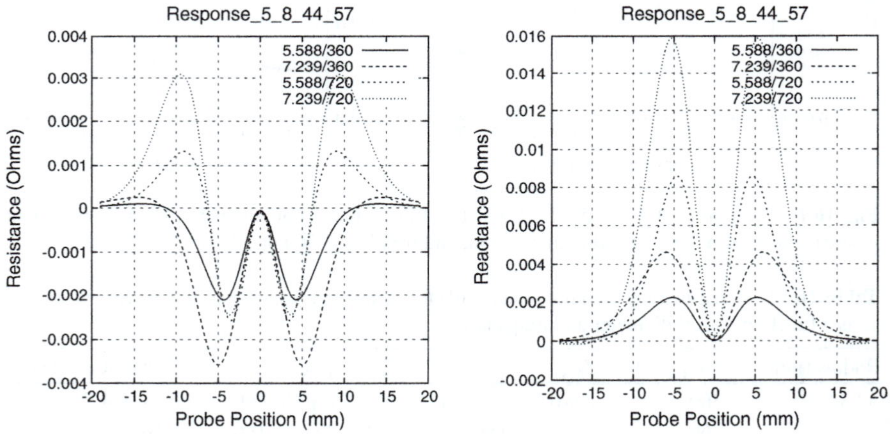

Fig. 16.18 Response of HAL_5_8_44_57. The title denotes coil diameter, 0.44 inch or 0.57 inch, whereas the legend denotes coil radius in mm, and frequency, 360 Hz or 720 Hz

diameter of 0.57 inch. This figure shows that the responses are well distinguished, which means that at this frequency, and using this coil configuration, we should be able to invert the measured impedance data and resolve the layer-two pits. For this reason we recommend the use of the 0.57-inch coil operating at 720 Hz.

Finally, note in all of these figures, especially in the reactive components, the significant dip that occurs when the coil is centered over the flaw. This is another manifestation of the response that we have seen earlier, in that a very large coil will have a very significant valley when the coil is so positioned. This valley will serve as a "pointer" to the pit and should help resolve nearby pits in the lateral direction.

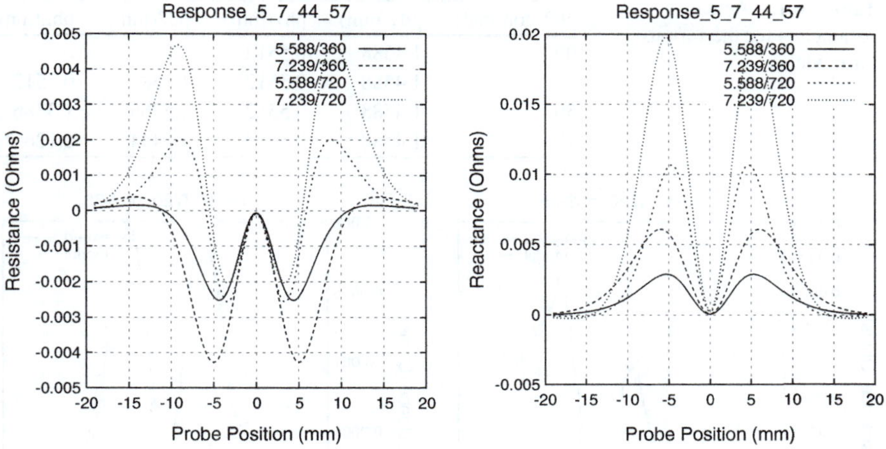

Fig. 16.19 Response of HAL_5_7_44_57. The title denotes coil diameter, 0.44 inch or 0.57 inch, whereas the legend denotes coil radius in mm, and frequency, 360 Hz or 720 Hz

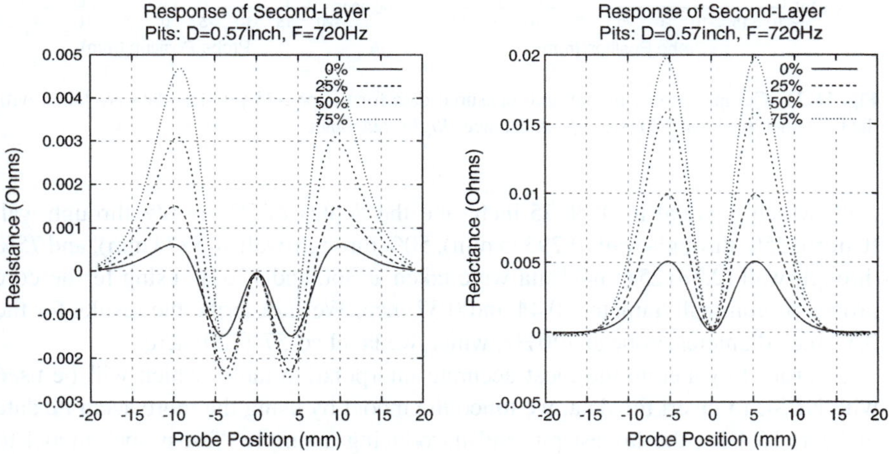

Fig. 16.20 Response of second-layer pits. The first-layer pit is fixed at a nominal value of 25% through-wall. The legend denotes percent through-wall of the second-layer pits. The coil diameter is fixed at 0.57 inch, and the frequency is fixed at 720 Hz

16.5.1 Inversion of Measured Data

Now, we return to the situation depicted in Fig. 16.15 and invert measured data taken on this system. We label the various configurations by their nominal values, in which the diameter of each pit is "fixed" at 0.0625 inch. In truth, the diameters vary with depth, as indicated in Sect. 16.4, but since the "β-scaling" is done using the usual nominal values, that's what we'll use. The depth of P1 is fixed at 25% through-

Table 16.13 Actual parameters for the various pit-models

ph2(nominal)	pd1(mm)	ph1(mm)	pd2(mm)	ph2(mm)
00	1.4588	0.5512		0
25	1.4588	0.5512	1.458	0.5512
50	1.4588	0.5512	1.5265	1.3386
75	1.4588	0.5512	1.5444	2.126

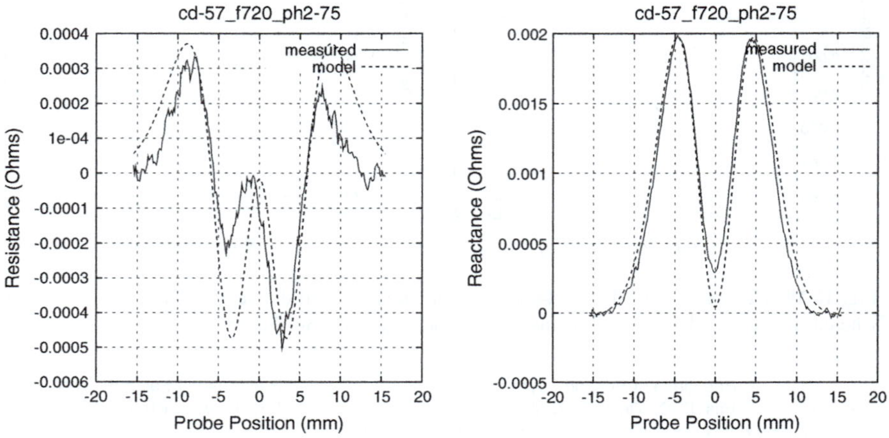

Fig. 16.21 Comparison of model and measured data for the ph2-75 pit. The data are taken with the 0.57-inch probe at 720 Hz. *Left*: resistance; *Right*: reactance

wall, which is equal to 0.79375 inch, and the depths of P2 are 0% through-wall (0 mm), 25% through-wall (0.79375 mm), 50% through-wall (1.5875 mm), and 75% through-wall (2.38125 mm). Data were taken at 360 and 720 Hz using ferrite-core probes of nominal diameters 0.44 and 0.57 inch. We'll describe the results for the 0.57-inch diameter probe at 720 Hz, which we label cd-57_f720, here.

In order to generate the most accurate interpolation table, which will be used with NLSE to invert the data, we tuned the model by using the more accurate data of Sect. 16.4 for the largest pit, and introducing a simple "fast-probe" model to facilitate generating the interpolation table. The inner radius of the probe is 0 mm, the outer radius is 5.588 mm, the height is 2.54 mm, and there are 1,075 turns. These parameters were chosen to give the best fit to the ph2-75 model and measured data. The actual parameters for the various pit-models are shown in Table 16.13.

Figure 16.21 shows the result of tuning the probe and ph2-75 model to the measured data. We emphasize that only the ph2-75 measured data were used in determining the fast-probe model. Figures 16.22–16.24 demonstrate that the other measured data were scaled consistently, so that they agree with the fast-probe model and the actual pit parameters tabulated in Table 16.13. No attempt has been made to model the asymmetries in the measured impedance data (especially in the resistance), as was done earlier when we "tilted" the probe model.

We assume that P1 is known (perhaps from UT data), so the problem is to reconstruct P2. We do this by first generating an interpolation table based on the grid

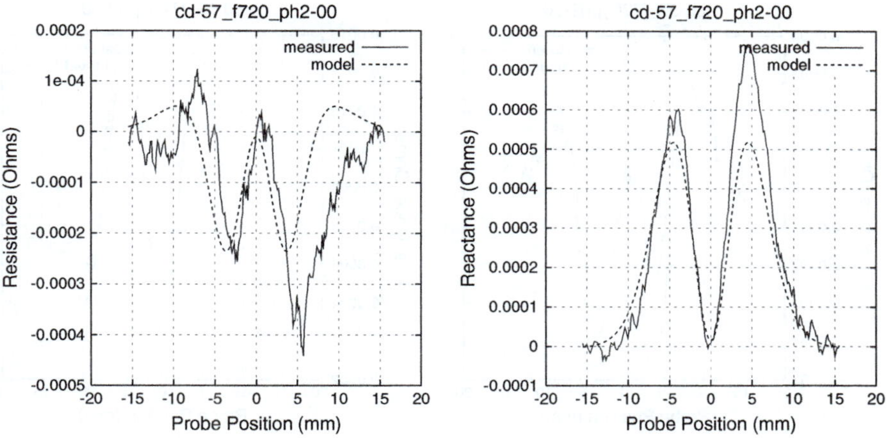

Fig. 16.22 Comparison of model and measured data for the ph2-00 pit. The data are taken with the 0.57-inch probe at 720 Hz. *Left*: resistance; *Right*: reactance

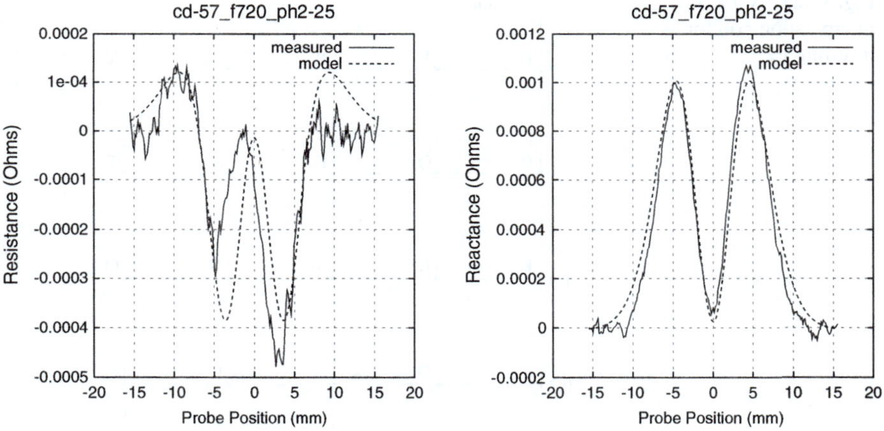

Fig. 16.23 Comparison of model and measured data for the ph2-25 pit. The data are taken with the 0.57-inch probe at 720 Hz. *Left*: resistance; *Right*: reactance

of Fig. 16.25. In creating this table, we fixed P1 using the parameters of Table 16.13, while allowing the P2 parameters to vary according to Fig. 16.25. The results are shown in Tables 16.14–16.17.

In order to validate our interpolation table, we inverted the model data of Figs. 16.23 and 16.24, and obtained the results shown in Tables 16.18 and 16.19, respectively. The results, especially for the model ph2-50 data, indicate that the table is correct. The discrepancies in the results for the ph2-25 data may be due to the fact that we did not put enough cells in the grid of the smaller pit while computing the response with **VIC-3D®**.

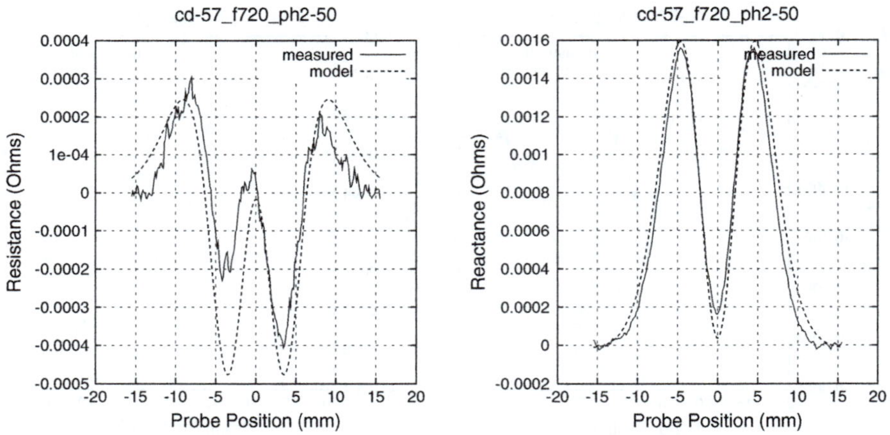

Fig. 16.24 Comparison of model and measured data for the ph2-50 pit. The data are taken with the 0.57-inch probe at 720 Hz. *Left*: resistance; *Right*: reactance

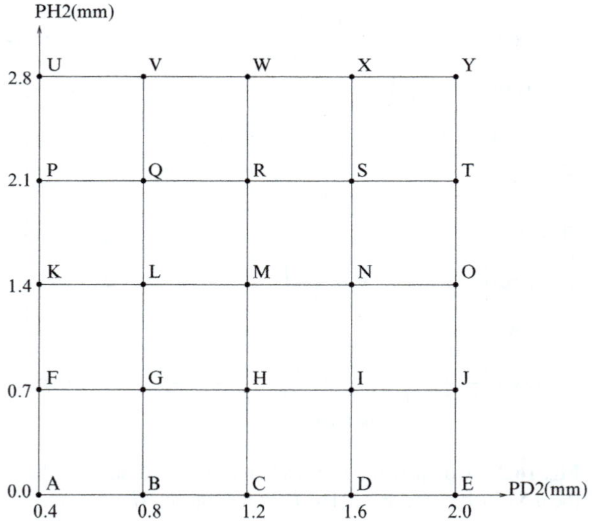

Fig. 16.25 A 5×5 interpolation grid for inverting P2

Table 16.14 Results for ph2-00_57_720. σ denotes the sensitivity of the parameter at the solution point

Order	Φ	PD2/σ	PH2/σ
1	0.2183248(−2)	1.200/0.9015	0.1514/0.4280
2	0.2179565(−2)	1.414/1.1797	0.1107/0.3084
3	0.2178009(−2)	1.473/1.2648	0.1024/0.2840
4	0.2172408(−2)	1.974/2.0086	0.0640/0.1757

Tables 16.20–16.23 display the results of inverting the measured data, but with the resistance data suppressed.

Table 16.15 Results for ph2-25_57_720. σ denotes the sensitivity of the parameter at the solution point

Order	Φ	PD2/σ	PH2/σ
1	0.1884036(−2)	2.0/0.3593	0.3222/0.1631
2	0.1877925(−2)	2.0/0.3631	0.3123/0.1574
3	0.1877159(−2)	2.0/0.3638	0.3107/0.1594
4	0.1875607(−2)	2.0/0.3706	0.3039/0.1601

Table 16.16 Results for ph2-50_57_720. σ denotes the sensitivity of the parameter at the solution point

Order	Φ	PD2/σ	PH2/σ
1	0.2083930(−2)	1.201/0.0749	2.2859/1.1356
2	0.2082139(−2)	1.198/0.0800	2.2972/1.1161
3	0.2082175(−2)	1.198/0.0793	2.2981/1.1196
4	0.2081406(−2)	1.196/0.0792	2.3090/1.1211

Table 16.17 Results for ph2-75_57_720. σ denotes the sensitivity of the parameter at the solution point

Order	Φ	PD2/σ	PH2/σ
1	0.2875837(−2)	1.492/0.1077	2.1000/0.7593
2	0.2878217(−2)	1.502/0.1049	2.0782/0.8968
3	0.2878120(−2)	1.502/0.1051	2.0783/0.9077
4	0.2877737(−2)	1.500/0.1050	2.0872/0.9029

Table 16.18 Results for the ph2-25 model data of Fig. 16.23

Order	Φ	PD2	PH2
1	0.9096408(−4)	1.565	0.5172
2	0.2538597(−4)	1.600	0.4860
3	0.2558663(−4)	1.604	0.4840
4	0.1026699(−4)	1.533	0.5176

Table 16.19 Results for the ph2-50 model data of Fig. 16.24

Order	Φ	PD2	PH2
1	0.1036414(−4)	1.554	1.3017
2	0.7781143(−6)	1.528	1.3393
3	0.1187602(−5)	1.526	1.3423
4	0.9500942(−6)	1.527	1.3401

Table 16.20 Results for ph2-00_57_720 with resistance data suppressed. σ denotes the sensitivity of the parameter at the solution point

Order	Φ	PD2/σ	PH2/σ
1	0.9401064(−3)	0.8000/0.2364	0.4476/0.4364
2	0.9390695(−3)	0.9116/0.2063	0.3310/0.3223
3	0.9387865(−3)	1.0314/0.2415	0.2536/0.2494
4	0.9373913(−3)	1.2189/0.3158	0.1756/0.1750

Table 16.21 Results for ph2-25_57_720 with resistance data suppressed. σ denotes the sensitivity of the parameter at the solution point

Order	Φ	PD2/σ	PH2/σ
1	0.1097437(−2)	1.2536/0.1016	0.7000/0.2057
2	0.1089299(−2)	1.8330/0.1851	0.3681/0.1080
3	0.1087329(−2)	2.0000/0.2094	0.3241/0.0960
4	0.1077701(−2)	2.0000/0.2113	0.3179/0.0957

Table 16.22 Results for ph2-50_57_720 with resistance data suppressed. σ denotes the sensitivity of the parameter at the solution point

Order	Φ	PD2/σ	PH2/σ
1	0.1307271(−2)	2.000/0.1152	0.7231/0.1479
2	0.1303683(−2)	2.000/0.1145	0.7222/0.1294
3	0.1303743(−2)	2.000/0.1144	0.7221/0.1289
4	0.1304370(−2)	2.000/0.1139	0.7220/0.1306

Table 16.23 Results for ph2-75_57_720 with resistance data suppressed. σ denotes the sensitivity of the parameter at the solution point

Order	Φ	PD2/σ	PH2/σ
1	0.1761228(−2)	2.000/0.1035	1.1302/0.1993
2	0.1741217(−2)	2.000/0.1027	1.1035/0.2017
3	0.1739535(−2)	2.000/0.1023	1.1035/0.2018
4	0.1742446(−2)	2.000/0.1011	1.1044/0.2011

16.5.2 Comments and Conclusions

The fourth-order result of Table 16.14 indicates that PH2 is quite small, which agrees with the actual parameter for ph2-00 in Table 16.13. The values of PD2 in Table 16.14 are not well determined, of course, because of the small value of PH2. The results for ph2-75, shown in Table 16.17, also agree quite well with the actual parameters listed in Table 16.13, for both PD2 and PH2.

The results for PD2 and PH2 for ph2-25 and ph2-50, given in Tables 16.15 and 16.16, respectively, though not accurate, display an interesting feature: the volume of each of the pits is reproduced accurately. Consider the fourth-order result for ph2-25 in Table 16.15; we have for the volume, $\mathrm{PH2} * \pi * \mathrm{PD2}^2/4 = 0.9547\ \mathrm{mm}^3$. The "true" volume for this pit is computed from Table 16.13 to be 0.9203, which differs by 3.6% from the "measured" volume. Similarly, the "measured" volume of ph2-50, as given by the fourth-order result of Table 16.16, is equal to $2.5940\ \mathrm{mm}^3$, whereas the true volume is equal to $2.45\ \mathrm{mm}^3$, and differs from the measured volume by 5.9%.

Tables 16.20–16.23 show that suppressing the resistance data (which are quite ugly when compared to the reactance data in Figs. 16.21–16.24) has a moderately stabilizing effect on PD2, in that all of the results cluster closely to each other (except for ph2-00), as is the case with the actual data of Table 16.13, but, of course, the measured PD2's are not close to the actual values. The PH2's follow an orderly ascent for ph2-00 to ph2-75, but for the latter four the values are off by a factor of about 2. So, it is difficult to conclude precisely the role played by the resistance data, except to say that it is useful, and it needs to be measured more accurately than it is now.

An alternative inverse problem is to choose a value of PD2 for each flaw, and then fix that parameter during the inversion and determine the optimum estimate of the height, PH2, of each flaw. This corresponds to our notion, expressed earlier in this section, of fixing a priori the lateral resolution of a simple or complex pit, and then determining the height of the pit. The question is how to determine the resolution. One way of doing this, if we simply don't want to postulate a lateral resolution, is to

Table 16.24 Results of applying the POCS algorithm to ph2-50_57_720. β_0 is the scale factor relating the impedance after the first iteration to the original data, β_1 is the scale factor relating the impedance after the second iteration to that after the first, and β_2 is the scale factor relating the impedance after the third iteration to that after the second. We do not go any further because β_2 is very close to unity. The "Error" is between the final computed geometry and the "Nominal" geometry

NLSE STEP	(PH2,PD2)	β
0	(2.052,1.24)	$0.98757 - j0.072423$
1	(1.7884,1.3052)	$0.98852 - j0.013285$
2	(1.5528,1.3660)	$0.99960 - j0.014508$
3	(1.2625,1.4891)	
Nominal	(1.3386,1.5265)	Error: (6.0%, 2.5%)

take, say, 500 random samples of PD2 and PH2, and determine Φ for each of these samples. Then we order Φ from smallest to largest in the ensemble, choose PD2 for, say, the smallest 10 values of Φ, and average the results to give us a prior estimate of PD2. The inversion is then completed as a one-dimensional problem for PH2. We will call this the "exhaustive-search" method.

We did this for ph2-25 and ph2-50, and got for ph2-25 values of PD2 $= 1.324$, PH2 $= 0.5997$, and for ph2-50, PD2 $= 1.24$, PH2 $= 2.052$. The results for ph2-25 are not in bad agreement with Table 16.13, being off by about 10% in diameter and about 9% in height. The errors in PD2 and PH2 for ph2-50 are, respectively, 23% and 53%. Clearly, the measured data for ph2-50 are not in good agreement with the data that should correspond to the pit whose dimensions are given in Table 16.13.

In general, however, we can conclude that the use of a 0.57-inch probe, operating at 720 Hz is well suited for characterizing second-layer pits in a system comprising two 0.125-inch panels.

16.5.3 Applying the POCS Algorithm

As we have stated many times before, the resistance data are the least reliable and most prone to producing poor reconstructions. This is certainly true of the data set ph2-50_57_720. The POCS algorithm of Fig. 14.3 was developed to ameliorate this problem and was applied to this data set.

The results of running the POCS algorithm on ph2-50_57_720 are listed in Table 16.24.

The input impedance data generated by the POCS algorithm are shown in Fig. 16.26. Note that the resistance data approach the "model" with each iteration. The reactances behave in a similar manner, but do not change as strongly from iteration to iteration. Only the measured reactance data are used in each iteration. It

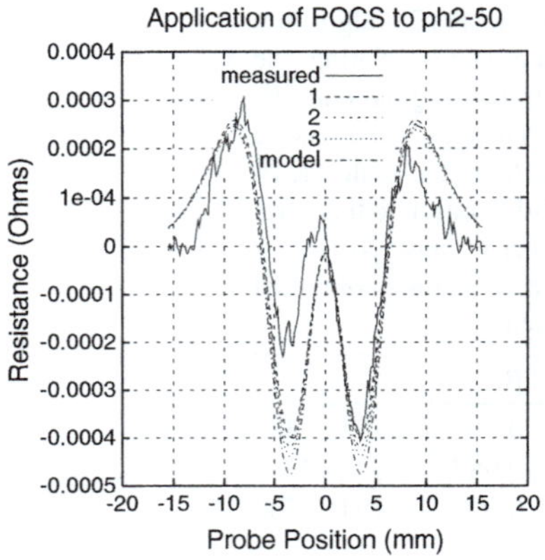

Fig. 16.26 Input impedance data generated by the POCS algorithm applied to ph2-50. The "measured" and "model" data are repeated from Fig. 16.24. The numbered data correspond to $(PH2, PD2)_1$, $(PH2, PD2)_2$ and $(PH2, PD2)_3$ given in Table 16.24. *Left*: resistance; *Right*: reactance

is clear that the improvement in the reconstruction is due to the improved resistance data during the iterations. Finally, we stress that we never use the "model" data because the model is not known to us a priori. Indeed, determining the model is the reason for doing the inversion in the first place.

16.6 Results of a 2D Circumferential Sweep Feature Extraction Algorithm

We have developed a new feature extraction algorithm for axisymmetric pits, in which averaging in the circumferential direction around the pit center is proposed to form the "feature vector" as a function of radial position. This method is made possible because all of the data are taken as 2D raster scans. Figure 16.27 compares the result of the earlier data for the ph2-25 pit, shown in Fig. 16.23, which we call "1D" with the new data, called "2D", together with the model results for the nominal ph2-25 pit dimensions shown in Table 16.13. Clearly, the resistance data are much improved.

Curiously, when we apply the 2D data to the usual inversion algorithm with the grid of Fig. 16.25, we get results that are not significantly different than those shown in Table 16.15. This may be due to some peculiar mathematical feature of the grid when applied to the ph2-25 data. We recall in Table 16.18 that even the model ph2-25 data did not produce as good an inversion as did the model ph2-50 data in Table 16.19.

If, however, we use the exhaustive-search algorithm, then we get a significant improvement when using the 2D data over the 1D data. For example, we ran the

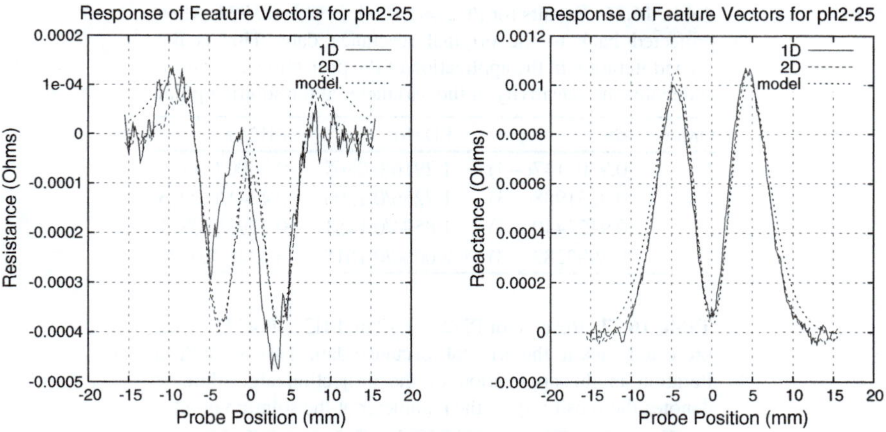

Fig. 16.27 Comparison of processed data with model data when 1D and 2D feature vectors are used. *Left*: resistance; *Right*: reactance

Table 16.25 Results for ph2-25_57_720_2DFV. This is the first iteration in the application of the projection algorithm. σ denotes the sensitivity of the parameter at the solution point

Order	Φ	PD2/σ	PH2/σ
1	0.1200470(−2)	1.2811/0.1072	0.7000/0.2088
2	0.1169658(−2)	1.8294/0.1840	0.3834/0.1126
3	0.1165389(−2)	2.0000/0.2086	0.3366/0.0995
4	0.1141903(−2)	2.0000/0.2082	0.3297/0.0982

random-PD2 algorithm twice, getting PD2 $=$ 1.4866 the first time, and PD2 $=$ 1.4758 the second. The results for PH2 and Φ in each case were PH2 $= 0.5444$, $\Phi = 0.1200699 \times 10^{-2}$ for the first, and PH2 $= 0.5508$, $\Phi = 0.1200684 \times 10^{-2}$ for the second. These results are in excellent agreement with the nominal values of PD2 $= 1.458$ and PH2 $= 0.5512$ that are listed in Table 16.13. The corresponding results for the 1D-data of Sect. 16.5.1 were PD2 $= 1.324$, PH2 $= 0.5997$.

Furthermore, when we force PD2 $= 1.458$ in NLSE, we get a value of PH2 $=$ 0.5193 when using the 1D-data, and PH2 $= 0.5618$ when using the 2D-data. The former error is 5.79%, whereas the latter error is 1.92%, so we see that the 2D-data are much better.

16.6.1 Application of the POCS Algorithm

Now we'll go back and apply the projection (POCS) algorithm of Fig. 14.3 to the 2DFV data. The result of the first iteration is shown in Table 16.25. We will use the second-order result for generating the input for the second step, because the

Table 16.26 Results for PD2 = 1.8294, PH2 = 0.3834, when projected back to the original reactance data. This is the second iteration in the application of the projection algorithm. σ denotes the sensitivity of the parameter at the solution point

Order	Φ	PD2/σ	PH2/σ
1	0.9301147(−3)	1.2916/0.0805	0.7000/0.1594
2	0.9131988(−3)	1.7256/0.1300	0.4241/0.0975
3	0.9121440(−3)	1.8594/0.1451	0.3793/0.0871
4	0.8992282(−3)	2.0000/0.1615	0.3348/0.0775

Table 16.27 Results for PD2 = 1.7256, PH2 = 0.4241, when projected back to the original reactance data. This is the third iteration in the application of the projection algorithm. σ denotes the sensitivity of the parameter at the solution point

Order	Φ	PD2/σ	PH2/σ
1	0.9289556(−3)	1.2921/0.0829	0.7000/0.1591
2	0.9128185(−3)	1.7188/0.1291	0.4270/0.0981
3	0.9118760(−3)	1.8409/0.1428	0.3852/0.0885
4	0.8996030(−3)	2.0000/0.1614	0.3350/0.0775

first-order result is "trapped" at the node PH2 = 0.7 of the interpolation grid of Fig. 16.25. This often happens with first-order results because the interpolation table is linear in this order, which means that there will be discontinuities in the derivative at each node. This discontinuity is suspected to be the culprit in trapping the solution at a node. The third- and fourth-order results are artificially constrained at the boundary PD2 = 2.0 of the grid, so we reject those solutions, as well.

When we use PD2 = 1.8294, PH2 = 0.3834 to generate our next input impedance, which is then projected back to the original reactance data and fed into NLSE, we get the results shown in Table 16.26.

The third iteration, which uses the results for PD2 = 1.7256, PH2 = 0.4241 as its input data (after projection), is shown in Table 16.27. Clearly, the 2nd-order results are converging slowly to the nominal values of PD2 = 1.458, PH2 = 0.5512.

16.6.2 Validation with Real Corrosion Pits

In addition to the examples using artificial pits as described in the earlier sections of this chapter, validation of this methodology has also been made using a sample with real corrosion pitting. The pits were generated by an accelerated corrosion process: ASTM B117 (salt fog test chamber at 95 °F, modified Exco solution, pH 3.2) in 7075-T6 aluminum. Figure 16.28 shows the methods of imaging these pits.

Figure 16.29 shows various images of pits as obtained by the methods shown in Fig. 16.28. One of the more significant pitting regions was selected from a set

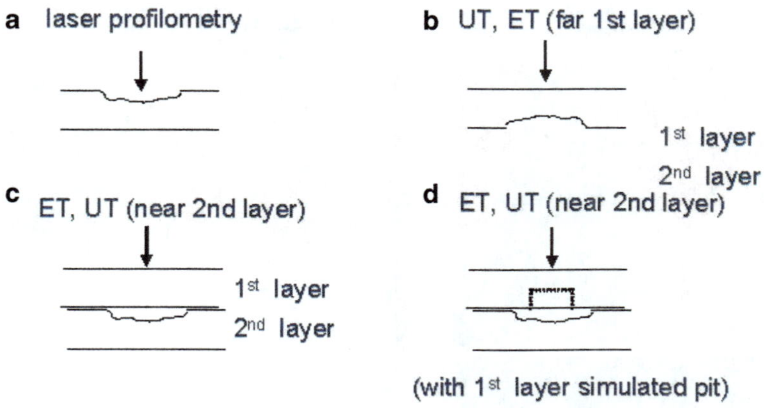

Fig. 16.28 Sketch illustrating the methods of imaging real corrosion samples. (**a**) Laser profilometry, (**b**) ultrasonic/eddy-current (far 1st layer), (**c**) eddy-current/ultrasonic (near 2nd layer), (**d**) eddy-current/ultrasonic (near 2nd layer with 1st layer simulated pit)

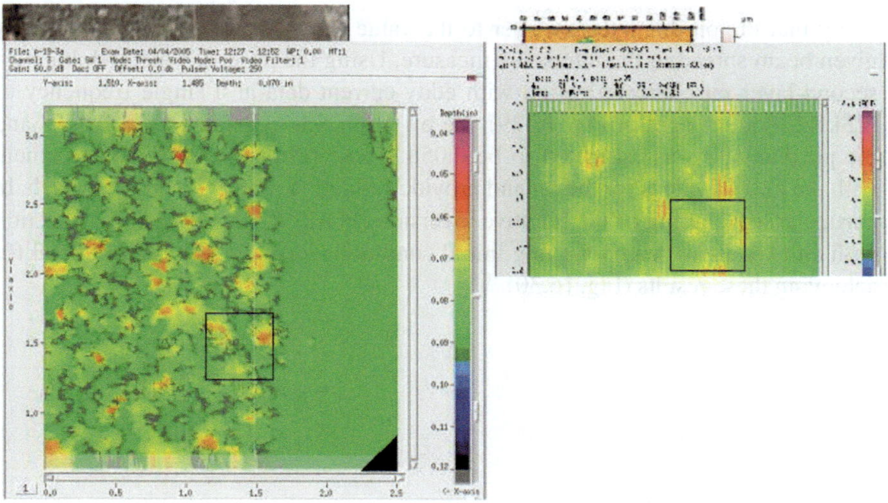

Fig. 16.29 Real corrosion pits. *Top left*: photograph, *Top right*: laser profilometry image, *Bottom left*: ultrasonic image (pits in far side of single layer), *Bottom right*: eddy-current image (pits in near side of second layer, no corrosion in first)

of samples evaluated using laser profilometry (top right of Fig. 16.29). Ultrasonic time-of-flight data were also acquired to evaluate the pit height and diameter (bottom left of Fig. 16.29). The pit height was determined to be 0.020″ by laser profilometry and 0.026″ using an ultrasonic method. However, there may be some error in the laser profilometry results due to the presence of corrosion by-product. From the image data for each method, an estimate of equivalent pit diameter was also made, varying from 0.040″ by laser profilometry and 0.100″ using ultrasonic method. It is

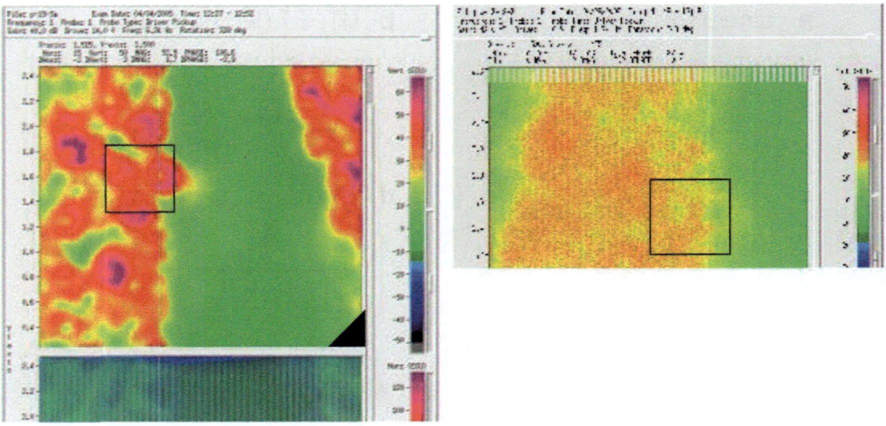

Fig. 16.30 Real corrosion images. *Left*: eddy-current (far 1st layer at 6.2 kHz), *Right*: eddy-current (near 2nd layer at 1.7 kHz)

likely that the pit diameter is closer to the value determined by laser profilometry given beam spread in the ultrasonic measure. Using the proposed methodology for second-layer pit characterization with eddy-current data at a single frequency of 6.2 kHz (bottom right of Fig. 16.29), the pit depth was estimated to be 0.029″ and the pit diameter was estimated to be 0.058. These results are in good agreement with the known values for the pit and provide validation of this general approach. In particular, the pairing of the iterative inversion (POCS) scheme for noise rejection with an exhaustive search varying initial conditions was found to be beneficial for achieving these results (Fig. 16.30).

Chapter 17
Applications to Aerospace Structures

17.1 Inspection of Fastener Sites in Aircraft Structures

A problem of particular interest in the field of nondestructive evaluation is the inspection of fastener sites in aircraft structures for fatigue cracks. An important class of structures comprises plane-parallel layered media representing joints in both fuselage and wing locations. The layer stackup typically consists of two to four panels and often includes thin layers of a nonconducting sealant or adhesive between the panels. Additional complexity for computational electromagnetics concerns the fastener shape (countersunk or buttonhead fastener) and material type (where ferrous materials are prevalent). Also, fatigue cracks of complex morphology provide a particular challenge for representation using numerical methods. Prior work has investigated the problem of modeling an eddy-current inspection of fastener sites in multilayered structures for fatigue cracks using a volume-integral equation approach [102]. Although good agreement was achieved with experimental results, simplications in the model representation were used. In this chapter, the capability of the volume-integral equation approach with spatial decomposition algorithms is demonstrated to fully address the problem of ferromagnetic fastener sites in multilayered structures with gaps between the layers and cracks emanating from the holes.

17.2 The Cessna Sandwich

To demonstrate the capability of the volume-integral equation method, the "Cessna Sandwich" example is presented (Amos, J., Pendse, V., Shashikiran, A.: Cessna Aircraft Co., private communication). Figure 17.1 illustrates the geometry of the Cessna-Sandwich, which consists of three layers of aluminum comprising two different aluminum alloys (conductivities), two "air gaps" representing the nonconducting sealant layers, and a titanium rivet-insert with a countersunk head

H.A. Sabbagh et al., *Computational Electromagnetics and Model-Based Inversion*, Scientific Computation, DOI 10.1007/978-1-4419-8429-6_17,

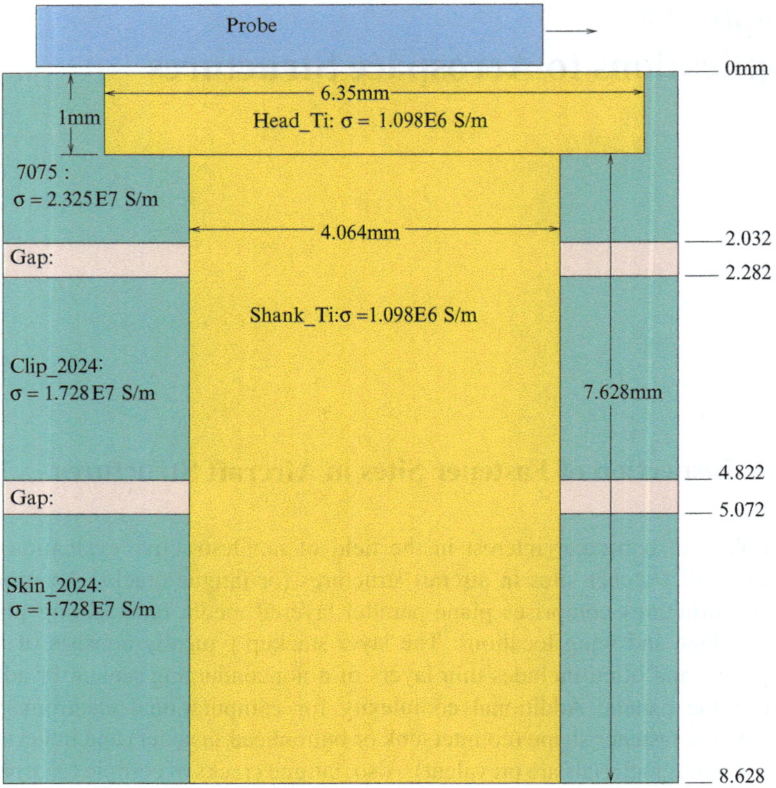

Fig. 17.1 The Cessna-Sandwich problem

and a shank connecting all layers [102]. In the model, the layers and gaps make up a uniform background or host (in x and y directions) with the fastener being an "anomaly" requiring a local volume element mesh. Because the anomaly extends through several layers of material with different electromagnetic properties, we must use **VIC-3D**® that has been augmented with the spatial-decomposition algorithm (SDA) of Chap. 10.

Two model calculations were run based upon Fig. 17.1, the first with both gaps filled with air, and the second with the top gap filled with 7075-aluminum ($\sigma = 2.325 \times 10^7$), and the other gap filled with 2024-aluminum ($\sigma = 1.728 \times 10^7$). The probe was a simple air-core coil with inner and outer radii of 3.02 and 5.14 mm, respectively, and a height of 2.48 mm. Both model calculations were run with a coil lift-off of 1.0 mm, and a frequency of 2,500 Hz. Five spatial-decomposition grids were used to represent the "anomaly" regions, each with $8 \times 8 \times 2$ cells, yielding a total of 1,920 unknowns. The results are shown in Figs. 17.2 and 17.3. The plot labeled "Gap" in Fig. 17.2 corresponds to the air-filled gaps of Fig. 17.1, and the plot labeled "NoGap" corresponds to the case in which the gaps are filled

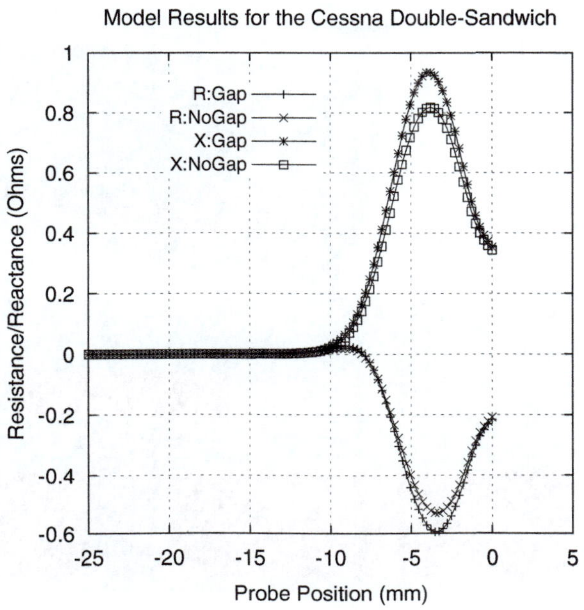

Fig. 17.2 Model results for the Cessna Double Sandwich of Fig. 17.1

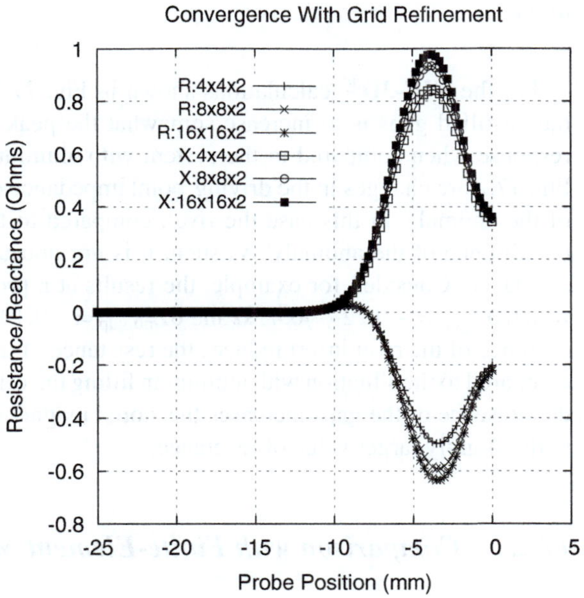

Fig. 17.3 Convergence of the Cessna-Sandwich Problem with grid refinement. The coarsest grid is $4 \times 4 \times 2$ cells per SDA grid, and the finest grid is $16 \times 16 \times 2$ cells per SDA grid. There are five grids in all. A linear extrapolation of the $4 \times 4 \times 2$ and $8 \times 8 \times 2$ results is within a percent or so of the $16 \times 16 \times 2$ results

with aluminum. The solution time for the "Gap" run with 26 probe scan points is about 6 min on an AMD/Athlon machine, whereas the "NoGap" run took about 2 min. Figure 17.3 shows convergence of the Cessna-Sandwich Problem with grid refinement.

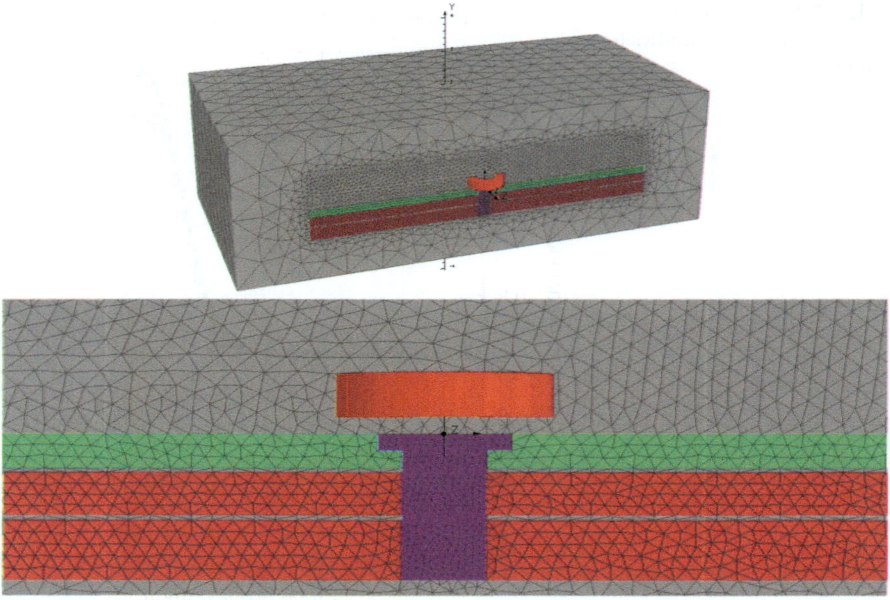

Fig. 17.4 Diagrams of fastener site FEM model with two gap layers between three aluminum panels

For the **VIC-3D®** calculations shown in Fig. 17.2, it is clear that the effect of the air-filled gaps is to increase somewhat the peak magnitude of the impedance response when compared to the system with aluminum-filled gaps. The results in Fig. 17.2 are changes in the driving-point impedance of the coil due to the presence of the anomaly, in this case the rivet, compared to the "host-only" impedance in the absence of the anomaly. As such, it is not unusual to have negative values of resistance. Consider, for example, the results at a probe position of -4.0 mm, for which $\delta Z_{\text{gap}} = -0.22 + j0.41\,\Omega$ and $\delta Z_{\text{no gap}} = -0.195 + j0.35\,\Omega$. In the former, the presence of the rivet insert reduces the resistance of the host structure with the gap, compared to the situation with aluminum filling the original gaps. On the other hand, the presence of the gaps increases the stored magnetic energy, which is manifested in the slightly larger value of reactance.

17.2.1 Comparison with Finite-Element Method Results

To provide a baseline for this challenge-problem concerning the performance the volume-integral equation method, an FEM model for the Cessna-Sandwich problem was implemented in Opera-3D© software package (V9.0) [103]. The numerical formulation of FEM is well established in the literature [104, 117]. A diagram of the model is shown in Fig. 17.4. Irregular meshes of tetrahedral elements were used

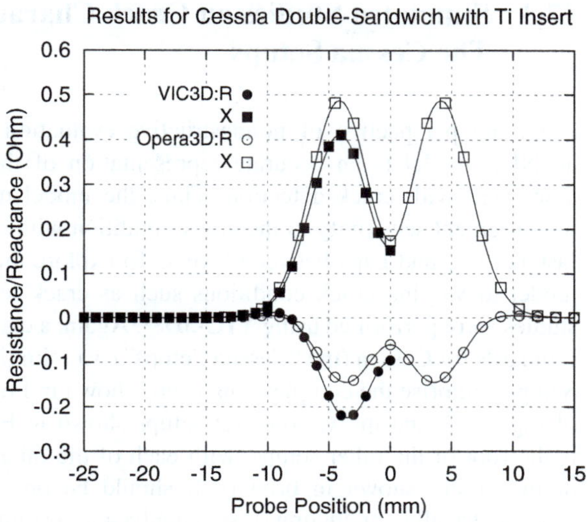

Fig. 17.5 FEM and **VIC-3D®** model results for the Cessna Sandwich with two airgaps

with the finite-element formulation to generally represent the complex geometries. The mesh for the model required at least two elements per skin depth in the fastener site region. The lateral dimensions of the plates in the model were extended well beyond the field generated by the eddy-current coil to minimize the effect of the part or model domain edges on the measurement response. Continuity conditions were maintained between all part/air interfaces in the model. Far from the coil and fastener site, the boundary conditions at the FEM model boundary were defined with the tangential magnetic field equal to zero. The output of this model is the electric and magnetic field intensities. Change in impedance can then be calculated using the change in resistance associated with dissipated energy in the region of the conductor and the change in inductance related to the stored energy in the whole solution domain. More information on constructing eddy-current models in Opera3D© can be found in [103].

Figure 17.5 displays the model results for the Cessna-Sandwich problem with both airgaps and compares these results with the **VIC-3D®** results of Fig. 17.2 labeled "Gap." Clearly, the FEM response shows the same characteristics as does the **VIC-3D®**-generated solution. Some differences found in the magnitude of the calculations are likely due to mesh-related error present in the models. Prior work has shown very good agreement between FEM and VIE for this class of problem [102]. Computation of the FEM model required 60 h on a 3.02 GHz Pentium 4 with 1 GB of memory. For this complex problem, the advantage of the compact formulation with volume-integrals (1,920 unknowns) over FEM is obvious.

17.3 Simulated Studies on Crack Characteristics: The Cessna Setups

From the perspective of nondestructive evaluation, the critical requirement of an NDE model is an accurate representation of the eddy-current measurement associated with crack detection. Thus, the model must accurately represent the measurement sensitivity to the crack condition with respect to features such as the fastener site and gaps between layers. To explore the sensitivity of the numerical model to varying crack conditions such as crack length and location, simulated studies were performed using **VIC-3D®**. Again, a class of problems based on setup standards for Cessna (the "Cessna Setups") was simulated in this study. The Cessna Setups comprise the complex ring probe shown in Fig. 17.6, the "hi-lock" pin rivet of Fig. 17.7, and the various test setups shown in Figs. 17.8–17.12. Test setup 0 is the root of the other setups, with each of the others differing by the placement of the crack (shown in brown). It should be observed that the setups contain both electrically conducting (lossy) media and ferromagnetic steels. The cracks are modeled as empty slots.

In modeling the setups, we use the spatial-decomposition algorithm with five grids, two above the air gap, and two below, with one within the gap. Each grid has $16 \times 16 \times 2$ cells, yielding a total of 15,360 electric and magnetic currents to be determined. The ring probe core was modeled with 12,288 electric and magnetic currents, so that the entire package of unknowns for each setup is 27,648. The models were run at a frequency of 1.0 kHz.

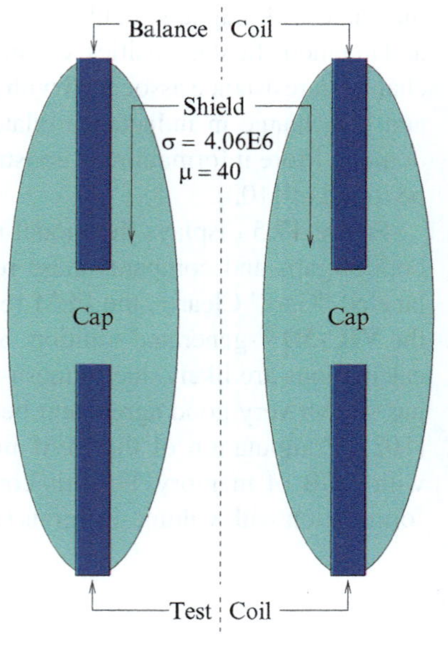

Fig. 17.6 The ring probe

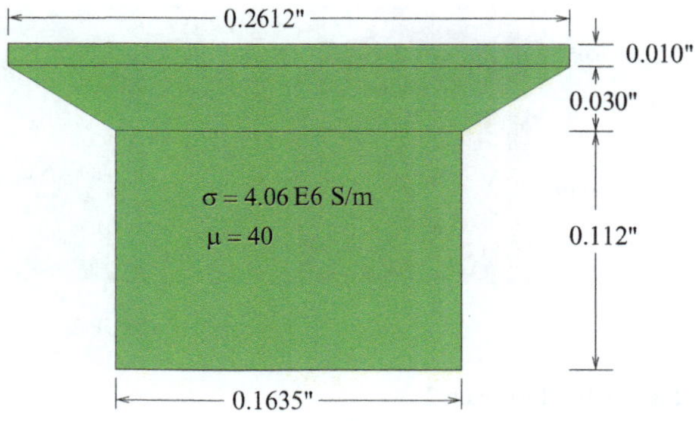

Fig. 17.7 The hi-lock pin rivet

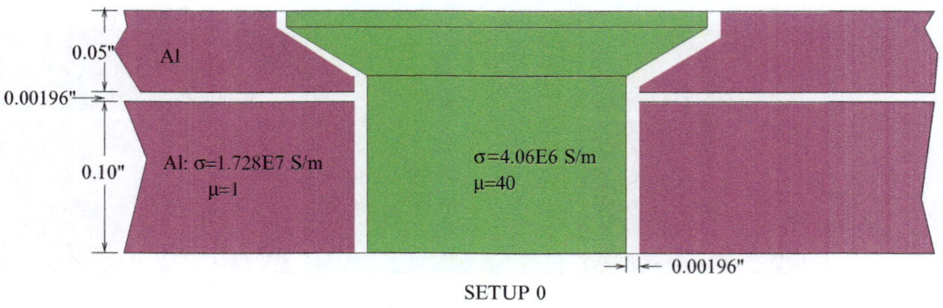

Fig. 17.8 Test setup 0

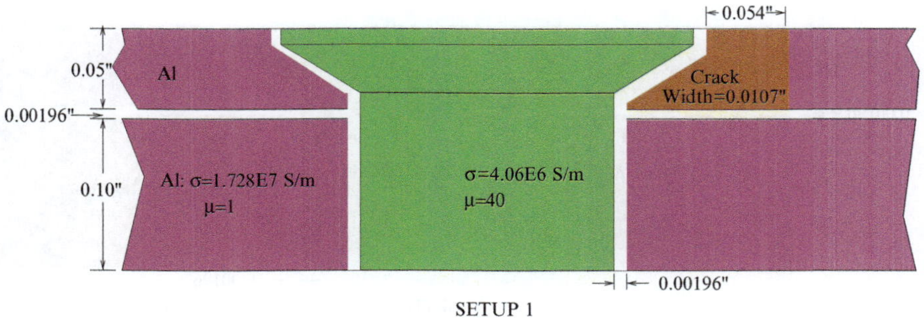

Fig. 17.9 Test setup 1

17.3.1 Analysis of the Ring Probe Circuit

The test (bottom) and balance (top) coils of the ring probe shown in Fig. 17.6 are connected in two arms of a bridge within the eddy-current test instrument, as shown in Fig. 17.13. Z_t is the driving-point impedance of the test coil in the presence of the workpiece (the "loaded impedance") and Z_b is the driving-point impedance of the balance coil. By the construction of the ring probe, as shown in Fig. 17.6, the cap

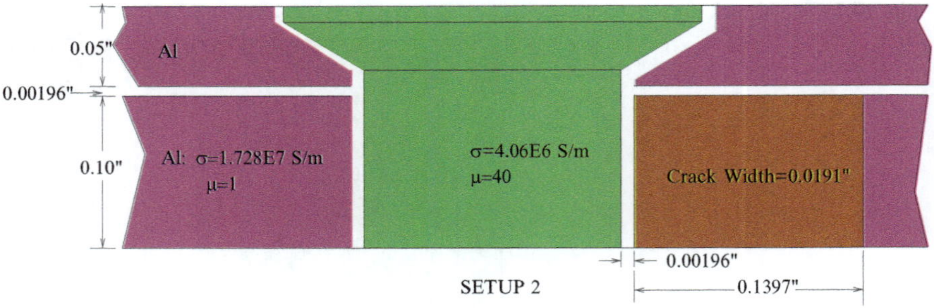

Fig. 17.10 Test setup 2

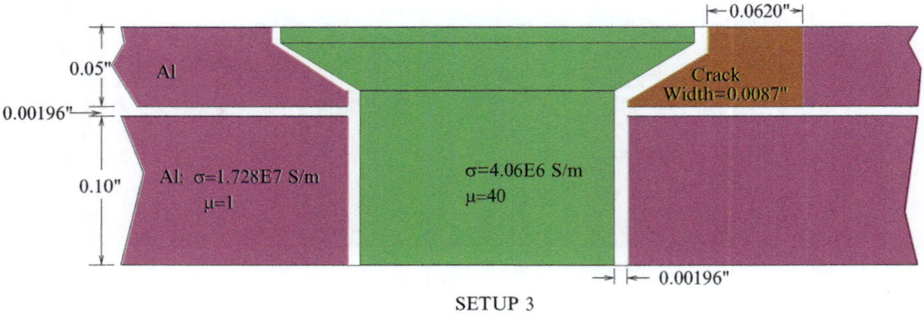

Fig. 17.11 Test setup 3

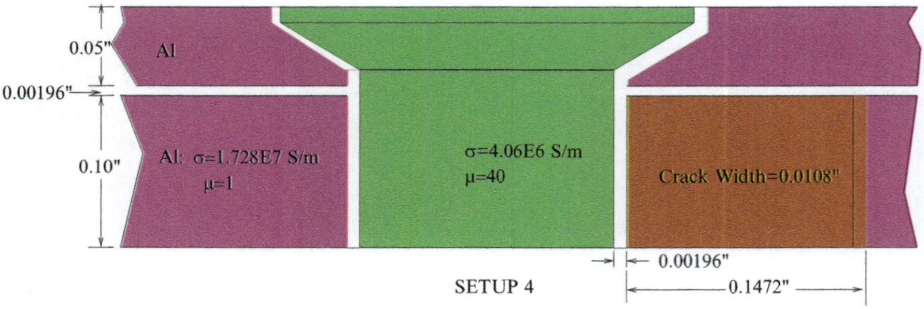

Fig. 17.12 Test setup 4

separating the test and balance coils isolates them electrically, so we assume that the mutual inductance between the two coils is small enough to ignore. That is, the cap is a "flux-shunt" that shunts the flux produced by either coil from linking the other coil. For the same reason, the balance coil is isolated from the workpiece, because the workpiece is beneath the cap. This means that the driving-point impedance of the balance coil will always be the freespace impedance, no matter where the probe is positioned.

Fig. 17.13 The eddy-current test instrument bridge circuit

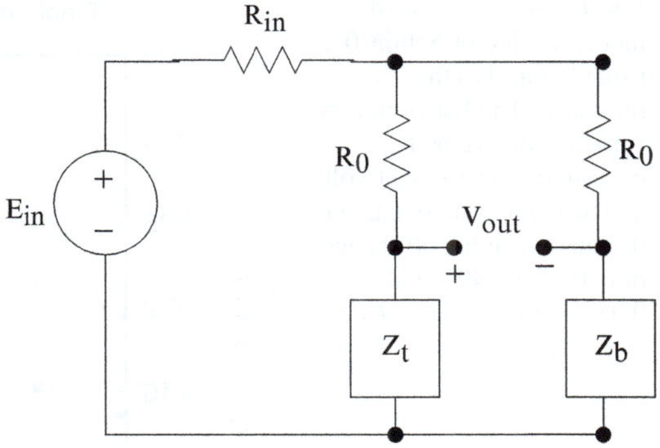

The bridge circuit of Fig. 17.13 is designed so that R_0 is much larger than either Z_t or Z_b. Under this condition, and the condition of zero mutual inductance between the two coils, it is a simple matter to compute the response of the bridge circuit to be

$$V_{\text{out}} \approx \frac{E_{\text{in}}}{R_0 + 2R_{\text{in}}} (Z_t - Z_b) \ , \tag{17.1}$$

which demonstrates that the bridge circuit acts as an impedance differentiator, and that the impedance measured is balanced against the freespace impedance of the test coil. This follows because the freespace impedance of the test and balance coils are equal, because the coils are identically situated when in freespace. The output of the test instrument is the "total signal" minus the "freespace signal," which, in the terminology of **VIC-3D®**, corresponds to the "host + flaw impedance."

The output of **VIC-3D®** is already balanced against both the freespace and host medium, so it appears that in order to compare the test-output with the **VIC-3D®** output, we would have to add the host-only impedance, which will be computed in a separate model. (All of the **VIC-3D®** calculations will be done without the need to model the balance coil.) Then we will scale the experimental data so that they agree with the **VIC-3D®**-generated model data and produce a scale-factor, β, which will then be used on all other problems at the same frequency.

The practical upshot of all of this is that because the subtraction (balancing) of the impedances is done analogically (in a bridge circuit), the probe requires two identical coils, with identical shielding. If the balancing were done numerically, one would need only the test coil, which would result in a much less expensive probe.

17.3.2 Model Results

Data for determining the presence of a flaw are obtained by varying the lift-off of the ring probe over the range 0, 0.25, 0.7493, and 0.9398 mm. By "lift-off" we mean the position of the bottom of the probe above the workpiece. The solution of the

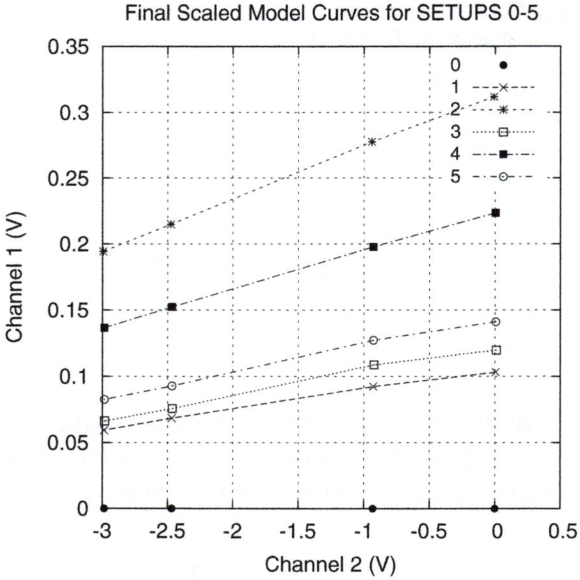

Fig. 17.14 Final scaled model results for Setups 0–5, using Setup 0 as the reference. The horizontal and vertical axes are both measured in instrument volts, V. The parametric points on the curves are lift-off values of 0, 0,25, 0.7493, and 0.9398 mm (*right to left*)

problem with 27,648 unknowns takes about 35 min for each lift-off, giving a total solution time of about 2.3 h for each setup. About 1Gb of storage for auxiliary files was required.

VIC-3D® computes impedances, but the industry typically uses analog instruments to measure data, so the final results of the computation are transformed into "instrument volts," as shown in Fig. 17.14. Channel 1 is called the vertical channel (on the oscilloscope) and indicates the difference signal between flawed and unflawed rivet. Channel 2 is the horizontal channel and corresponds to lift-off. The markers correspond to the four lift-off values of 0(mm), 0.25, 0.7493, and 0.9398, from right to left. Both channels (axes) are measured in instrument volts.

We showed in Chap. 15 how impedance calculations can be transformed into voltage measurements using calibration data with known samples (see also [102]). Setup 0 is the reference, so it is zero for each lift-off along the negative real axis. The remaining curves indicate the presence of a flaw and yield information that can be used in an inversion procedure.

It is clear that Setup 2 produces the largest signal, even though the flaw is in the second layer. This is due to the fact that the flaw in this setup has a larger volume than any of the other flaws. Furthermore, it is clear that the flaws in Setups 1 and 3, even though they are in the upper layer, are more difficult to distinguish than the other two flaws. This is due to the fact that these two flaws have a smaller volume than those in Setups 2 and 4. Setup 5 is the same as Setup 4, except that the length of the crack is 0.0986 in., instead of 0.1472 in. The response of Setup 5 is about 0.67 times the response of Setup 4, which is consistent with the fact that the response is roughly proportional to the volume of the defect.

Table 17.1 Convergence of Half-Sandwich No. 1 with σ_3

σ_3	Iterations	δR	δX	δZ
10^5	106	−0.85638	1.8332	$2.0234\angle 115.04$
10^4	186	−0.86202	1.8394	$2.0314\angle 115.11$
10^3	361	−0.86425	1.8425	$2.0351\angle 115.13$
100	422	−0.86506	1.8438	$2.0366\angle 115.14$
10	428	−0.86513	1.8439	$2.0368\angle 115.14$
1	431	−0.86516	1.8439	$2.0368\angle 115.14$

These studies indicate that we are capable of modeling and inspecting fastener-crack problems without removing the fastener, in contrast to the standard practice which uses a split-D probe, of the type shown in Fig. 6.16 of Chap. 6, that is inserted within the hole after the fastener is removed. Problems of this type, in which the fastener is removed before inspection, are called bolt-hole eddy-current (BHEC) problems, and we have already seen two examples of them in Figs. 10.23 and 10.24 of Chap. 10. In those examples, of course, the probe was a simple coil that was scanned along the surface, and not inserted within the hole.

Lift-off data are generally too sparse to allow a good inversion. At the time that the manuscript for this book was submitted to the publisher, the authors began a program that will lead to the validation of inversions of fastener-flaw systems without the need to remove the fastener, and using a surface probe of the type shown in Figs. 10.23 and 10.24. The anticipated success of this project would lead to a considerable simplification, with significant cost-savings, of the fastener-flaw inspection problem.

17.4 Further Tests with the SDA and "Weak-Host" Layers

17.4.1 Half-Sandwich No. 1

The setup for this exercise is shown in Fig. 17.15. The insert can be ferromagnetic, and we have run tests with $\sigma = 1.79 \times 10^6$ S/m and $\mu = 200$. In those runs, we used three SDA grids, each with $32 \times 32 \times 4$ cells, yielding over 62,000 unknowns. The convergence to 1 % took 251 iterations. With a nonpermeable insert, the number of unknowns is reduced to 33,024, and convergence to 1 % required 431 iterations. Apparently the permeability yields a better conditioned system. In those runs, the middle layer conductivity was $\sigma_3 = 0$, which made it a "weak-host" layer.

Our interest, now, is in testing the effect of varying σ_3 over several decades in order to determine the effect of a nonzero conductivity in speeding up the convergence of a problem in which $\sigma = 0$. For this exercise we made the insert nonpermeable, keeping the conductivity equal to 1.79×10^6 S/m, and used a convergence criterion of 1 %. The convergence results with respect to σ_3 are shown in Table 17.1.

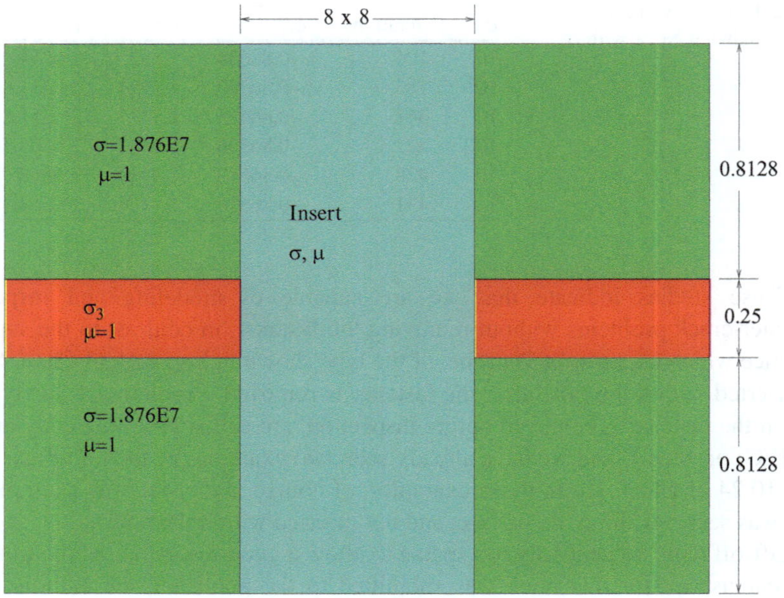

Fig. 17.15 Setup for Half-Sandwich No. 1. All dimensions are in mm

The immediate observations are that the model is robust and consistent, in the sense that there is a monotonic increase in the number of iterations required to achieve 1 % convergence. This is an indication that the problem becomes slightly less well conditioned as σ_3 decreases, which is to be expected. Further indication of the consistency in the results is that the magnitude of δZ increases monotonically with a decrease in σ_3, as we would expect because the electrical contrast of the anomaly increases with a decrease in σ_3 in the middle slab.

The most important observations, however, are that the phase of δZ is virtually independent of the value of σ_3, which means that the magnitude of δZ, alone, is sufficient to determine the effects of changing the conductivity of the slab, and furthermore, the magnitudes change by less than 1 %, which means that we could safely replace a slab of air with one of $\sigma_3 = 10^5$ S/m and get the same answer, while achieving a significant reduction of computation time. This could be extremely important when doing a large scan. We shall see how this can be very effective in the next example.

17.4.2 Half-Sandwich No. 2

This sandwich is shown in Fig. 17.16. Each of the three SDA grids has $16 \times 16 \times 2$ cells, which produces a total of 768 electric unknowns and 498 magnetic unknowns.

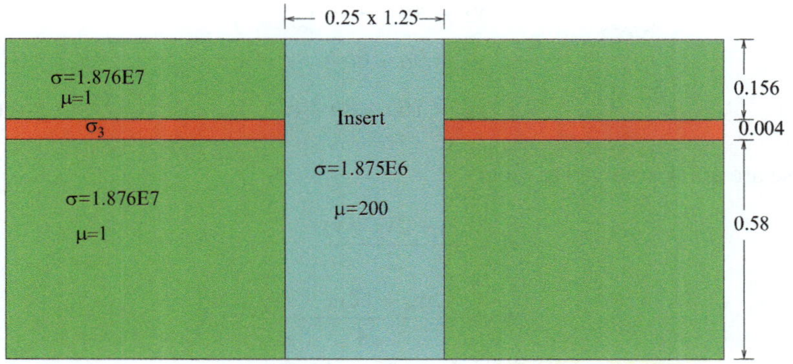

Fig. 17.16 Setup for Half-Sandwich No. 2. All dimensions are in inches

Table 17.2 Convergence of Half-Sandwich No. 2 with σ_3. The first four entries are computed, and the fifth is extrapolated using the first three entries

σ_3	Iterations	$\delta R \times 10^{-5}$	$\delta X \times 10^{-5}$	$\delta Z \times 10^{-5}$
10^4	520	−5.012	−3.8032	$6.2916\angle -142.81$
10^3	598	−5.4627	−4.1561	$6.864\angle -142.74$
100	1,332	−5.9106	−4.4743	$7.4132\angle -142.87$
0	4,268	−5.2723	−4.0658	$6.6579\angle -142.36$
1(Extrap.)		−6.798	−5.0066	$8.4427\angle -143.63$

We do another convergence study with respect to σ_3, with the results shown in Table 17.2. We note that even with a much smaller number of unknowns than in Half-Sandwich No. 1, the convergence of the iterative solver to a precision of 1 % requires a significantly larger number of iterations. We attribute this to the very small transition layer depth compared to the top and bottom of the sandwich (this conjecture will be supported in the next example: "Half-Sandwich No. 3").

The first four entries in Table 17.2 are computed, while the fifth is extrapolated using a quadratic algorithm that uses the first three entries. We'll discuss that shortly, but we want to first comment on the nature of the computed results. The first three entries show a consistent monotonic increase in magnitude, while their phases remain virtually equal, as we saw in Half-Sandwich No. 1. The fourth result, in which $\sigma_3 = 0$, does not fit the pattern, and that is due to the fact that our convergence criterion of 1 % is too stringent, given the relatively poor conditioning of the operator (the large number of iterations required to attain 1 % convergence indicates this). Tightening the criterion to 0.5 % or smaller to get consistency would defeat the purpose of extrapolating the better-conditioned models to obtain the desired model.

The extrapolation algorithm used to obtain the final result is quadratic and will be developed here. Let $y = ax^2 + 2bx + c$ be the quadratic relation between the impedance component and $x = \log \sigma_3$, and let y_2, y_3, and y_4 be values of the impedance at x_2, x_3, and x_4, respectively (the subscript denotes the logarithm of σ_3). We then have

$$
\begin{aligned}
y_2 &= 4a + 4b + c \\
y_3 &= 9a + 6b + c \\
y_4 &= 16a + 8b + c.
\end{aligned}
\tag{17.2}
$$

These are quickly solved to yield

$$
\begin{aligned}
a &= \frac{y_4 - 2y_3 + y_2}{2} \\
b &= \frac{-5y_4 + 12y_3 - 7y_2}{4} \\
c &= 6y_2 - 8y_3 + 3y_4,
\end{aligned}
\tag{17.3}
$$

where c is the sought-for solution for $\sigma_3 = 1$, which is close enough to zero for our purposes.

Referring to Table 17.2, we take y_4 to be the entry for either δR or δX for $\sigma_3 = 10^4$, and so on for y_3 and y_2. Using these values, we compute the entries for δR, δX, and δZ in the fifth row "(Extrap.)". Our strategy, then, is to do the **VIC-3D®** model scans using $\sigma_3 = 10^4$ S/m, and then multiplying these results by the ratio

$$
\begin{aligned}
\frac{8.4427e^{-j143.63}}{6.2916e^{-j142.81}} &= 1.3419e^{-j0.82} \\
&= 1.3418 - j0.0192\,.
\end{aligned}
\tag{17.4}
$$

17.4.3 Half-Sandwich No. 3

This is the same as Half-Sandwich No. 1, except that the gap thickness is 0.025 mm, and the top and bottom halves of the sandwich are 0.9253 mm thick, yielding the same overall height of the sandwich of 1.8756 mm. Our interest is in determining the effect of gap thickness on convergence. For Half-Sandwich No. 3, after 1,001 iterations, the convergence criterion was reduced to only 6.559 %, which clearly shows the deleterious effect that a very thin gap has on convergence. This supports our contention that the slow convergence of Half-Sandwich No. 2 is largely due to the very thin airgap.

17.5 Comments and Conclusions

We have shown that the volume-integral approach has significant advantages over other numerical methods such as the finite-element method, in that the formulation of the numerical model is much simpler in the former than in the latter. Furthermore, the solution time for **VIC-3D®** is extremely short for many problems in NDE, because the formulation in terms of the Galerkin variant of the method of moments on a regular grid results in operators that have very special structures; they are either three-dimensional convolutions, or two-dimensional convolutions and one-dimensional correlations, which means that we can use three-dimensional FFT's to evaluate them in a conjugate-gradient search algorithm. The use of a highly irregular mesh in the finite-element technique does not allow a similar advantage in the solution process.

This advantage accrues from the very different nature of the physics that goes into the formulation of the mathematical models. In volume-integral equations, as well as boundary-integral equations, the unknowns are anomalous currents that are supported in a compact domain, namely the domain of the anomaly; in the example of the Cessna Series, the anomaly is the rivet or rivet and crack. In finite-element or finite-difference methods the unknowns are the electric and magnetic fields, which extend to infinity. This has two disadvantages; it increases the number of unknowns and requires some method of approximating the "boundary-at-infinity" in order to truncate the problem domain. This increases the complexity in simply defining the model and presents an extreme challenge to prospective users who are not skilled in computational electromagnetics. Furthermore, the FEM method is not particularly well suited to solve typical problems in NDE, because the anomalies, such as rivets or cracks, require a very complicated mesh, with a large number of very irregular cells.

Finally, we conclude from the model calculations of the Cessna Series that the spatial decomposition algorithm is a very efficient method of solving problems in which an anomaly extends through layers with different electric and magnetic properties. The spatial decomposition algorithm formulation was also demonstrated to be valid for cases where a conducting anamoly is present in a nonconducting layer such as air. This approach greatly expands the capability of volume-integral equation methods for complex problems in computational electromagnetics, and in particular nondestructive evaluation. Lastly, rapid solution of the forward problem for computational electromagnetics will also be beneficial for the practical application of advanced inverse method techniques for quantitative nondestructive evaluation of material discontinuities.

Chapter 18
Applications to Nuclear Power

The nuclear-power industry faces the serious challenge of convincing a skeptical public and regulatory agencies that it can operate safely and efficiently. Nondestructive evaluation (NDE) plays a significant role in this task, and computer modeling is playing a significant role in NDE. The industry now realizes the value of using such modeling to replace expensive experimental tests, as well as to design equipment, and interpret results. Eddy-currents have a traditional place in the inspection of steam generator tubing, and the industry seeks improved tools for such inspections.

In this chapter we demonstrate the value of model-based inversion in providing that tool. The problems that we solve are similar to those in the preceding chapters of this book, with the main difference being the need to model workpieces with cylindrical geometries rather than planar. The tools for carrying out this model were developed in Chap. 9.

We start the chapter with a discussion of an extremely important problem in the NDE of heat-exchanger tubes in nuclear power plants: the ability to model and reconstruct pitting and corrosion phenomena on the inner diameter of these tubes. We then extend our study to other classes of anomalies in tubes, while continuing to emphasize the role of model-based inversion in solving these problems.

18.1 Modeling Pitting and Corrosion Phenomena in Heat-Exchanger Tubes

18.1.1 An Ellipsoidal Model for Pits

The photomicrographs shown in Fig. 18.1 illustrate the results of long-term, under deposit pitting corrosion on the inner-diameter of a section of a component

H.A. Sabbagh et al., *Computational Electromagnetics and Model-Based Inversion*, Scientific Computation, DOI 10.1007/978-1-4419-8429-6_18,

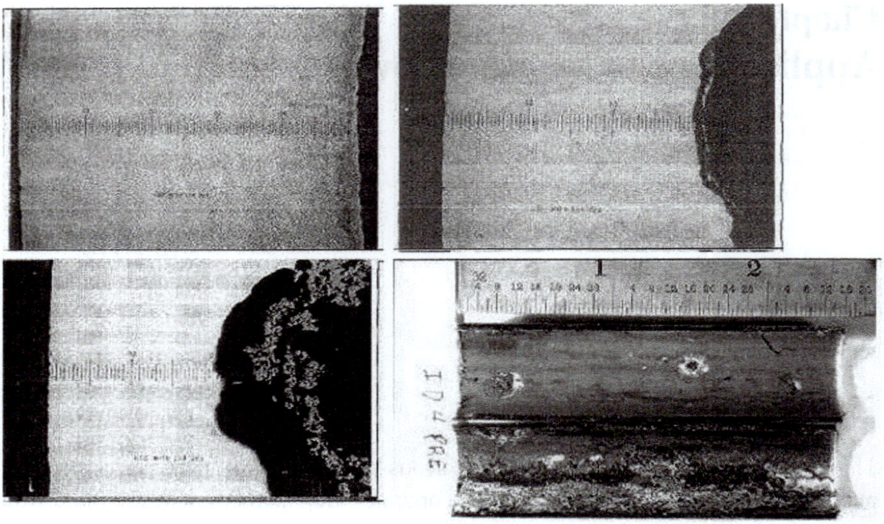

Fig. 18.1 Photomicrographs showing the effects of long-term, under deposit pitting corrosion of a heat exchanger tube whose nominal wall thickness was 49–51 mils. *Upper-left*: little corrosion; *Upper-right*: remaining wall of 29 mils; *Lower-left*: remaining wall of 20 mils; *Lower-right*: actual pits, with an indication of 88 % through-wall within the middle third of the sample shown, based on nominal 0.049 inch

cooling heat exchanger tube whose nominal wall thickness is 49–51 mils.[1] They further suggest that we can construct a reasonable model of corrosion pits as three-dimensional semiellipsoids that originate in the inner-radius of tube walls. Figure 18.2 illustrates the model, together with its three defining parameters, the semiaxes, A, B, and C. By modeling the pit in this manner, we cast the problem into the model-based inversion paradigm. In this case, we hope to determine the three semiaxes, thereby solving the inverse problem. Clearly, C determines the depth of the pit.

Furthermore, the photographs give us insight into the sizes of typical pits, which can then be translated into values of A, B, and C for typical pits. Indeed, from these photographs we see that the pits appear to be circular, when looking into the inner surface of the tube, with a nominal radius of 0.0625 inch. Hence, we model these pits by setting A and B equal to 0.0625. We then choose C to be equal to the difference between 0.049 inch (the nominal wall thickness) and the minimum measured wall thickness, as given in the photomicrographs. We assume that the pits are empty, but, it would be quite easy to model pits filled with corrosion deposits, whether conducting or magnetic (or both).

[1] These photographs and the data concerning them are courtesy of Exelon PowerLabs.

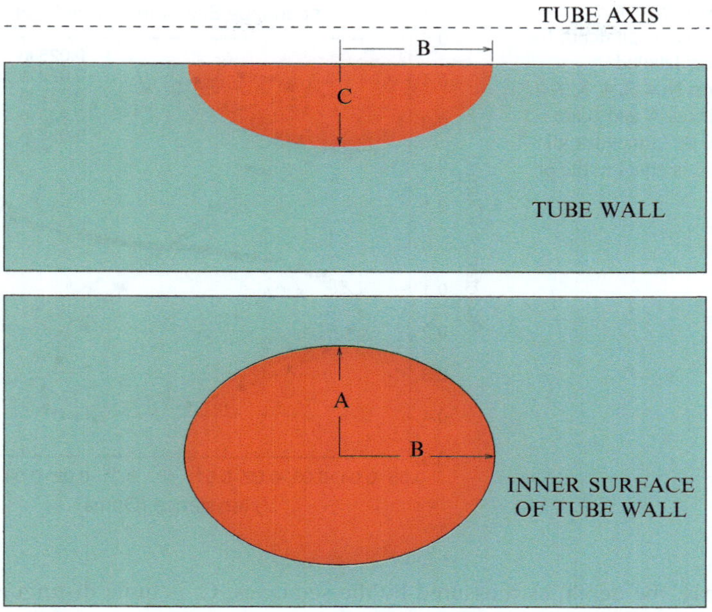

Fig. 18.2 Illustrating a model of pits as three-dimensional semiellipsoids that originate in the inner-radius of a tube wall. The ellipsoid is defined by its three semiaxes, A, B, and C

Table 18.1 Measured pit data for a component cooling heat exchanger tube

Indication	Minimum measured remaining wall (in)	Pit depth (in)	Calculated % thru-wall based on nominal 0.049 in	Eddy current % thru wall
No. 3	0.023	0.026	53	59
No. 4	0.015	0.034	69.4	88
No. 5	0.020	0.029	59.2	89
No. 7	0.016	0.033	67.3	65
No. 8	0.015	0.034	69.4	97
No. 9	0.017	0.032	65.3	100

18.1.2 Computing Impedance Signatures with VIC-3D®

The pit data of Table 18.1 are measurements taken on the component cooling heat exchanger tubes of Fig. 18.1. The first calculations that we make using **VIC-3D®** are to determine the impedance signature versus depth for the various pits labeled "Indication Nos. 3–9," omitting No. 8, because it has the same depth as No. 4. In all cases, we fixed A and B to be equal to 0.0625 inch, as explained above. The results are shown in Fig. 18.3.

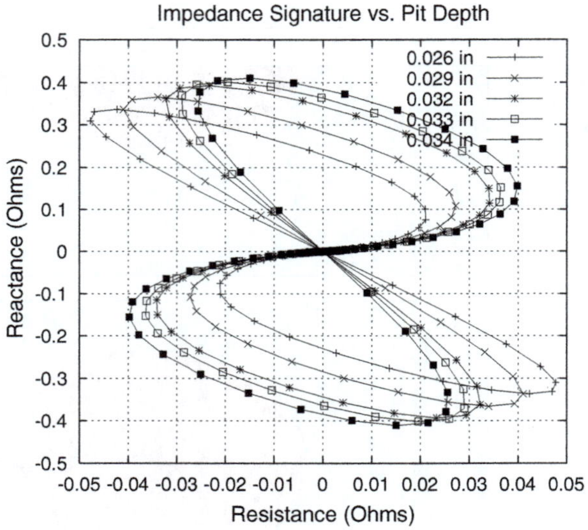

Fig. 18.3 Impedance signature versus pit depth for the five pits labeled "Indication Nos. 3, 4, 5, 7, 9." The semiaxes A and B are set equal to 0.0625 inch for all five, and C is equal to the pit depth

Clearly, the depth, as measured by the semiaxes, C, is quite distinguishable in these curves because the slope of the corresponding Lissajous figure[2] changes with C. This is the standard eddy-current inspection technique and is based on visual observation of the curve produced in an oscilloscope of an analog instrument. It suffers from two disadvantages: (a) noise will usually obliterate the curve, making it impossible to define a clear slope for the curve and (b) without having prior knowledge of the lateral dimensions of the pit, the A and B semiaxes, it is impossible to precisely define C. This is clear from Table 18.1, wherein the eddy-current technique generally predicted a pitting depth that was larger than true.

In carrying out the model calculations, we assume that the tube is made of 90–10 Copper–Nickel, whose electrical conductivity is $\sigma = 5.277 \times 10^6$ S/m (9.1 % IACS), and is nonmagnetic. The outer diameter of the tube is 0.75 inch, and the nominal wall thickness is 0.049 inch. The coil parameters are given in the table and the frequency of operation is assumed to be 70kHz.

Coil spacing	0.060 inch
Coil height	0.055
Coil width	0.060
Coil OD	0.576
Coil ID	0.466
Coil turns	156

[2]Lissajous figures are normally associated with "differential-bobbin" probes, as described in connection with Fig. 18.19.

Table 18.2 Estimate of A, B, and C for the semiellipsoidal representation of the pits labeled Indication Nos. 3–9

Indication	Original model data (in)	Reconstructed (inverted) data (in)
No. 3	A = 0.0625, B = 0.0625, C = 0.026	A = 0.0643, B = 0.0634, C = 0.0263
No. 4	A = 0.0625, B = 0.0625, C = 0.034	A = 0.0644, B = 0.0639, C = 0.0325
No. 5	A = 0.0625, B = 0.0625, C = 0.029	A = 0.0645, B = 0.0637, C = 0.0285
No. 7	A = 0.0625, B = 0.0625, C = 0.033	A = 0.0645, B = 0.0639, C = 0.0316
No. 8	A = 0.0500, B = 0.0700, C = 0.034	A = 0.0550, B = 0.0686, C = 0.0328
No. 9	A = 0.0625, B = 0.0625, C = 0.032	A = 0.0645, B = 0.0639, C = 0.0309

18.1.3 The Inversion Problem

Now we consider the inversion problem in detail. We take the model data of Fig. 18.3 and use them to simulate measured impedance data, which are then submitted to NLSE. The output of the estimator is an estimate of the semiaxes, A, B, and C of the ellipsoidal pit, as displayed in Table 18.2. It only takes a few minutes to compute a table for the inversion algorithm, which is then used for all pits. Subsequently, it takes about 2–4 s to compute the results for each pit.

Indication No. 8 was handled differently from the others. Because the X-Y cross-section of this pit was not circular, we found it more reliable to acquire data for it using a surface pancake probe scanned in a two-dimensional raster pattern. These data were then submitted to the estimator exactly as were the other five.

The results displayed in Table 18.2 indicate that the algorithm for nonlinear estimation is reliable and robust. The maximum error occurs in A for Indication No. 8, and is only 10 %, which is quite tolerable for this dimension (the width of the pit). The important point to be made from this study is the reliability in reconstructing C, the semiaxes that determines the depth of the pit. Indeed, the results of the algorithm indicate that the computation is more sensitive to depth then to the lateral dimensions, which is exactly what we're looking for!

18.1.4 Inversion of a Complex Pit

Figure 18.4 illustrates a complex pit, comprising three intersecting semiellipsoids. The A and B semiaxes are 0.0625 inches, as before, and the C semiaxes are 0.029, 0.026, and 0.034 inches. In this exercise, we assume that the A and B semiaxes are known, and the job is to determine the profile-in-depth of the pit, that is, to determine the three C-axes. The input data to NLSE at six frequencies, 10, 20, 40, 70, 200, and 700 kHz are shown in Fig. 18.5. The results of applying NLSE at these frequencies are shown in Table 18.3. The interesting result is the average over the six frequencies. Clearly, the reconstruction in depth is robust, especially when frequency scanning is incorporated along with spatial scanning. The middle lobe, which is effectively lost in the presence of its two larger neighbors, is still quite accurately reconstructed.

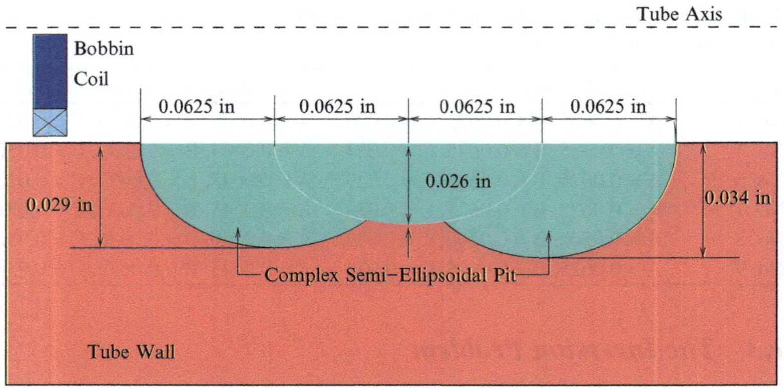

Fig. 18.4 Illustrating a complex pit that comprises three intersecting semiellipsoids, whose dimensions are shown

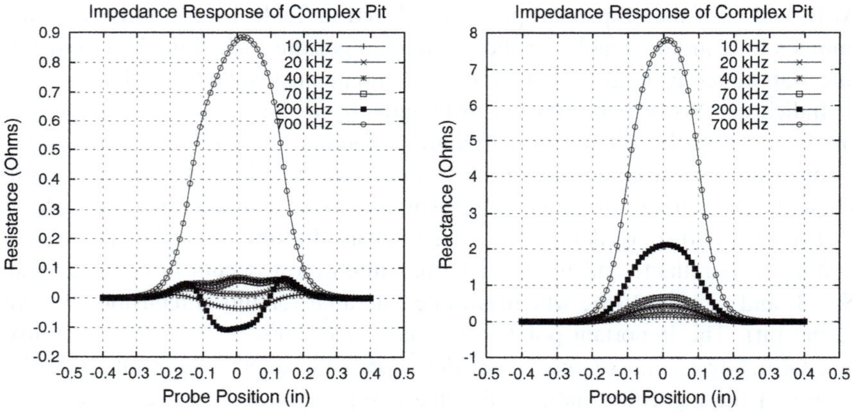

Fig. 18.5 The model input impedance data to NLSE at 10, 20, 40, 70, 200, and 700 kHz: resistance (*left*), reactance (*right*)

Table 18.3 Estimation of three C-axes for the complex semiellipsoidal pit model. The row labeled "Freq. Avg." is the average of the results over the six frequencies

Freq (kHz)	Result
10	(0.026, 0.022, 0.032)
20	(0.026, 0.023, 0.032)
40	(0.026, 0.023, 0.032)
70	(0.028, 0.023, 0.033)
200	(0.032, 0.026, 0.036)
700	(0.033, 0.030, 0.037)
Freq. Avg.	(0.029, 0.025, 0.034)
True	(0.029, 0.026, 0.034)

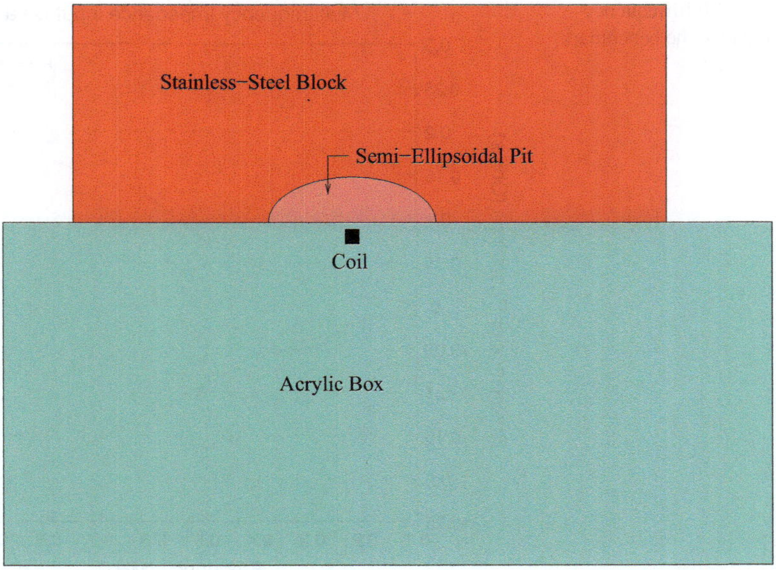

Fig. 18.6 A multifrequency validation test. The coil is excited at 10 frequencies between 0.1 and 1.0MHz

18.1.5 A Multifrequency Benchmark Inversion Test

In order to validate our inversion algorithm that utilizes NLSE, we perform a multifrequency experiment on a block of stainless steel that contains a machined semiellipsoid, whose semiaxes were nominally $0.0625 \times 0.0625 \times 0.026$ in. The data are obtained by simply laying the stainless-steel block over a coil that is embedded in an acryllic block, as in Fig. 18.6, and then exciting the coil over the frequency range of 0.1–1 MHz, in 100 kHz steps. The "pit" is positioned as accurately as possible to be directly above the coil, whose dimensions are 15 mils inner radius, 48 mils outer radius, 7 mils height, and is wound with 19 turns over two layers. The measured dc resistance of the coil is 0.741 Ω, and the measured dc inductance is 0.51092 μH.

The first task is to characterize the host material and to determine the lift-off of the coil over the host. This is done by taking multifrequency data when the coil is over the unflawed workpiece, and then introducing these data into NLSE. The results are that $\sigma_{\text{host}} = 1.1763 \times 10^6$ S/m, lift-off = 10.201 mils. These, then, are the "host parameters" for the pit-inversion process.

The multifrequency input (impedance) data obtained when the pit is over the coil are shown in Fig. 18.7, and these data are presented to NLSE in order to determine the X,Y coordinates of the center of the coil from the center of the pit, as well as the values of the three semiaxes. The result of the inversion gives X and Y coordinates of 6.711 mils and 6.695 mils, respectively, for the location of the probe from the center

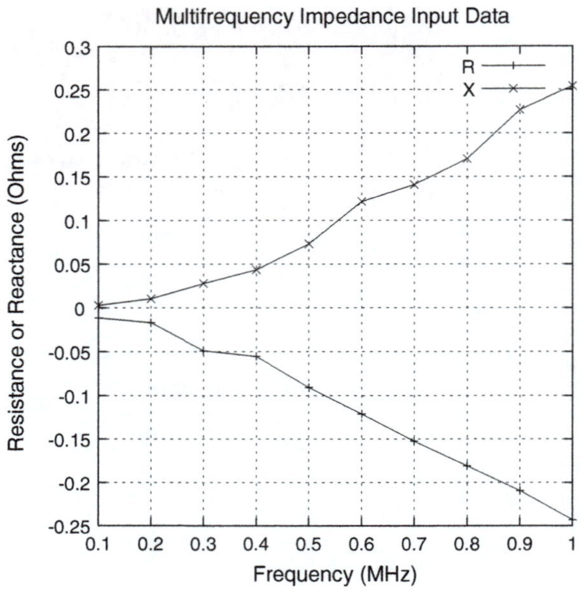

Fig. 18.7 Multifrequency input data for the benchmark inversion test

of the pit. The dimensions of the pit semiaxes are: A = 0.0737 in., B = 0.0737 in., C = 0.0279 in. The solution is least sensitive to the lateral coordinates of the probe relative to the pit, and most sensitive to the depth of the pit, as we noted before.

This validates the NLSE inversion process, as it is applied to semiellipsoidal pit models. Further, this process illustrates another method for obtaining data when a physical scan is difficult or impossible. It would be impossible to infer the solution for five parameters by using analog instruments and checking traces on an oscilloscope. This, again, demonstrates the value of model-based inversion in eddy-current NDE.

18.1.6 Modeling Wall-Thinning Effects

Wall-thinning is often associated with corrosion effects, and it is necessary to distinguish wall-thinning from such things as pitting. Following the same strategy used in the complex pit model, we can model wall-thinning in a tube by stacking axisymmetric rings contiguous to each other, as shown in Fig. 18.8. In this case, we agree that the axial resolution for wall-thinning will be 0.2 inch, and we take enough rings to cover the region of interest. The radius of each ring determines the skyline profile of the inner radius of the tube. Because we assume that bobbin coils are being used in gathering data, it follows that we can only assume the wall-thinning to be axisymmetric. In the model shown in Fig. 18.8 the inverse problem

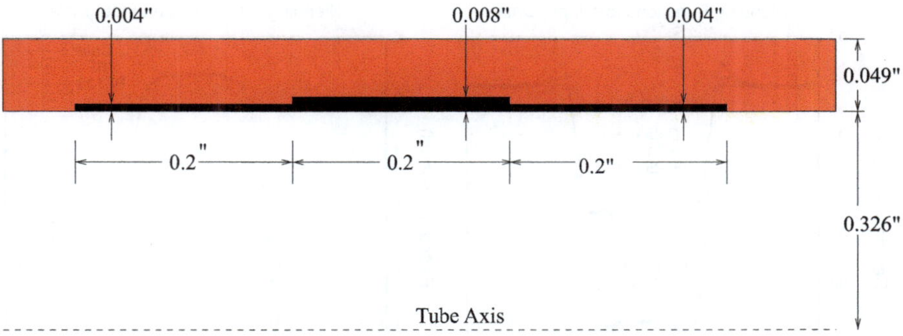

Fig. 18.8 Modeling localized wall-thinning by using a series of axisymmetric rings

Fig. 18.9 Input data for inverse problem of wall-thinning

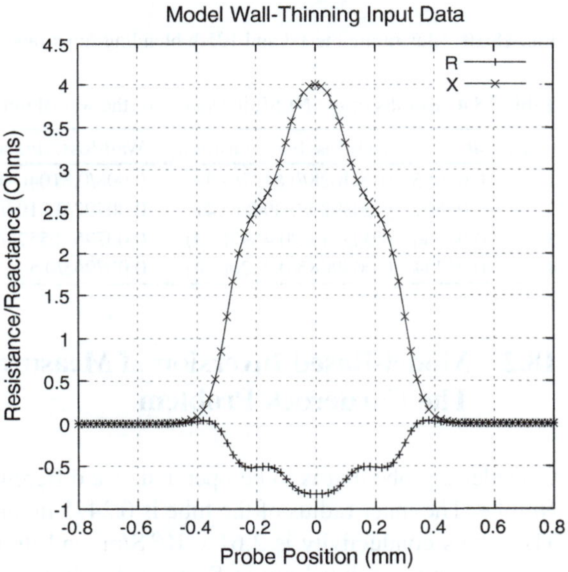

is to determine the radii of the three rings. The model input data are the scanned impedances associated with the configuration shown in the figure, and are shown in Fig. 18.9.

In order to invert these data using up to a fourth-order polynomial, we first establish a $5 \times 5 \times 5$ interpolation table, in which each wall-loss variable has nodes at 0.0005, 0.003, 0.0055, 0.008, and 0.0105 inch. The blending functions associated with the 1st and 125th nodes are shown in Fig. 18.10, along with the input data of Fig. 18.9.

The results of the NLSE inversion are tabulated in Table 18.4.

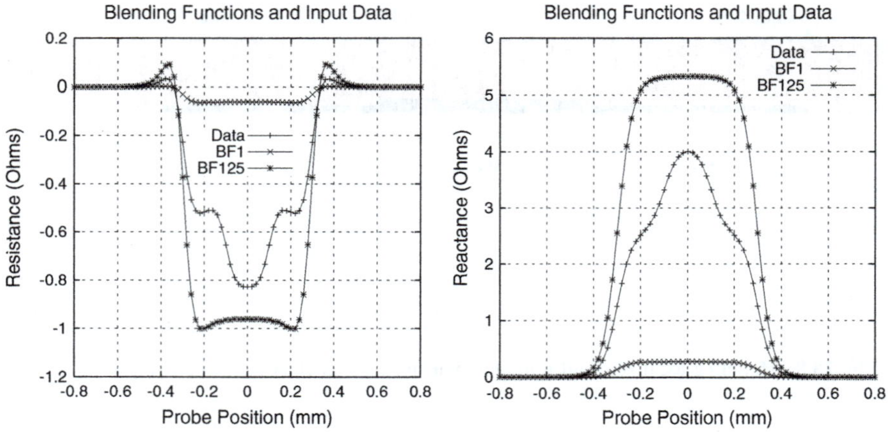

Fig. 18.10 Illustrating the 1st and 125th blending functions, together with the input data

Table 18.4 Results of the NLSE inversion for the wall-thinning problem

Order	Φ	Wall-loss1(in)/sensitivity	Wall-loss2(in)/sensitivity	Wall-loss3(in)/sensitivity
1	0.05056	0.004028/0.6127(−4)	0.008/0.8104(−4)	0.004028/0.6127(−4)
2	0.03609	0.003939/0.3948(−4)	0.00797/0.5103(−4)	0.003939/0.3948(−4)
3	0.04160	0.0039175/0.4593(−4)	0.00795/0.5578(−4)	0.0039175/0.4593(−4)
4	0.04734	0.0039085/0.5220(−4)	0.007945/0.5905(−4)	0.0039085/0.5220(−4)

18.2 Model-Based Inversion of Measured Impedance Data: The Carderock Problem

Consider a probe that is to be operated as a differential-bobbin within a 70/30 Cu-Ni tube. The inner radius of the tube is 0.2475 in and the outer radius is 0.3125 in. The tube's conductivity is 2.61×10^6 S/m, and its relative magnetic permeability is unity. As we pointed out in Sect. 15.3.1, in modeling such a probe, we simply take the response to a single (absolute) bobbin coil, and then interpolate within that response to compute the response of the second coil, from with the differential response is computing by subtraction.

When the absolute coil and its connecting cable are scanned over a 0.25-inch, through-wall round hole in the same tube at 88 kHz, the resulting measured data are shown in Fig. 18.11, along with the results from a **VIC-3D**®-model of the problem.[3] Except for the host-only offset in the two data sets, the results are in excellent agreement, the l_2-norm error being 1.55 %. The host-only offset is probably attributable to uncertainty in the original data of the probe in freespace,

[3]The measured data and parameters of this experiment were provided by J. Liu and N. Trepal of the Naval Surface Warfare Center, Carderock Division.

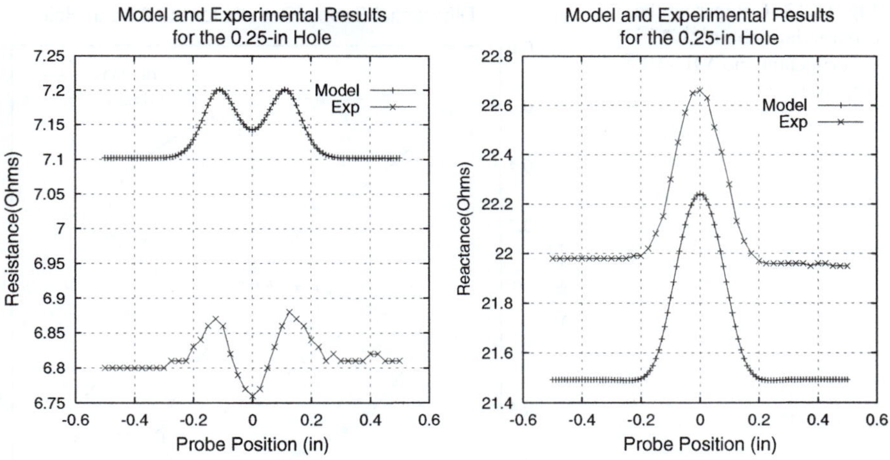

Fig. 18.11 Model and experimental data for a 0.25 inch, through-wall round hole model of a pit at 88 kHz

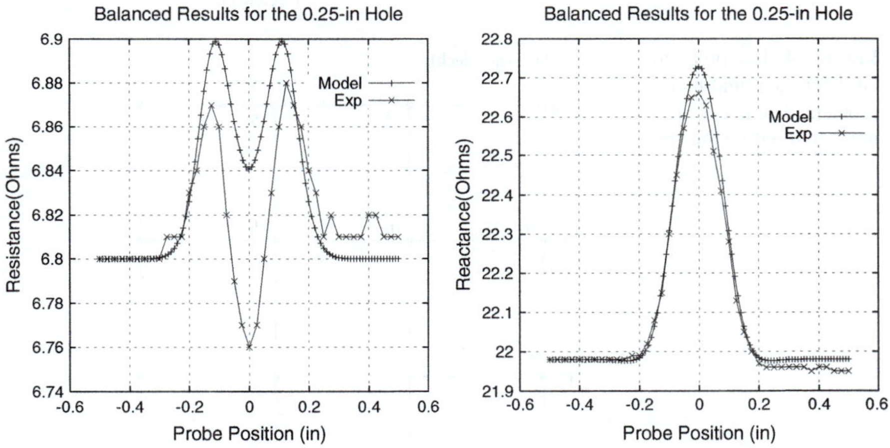

Fig. 18.12 Model and experimental data for a 0.25 inch, through-wall round hole model of a pit at 88 kHz after the host-only offset has been removed

without the cable.[4] Figure 18.12 compares the responses when the model results are "balanced" against the measured data by forcing the host-only offsets to be removed. Now the l_2-norm error is less than 0.1 %.

A comparison of the differential-bobbin responses is shown in Fig. 18.13. It is clear from this figure that noise in the measured data will produce a Lissajous figure

[4]See Appendix A.1 for a discussion of probe + cable effects.

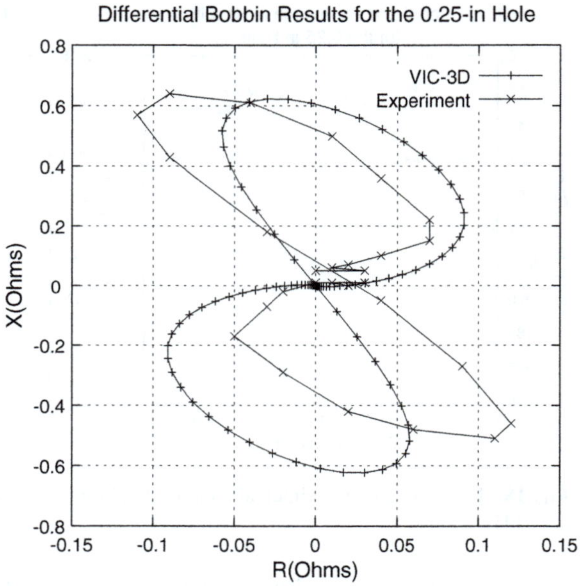

Fig. 18.13 Comparing the differential-bobbin experimental and **VIC-3D®** responses

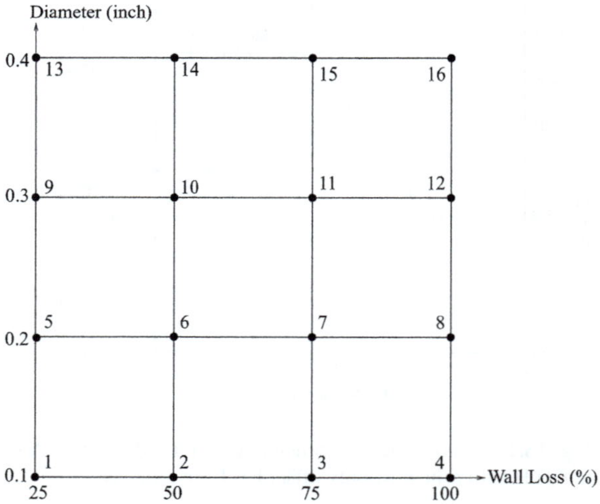

Fig. 18.14 Interpolation table for the round-hole problem. The nodes are numbered within the grid

that will not allow a very accurate assessment of the size and depth of the pit. This is the reason for doing model-based inversion, as we alluded to in the preceding section.

The first step in applying model-based inversion to this problem is to develop an interpolation table for the variables of interest. Figure 18.14 illustrates one such table for the variables of diameter and wall-loss (depth) of a round-hole. We use **VIC-3D®** to generate the standards (blending-functions) associated with each node of the interpolation table. These functions are shown in Figs. 18.15–18.18.

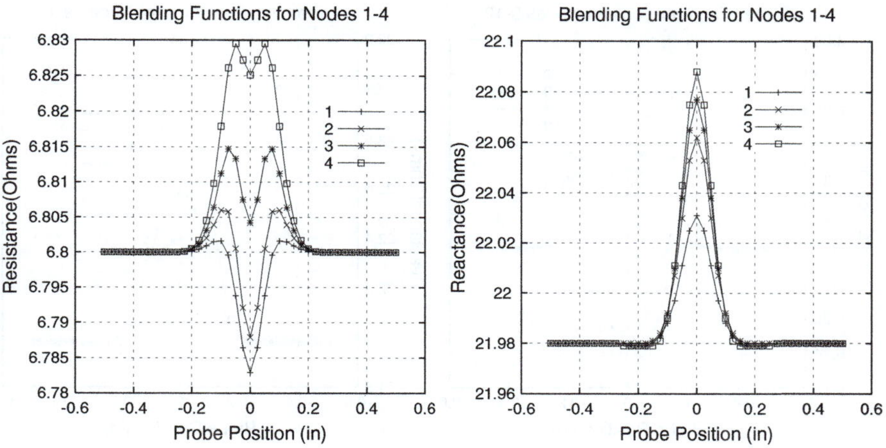

Fig. 18.15 Blending functions for nodes 1–4

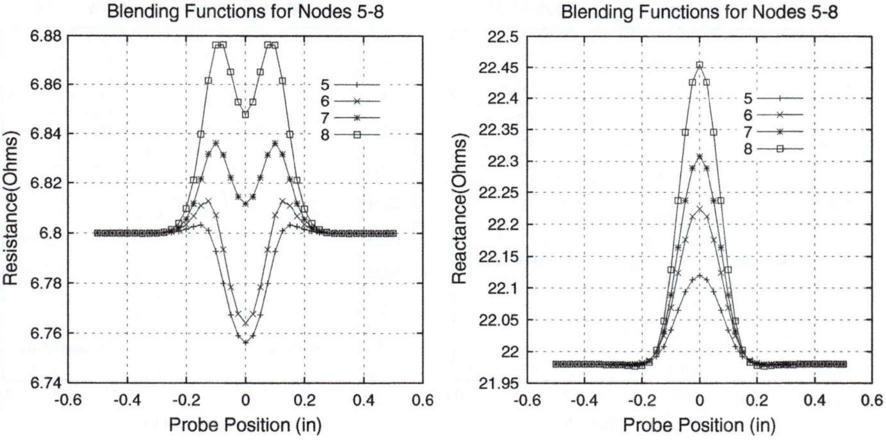

Fig. 18.16 Blending functions for nodes 5–8

The results when the experimental data of Fig. 18.12 are used as the input to NLSE are shown in Table 18.5. The heading labeled "Order" denotes the order of the interpolating polynomial and Φ denotes the norm of the residuals. Recall that "sensitivity" is the parameter that determines the sensitivity of Φ to that unknown—the smaller the value of "sensitivity" the more sensitive to that parameter. In addition to Φ, the sensitivity parameter is important in determining the quality of the solution, especially in making a judgement as to the importance of an unknown parameter, or in determining the confidence one has in the value of that parameter at the solution.

The corresponding results when the model data of Fig. 18.12 are used as the input to NLSE are given in Table 18.6. The purpose of showing these results is to

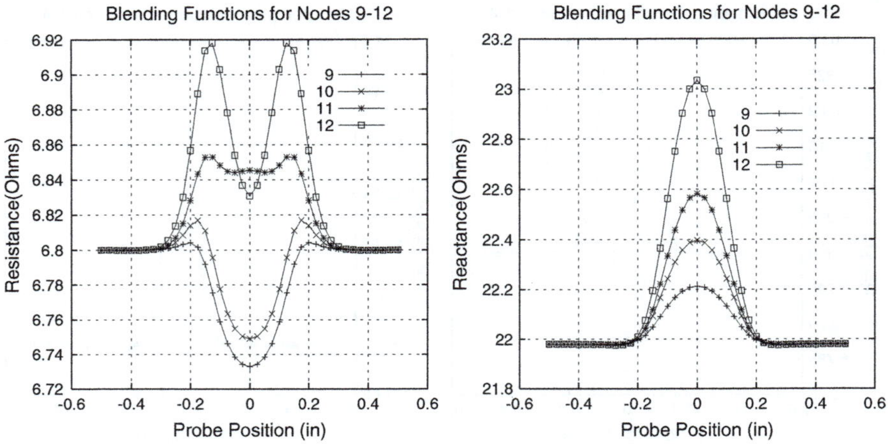

Fig. 18.17 Blending functions for nodes 9–12

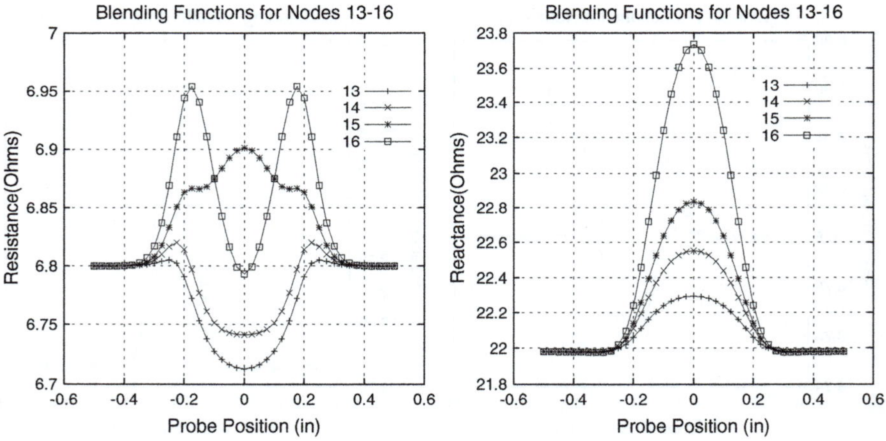

Fig. 18.18 Blending functions for nodes 13–16

Table 18.5 Response of NLSE when experimental data are used

Order	Φ	Diameter/sensitivity	Depth/sensitivity
1	0.2116	0.2479/0.1466(−1)	0.0601/0.4877(−2)
2	0.2123	0.259/0.1487(−1)	0.0588/0.4355(−2)
3	0.2163	0.2594/0.1523(−1)	0.05895/0.4467(−2)
Nominal		0.25	0.065

establish a standard for comparison. The fact that Φ in Table 18.5 is only about an order of magnitude larger than that in Table 18.6 indicates that the results shown in Table 18.5 are acceptable.

Table 18.6 Response of NLSE when model data are used

Order	Φ	Diameter/sensitivity	Depth/sensitivity
1	0.6894(−1)	0.2443/0.4083(−2)	0.065/0.1655(−2)
2	0.1767(−1)	0.2509/0.1018(−2)	0.065/0
3	0.181(−1)	0.2513/0.1054(−2)	0.065/0.3913(−3)
Nominal		0.25	0.065

Table 18.7 Response of NLSE after first iteration of POCS

Order	Φ	Diameter/sensitivity	Depth/sensitivity
1	0.127	0.2536/0.7522(−2)	0.065/0.3393(−2)
2	0.1230	0.2448/0.7617(−2)	0.06386/0.2997(−2)
3	0.1231	0.2477/0.7865(−2)	0.06293/0.2799(−2)
Nominal		0.25	0.065

Even though these results are acceptable, we can improve them by performing the iterative-refinement algorithm that is based upon projections onto convex sets (POCS) (recall Chap. 14, Fig. 14.3). We start the next iteration of POCS by computing the impedance response to a hole that has a diameter of 0.2479 inch and a depth of 0.0601 inch. These are the estimated parameters shown in Table 18.5. Then we take the computed resistance as the next estimate of the "correct" resistance, but replace the computed reactance by the original measured reactance. This keeps us in touch with the original measurements. Using the "corrected resistance" and the original reactance, we rerun NLSE and get the new inversion results shown in Table 18.7. Clearly, this is a significant improvement over the previous results of Table 18.5, both with respect to a smaller Φ, and for a more accurate estimate of the diameter and depth of the hole, with smaller sensitivity values. Note that in all of our inversion experiments, we find that the sensitivity in the depth parameter is better (smaller) than in the diameter. This is typical for problems, such as corrosion modeling, in which these are the two parameters of interest.

18.3 Model-Based Inversion of Measured Instrument Data: The EPRI ID Pits Benchmark Test

Figure 18.19 illustrates the model of the EPRI ID Pits test problem.[5] The parameters of the coil are estimates, only. Each coil is a bobbin coil, and their individual responses are called "absolute." When the two coils are connected in a bridge circuit, they form a differential bobbin, whose response is essentially the derivative of the

[5]The test setup and measured data for this section were supplied by K. Krzywosz of the Electric Power Research Institute (EPRI) as part of its benchmark validation test for sizing inner-diameter (ID) pit models.

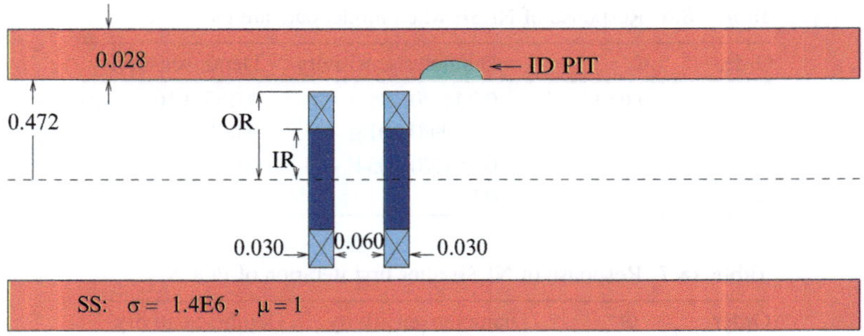

Fig. 18.19 Schematic of the EPRI ID Pits test problem. The parameters of the coil are estimates, only

Table 18.8 Labeling the 999-series of round-bottom pits

Pit no.	Diameter (in)	Height (in)	Wall loss (%)
1	0.0625	0.007	25
2	0.0625	0.014	50
3	0.0625	0.021	75
4	0.0625	0.028	100
5	0.09375	0.007	25
6	0.09375	0.014	50
7	0.09375	0.021	75
8	0.09375	0.028	100
9	0.125	0.007	25
10	0.125	0.014	50
11	0.125	0.021	75
12	0.125	0.028	100

absolute response. In **VIC-3D®** we simulate the differential-bobbin response in a post-processing filter, without the need to actually create a physical differential-bobbin probe with the accompanying bridge circuit, or other circuit that performs the differencing operation. The results of this section are obtained by means of the simulated differential-bobbin probe.

Calibration standard data are taken at 1 MHz, 800 kHz, 700 kHz, 600 kHz, 500 kHz, 400 kHz, 300 kHz, and 100 kHz in differential and absolute modes on a series of flaws with 25 %, 50 %, 75 % and 100 % wall loss. The 999-series comprises round bottom pits of 1/16″, 3/32″' and 1/8″' diameters, the 888-series comprises elongated pits 1/32″W×3/16″L, the 777-series comprises elongated pits 1/16″W×3/16″L, and the 666-series comprises elongated pits 1/32″W×1/8″L.

The 999-series is identified in Table 18.8. We will apply **VIC-3D®** to develop model-based "equivalent pillbox" standards for this series at 1 MHz, 800 kHz, 600 kHz, and 400 kHz. By an equivalent pillbox, we mean a truncated right-circular cylinder that is defined by its diameter and height, and that approximates the true

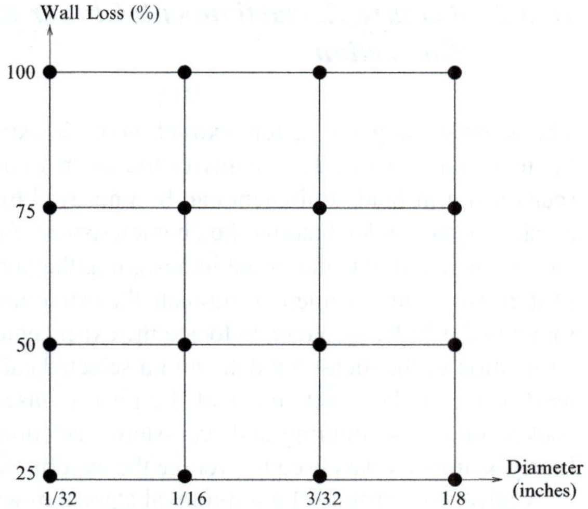

Fig. 18.20 A 4×4 interpolation table for the 999-series at 1 MHz, 800 kHz, 600 kHz, and 400 kHz. This table will allow us to a third-order (cubic-spline) interpolation

"round-bottom" circular pit in a manner to be described later. We add another standard for a 1/32"-diameter pit in order to generate at 4×4 interpolation table for inverting data using these model-based standards. Such a table (see Fig. 18.20) will allow us to do up to a third-order (cubic-spline) interpolation.

18.3.1 Scaling the Measured Instrument Data

The measured data are instrument voltages, which means that they must be transformed into impedances in order to be used as a source for inversion with **VIC-3D**®. Following the discussion in Chap. 15, we accomplish this by using a general linear filter,

$$Z = \beta_1(C_1 + jC_2) - \beta_2, \tag{18.1}$$

relating the impedance, Z, to the voltages, C_1 and C_2, of instrument channels 1 and 2, respectively.

The four parameters given by the two complex numbers, β_1 and β_2, are frequency dependent and are determined by fitting the instrument voltages measured for pit 12 at each frequency to the impedances calculated by **VIC-3D**® for nominal values of pit 12 dimensions. A value of $\beta_2 = 0$ was found to give a good match between channel data and computed impedances at all frequencies. Thus, we have a two-parameter transformation that simply scales the complex voltage, $C_1 + jC_2$, and rotates it in the complex plane. These two operations account for the amplification and phase shift that the instrument applies to the probe signal.

18.3.2 *Feature Extraction with Clutter Removal and Scan-Step Correction*

The preprocessing and feature extraction of the experimental data are a key part of the inversion process. The quality of the inversion results from NDE measurement scans using in-field equipment can be improved by using the appropriate preprocessing steps. Understanding the characteristics of the measurement data and how they were acquired is important in designing the preprocessing steps. Here, a series of steps were implemented to automate the extraction and filtering of the pit features found in the EPRI scan data to locate the exact center of the pit responses. A cross-correlation of the measured data with a selected calibration pit response curve was used to locate the exact center of the pit responses. Subsequent steps addressing background noise filtering and scan-step correction are presented in detail below. Lastly, symmetry was used to average the experimental response about the pit.

A novel background clutter removal algorithm was developed to compensate for systematic measurement noise in scan data. Typical sources for clutter and noise found in eddy-current scans are probe lift-off variation, local changes in material properties, sensitivity to sub-surface structures, thermal variation of measurement components, and electrical noise. Figure 18.21a displays an example where the background noise of the original experimental response is evaluated, using the procedure described in Chap. 15, providing efficient removal of background clutter from the experimental data. Compensation for the systematic background noise will reduce the role that the tails of the pit response play in the least-squares error minimization and emphasize the center pit region achieving a more accurate model fit.

A method was also developed to correct for the significant variation in the scan step size found between different pits in a scan. Since the data were acquired using a constant rate of acquisition (1,000 Hz), it was observed that the pull velocity of the probe through the tube actually varied by as much as 30 % with respect to the mean rate. This resulted in uncertainty in the actual scan step size in the neighborhood of each pit in the test sample. Thus, it was concluded that the local scan step size must be estimated for each pit. To determine the local scan step size, an inversion process was designed that iteratively fit both the equivalent depth and diameter of the pit and the scan step size. Figure 18.21b displays the reactance response for pit 7 for the measured data with and without scan-step correction and the corresponding simulated data. Since the scan step size more greatly affects the lateral fit of the data, while the depth and diameter of the pit more greatly impact the fit in both magnitude and at the pit center, the inversion of these three parameters simultaneously was found to be well-posed over the parameter range tested in the EPRI study.

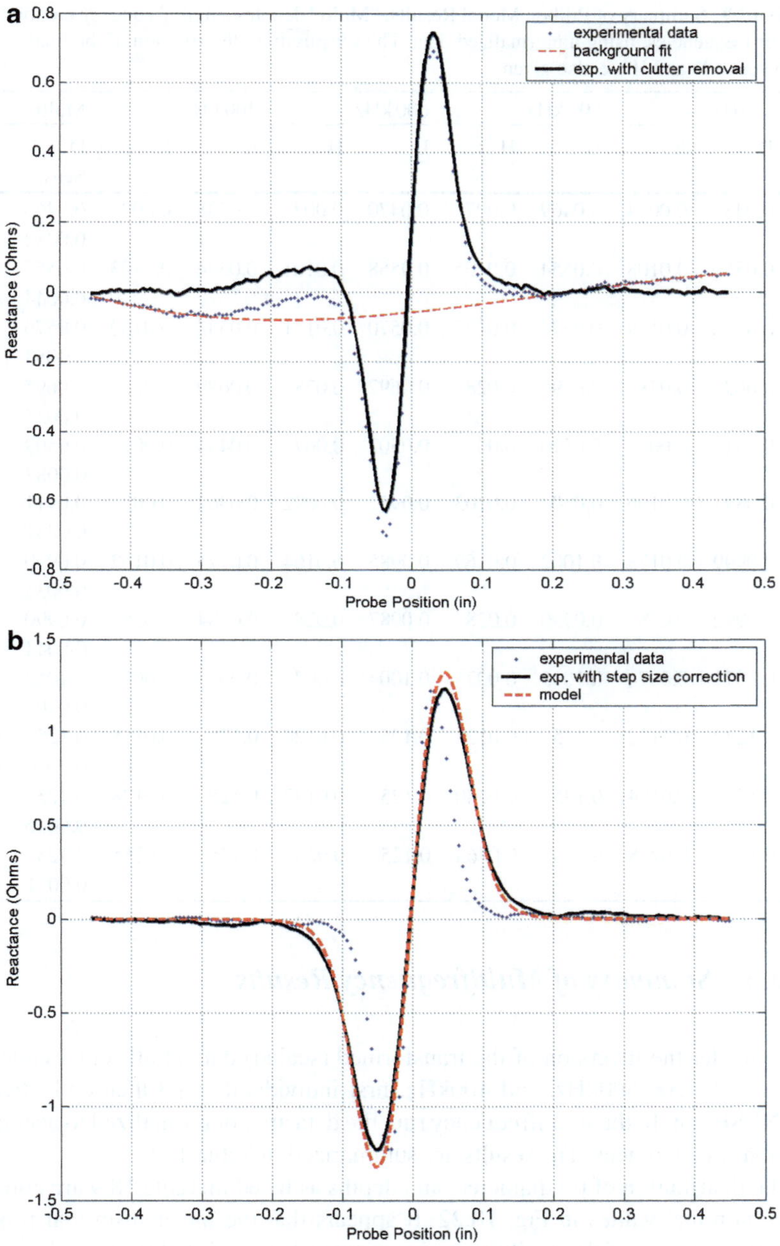

Fig. 18.21 (**a**) Reactance response for pit 6 demonstrating background clutter removal. (**b**) Reactance response for pit 7 demonstrating scan-step correction (for 008–999 at 1.0 MHz)

Table 18.9 Summary of Pillbox Model Results. "Multi" denotes a multifrequency combination of all four frequencies, using unnormalized data. The sensitivity to the solution of the final ("Multi") estimates of D and H are also given

	1 MHz		800 kHz		600 kHz		400 kHz		Multi	
Pit	D	H	D	H	D	H	D	H	D Sensit	H Sensit
1	0.0433	0.0084	0.0469	0.007	0.0470	0.007	0.0438	0.007	0.0463 0.0082	0.007 0.0028
2	0.0550	0.0108	0.0554	0.0105	0.0558	0.0101	0.0534	0.0103	0.0557 0.0044	0.0103 0.0015
3	0.0578	0.0179	0.0574	0.0175	0.0570	0.0171	0.0538	0.0180	0.0576 0.0053	0.0173 0.0029
4	0.0699	0.028	0.0695	0.028	0.0692	0.028	0.0689	0.028	0.0695 0.0033	0.028 0.0022
5	0.0513	0.007	0.0510	0.007	0.0502	0.007	0.0484	0.007	0.0509 0.0083	0.007 0.0024
6	0.0897	0.0106	0.0904	0.0103	0.0952	0.0092	0.0964	0.0087	0.0911 0.0051	0.0101 0.0010
7	0.0999	0.0173	0.1002	0.0169	0.0985	0.0164	0.0975	0.0162	0.0990 0.0033	0.0166 0.0010
8	0.0992	0.028	0.0990	0.028	0.0987	0.028	0.0984	0.028	0.0990 0.0044	0.028 0.0017
9	0.1021	0.007	0.1015	0.007	0.1004	0.007	0.0973	0.007	0.1012 0.010	0.007 0.0014
10	0.1244	0.0132	0.125	0.0128	0.125	0.0124	0.125	0.0118	0.125 0.0036	0.0127 0.0006
11	0.125	0.0184	0.125	0.0180	0.125	0.0177	0.125	0.0178	0.125 0.0046	0.0180 0.0009
12	0.125	0.0268	0.125	0.0268	0.125	0.0267	0.125	0.0265	0.125 0.0071	0.0267 0.0015

18.3.3 Summary of Multifrequency Results

We consider the inversion of the transformed (scaled) data at all four frequencies, 1MHz, 800kHz, 600kHz, and 400kHz, first individually, and then collectively in one NLSE run. In the multifrequency run, the data are not normalized to account for frequency differences. The results are summarized in Table 18.9.

The distribution of pit diameters and depths as listed in Table 18.9 are compared to the nominal values in Fig. 18.22. It appears that the mechanism that produces the round-bottom of the shallowest pits, namely 1, 5, and 9, does not allow a fully developed pit to form, and this may account for the smaller than expected diameters of these pits. This is reminiscent of the need for a "new flaw model" for pits, as described in Sect. 16.4.

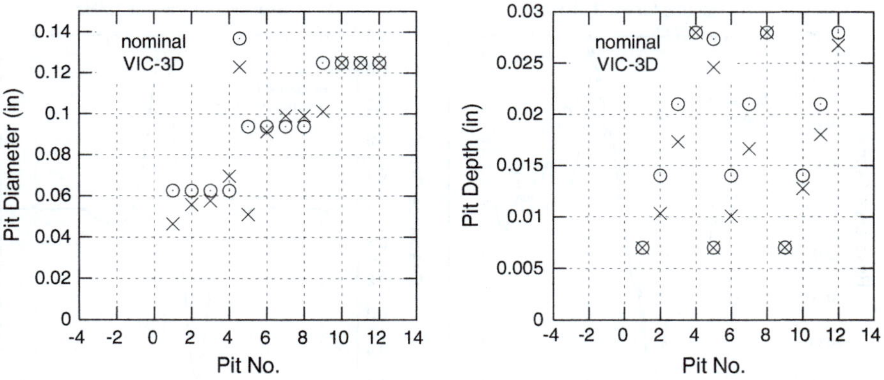

Fig. 18.22 Distribution of computed and nominal pit diameters and depths

The final results for each pit response at 1 MHz, when the diameter and height shown in Table 18.9 under "Multi" are used, are plotted in the next figures.

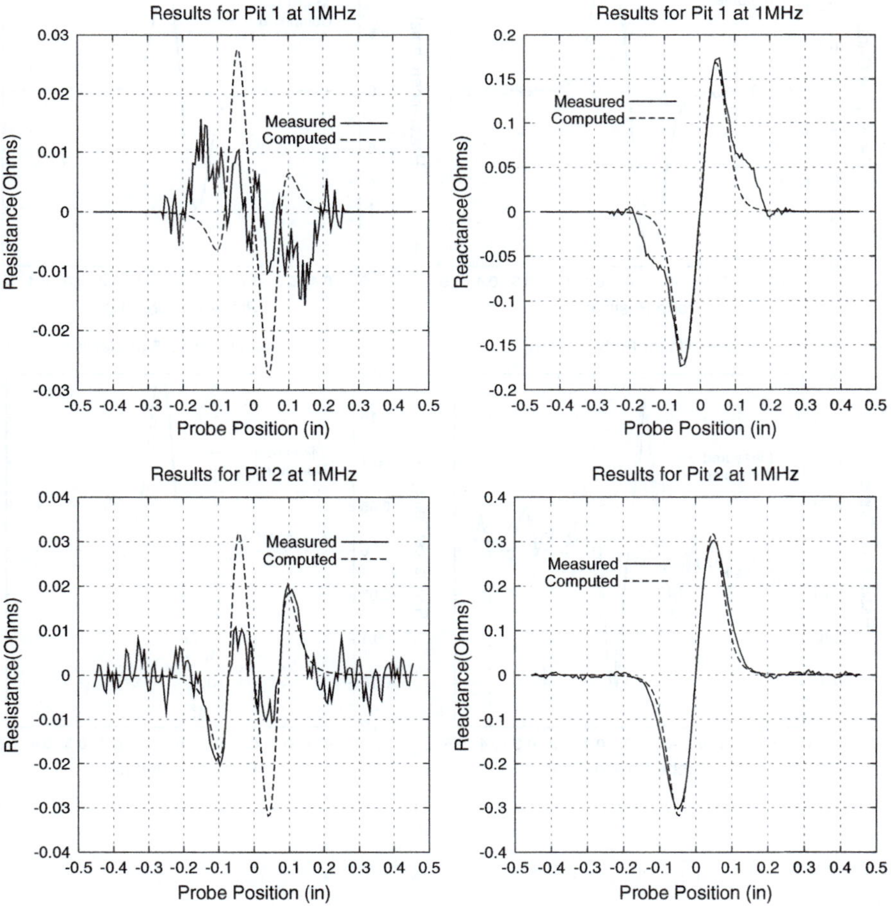

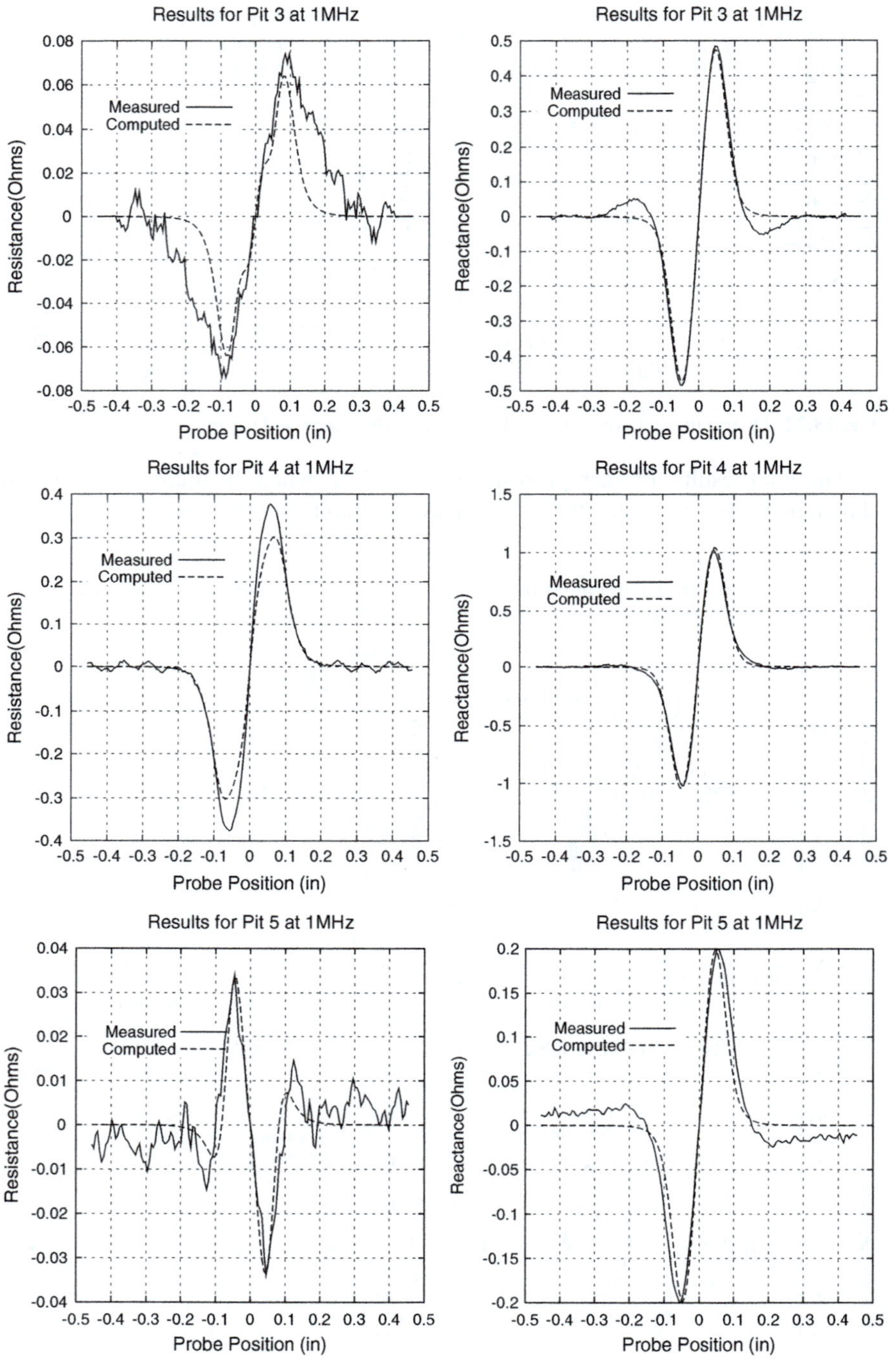
Results for Pit 3 at 1MHz
Measured
Computed
Resistance(Ohms)
Probe Position (in)
Results for Pit 3 at 1MHz
Measured
Computed
Reactance(Ohms)
Probe Position (in)
Results for Pit 4 at 1MHz
Measured
Computed
Resistance(Ohms)
Probe Position (in)
Results for Pit 4 at 1MHz
Measured
Computed
Reactance(Ohms)
Probe Position (in)
Results for Pit 5 at 1MHz
Measured
Computed
Resistance(Ohms)
Probe Position (in)
Results for Pit 5 at 1MHz
Measured
Computed
Reactance(Ohms)
Probe Position (in)

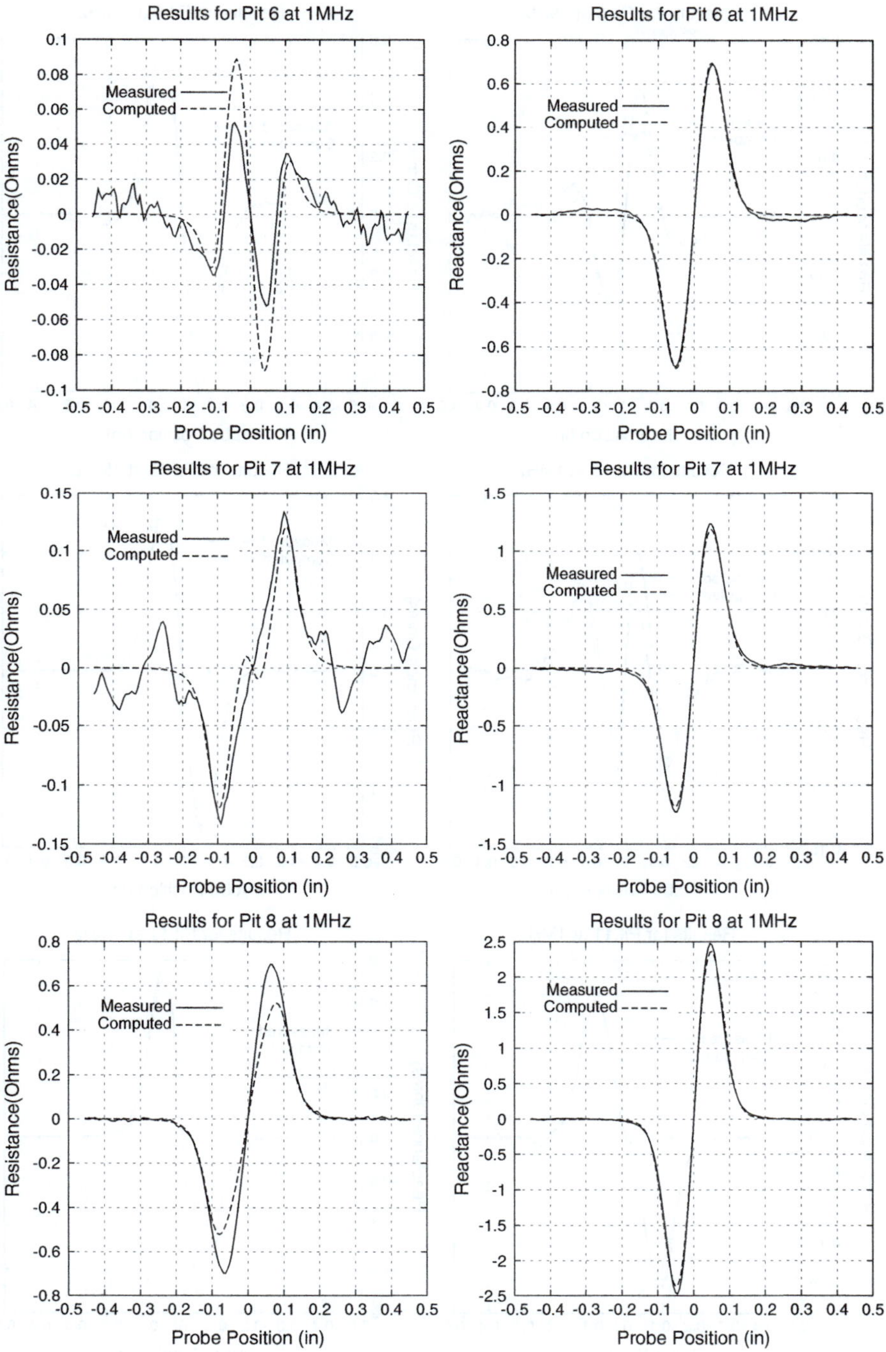
Results for Pit 6 at 1MHz
Measured
Computed
Resistance(Ohms)
Probe Position (in)
Results for Pit 6 at 1MHz
Measured
Computed
Reactance(Ohms)
Probe Position (in)
Results for Pit 7 at 1MHz
Measured
Computed
Resistance(Ohms)
Probe Position (in)
Results for Pit 7 at 1MHz
Measured
Computed
Reactance(Ohms)
Probe Position (in)
Results for Pit 8 at 1MHz
Measured
Computed
Resistance(Ohms)
Probe Position (in)
Results for Pit 8 at 1MHz
Measured
Computed
Reactance(Ohms)
Probe Position (in)

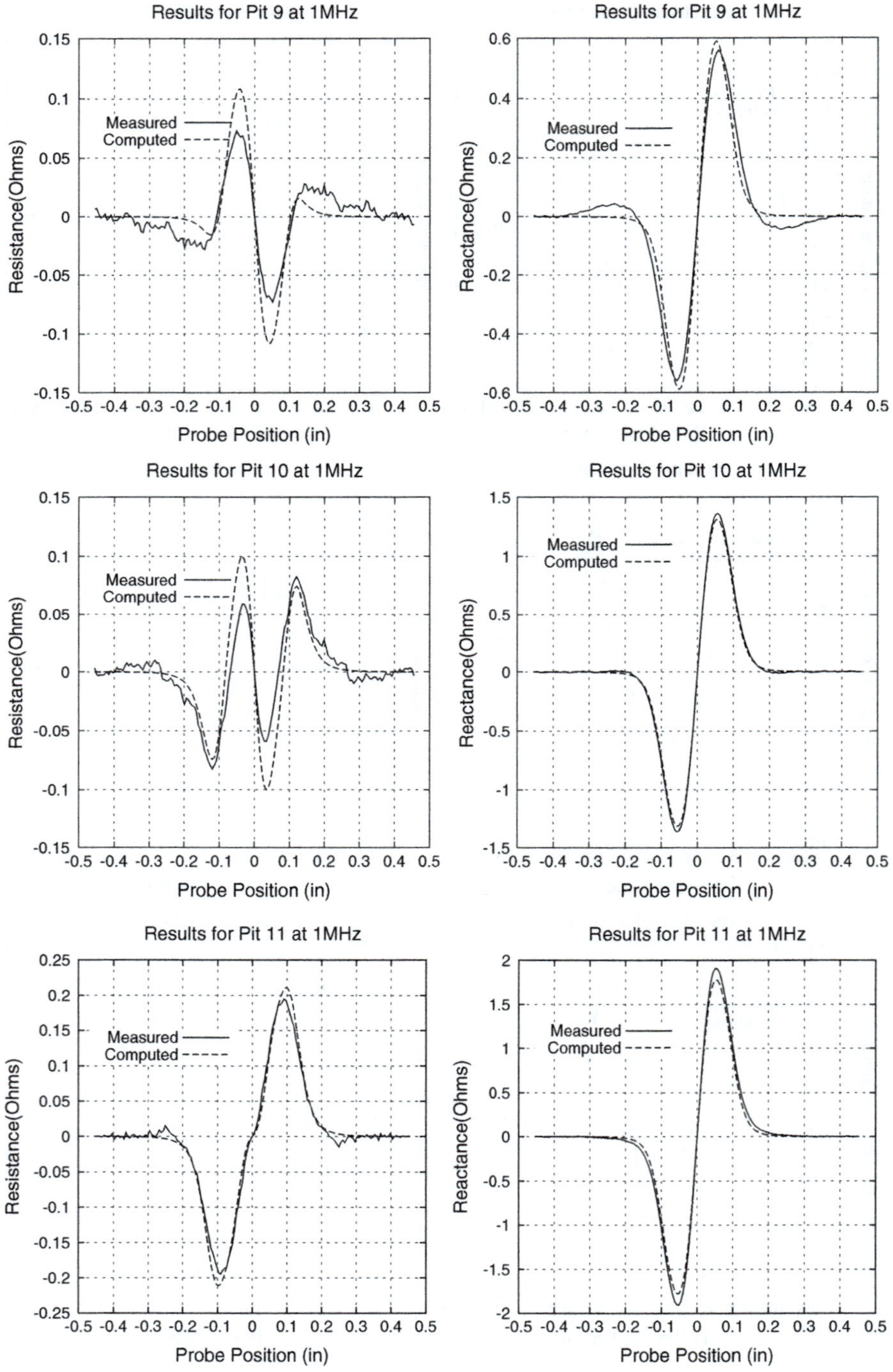
Results for Pit 9 at 1MHz
Measured
Computed
Resistance(Ohms)
Probe Position (in)
Results for Pit 9 at 1MHz
Measured
Computed
Reactance(Ohms)
Probe Position (in)
Results for Pit 10 at 1MHz
Measured
Computed
Resistance(Ohms)
Probe Position (in)
Results for Pit 10 at 1MHz
Measured
Computed
Reactance(Ohms)
Probe Position (in)
Results for Pit 11 at 1MHz
Measured
Computed
Resistance(Ohms)
Probe Position (in)
Results for Pit 11 at 1MHz
Measured
Computed
Reactance(Ohms)
Probe Position (in)

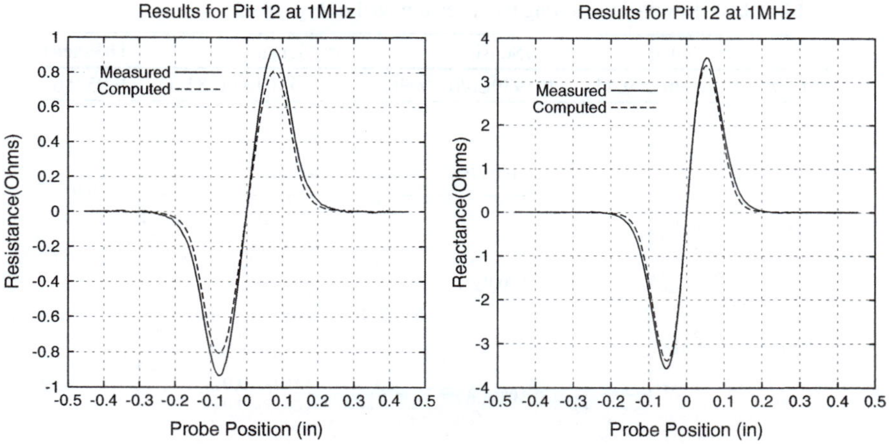

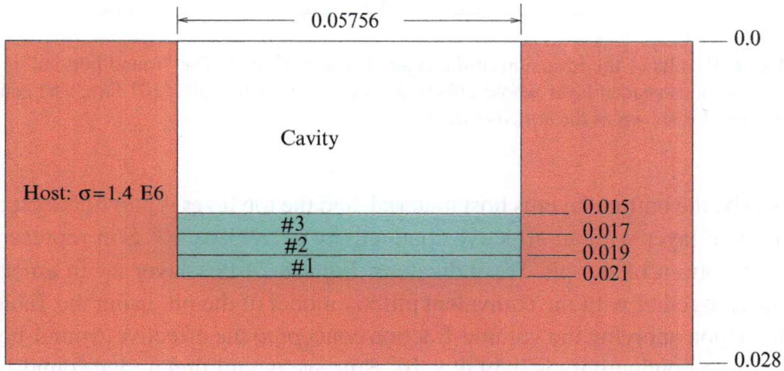

Fig. 18.23 Layered model of pit 3

18.3.4 The Morphology of "Round-Bottom" Pits

We will develop a layered model[6] of pit 3 in order to gain some insight into the morphology of round-bottom pits. Figure 18.23 illustrates a three-layer model of pit 3, with a resolution of 2 mils per layer. We used the data of Table 18.9 to establish the size of the cavity and the position of the three layers. The solution of the multifrequency inverse problem for the conductivities of the layers is shown in Table 18.10.

---

[6]See Appendix A.2 for a discussion of layered models of pits.

Table 18.10 Results of inverting the layered model of Fig. 18.23

Φ	σ_1/Sensit	σ_2/Sensit	σ_3/Sensit	D(Fixed)
0.4788	1.4E6/0.2192(1)	0.5939E6/0.2099(1)	0.1151E1/0.1781(1)	0.05756

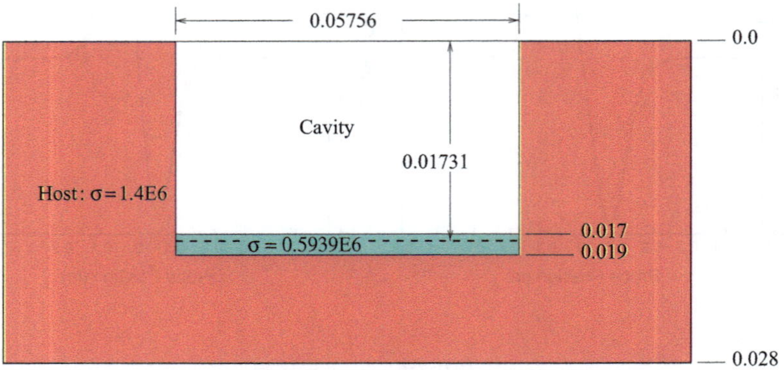

Fig. 18.24 Results of the inversion of the layered model of pit 3. The "round-bottom" is shown as the amorphous shaded layer whose effective conductivity is 0.5939×10^6 S/m. The equivalent pillbox model is shown as the *dotted outline*

Clearly, the bottom layer is host material, and the top layer is part of the pit cavity. The middle layer with an effective conductivity of 5.939×10^5 S/m represents the rounded bottom of the pit. We indicate in Fig. 18.24 this layer as an amorphous structure, together with the equivalent pillbox model of the pit, using the Table 18.9 results. Upon applying the volume-fraction concept to the effective "round-bottom" layer whose conductivity is 0.5939×10^6 S/m, we would find a total round-bottom + cavity volume that is virtually equal to the cavity volume of the equivalent pillbox model. The important thing to note here is that the "rounded-bottom" that we have just computed gives us a net depth of 0.019 inch for pit 3, which is in better agreement with the nominal value listed in Table 18.8. (See, also, Fig. 18.22). This shows the power and flexibility of model-based inversion in characterizing pits by choosing appropriate parameters for modeling the pit.

Using the "Multi" results for D and H of pit 3 from Table 18.9, and the results for the round-bottom model of pit 3 from Fig. 18.24, we plot in Fig. 18.25 the corresponding impedances at 1MHz. The round-bottom model gives a more accurate estimation of the peak and slope through the origin of the resistance scan. This is what we would expect, because the resistance data are important in determining the depth of the pit. We have not applied this algorithm to the shallow pits, 1, 5, and 9, because our interest was in determining the origin of the depth-discrepancies of the deeper pits when the pillbox model was used.

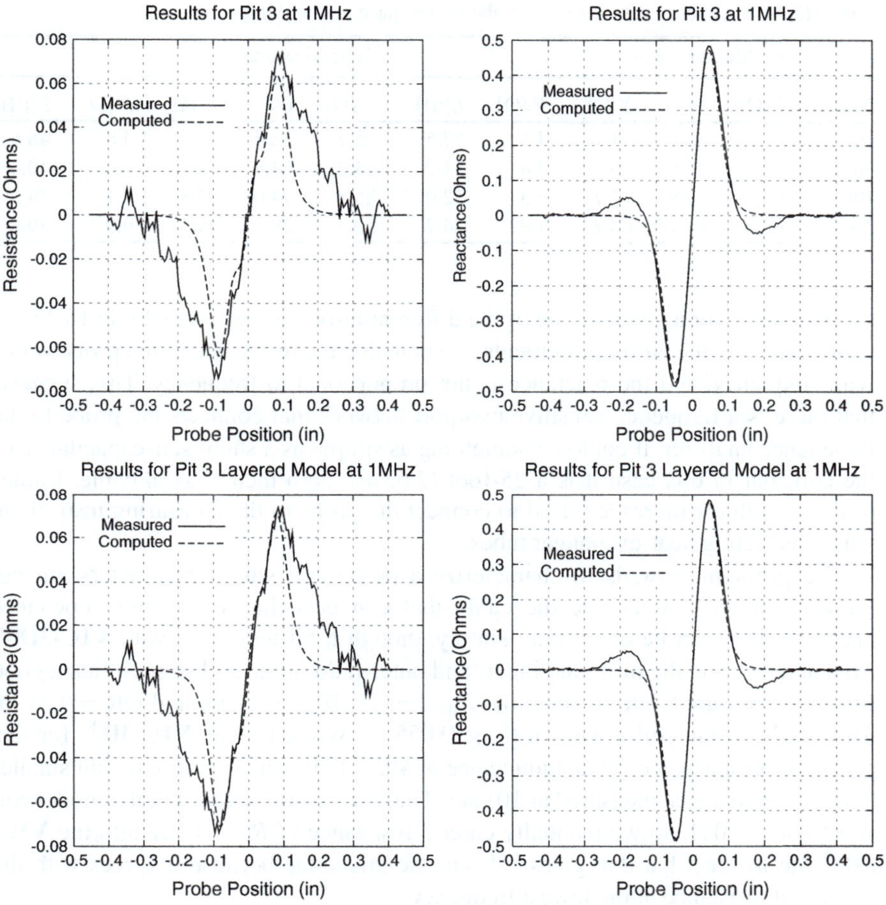

Fig. 18.25 Final results for pit 3 at 1MHz. *Top*: pillbox model. *Bottom*: layered (round-bottom) model

Appendix

A.1 Modeling Probes + Cables

In order to characterize a probe for the purpose of modeling probe-flaw responses and validating benchmark tests, it is necessary to consider the effects of real conditions under which the ideal probe operates. Consider a probe that is to be operated as a differential-bobbin within a 70/30 Cu-Ni tube with no defects. The inner radius of the tube is 0.2475 in and the outer radius is 0.3125 in. The tube's conductivity is 2.61×10^6 S/m, and its relative magnetic permeability is unity. Almost nothing is known about the probe except some external dimensions.

Table 18.11 Frequency response of a probe in freespace and within a tube

	Air Measurements					Tube Measurements				
f(khz)	Z (Ω)	Θ(°)	X (Ω)	R (Ω)	L(μH)	Z (Ω)	Θ(°)	X (Ω)	R (Ω)	L(μH)
10	5.6	36.5	3.3	4.5	52.6	5.7	32.4	3.1	4.8	48.6
50	17.0	73.5	16.3	4.8	51.7	14.2	64.8	12.8	6.0	40.6
100	33.1	80.8	32.7	5.3	52.0	25.8	74.0	24.8	7.1	39.5
200	69.1	84.1	68.7	7.0	54.7	50.7	78.7	49.7	9.9	39.6

Data taken for a single probe of the differential-pair are given in Table 18.11.

Referring to the air measurements, we note that the resistance of the probe varies with frequency, and the reactance is not proportional to frequency. This suggests that there is a frequency-sensitive two-port network that connects the probe to the impedance analyzer. It could be something as simple as a shunt self-capacitance of the coil, but in this case it is a 25-foot (7.62 m), 5/16-inch coaxial cable. Cables of this length (or more) are used to connect the probe to the measuring instrument when inspecting heat-exchanger tubes.

The problem, now, is to characterize a probe coil whose parameters are not known, and to characterize the cable that connects the coil to the impedance analyzer. This can be done quite simply, though a bit tediously, with **VIC-3D®**. Trial and error on the turns and inner- and outer-radii of the coil suggest that a good fit to the measured data is given by: $N_{\text{turns}} = 55$, IR = 0.19 in. and OR = 0.21 in. The height of the coil is known to be 0.055 in. When we run **VIC-3D®** for this probe in freespace, we get an inductance of 49.55 μH, which is close to, but smaller than, the inductance measured at 10 kHz. The freespace resistance of the model coil is, of course, 0 Ω, so we manually enter a resistance of $R_0 = 4.5\,\Omega$ into the **VIC-3D®** file to see what that gives. We choose this value because it agrees with the measured resistance at the lowest frequency.

Cable effects can be accounted for by defining four parameters: characteristic impedance, capacitance per unit length, attenuation in dB/m and length. This follows from transmission-line theory [98]. Let Z_L be the load (terminating) impedance of a coaxial cable, and Z_{in} be the driving-point (input) impedance of the loaded cable. Then

$$
\begin{aligned}
Z_{\text{in}} &= Z_0 \frac{Z_L + Z_0 \tanh(j\beta l + \alpha l)}{Z_0 + Z_L \tanh(j\beta l + \alpha l)} \\
Z_0 &= \sqrt{(L/C)} = \text{characteristic impedance of the cable} \\
\beta &= \frac{\omega}{v_p} \\
v_p &= \frac{1}{\sqrt{(LC)}} \approx 2 \times 10^8\ \text{m/s}.
\end{aligned}
\tag{18.2}
$$

Table 18.12 Freespace impedance response of the coil + cable

	Frequency			
α (dB/m)	10^4	5×10^4	10^5	2×10^5
0.0	$4.5014+j3.2325$	$4.5344+j16.221$	$4.6399+j32.809$	$5.1008+j68.743$
0.005	$4.7197+j3.2299$	$4.7749+j16.208$	$4.9519+j32.782$	$5.7317+j68.677$
0.010	$4.9379+j3.2272$	$5.0151+j16.194$	$5.2628+j32.752$	$6.3612+j68.604$
0.015	$5.1559+j3.2243$	$5.2551+j16.179$	$5.5735+j32.721$	$6.989+j68.524$

Table 18.13 In-tube impedance response of the coil + cable

f(Hz)	Z (Ω)
10^4	$4.8594+j2.9665$
5×10^4	$6.1782+j12.567$
10^5	$7.3656+j24.325$
2×10^5	$10.117+j48.48$

L and C are, respectively, inductance and capacitance per unit length of the cable. The value for phase velocity, v_P, is a reasonable approximation for typical transmission lines.

Typical values for coaxial cables that are used in eddy-current NDE are 50 Ω for the characteristic impedance, and 100 pF/m for the distributed capacitance. Our problem calls for a cable length of 7.62 m, which leaves only the attenuation to be determined. Attenuation, α, is frequency dependent, but **VIC-3D**® assumes it to be constant, so it becomes necessary to run several tests to determine the frequency response of the cable.

The results for tests of the probe+cable in freespace are shown in Table 18.12. The diagonal entries in this table agree with the corresponding freespace results of Table 18.11 within 2 %. This indicates that the attenuation varies with frequency as shown in the left-hand column. This variation agrees with theory, in that attenuation always increases with frequency.

When modeling the probe within the tube, we use the same cable parameters as above, except that we use $\alpha = 0.0$ only for $f = 10^4$ Hz, $\alpha = 0.005$ only for $f = 5\times10^4$, etc. The results for the model calculations for the coil within the tube are given in Table 18.13. The differences between these values and those shown in Table 18.11 are 3 %, or less, except for X_{10^4} and R_{10^5}.

It is difficult to determine an outer radius of a many-turn bobbin coil when it is within a tube, because the windings on the outer layer of the coil nearest the tube will not lie smoothly in that layer, as shown in Fig. 18.26. This gives the appearance of an uneven spacing between the outer radius of the coil and the inner tube wall. This is not as serious a problem when the coil is in the "pancake" aspect.

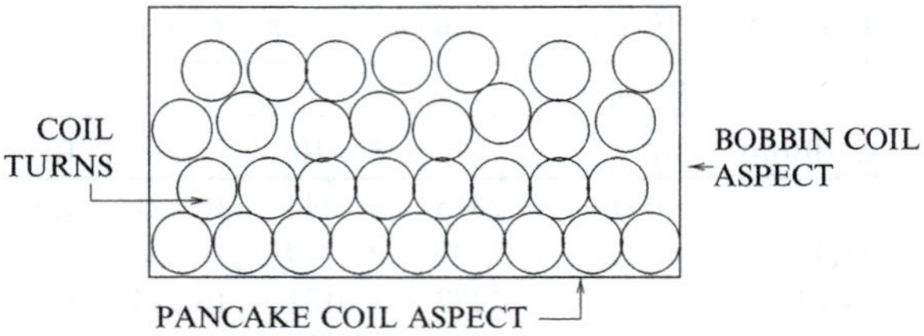

Fig. 18.26 The nonuniform distribution of turns within a typical real coil

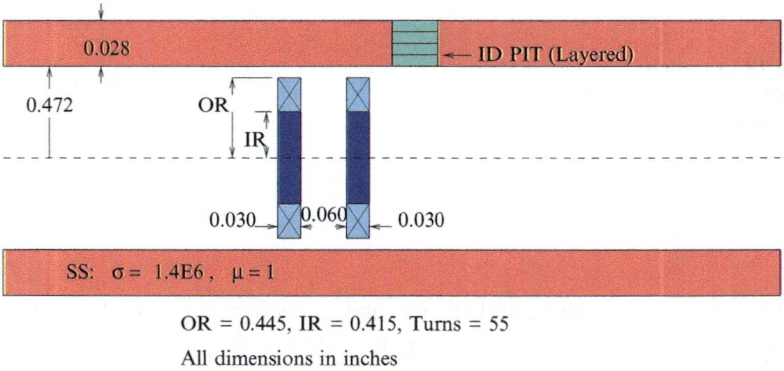

Fig. 18.27 A layered model of the ID pit shown in Fig. 18.19

A.2 A "Layered" Model of Corrosion Pits

One possible scheme for determining the shape of a corrosion pit that is assumed to have a fixed morphology would be to use the layered-pit model shown in Fig. 18.27.

The objective is to determine the conductivity of each layer, and from that result infer the size, and perhaps shape, of the pit. We assume that the pit has a circular cross-section, as before, so that the only parameter that defines the model is the radius of the anomalous region. If the layer has a conductivity of 0S/m, then clearly that layer is entirely filled by the pit, whereas if the layer has a conductivity of 1.4×10^6S/m, then the layer is entirely free of the pit, being host material, only. Anything in between requires further consideration, as we show next.

The unknowns in the inversion process are $\sigma_1, \ldots, \sigma_4$, the conductivities of layers 1–4, with 1 corresponding to the inner wall and 4 to the outer wall, as well as the diameter of the anomalous region. Once the conductivity of a layer has been estimated by NLSE, we apply a simple volume-fraction computation to determine the relative volume of cavity to host material within that layer. For example, suppose layer L has a conductivity of 1.135×10^6 S/m. Then to see how much of the Lth layer is occupied by the cavity, we compute the volume-fraction of the cavity: $\mathrm{VF}_L = 1 - \dfrac{1.135}{1.40} = 0.1893$. Thus, the cavity occupies less than 19 % of the Lth layer.

Chapter 19
Coupled Problems in Heat-Exchanger Tubes

19.1 Introduction

The preceding chapters of this book dealt largely with problems in which the host material and any anomalies were pure electrical conductors. The presence of magnetic permeabilities was merely an interesting side effect to the creation of the **VIC-3D®** model and its solution. There are, of course, many problems in which magnetic permeability is present with electrical conductivity and must be accounted for, not only in creating the model but also in understanding the physics of the solution and its effect on the NDE process. In this chapter we consider magnetic effects in heat-exchanger tubes, focusing later on ferritic tubes, which are becoming of increasing importance in the nuclear power industry.

19.2 Reconstruction of a Semiellipsoidal Wear Scar and Permeable Crust in a Heat-Exchanger Tube

In Figs. 19.1 and 19.2, we model "fretting" damage in a heat-exchanger tube [57]. The damage manifests itself as a "wear scar" surrounded by a "crust." The crusts of both figures are slightly magnetic, with that of Fig. 19.1 highly conducting, relative to the host tube, and that of Fig. 19.2 having the same conductivity as the host tube. The scars and crusts are modeled as semiellipsoids.

We model a seventeen-point scan over the anomalous region shown in Fig. 19.2 at three frequencies, 100 kHz, 400 kHz and 1.6 MHz, to reconstruct the conductivity, σ, and permeability, μ, of the crust. The results are shown in Table 19.1. In all cases, the result is more sensitive to μ than to σ. The computed impedance trajectories that are the input to the inversion algorithm are shown in Fig. 19.3. The rotation of these trajectories as a function of frequency is clearly shown.

H.A. Sabbagh et al., *Computational Electromagnetics and Model-Based Inversion*, Scientific Computation, DOI 10.1007/978-1-4419-8429-6_19,

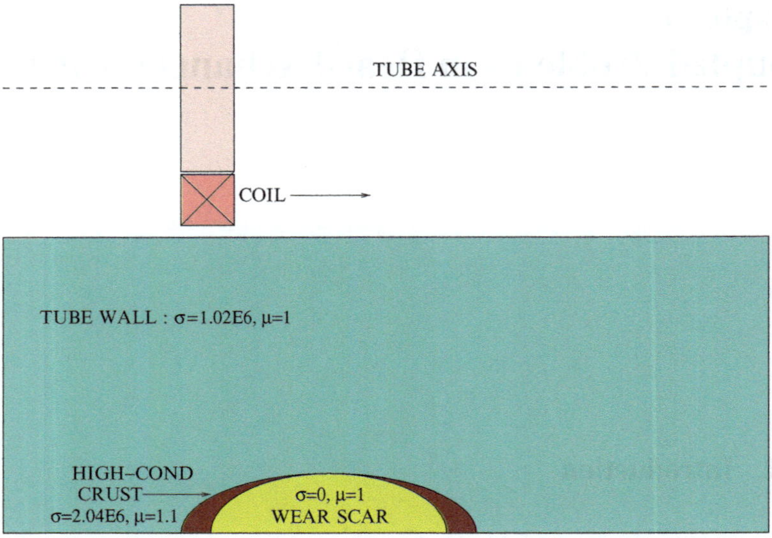

Fig. 19.1 A semiellipsoidal "wear scar" and highly-conducting permeable "crust" in the outer surface of a tube

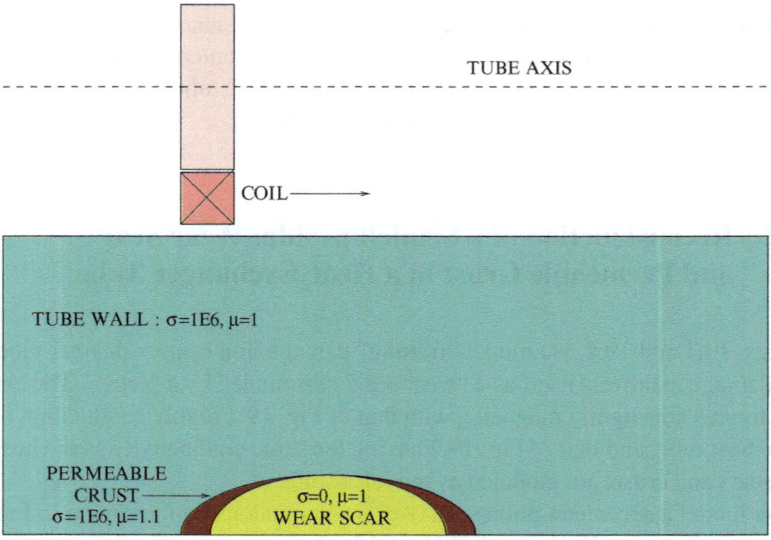

Fig. 19.2 A semiellipsoidal "wear scar" and permeable "crust" in the outer surface of a tube

Table 19.1 Results (μ, σ) for wear scar and permeable crust

Freq (kHz)	Computed	Original
100	(1.056, 0.84E6)	(1.1, 1E6)
400	(1.105, 0.85E6)	(1.1, 1E6)
1,600	(1.106, 0.86E6)	(1.1, 1E6)

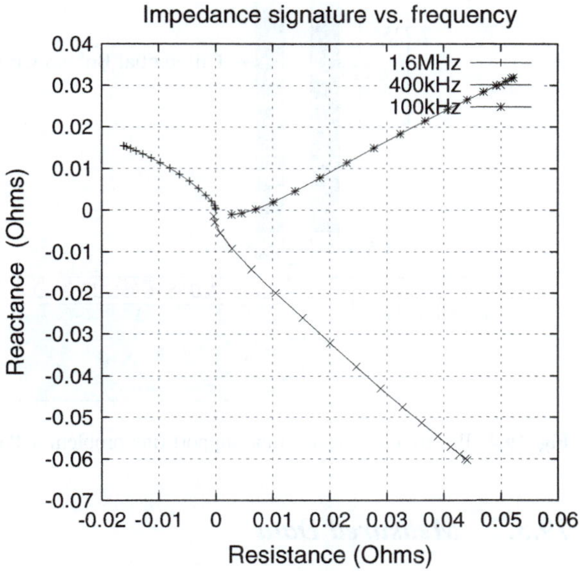

Fig. 19.3 Impedance-plane results for "wear scar" and permeable "crust" at three frequencies

19.3 Modeling Tube-Support Rings (TSR)

19.3.1 Statement of the Problem

In this section and the next we model the tube support ring problem of Fig. 19.4 that is quite common in the inspection of nuclear power plants. The major difference between the problem here and in the next section is the nature of the tube, itself. Here, it is inconel, which is nonmagnetic, while in the next section it is ferritic, which introduces additional interesting challenges. In both cases, however the TSP+tube structure is axisymmetric, which allows us to use the axisymmetric model in cylindrical geometries that was developed in Chap. 9.

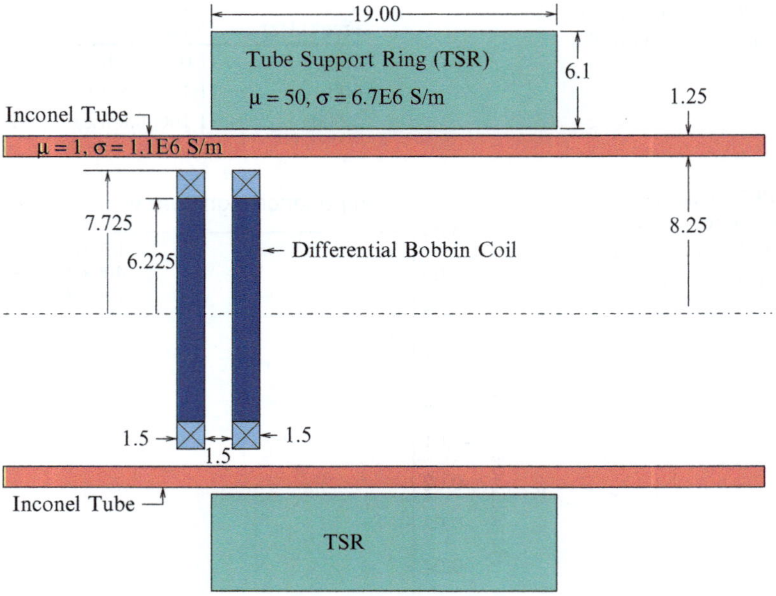

Fig. 19.4 Illustrating a typical tube support ring problem. All dimensions are in millimeters

19.3.2 Measured Data

The measured data for the problem are taken at 100, 200, 300 and 400 kHz, and are presented here for convenience in Fig. 19.5.[1] Note that the data are plotted in the impedance plane (real vs. reactive), even though they are instrument A/D counts (or "voltage") in two channels that are orthogonal in time; i.e., they are ninety degrees out of phase with each other. The "Lissajous figure" closed-loop curve is typical of data that are taken with a differential-bobbin probe. The user of the instrument is usually instructed to "rotate the impedance-plane plot" so that it satisfies certain criteria, and this makes the interpretation of the measured data more challenging. We will use a linear filter to transform the measured voltage data into impedances in a manner similar to that which was done in the EPRI ID pits problem of Sect. 18.3.1. There are several important features to glean from this figure: note the clockwise rotation of the loops with frequency and note that the loops get smaller with frequency. Indeed, the loop for 400 kHz, actually fits within the loop for 100 kHz.

---

[1]The statement of the problem described in this section, and the original measured data were supplied by Prof. S-J. Song of Sungkyunkwan University, South Korea.

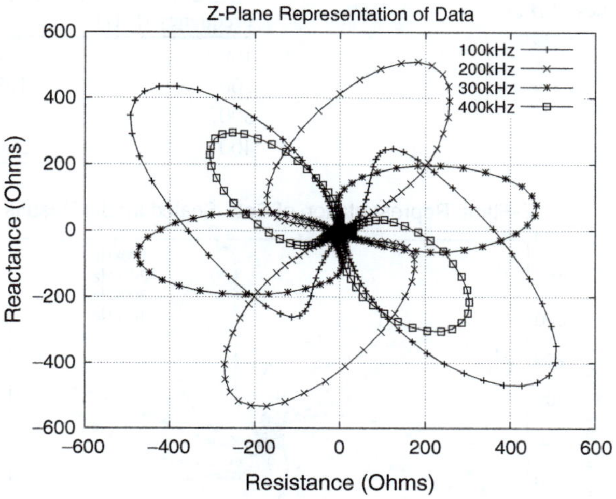

Fig. 19.5 The measured data at the four frequencies

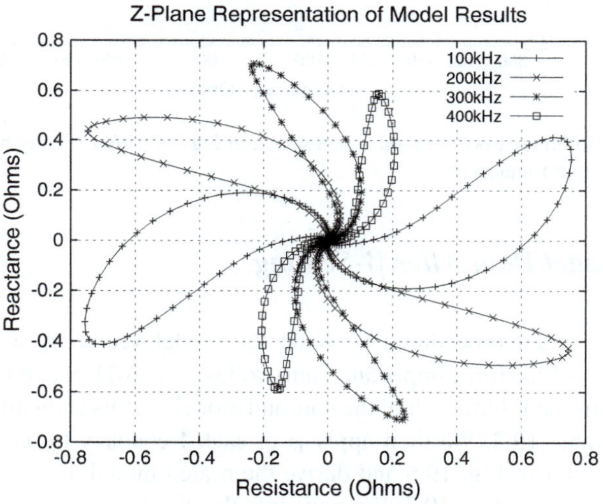

Fig. 19.6 The differential-bobbin model data at the four frequencies

19.3.3 Model Data Before β-Scaling

The model results (impedances) for Fig. 19.4 are shown in Fig. 19.6. We have not yet applied the β-scaling that will bring the impedances to voltage or A/D-counts, but the important features described above are already apparent. The loops decrease in size with frequency and are rotated clockwise.

Table 19.2 Values of β at each frequency

Frequency (kHz)	β
100	$-384.6 + j464.63$
200	$-197.7 + j530.7$
300	$3.4408 + j587.07$
400	$260.09 + j562.31$

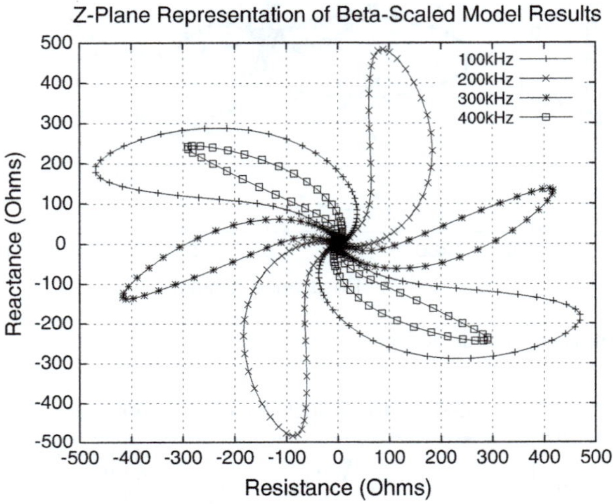

Fig. 19.7 The differential-bobbin model data shown in Fig. 19.6 after scaling by the appropriate β at each of the four frequencies

19.3.4 Model Data After β-Scaling

Following the usual procedure, we scale the model vector by a factor, β, that will allow us to transform impedance into voltage (or A/D-counts), or vice-versa. We have determined β for each of the data and model vectors at the four frequencies, as listed in Table 19.2. We then apply β at each frequency to the corresponding model data vector of Fig. 19.6 and derive the scaled-model data vectors shown in the complex plane of Fig. 19.7. We now note the similarities to the measured data of Fig. 19.5: the loops rotate clockwise and shrink with frequency, and the loop for 400 kHz fits within that for 100 kHz. Hence, we can conclude that there is consistent agreement between model and measured data.

19.3.5 Scaling in the Opposite Direction

Up to this point, we scaled the model data of Fig. 19.6 to agree as closely as possible (in the least-squares sense) with the measured data of Fig. 19.5. The result is shown in Fig. 19.7.

Table 19.3 Values of β_1 to β_4 at each frequency

Frequency (kHz)	β_1	β_2	β_3	β_4
100	−1.5305(−3)	2.7096(−4)	−3.1574(−4)	−9.0284(−4)
200	−2.1057(−4)	1.5131(−3)	−9.0218(−4)	−3.3706(−4)
300	3.4058(−4)	4.2174(−4)	−3.2197(−4)	−1.5568(−3)
400	−1.0023(−4)	5.7554(−4)	1.1822(−3)	−9.4413(−4)

Now, we want to do the opposite and scale the measured data to agree as closely as possible (in the least-squares sense) with the model data. Furthermore, we are going to use a more general scaling approach in which four parameters, β_1, β_2, β_3, and β_4, are used instead of only β_1 and β_2. The four parameters are defined by

$$\begin{aligned} R &= \beta_1 C_1 + \beta_2 C_2 \\ X &= \beta_3 C_2 + \beta_4 C_1, \end{aligned} \tag{19.1}$$

where R and X are the vectors of the model results and C_1 and C_2 are vectors of the measured data in each channel of the instrument (which are supposed to be orthogonal in time).

Upon taking the dot-product of the first equation by, respectively, C_1 and C_2, and similarly the second equation, we get the two linear systems

$$\begin{aligned} \Gamma_{R1}^{(VM)} &= \beta_1 \Gamma_{11}^{(M)} + \beta_2 \Gamma_{21}^{(M)} \\ \Gamma_{R2}^{(VM)} &= \beta_1 \Gamma_{12}^{(M)} + \beta_2 \Gamma_{22}^{(M)} \\ \Gamma_{X1}^{(VM)} &= \beta_4 \Gamma_{11}^{(M)} + \beta_3 \Gamma_{21}^{(M)} \\ \Gamma_{X2}^{(VM)} &= \beta_4 \Gamma_{12}^{(M)} + \beta_3 \Gamma_{22}^{(M)}, \end{aligned} \tag{19.2}$$

where, if these were random processes, the Γs would be auto- and cross-correlation functions. For our purposes, they are simply real numbers that are obtained by taking the dot products of the appropriate vectors.

The solutions of (19.2) are tabulated for each frequency in Table 19.3.

An impedance-plane plot is shown in Fig. 19.8. Once again we see the strong similarity between the scaled data and the model data of Fig. 19.6; the curves rotate clockwise with frequency, and they become smaller and narrower.

19.3.6 An Inverse Problem

When we compare the Lissajous figures of Figs. 19.5 and 19.7, we are liable to draw the conclusion that the model data cannot be used to characterize the system, if the only data that are known for the system are given by the measuring instrument,

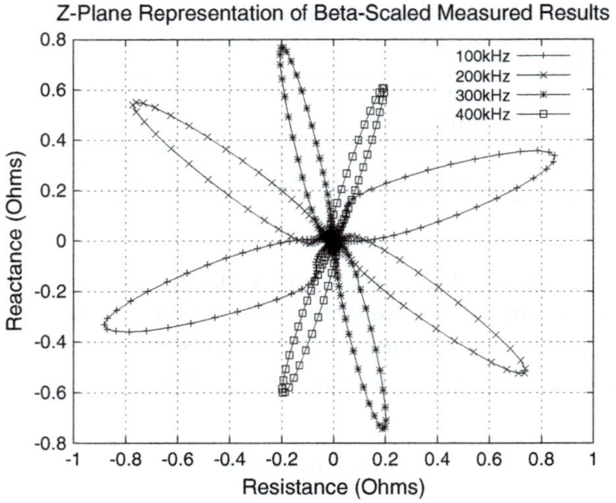

Fig. 19.8 The differential-bobbin measured data shown in Fig. 19.5 after scaling by the appropriate β_1–β_4 at each of the four frequencies

Table 19.4 Results for the two variables, σ and μ, of the TSR

Freq(kHz)	σ/sens	μ/sens	Φ
100	6.2224E6/4.36	40/20.08	1.67716
200	6.4227E6/3.84	40/28.13	1.28258
300	6.5032E6/3.72	40/43.14	0.93008
400	6.5410E6/3.81	40/30.37	0.68515

because they are not "perfectly" matched. We will show that the model data and measured data, after either has been scaled to agree with the other, can be used together to characterize the system. We will do this by using both in an inverse problem whose goal is to determine the conductivity and magnetic permeability of the tube-support ring. The inverse problem will be solved using NLSE in impedance space, which means that we will use the β-scaled measurements of Fig. 19.8.

When we have two unknowns, the conductivity and magnetic permeability, and apply NLSE to the scaled measured data at each of the four frequencies, we get the results shown in Table 19.4. The term "sens" denotes the sensitivity of the parameter at the solution point: the smaller, the better. Φ is the norm of the residual vector at the solution point: the smaller the better, again.

Note that the conductivity estimate increases somewhat with frequency. Because conductivity is more accurately estimated at higher frequencies (see [57] as well as Chap. 11 of this book) we will take $\sigma = 6.55 \times 10^6$ S/m as our estimate of the conductivity.

The estimate of the magnetic permeability, however, is somewhat suspect, because the solution is rather insensitive to the parameter. This follows from the large values of *sens* associated with μ that are shown in the table. Hence, we will

Table 19.5 Results for the single variable, μ, of the TSR when the conductivity is fixed at $\sigma = 6.55 \times 10^6$

Freq (kHz)	μ/sens	Φ
100	42.517/0.3345	1.5871
200	41.603/0.2763	1.2401
300	41.206/0.2434	0.9072
400	41.007/0.2239	0.6710

redo the inversion, this time fixing $\sigma = 6.55 \times 10^6$, and solving for only the single variable, μ. When we use NLSE for this one-variable problem, we get the results shown in Table 19.5.

Now, we see that Φ is reduced, meaning that the solution is in better agreement with the measured data, and the sensitivity of the solution for μ is much better. We also see that the estimated value of μ decreases slightly with frequency. Because μ is more accurately estimated at lower frequencies (see Sect. 11.4), we choose $\mu = 42.517$ for our best estimate. Thus, both σ and μ are well estimated, given their nominal values of 6.7×10^6 and 50, respectively.

In performing this inversion, it is important to understand that we did not rely upon any particular "feature" of the Lissajous figure at any one frequency. The approach that utilizes features of measured data is likely to be subject to significant errors, given the subjective nature of "observing" such features. This is especially true if the Lissajous figure is corrupted by noise or is embedded in significant background clutter. The mathematical approach that we have indicated is much more reliable and objective and can be further enhanced by the application of data-smoothing and clutter-rejection algorithms.

19.4 Modeling Direct and Inverse Problems in Ferritic Tubes

19.4.1 The Model Problem

Ferritic stainless steels, such as Type 439 or SEACURE, are being increasingly used in heat-exchanger tubes because of the increased resistance to chloride stress corrosion and intergranular attack when compared to older alloys, such as Type 304. This presents interesting modeling opportunities for the eddy-current nondestructive evaluation of these tubes because their magnetic permeability is ~60–100.[2] In this section,[3] we present examples of direct and inverse problems involving such tubes,

[2]The magnetic permeability is a function of the stress-state of the tube, as well as the frequency of operation.

[3]Details of the experiments that are described in this section are given in [112]. Related work on ferritic tubes can be found in [113].

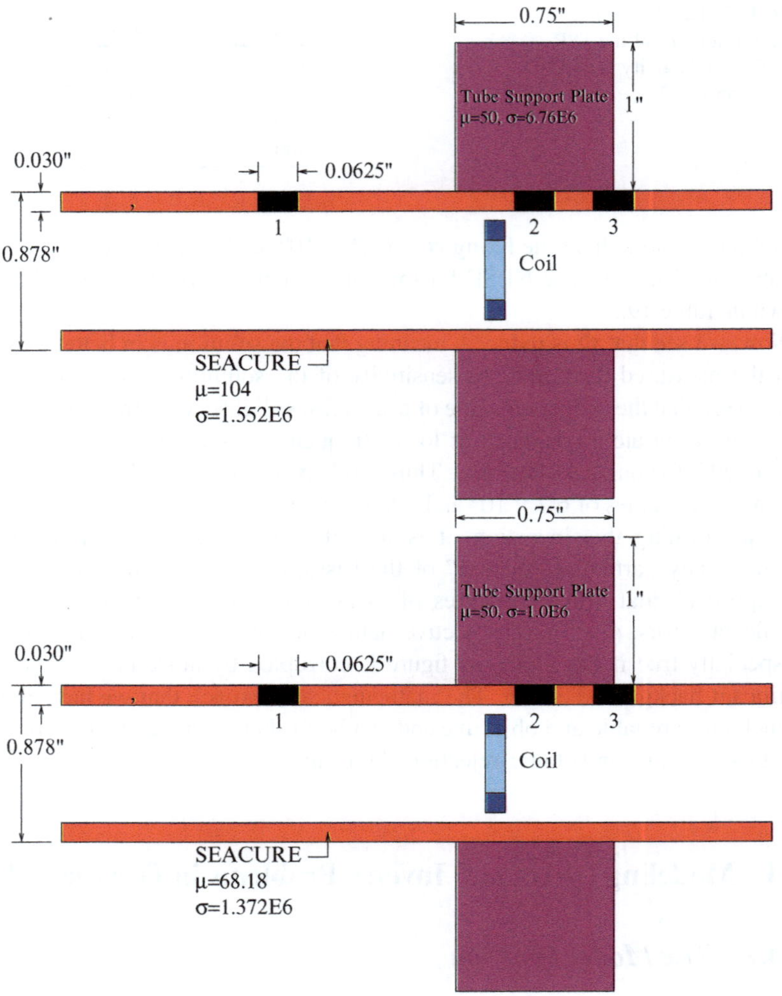

Fig. 19.9 Illustrating a ferritic tube, such as SEACURE, with a tube-support plate (TSP), and three 0.0625-inch through-wall holes. Hole 1 is well away from the TSP, hole 2 is centered under the TSP, and hole 3 is centered at edge of the TSP. *Top*: High-frequency model. *Bottom*: Low-frequency model. See text for details on parameters

then demonstrate conditions that are peculiar to ferritic tubes, and give insight into the optimum methods for characterizing the tubes and flaws within them.

The problem configuration is shown in Fig. 19.9. It is similar to the model shown in Fig. 19.4, except that the tube is ferritic, and there are three holes, one located away from the tube support plate (TSP),[4] one located under the center of the TSP,

[4]Tube support plate and tube support ring are synonymous terms.

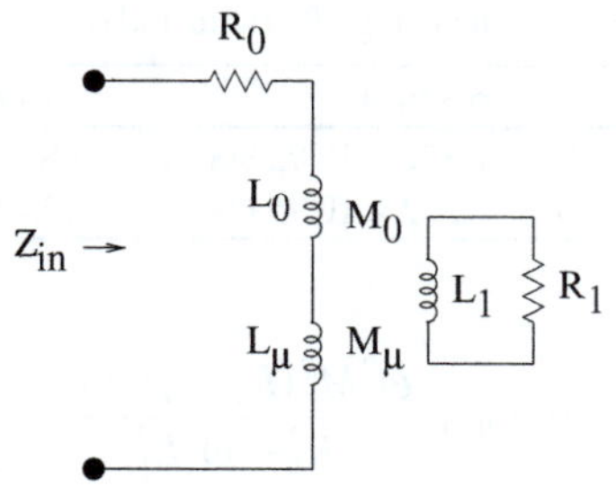

Fig. 19.10 A coupled-circuit model of the coil in the presence of the ferritic tube. R_0 and L_0 are, respectively, the resistance and self-inductance of the coil in freespace, and L_μ is the increased inductance of the coil due to the permeability of the tube. L_1 is the "virtual" secondary inductance that accounts for induction effects within the tube, M_0 is the mutual inductance between L_0 and L_1, and M_μ is the mutual inductance between L_μ and L_1. R_1 is the effective "secondary resistance" that is due to the electrical conductivity of the tube

and one located at the edge of the TSP. We will characterize the tube and then the composite structure, including the TSP and holes in the following subsections.

19.4.2 A Coupled-Circuit Model

We have seen several times that there is an advantage to working in the lower-frequency range when trying to discriminate between conductivity and permeability changes, because the correlation between these two parameters is smallest there. We can see this again by referring to the coupled-circuit model of the coil and tube shown in Fig. 19.10. R_0 and L_0 are, respectively, the resistance and self-inductance of the coil in freespace, and L_μ is the increased inductance of the coil due to the permeability of the tube. L_1 is the "virtual" secondary inductance that accounts for induction effects within the tube, M_0 is the mutual inductance between L_0 and L_1, and M_μ is the mutual inductance between L_μ and L_1. R_1 is the effective "secondary resistance" that is due to the electrical conductivity of the tube.

Elementary coupled-circuit theory yields an expression for the driving-point impedance of the loaded coil:

$$Z_{\text{in}} = R_0 + j\omega L_0 + j\omega L_\mu + \frac{\omega^2 M_\mu^2}{R_1 + j\omega L_1} + \frac{\omega^2 M_0^2}{R_1 + j\omega L_1} . \tag{19.3}$$

From this we get the change in impedance due to the presence of the tube:

$$\begin{aligned}\delta Z_{\text{in}} &= Z_{\text{in}} - R_0 - j\omega L_0 \\ &= j\omega L_\mu + \frac{\omega^2 M_\mu^2}{R_1 + j\omega L_1} + \frac{\omega^2 M_0^2}{R_1 + j\omega L_1}\end{aligned}$$

Table 19.6 Inversion results using 100 Hz to 1 kHz data

Location	Φ	σ/sensit	μ/sensit	No. pts
A	0.188(−2)	1.372×10^6/2.90(−2)	68.18/3.76(−3)	500
B	0.397(−2)	1.552×10^6/5.66(−2)	73.43/8.35(−3)	500

$$= j\omega L_\mu + \frac{\omega^2 M_c^2 (R_1 - j\omega L_1)}{R_1^2 + \omega^2 L_1^2}$$

$$= \frac{\omega^2 M_c^2 R_1}{R_1^2 + \omega^2 L_1^2} + j\omega \left(L_\mu - \frac{\omega^2 L_1 M_c^2}{R_1^2 + \omega^2 L_1^2} \right)' \tag{19.4}$$

where $M_c^2 = M_0^2 + M_\mu^2$. Again, we see that at low frequencies the effects of conductivity and permeability of the tube are uncoupled. The first term in (19.4) is the effective coupled resistance as a function of frequency, whereas the second term displays the coupled inductance as a function of frequency. Note, in particular, the quadratic dependence on ω of the resistance at low frequencies, $\omega << R_1/L_1$. Furthermore, when ω is very small the coupled inductance is simply L_μ, the inductance of the coil due solely to permeability. The coupling to R_1, which corresponds to conductivity, is virtually nil. Depending upon the values of the various parameters, the coupled inductance, while starting out as a positive number, may go negative. If the permeability of the tube were that of freespace, then $L_\mu = 0$, and the change in coupled inductance would be negative at all frequencies, in accordance with Lenz' law.

19.4.3 Characterizing the Tube: Outer-Coil Experiments

19.4.3.1 Inversion Using 100 Hz to 1 kHz Data

We perform a series of experiments using an outer coil and an inner coil to characterize the tube. We inverted data taken at two locations with the outer coil: location A, near one end of the tube, and location B, near the center. The results of the inversions are shown in Table 19.6. These results have a residual norm that is quite small, with sensitivities that are consistently small for each estimated variable. We attribute this to "unraveling" the confounding variables, σ and μ, at low frequencies, exactly as our analysis above indicated. The fifth column in Table 19.6 gives the number of points, out of the original distribution of 500 (recall Sect. 12.2.2), that are mapped into the global minimum.

The results of the inversion are compared with the original data in Fig. 19.11. It is clear that the inversion not only gives a good fit to the data, but that the quadratic depence of δR on frequency, for low frequencies, holds, as derived in (19.4). Furthermore, the reactance, δX, varies linearly with frequency, which means that

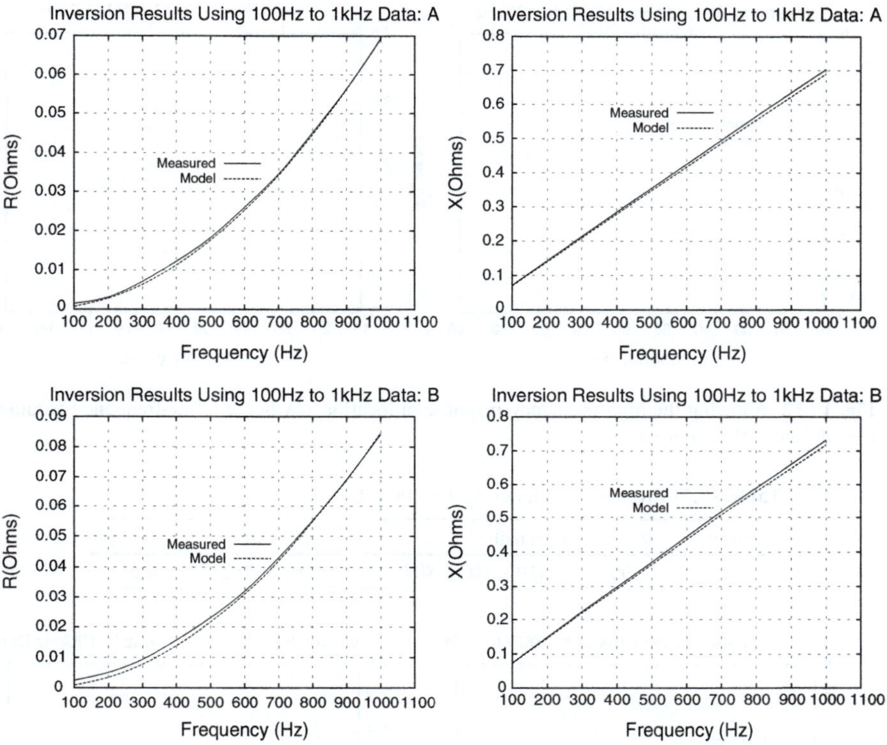

Fig. 19.11 Comparing the model results, based on the results of Table 19.6, with the original measured data over the frequency range of 100 Hz to 1 kHz. *Top*: location A. *Bottom*: location B

we are seeing the effects of L_μ, only, with no coupling to the conductivity in the tube. This further confirms our model results and demonstrates the advantage of working at low frequencies to distinguish conductivity and permeability effects.

19.4.3.2 Inversion Using 10–100 kHz Data

When we use the inverted results of Table 19.6 for location A, and generate a model response over the frequency range 10–100 kHz, we get the results shown in Fig. 19.12. Because the resistance values are in good agreement over this frequency range, but the reactances differ considerably, it is clear that L_μ in (19.4) is underestimated, which in turn suggests that μ is too small. We do another inversion to determine σ and μ, this time using the measured data for location A shown in Fig. 19.12. The results of this high-frequency inversion are shown in Table 19.7 and are plotted in Fig. 19.13.

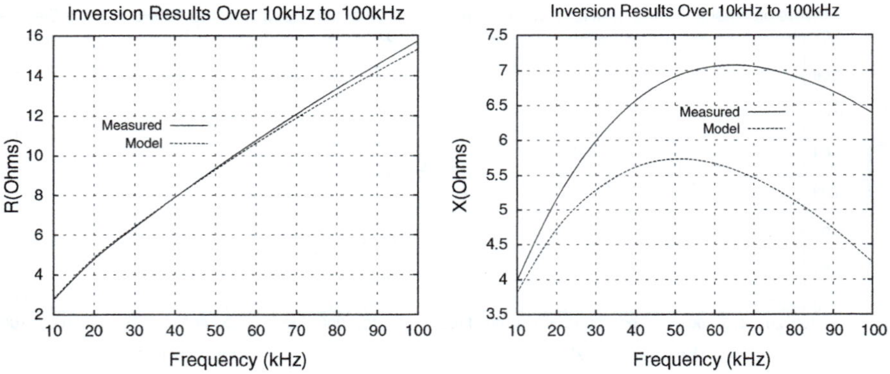

Fig. 19.12 Showing the high-frequency response at location A when the low-frequency parameters of Table 19.6 are used

Table 19.7 Inversion results using 10–100 kHz data

Location	Φ	σ/sensit	μ/sensit	No. pts
A	1.399	$1.319 \times 10^6/7.803(-2)$	78.636/0.232	426

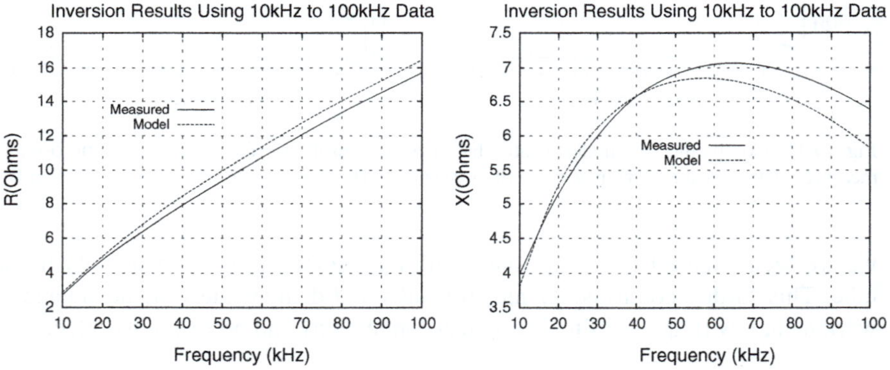

Fig. 19.13 Comparing the model results, based on the results of Table 19.7, with the original measured data for location A over the frequency range of 10–100 kHz

The fact that the norm of the residuals, Φ, is much larger here than in Table 19.6 is not surprising, given the much larger data values here, nor is it surprising that the sensitivity coefficient for μ is much larger, given our observation that the two variables have the smallest correlation at low frequencies. The important point to note is that σ is changed by only 4 %, which is probably in reasonable agreement with the data, but that μ increases by 15 %. It seems likely that the origin of the ferromagnetic response is dispersive, which yields a permeability that is frequency dependent. This frequency dependence is reasonable, given that dipolar effects, whether electrical or magnetic, have their origins in electrons that are bound to the

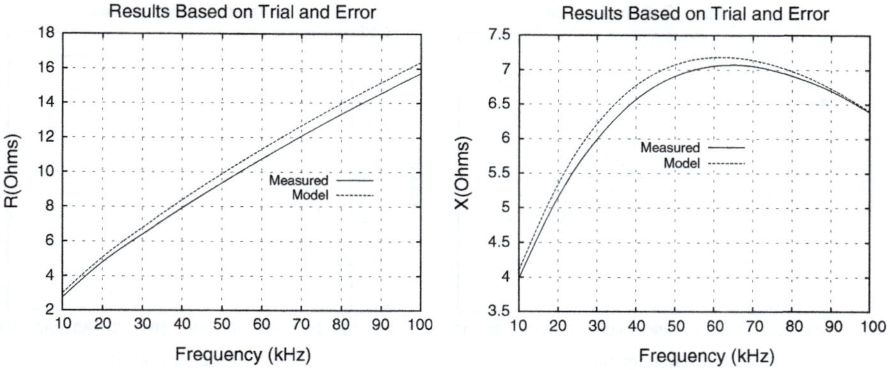

Fig. 19.14 Comparing measured and model results over 10–100 kHz for location A, when the model comprises $\sigma = 1.372 \times 10^6$ S/m, as in the low-frequency regime, and $\mu = 85$, chosen by trial and error

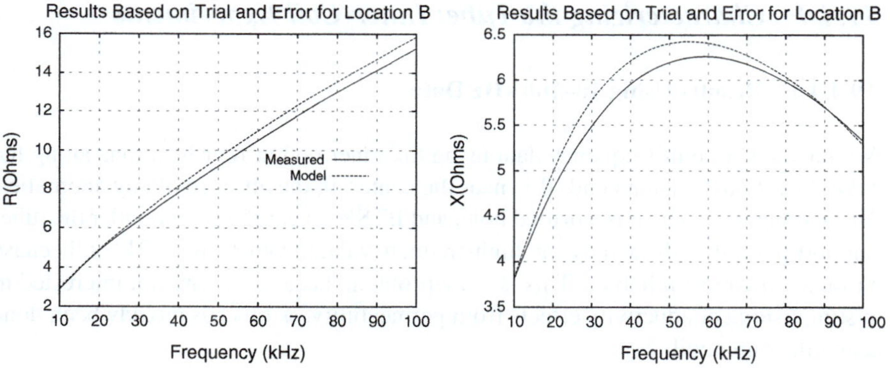

Fig. 19.15 Comparing measured and model results over 10–100 kHz for location B, when the model comprises $\sigma = 1.552 \times 10^6$ S/m, as in the low-frequency regime, and $\mu = 86$, chosen by trial and error

host lattice, and are not free to move as are those electrons in the conduction band [101]. This suggests that dipoles should share the frequency response of the host lattice, including resonance and similar dispersive phenomena.

An alternative strategy would be to keep the host conductivity constant at $\sigma = 1.372 \times 10^6$ S/m for all frequencies for location A, and then determine a value of μ that is consistent with the measured data simply by trial and error. We have done this, and found that $\mu = 85$ gives excellent results, as shown in Fig. 19.14. Figure 19.15 shows the corresponding results for location B, using $\sigma = 1.552 \times 10^6$ S/m and $\mu = 86$. In either case, it is clear from Figs. 19.14 and 19.15 that the high-frequency reactance demonstrates the nonlinear behavior shown in (19.4) that is due to Lenz' law for coupled circuits.

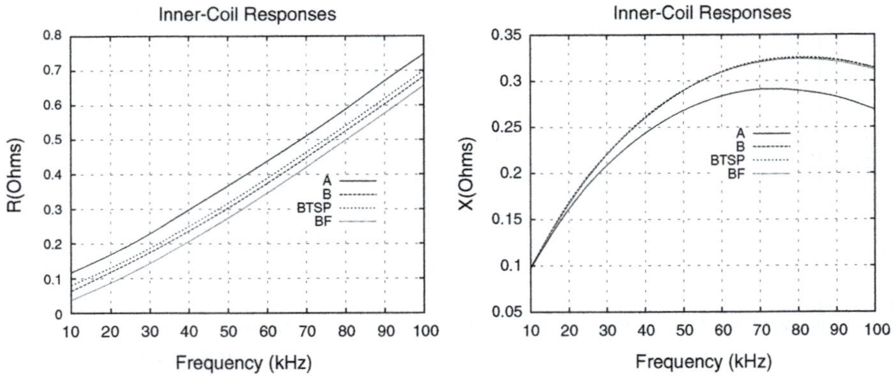

Fig. 19.16 Inner-coil responses at four probe locations. A: At end of tube. B: Near center of tube. BF: At center with flaw *(drilled hole)*. BTSP: Near center with tube support plate over it

19.4.4 Characterizing the Tube: Inner-Coil Experiments

19.4.4.1 Results Using 10–100 kHz Data

We have taken multifrequency data using the inner coil at four locations along the tube. Location A is at an end, B is near the center, in the absence of any anomalies, BF is at the flaw, which is a drilled hole, and BTSP is near the center under the tube-support plate. We show only the high-frequency data from 10 to 100 kHz, because those are the most reliable.[5] This is not a problem, because we are not interested in distinguishing conductivity effects from permeability, as this has already been done using the outer coil.

Figure 19.16 shows the impedance responses at these four locations. It is clear that the response at A is distinguished because of its location near the end of the tube. It is also clear from the resistance response that the tube-support plate is not as clearly distinguished from its background (B) as is the flaw. This is reasonable, given that the skin-effect is pronounced at these frequencies, which effectively shields the TSP from the coil. The flaw, being a surface-breaking inner-diameter hole, is not similarly shielded.

We have noted that over the frequency range of 10–100 kHz the permeability does not remain fixed at its low-frequency value. By setting $\mu = 104$ over this frequency range, and letting the conductivity retain its low-frequency value of

[5]The inner coil is much smaller than the outer, with fewer turns, which means that it has a smaller inductance, and is, therefore, prone to produce noisier low-frequency data.

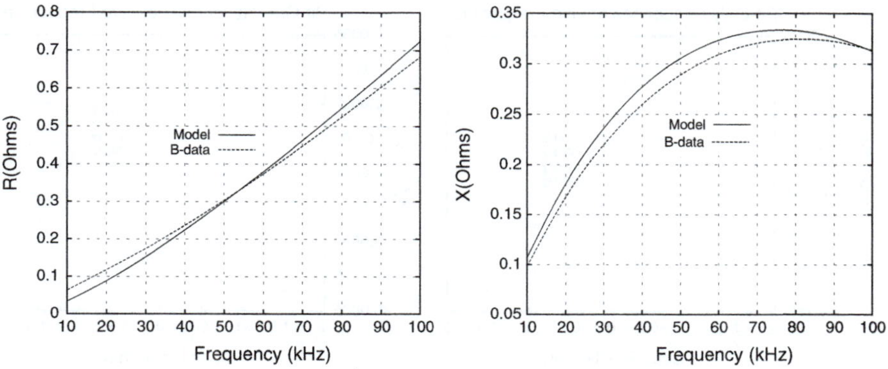

Fig. 19.17 Showing the model inner-coil response over the frequency range of 10–100 kHz, together with the measured data for location B. The model parameters used to obtain this result are $\sigma = 1.552 \times 10^6$ S/m, and $\mu = 104$

$\sigma = 1.552 \times 10^6$ for location B, we generate the model response for the inner coil shown in Fig. 19.17, along with the measured data for location B. The responses are in good agreement within a few percent over this frequency range.[6]

19.4.5 Characterizing the Composite Structure

19.4.5.1 Model Results

Now that the host ferritic tube has been characterized, we proceed with the main thrust of this section: characterizing the composite structure consisting of the inner coil, tube, through-wall hole, and tube support plate. We'll start with some model calculations, because these results will allow us to interpret the measured data that will be described in the next section. The configuration that produces the model results of this section is the same as that shown in the top of Fig. 19.9. It uses the same values of $\sigma = 1.552 \times 10^6$ S/m and $\mu = 104$ for the host parameters as in Fig. 19.17.

In Fig. 19.18, we show the results of a scan past the isolated hole at five frequencies, 1, 10, 50, 75, and 100 kHz. The important feature here is the monotonic increase (in the negative direction for *R*) of the responses with frequency. These results should be contrasted with the corresponding ones for the isolated TSP, shown in Fig. 19.19. The frequency responses of the TSP are much more convoluted, in

[6]One possible explanation for the difference in the values of the permeabilities of Figs. 19.15 and 19.17 is due to material stress inhomogeneities through the wall thickness. See [101] for a discussion of magnetoelastic effects on permeability.

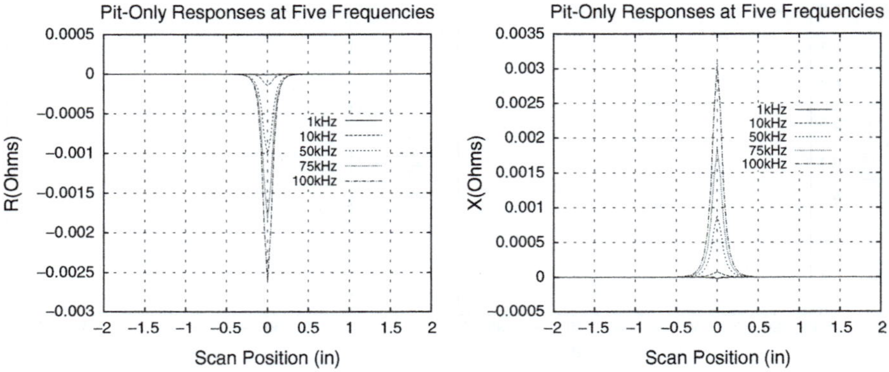

Fig. 19.18 Inner-coil responses at five frequencies for the isolated through-wall hole

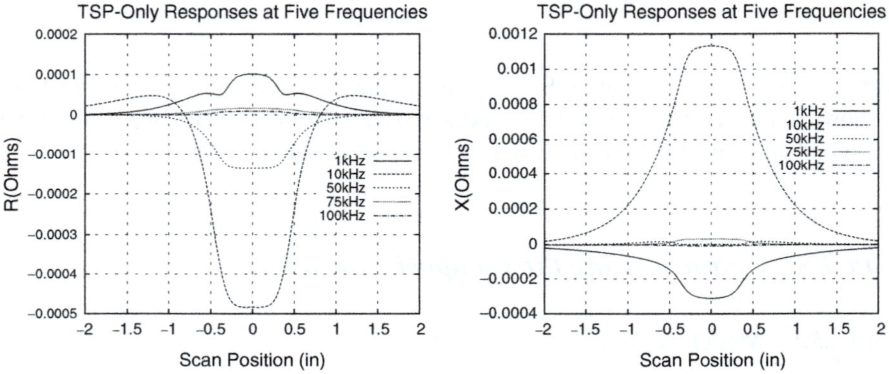

Fig. 19.19 Inner-coil responses at five frequencies for the isolated tube-support plate

the sense that they “flip” at low frequencies. This is typical of an outer-diameter anomaly and is due to the “competition” between μ and σ in the ferritic tube wall, as we have seen before. Furthermore, the TSP becomes invisible at higher frequencies (at least as far as the inner coil is concerned), due to the reduced skin depth. This reduction with frequency is pronounced due to the large magnetic permeability of the tube.

The important contrast between Figs. 19.18 and 19.19 is in the low- and high-frequency responses. At low frequencies, 1–10 kHz, the TSP response overwhelms the hole response, but the situation is completely reversed in the high-frequency regime at 50 kHz and above, for which the hole dominates the TSP. This suggests that any protocol for detecting pits in the presence of tube-support plates will necessarily require choosing the right frequency range in which to operate.

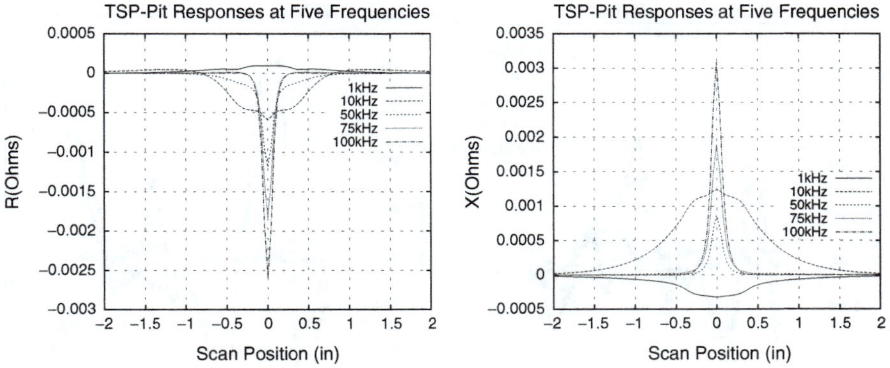

Fig. 19.20 Inner-coil responses at five frequencies for the tube-support plate with the hole centered beneath it

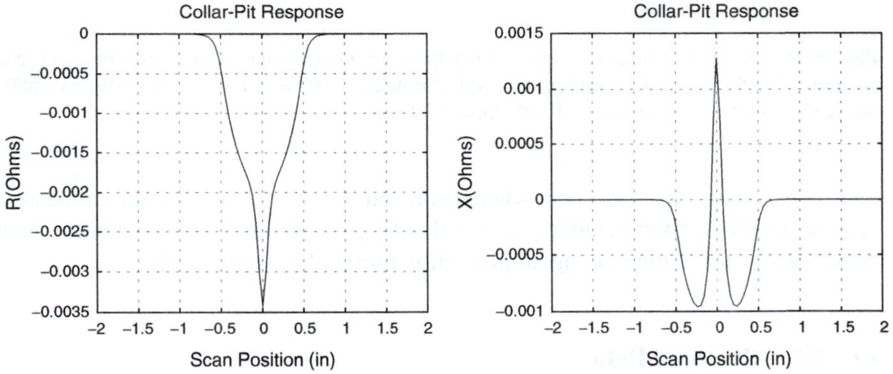

Fig. 19.21 Inner-coil response at 100 kHz for the hole of Fig. 19.18 surrounded by a through-wall collar whose conductivity is 1.552×10^6 S/m and permeability is 100. The diameter of the collar is one inch

We plot the combined response of a hole centered under the TSP in Fig. 19.20.

It is clear that the TSP completely dominates the hole response at frequencies of 10 kHz and lower, but that the hole dominates the high-frequency response, in accordance with the isolated results of Figs. 19.18 and 19.19.

Later, when interpreting measured scan data, we will make use of a model, whose result is shown in Fig. 19.21, comprising the through-wall hole of Fig. 19.18 surrounded by a through-wall collar whose permeability is 100 and conductivity is 1.522×10^6 S/m. The purpose of this model is to illustrate the significant effect that even a small material inhomogeneity, such as magnetic permeability, can have on the impedance response. Contrast, for example, the reactance response at 100 kHz in Fig. 19.18 with the reactance response in Fig. 19.21. The former

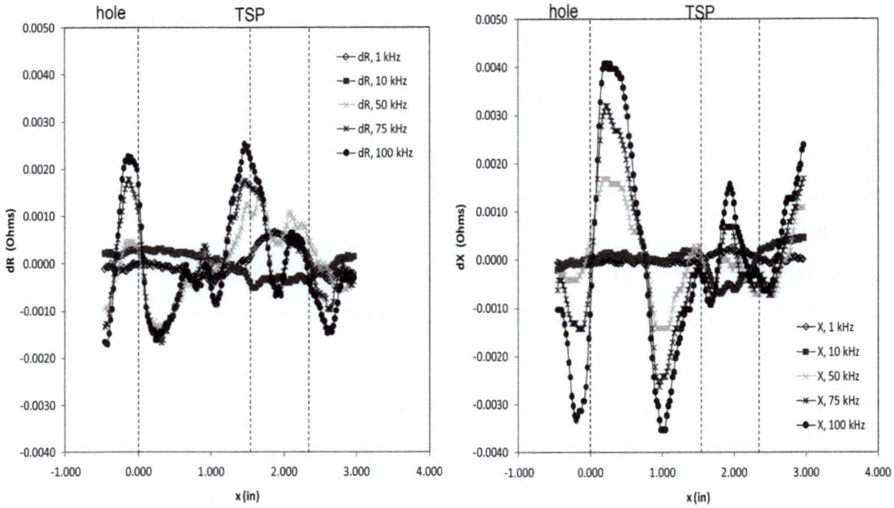

Fig. 19.22 Change in resistance (*left*) and reactance (*right*) as a function of frequency and scan position when the probe is scanned past the hole, located at 0.00", and TSP, located between 1.580" and 2.30". The frequencies are 1, 10, 50, 75, and 100 kHz

does not change sign with scan, whereas the latter does. As we will see in the next section, the background clutter associated with permeability inhomogeneities must be accounted for in developing an inspection protocol for ferritic tubes.

19.4.5.2 Measured Data

The model results presented in the last section are useful in interpreting measured data in which the inner-coil is scanned past the hole and TSP. Figure 19.22 shows the response of the inner coil when it is scanned past the hole, located at 0.00", and the TSP, which starts at 1.580" and ends at 2.300". The frequencies are 1, 10, 50, 75, and 100 kHz.

These results, while exhibiting considerable coherent background clutter, still show qualitative agreement with our model computations. For example, consider the reactance function shown in the right-hand side of Fig. 19.22. The hole portion of this response shows two negative-going "descenders" bracketing the upward-going response over the hole. This agrees with the model reactance response shown in Fig. 19.21, which suggests, but certainly does not prove, that the source of the coherent clutter could be an inhomogeneous distribution of magnetic permeability. Furthermore, note that the reactance response associated with the hole increases monotonically with frequency, exactly as we showed in Fig. 19.18.

The reactance function response of the TSP in the right-hand side of Fig. 19.22 is also in agreement with the model calculations; the 1 and 10 kHz responses have opposite signs, which agrees with the results shown in Fig. 19.19. The fact that each of these two measured responses has the opposite sign of the corresponding model responses is due to choosing the model TSP parameter, σ_{TSP}, arbitrarily to be 6.76×10^6 S/m; a more reasonable value of 1.0×10^6 S/m will be demonstrated shortly. Furthermore, note that the two lowest measured TSP signals (1 and 10 kHz) are much larger than the measured hole signals at 1 and 10 kHz and is the reason that we must not use such low frequencies to try to locate a hole in the presence of the TSP. This, too, was predicted in the model calculations, as shown in Figs. 19.18–19.20. It is clear that the higher frequency signals, namely 50, 75, and 100 kHz, are associated with an inner-diameter anomaly, since they monotonically increase with frequency. In fact, judging from the shape of these signals—they are miniversions of the corresponding through-wall hole signals—the anomaly may well be a small pit. It is likely, however, that these signals are simply part of the background clutter, whose source needs to be studied further.

Further validation is given in the series of responses shown in Figs. 19.23–19.26, which display the results for the isolated hole, the hole centered under the TSP, and the hole placed at an edge of the TSP. These data were taken with an inner coil that had more turns, and was somewhat larger than that used in the earlier experiments. Thus, with the increased sensitivity of the coil, the responses are larger than those shown previously, but retain the same qualitative features. In all of these figures, the hole is located at 0.00 inch.

The results shown in these figures confirm our previous conclusions. Note that the high-frequency response of the hole + TSP system approaches the isolated hole response, no matter where the hole is located relative to the TSP. On the other hand, the low-frequency response is dominated by the TSP, no matter where the hole is located, and furthermore, the low-frequency response exhibits the usual "flipping" phenomenon that we observed earlier. This is made quite clear in Figs. 19.25 and 19.26.

19.4.6 An Improved Low-Frequency Model

In our discussion of the results of Figs. 19.19 and 19.22, we determined that the permeability of the TSP was in error, and this was due to the fact that we had not characterized the TSP, but had simply used a previous model, as shown at the top of Fig. 19.9. In order to rectify this situation, we reran some of the models with the TSP and hole at low frequencies, namely 1, 10, and 19 kHz. The new runs were performed with low-frequency permeabilities of 68.18 and 73.43, instead of the high-frequency permeability of 104. Furthermore, we ran different host conductivities, namely 1.372×10^6 S/m and 1.552×10^6 S/m, which we computed earlier to be correct for two measurement points with the outer coil.

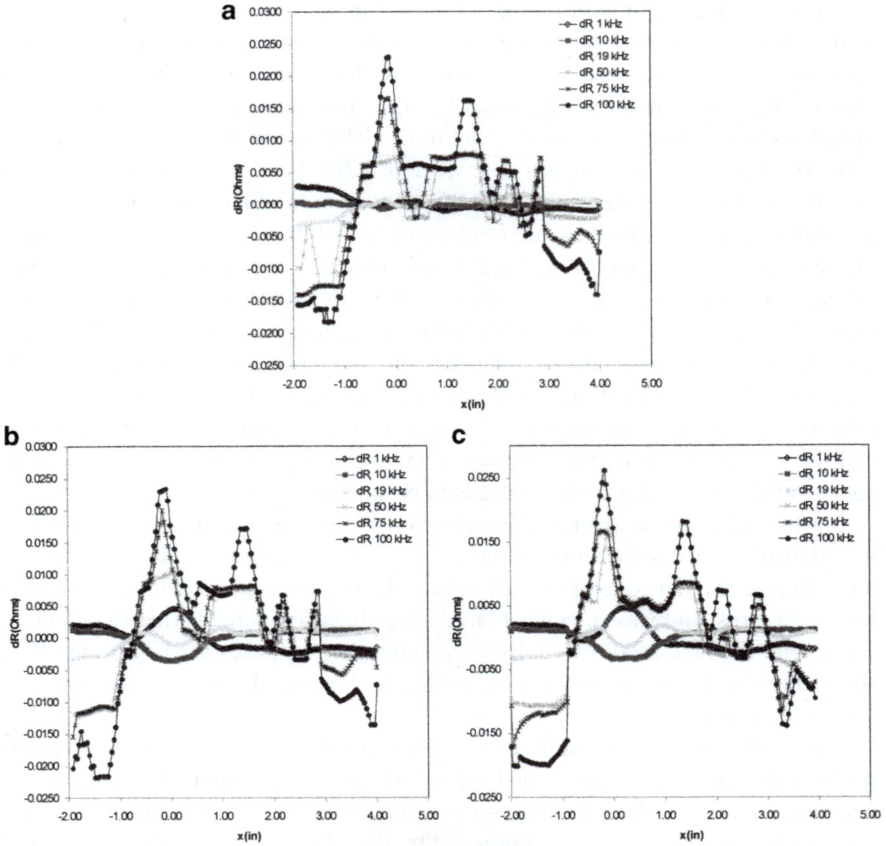

Fig. 19.23 Change in resistance as a function of frequency and scan for (**a**) no TSP, (**b**) TSP over hole center, (**c**) TSP over hole edge

Because we had not characterized the TSP independently of the tube, we varied the TSP parameters during these trials, using combinations of $\mu_{\text{TSP}} = 50$, and 100, together with $\sigma_{\text{TSP}} = 6.76 \times 10^6$, 1×10^6, 5×10^5 S/m. The combination of all of these parameters that gave a good qualitative fit to the measured data is shown in Table 19.8. These parameters are used in the lower configuration shown in Fig. 19.9.

The results of the runs are shown in the top row of Fig. 19.27, and when they are compared with the corresponding measured data shown in the bottom row, we see that there is a good qualitative agreement and that the relative magnitudes of the reactance responses are in reasonable agreement with the measured data.

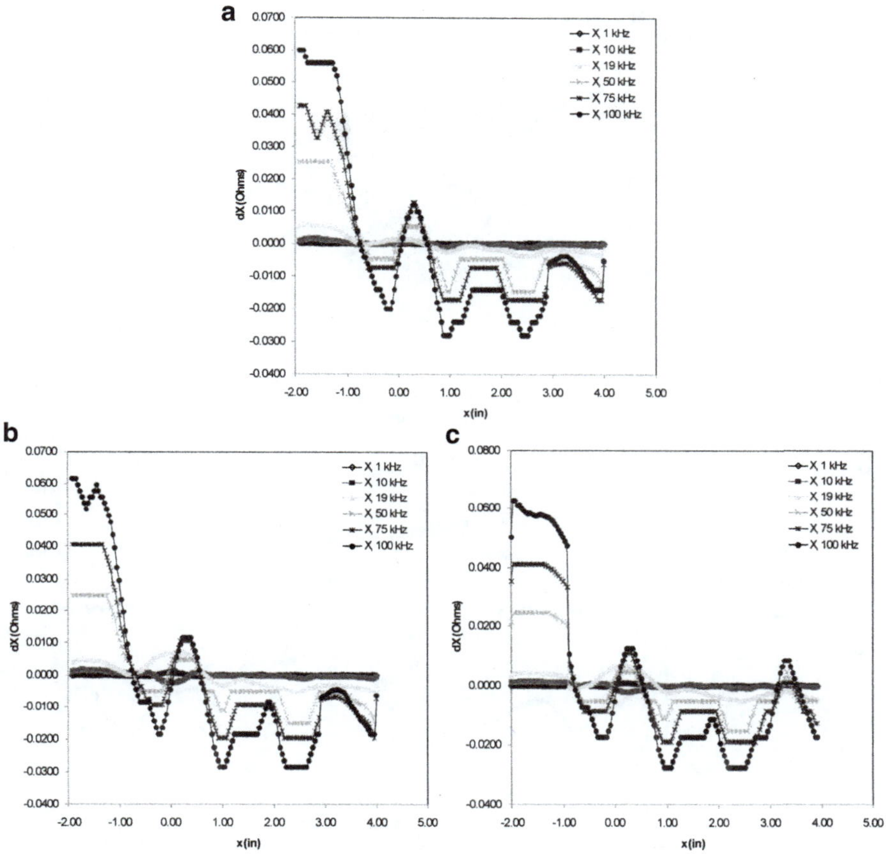

Fig. 19.24 Change in reactance as a function of frequency and scan for (**a**) no TSP, (**b**) TSP over hole center, (**c**) TSP over hole edge

19.4.7 Comments and Conclusions

The problem of detecting and characterizing anomalies in ferritic tubes can be solved using conventional eddy-current technology, with the use of multifrequency scanned impedance data being the key to success. It is necessary to first characterize the tube in order to determine the optimum operating frequencies to distinguish classes of anomalies. Typically, one will want to operate at "high" frequencies in order to reduce the effects of any external anomalies, such as a tube-support plate, on ID or through-wall anomalies, such as flaws or pits.

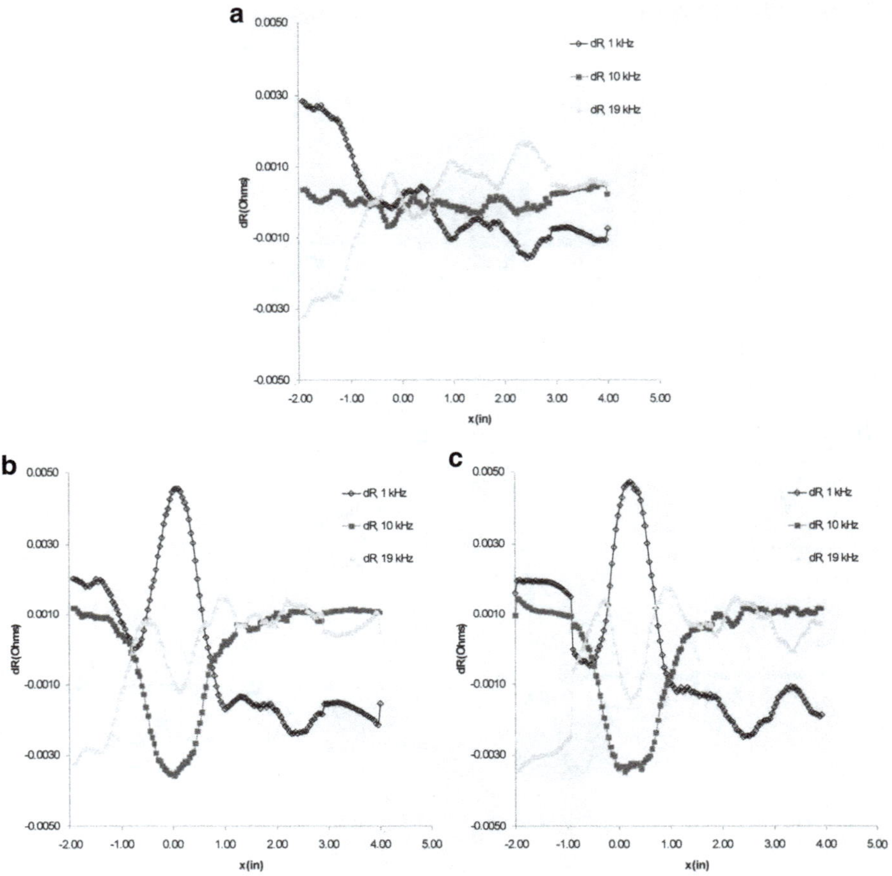

Fig. 19.25 Change in resistance as a function of frequency and scan for (**a**) no TSP, (**b**) TSP over hole center, (**c**) TSP over hole edge (low frequencies, only)

The presence of a large, coherent, background clutter, however, remains an interesting challenge. Based on a comparison of the model results and measurements, the metallurgical origin of this clutter could be inhomogeneities in the magnetic permeability of the tube. This suggests that a determined effort be made to study the statistical nature of this clutter, with the intention of developing a protocol to mitigate its effects. Methods for doing this will be studied in the second volume of this series.

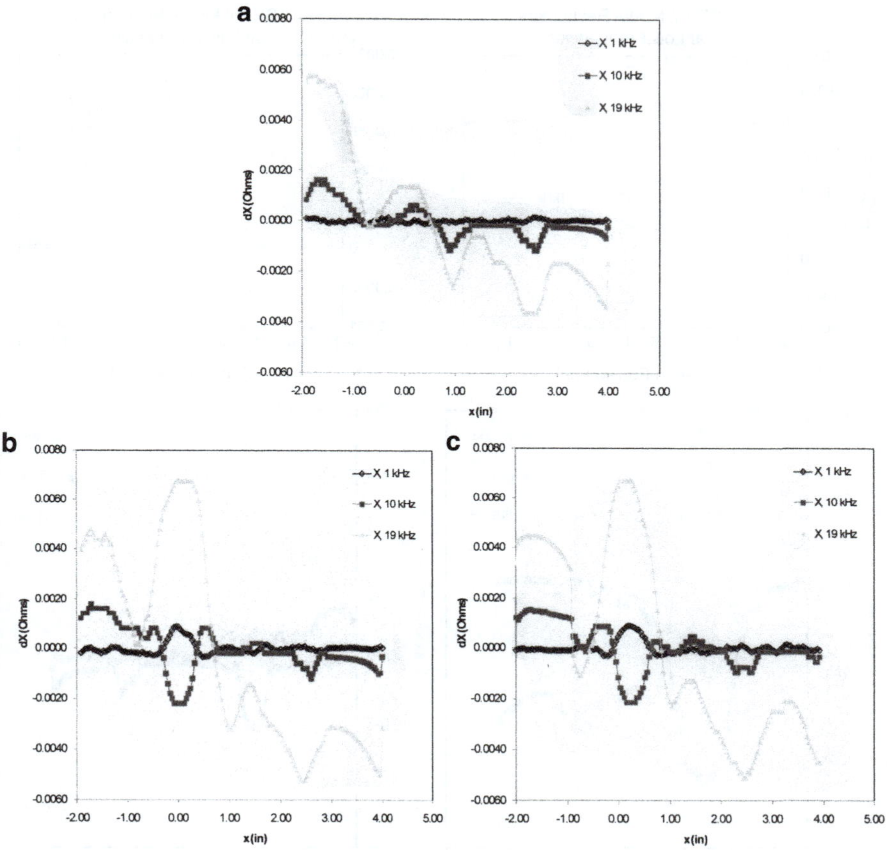

Fig. 19.26 Change in reactance as a function of frequency and scan for (**a**) no TSP, (**b**) TSP over hole center, (**c**) TSP over hole edge (low frequencies, only)

Table 19.8 Optimum parameters for low-frequency TSP-hole models

σ_{host}	μ_{host}	σ_{TSP}	μ_{TSP}
1.372×10^6	68.18	1×10^6	50

Appendix

A.1 Inverse Method Quality Metrics for the Ferritic Tube

Following the discussion of inverse method quality metrics and the Cramer–Rao Lower Bound in Chap. 12, we optimize the estimation of conductivity and permeability of the ferritic tube. Figure 19.28 presents changes in the normalized impedance plane for varying permeability (red) and conductivity (black) levels at

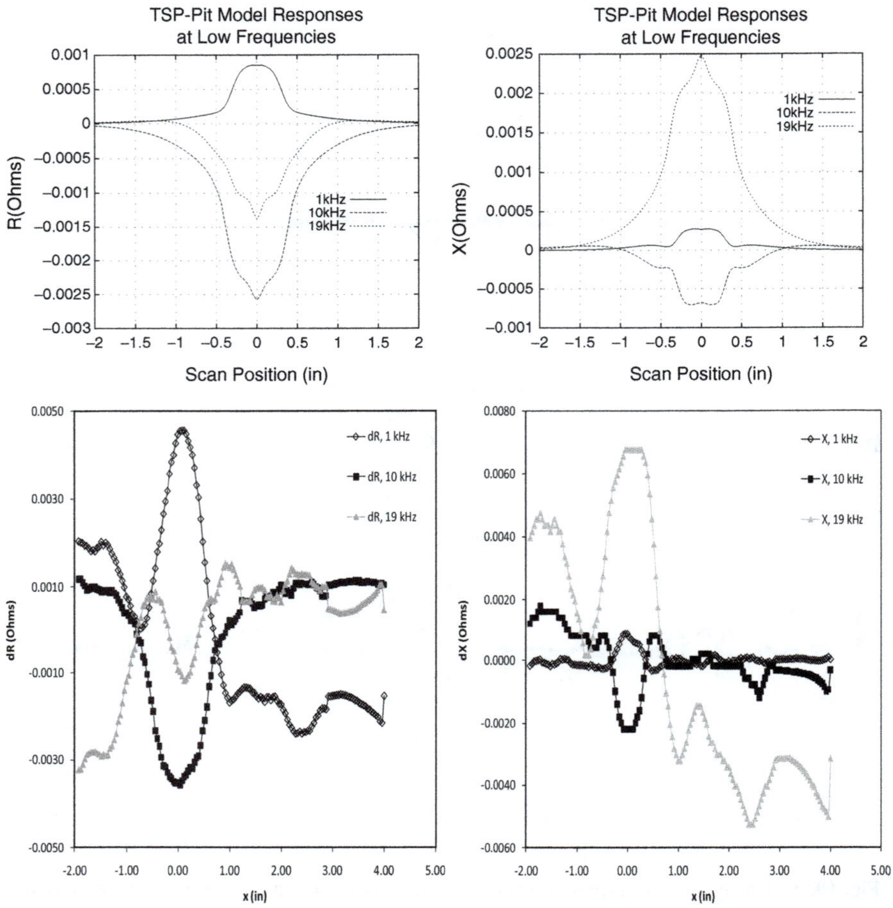

Fig. 19.27 Low-frequency model response of the TSP and hole when the data in Table 19.8 are used. *Top*: Model. *Bottom*: Measured

four frequencies: (a) 100 Hz, (b) 1.0 kHz, (c) 10 kHz, and (d) 100 kHz. Several observations can be made from these plots. First, it is helpful to increase frequency in order to get sensitivity to conductivity changes. At very low frequencies, sensitivity to changes in permeability is dominant in the reactance term. As the frequency approaches 100 kHz, it is impossible to distinguish changes in permeability and conductivity in the eddy-current measurements.

Figure 19.29 presents a comparison of four different estimation metrics for this inversion problem: (a) CRLB for variation in conductivity estimation, (b) CRLB for variation in permeability estimation, (c) condition number, and (d) correlation. Based on the CRLB for optimal sensitivity to permeability, lower frequencies are better. Note, there is really not much benefit from using 100 Hz versus 1.0 kHz.

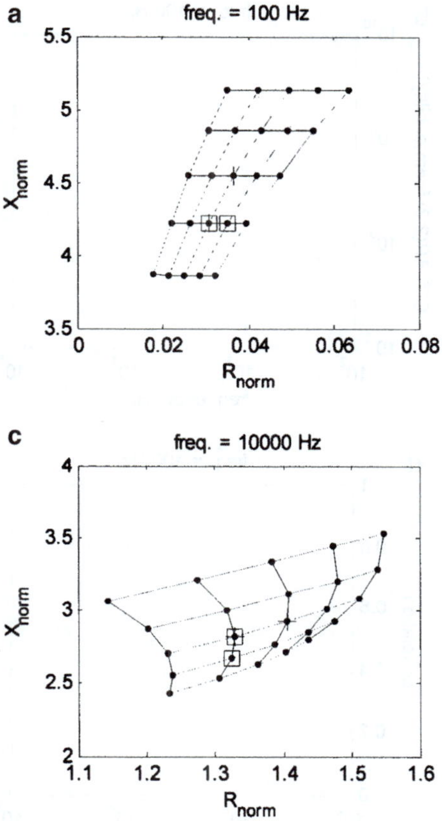

Fig. 19.28 Changes in normalized impedance plane for varying permeability and conductivity levels at four frequencies: (**a**) 100 Hz, (**b**) 1.0 kHz, (**c**) 10 kHz, and (**d**) 100 kHz. Center values considered in the study were $\sigma = 1.4 \times 10^6$ S/m and $\mu = 60$. The plus sign denotes a change in response with respect to a change in permeability, and the box denotes a change in response with respect to a change in conductivity

However, there is clearly an optimal region for estimating conductivity. Based on the CRLB for variation in conductivity estimation, 6.0 kHz was found to be optimal. This result was found to be in almost near agreement with the condition number maximum of 5.0 kHz. Lastly, the correlation between the two estimation parameters was found to be a problem (approaching 1) at 20 kHz and beyond. Thus, it is recommended that inspectors work in a frequency range of 10 kHz and below for simultaneously estimating permeability and conductivity.

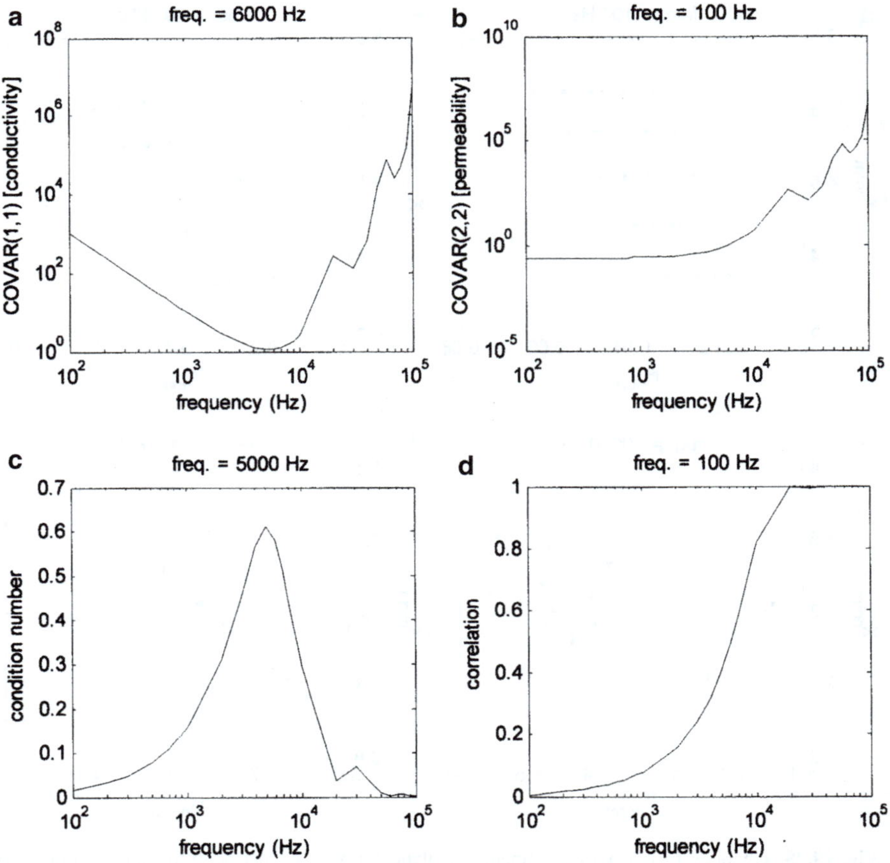

Fig. 19.29 Comparison of estimation metrics: (**a**) CRLB for variation in conductivity estimation, (**b**) CRLB for variation in permeability estimation, (**c**) condition number, and (**d**) correlation

Chapter 20
Applications to NDE of Coatings

20.1 Assessing Thermal Barrier Coatings

Characterizing surfaces and coatings has long been a staple of eddy-current technology, but with the advent of inverse methods that technology has become even more powerful. In this chapter we describe a case in which model-based inversion, coupled with computational electromagnetics, can be effectively used to solve complex coating problems.

The objective is to take measured impedance data and pass them through the inversion algorithm, NLSE, and thereby reconstruct the layer(s) of a coating, i.e., determine the electromagnetic properties of each layer. We have applied **VIC-3D®** to a similar problem of assessing the PWA286 thermal barrier coating by eddy-current inversion [99, 100]. Such coatings are important in maintaining the life of new, high-performance gas turbines that are increasingly being used in small power plants. Because the work reported in [99] and [100] involved actual measured laboratory data, and because the techniques used to assess these coatings are identical to what we are proposing in this book, it is useful to include this material as a benchmark validation.

Advanced turbines are used in applications ranging from aerospace to land-based power generators. These turbines are fired at higher temperatures (1,850–1,950°F) and utilize optimum cooling of hot section components. Because of the higher operating temperature, the performance and durability of the first-stage blades has become one of the prime life-limiting factors. Individual blades are nickel-based GTD 111 alloys that are protected by sacrificial metallic coatings to extend service life. The first-stage blades are especially important, and it is desirable to develop an in situ NDE system to monitor, evaluate, and predict remaining coating life. The coatings used on the turbine blades include CoCrAlY and NiCoCrAlY, with a top aluminide coating (GT29+, GT33+, respectively), and a NiCoCrAlY coating, called PWA 286. Figure 20.1 illustrates an as-coated PWA286 coating on a GTD111 substrate and the same coating after aging.

H.A. Sabbagh et al., *Computational Electromagnetics and Model-Based Inversion*, Scientific Computation, DOI 10.1007/978-1-4419-8429-6_20,

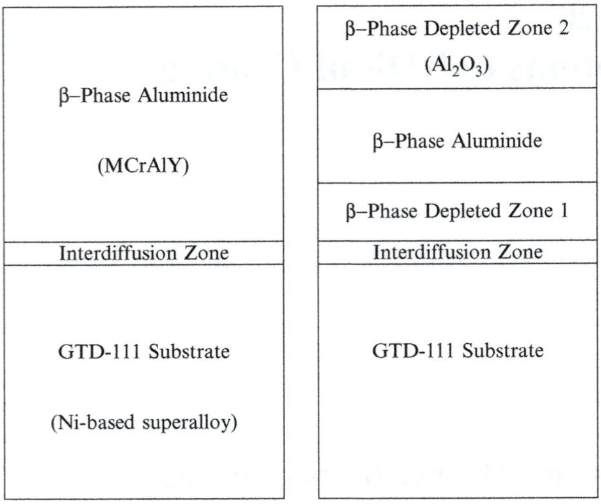

Fig. 20.1 An as-coated PWA286 coating on a GTD111 substrate (*left*) and the same coating after 2,400 h (*right*)

The primary objective is to estimate the equivalent thickness of the aluminum beta-phase content of the PWA286 coating. This information is essential to maintaining the integrity of blades, because it allows the timely repair or refurbishment of coatings to extend the service-life of operating blades. Further, it is desirable to obtain the interdiffusion layer thickness, since this information indicates the level of blade exposure to service temperature. The overall remaining coating thickness indicates the reduction of the coating thickness caused by the oxidation-induced degradation of the top beta-phase depleted layer.

As the coating evolves with use, aluminum diffuses out to reform the protective oxide coating and also diffuses into the substrate, causing a transformation of beta phase NiAl into a gamma matrix of solid nickel solution. This gamma matrix manifests itself as a beta-phase depleted zone (called "Zone1"), resulting in a "standard model" shown in Fig. 20.2. Further, while the conductivity of the various zones remains fixed over time, the layer-thicknesses of these zones changes, and it is our job to determine the thicknesses of each layer, because this gives us an indication of the remaining life of the coating.

20.1.1 Inversion of Impedance Data

Using an HP 3577A Network Analyzer, we took laboratory data on a number of coupon-samples of PWA286 coatings that had undergone various levels of deterioration. The measured impedances, from 1 to 50 MHz, are shown in Fig. 20.3. We then applied the eight-layer algorithm, shown in Fig. 20.4, to invert these data.

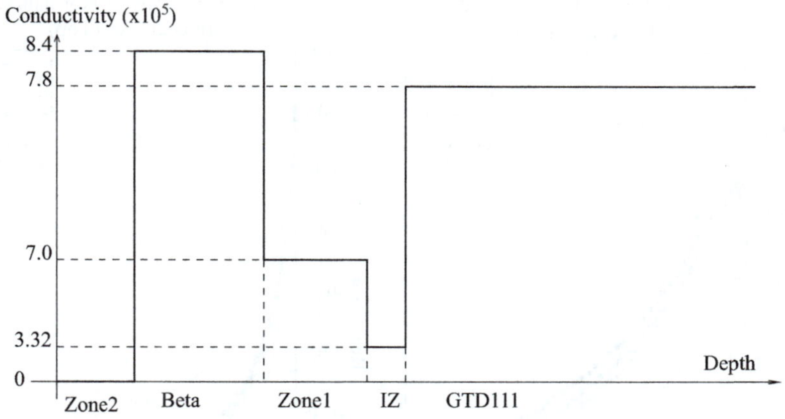

Fig. 20.2 Standard model of the PWA286 thermal barrier coating. The conductivities are determined from measurements on known samples

Inverting the impedance data via the eight-layer algorithm means assigning a value to L_0, the thickness of the top layer of Fig. 20.4, and assigning conductivities to the remaining seven layers, each of which has the same thickness, L, that defines the resolution of the reconstruction. This is done by a process of nonlinear least-squares (NLSE), in which the eight variables are chosen to give the best fit to the impedance data. The inversion is done in several steps, starting with the computation of a table of data, from which the final solution is determined by interpolation. The process is quite fast and is quite conservative in computational resources.

The results of the inversion process are fitted to the "standard model" of the PWA286 TBC, that is shown in Fig. 20.2. The values of the conductivity of the various layers are determined from inversions on known samples. In this manner, we are able to determine the thickness of each zone. The results are tabulated in Tables 20.1 and 20.2.

In several cases, the final computed conductivity profiles were obtained by a process of "focussing" the eight-layer inversion algorithm. This is accomplished by starting with a rather coarse grid, which allows the Beta Aluminide Zone and Interdiffusion and Inner Beta Depleted Zone 1 to be generally located relative to the GTD111 Substrate. Then we refine the grid between the Beta Aluminide Zone and GTD111 Substrate to get a more precise value for the various zone boundaries, as well as a more precise value for the conductivity of Zone 1 and the Interdiffusion Zone. In this manner, we were able to locate the Interdiffusion Zone and determine its conductivity for the "as-coated" samples, B1B and B1T. This is a fairly difficult computation to carry out without such a "multigrid rezoning" technique.

The tabulated results confirm that the inversion algorithm produces excellent reconstructions, with very good resolutions. In particular, we note the good agreement between the computed and measured thickness of the beta-zone of each sample. This is the critical datum, when it comes to determining the remaining life of the thermal barrier coating.

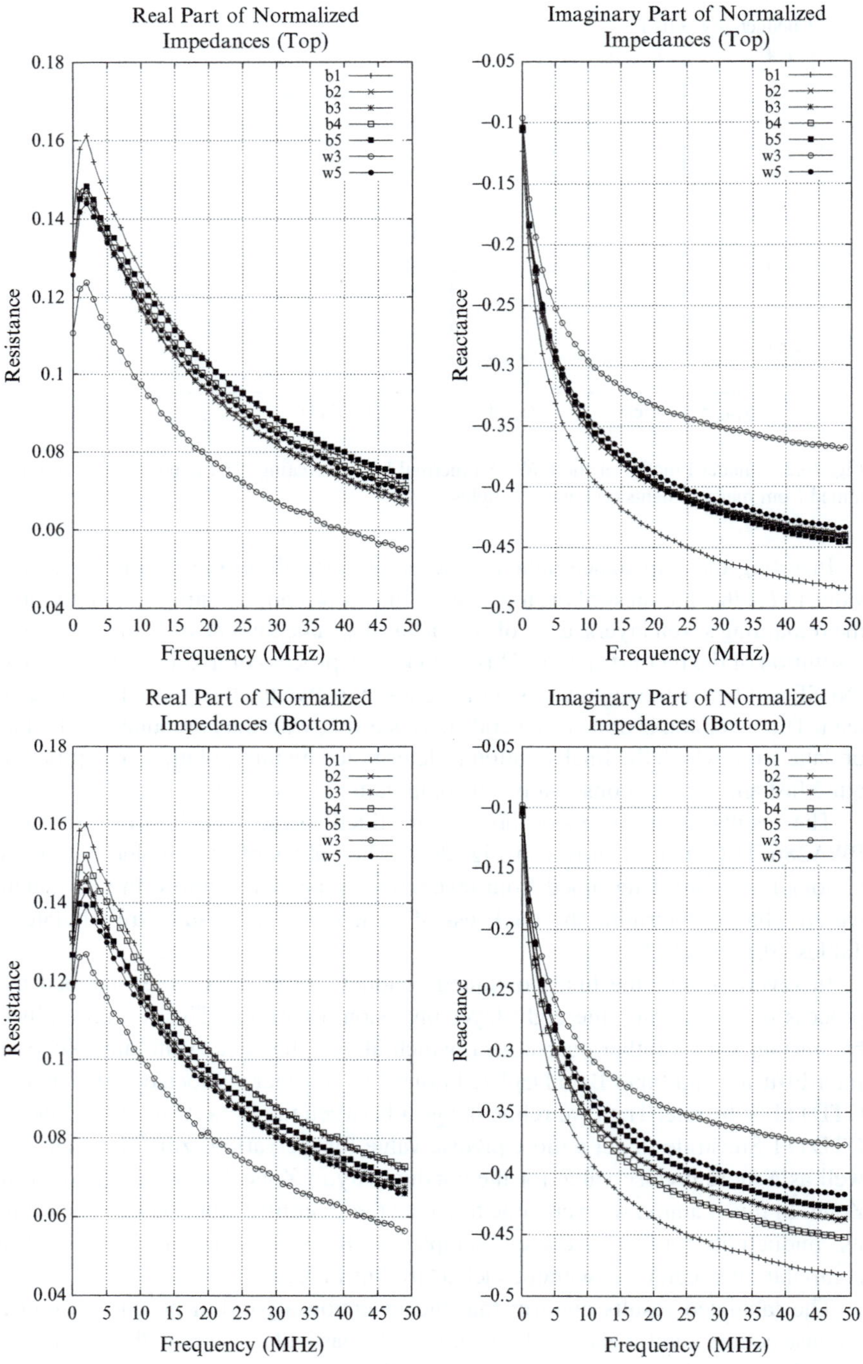

Fig. 20.3 Real (*left*) and imaginary (*right*) parts of the normalized change in impedance for the "top" PWA286 samples (*upper two figures*) and "bottom" PWA286 samples (*lower two figures*)

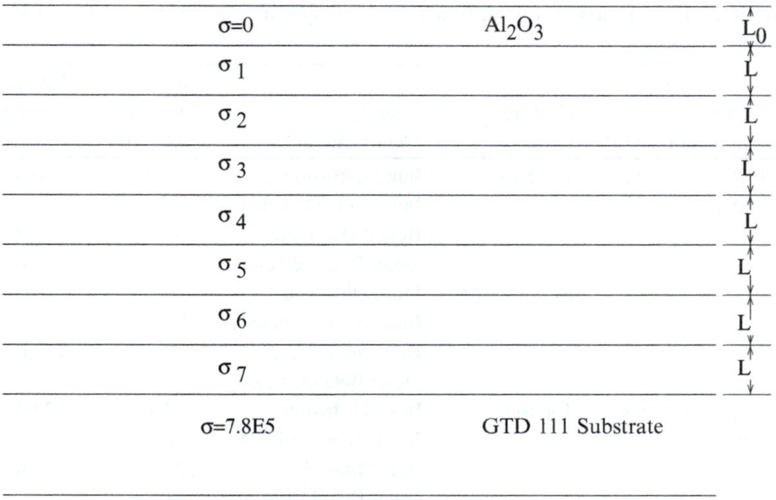

Fig. 20.4 The eight-layer inversion algorithm, in which L is the given resolution, and the objective is to determine L_0 and σ_1 through σ_7

20.1.2 *Sample Calculation: White-5 Top*

We'll demonstrate the eight-layer algorithm on the White-5 Top ("w5t") sample. The numbers that we get may be a little bit different than what are shown in Table 20.2 because we'll follow slightly different paths to the final solution.

The data for the problem are: coil IR = 12.5 mils, coil OR = 45 mils, coil HT = 8 mils, and coil turns = 23. The position of the coil within the probe coordinate system is $[0,0,5.7874]$ mils. **VIC-3D®** requires a minimum value for the z-coordinate of 4 mils, and the coil was recessed 1 mil beneath the surface of its container. The remaining 0.7874 mils is numerical fitting to give the best response to known lift-off conditions. The probe origin position is $[0,0,\text{Zone2}]$; i.e., we model the Zone 2 ("Outer Beta Depleted") height by the z-coordinate in the probe origin position. We use the 21 even-valued frequencies between 10 and 50 MHz for the model and measured input data.

Our first calculation uses Fig. 20.4 with $L = 20$ microns, so that the grid extends 140 microns into the coating. The result of the inversion produced a value of Zone 2=48.8 microns, but the conductivities of the seven layers were not well determined.

The second calculation uses an eight-layer grid, but now with all eight layers lying within the structure. Since each layer is 20 microns tall, the grid extends 160 microns into the structure. The coil coordinate system is as before, but the probe coordinates are fixed at $[0,0,48.8]$ microns. The results of this inversion are: $\sigma_1 = 897969.603411648$, $\sigma_2 = 748852.078029096$, $\sigma_3 = 576648.447023520$, $\sigma_4 = 821222.023798830$, $\sigma_5 = 888060.286078721$, $\sigma_6 = 300016.292476106$, $\sigma_7 = 300026.618414368$, $\sigma_8 = 796133.184675437$.

Table 20.1 Comparison of measured and computed coating thicknesses for the PWA286 test

Sample Identification	Coating Location	Coating Identification	Average Thickness (microns)	Computed Thickness (microns)
1950F-"as coated" (Blue-1)	Top Side	Inner Diffusion	10.6	10.0
		Inner Beta Depleted	–	–
		Beta Phase Zone	132.0	132.0
		Outer Beta Depleted	–	–
	Bottom Side	Inner Diffusion	10.1	10.0
		Inner Beta Depleted	–	–
		Beta Phase Zone	140.9	142.0
		Outer Beta Depleted	–	–
1950F-250 cycles (Blue-2)	Top Side	Inner Diffusion	22.4	20.0
		Inner Beta Depleted	8.9	9.4
		Beta Phase Zone	122.4	120.0
		Outer Beta Depleted	14.5	17.7
	Bottom Side	Inner Diffusion	22.2	22.6
		Inner Beta Depleted	8.1	9.1
		Beta Phase Zone	114.2	112.5
		Outer Beta Depleted	11.6	20.0
1950F-500 cycles (Blue-3)	Top Side	Inner Diffusion	24.4	26.3
		Inner Beta Depleted	14.9	14.4
		Beta Phase Zone	94.6	93.0
		Outer Beta Depleted	17.9	19.2
	Bottom Side	Inner Diffusion	24.8	28.2
		Inner Beta Depleted	16.8	15.3
		Beta Phase Zone	98.0	94.5
		Outer Beta Depleted	19.7	19.7

Even though there is considerable oscillation in the results (as is typical for these problems), we use an averaging procedure, together with mixture theory, to make sense of them. One reason for using a large number of layers is to allow averaging. Note that the average of σ_3 and σ_4 is 6.99×10^5 S/m, which means that Zone 1 covers layers 3 and 4 (at least) (see Fig. 20.2). Furthermore, when we get such agreement, it means that Zone 1 probably ends at a depth of 80 microns.

We can get more information about the start of Zone 1, and all of Beta by using mixture theory on σ_1 and σ_2. Let L be the volume-fraction associated with $\sigma = 8.4 \times 10^5$ S/m. Then the law for combining Beta and Zone 1 is: $8.9797 \times 10^5 + 7.4885 \times 10^5 = 16.4682 \times 10^5 = 16.8 \times 10^5 L + 14 \times 10^5(1-L)$, or $L = 0.8815$. Since the total distance covered is two layers, the length of Beta is $0.8815 \times 40 = 35.26$ microns. The remaining 4.74 microns of the second layer belong to Zone 1, which means that Zone 1 = 4.74 + 40 = 47.4 microns. Because of the oscillations in σ_5 through σ_7, we must use a third calculation to compute IZ. It is pretty clear that σ_8 is associated with the GTD111 substrate. Another thing that makes us suspicious of the results

Table 20.2 Comparison of measured and computed coating thicknesses for the PWA286 test

Sample Identification	Coating Location	Coating Identification	Average Thickness (microns)	Computed Thickness (microns)
1950F-1000 cycles	Top Side	Inner Diffusion	25.8	26.4
(Blue-4)		Inner Beta Depleted	34.0	33.0
		Beta Phase Zone	62.6	63.8
		Outer Beta Depleted	37.5	41.5
	Bottom Side	Inner Diffusion	23.7	25.0
		Inner Beta Depleted	38.1	40.1
		Beta Phase Zone	52.6	53.0
		Outer Beta Depleted	33.2	34.8
1950F-1938 cycles	Top Side	Inner Diffusion	23.6	24.0
(Blue-5)		Inner Beta Depleted	118.7	119.0
		Beta Phase Zone	–	–
		Outer Beta Depleted	–	34.3
	Bottom Side	Inner Diffusion	28.4	25.0
		Inner Beta Depleted	61.5	64.0
		Beta Phase Zone	15.2	16.0
		Outer Beta Depleted	48.5	46.3
1850F-3500 cycles	Top Side	Inner Diffusion	13.5	13.7
(White-5)		Inner Beta Depleted	46.0	45.2
		Beta Phase Zone	35.2	34.8
		Outer Beta Depleted	50.2	44.6
	Bottom Side	Inner Diffusion	14.2	15.1
		Inner Beta Depleted	45.9	45.5
		Beta Phase Zone	27.6	29.6
		Outer Beta Depleted	46.9	53.1

for σ_5 through σ_7 is that σ_6 and σ_7 are very close to the lower nodal value in the two-point interpolation table. Thus, the NLSE algorithm may have "hit the stops" for these two values.

20.1.3 Application of β-Scaling to the Thermal Barrier Coating Problem

At this point, we have concluded that Zone 2 is 48.8 micrometers thick, the Beta Zone is 35.26 micrometers thick, and Zone 1 is 44.74 micrometers thick. We will apply the linear-filter algorithm to determine the thickness of the remaining zone, IZ, and to do this we start with the four-layer model shown in Fig. 20.5.

We first define the 16 nodes of an interpolation table in Table 20.3, which shows the conductivities (divided by 10^5) of layers 1–4, respectively.

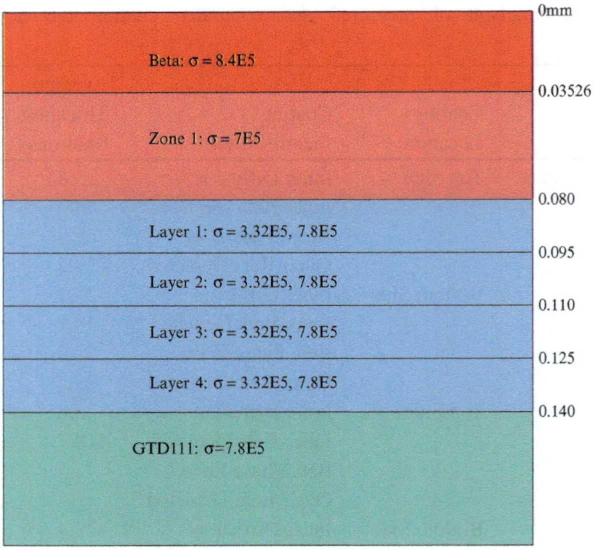

Fig. 20.5 Illustrating the four-layer algorithm that will be applied to the determination of IZ of the thermal barrier coating

Table 20.3 Interpolation table for applying the four-layer algorithm of Fig. 20.5. The conductivities (divided by 10^5) of layers 1–4, respectively, are shown

Node	Conductivity				Node	Conductivity			
1	3.32	3.32	3.32	3.32	9	3.32	3.32	3.32	7.8
2	7.8	3.32	3.32	3.32	10	7.8	3.32	3.32	7.8
3	3.32	7.8	3.32	3.32	11	3.32	7.8	3.32	7.8
4	7.8	7.8	3.32	3.32	12	7.8	7.8	3.32	7.8
5	3.32	3.32	7.8	3.32	13	3.32	3.32	7.8	7.8
6	7.8	3.32	7.8	3.32	14	7.8	3.32	7.8	7.8
7	3.32	7.8	7.8	3.32	15	3.32	7.8	7.8	7.8
8	7.8	7.8	7.8	3.32	16	7.8	7.8	7.8	7.8

In calculating β according to (15.6), we use the "White 5 Top" data of Fig. 20.3 as the measured (input) data for Z_1, and the model data corresponding to each of the nodes in Table 20.3 for Z_2. The results are shown in Table 20.4.

The data in the third column of Table 20.4 are entered into an interpolation table for NLSE. The nonlinear least-squares problem now becomes: $\min_{p_i} |\beta(p_i) - 1|$, where $p_i, i = 1,..,4$ are the four conductivities of Fig. 20.5. The result of the NLSE inversion is shown in Table 20.5. The results agree well with our standard model of Fig. 20.2, except that $p_4 = 3.754 \times 10^5$ is too small. Because of the relatively large value of σ_4, however, we conclude that the solution is fairly insensitive to this parameter, so that we can set its value to 7.8×10^5 to agree with the standard model.

Table 20.4 Results for β corresponding to the nodes of Table 20.3. The measured data are the "White 5 Top" data of Fig. 20.3

Node	β	$\lvert\beta-1\rvert$	Node	β	$\lvert\beta-1\rvert$
1	$1.0032+j0.0071912$	0.00787	9	$1.0014+j0.005348$	0.005528
2	$1.0025+j0.0009032$	0.002658	10	$1.0008-j0.0005944$	0.000997
3	$1.0018+j0.00266$	0.003212	11	$1.0001+j0.0011611$	0.001165
4	$1.0009-j0.0030114$	0.00314	12	$0.99929-j0.0042202$	0.00428
5	$1.0015+j0.0041487$	0.004412	13	$0.99975+j0.0026443$	0.002656
6	$1.0008-j0.0016615$	0.00184	14	$0.99912-j0.0028707$	0.0030
7	$1.000+j0.00012423$	0.000124	15	$0.99832-j0.0011249$	0.00202
8	$0.99912-j0.0051704$	0.005245	16	$0.99756-j0.0061376$	0.0066

Table 20.5 Results of $\min_{p_i} |\beta(p_i)-1|$. σ denotes the sensitivity of the solution to the conductivity. The conductivities are normalized to 10^5

Φ	p_1/σ_1	p_2/σ_2	p_3/σ_3	p_4/σ_4
0.4899173(−3)	3.334/0.4589	7.8/0.5568	7.535/0.8178	3.754/1.3211

Table 20.6 Results for the "White 5 Top" coating

Layer	Average of Measurements (μ)	Computed Results (μ)
IZ	13.5	15
Z1	46.0	44.74
Beta	35.2	35.26
Z2	50.2	48.8
Total	144.9	143.8

This implies that IZ exactly occupies Layer 1 of Fig. 20.5, and the remaining three layers belong to the GTD111 substrate. Hence, IZ is 15 micrometers thick. We tabulate the computed results for "White 5 Top" along with the average of the experimentally measured results in Table 20.6. The results computed by the β-scale method agree well with the measurements and the results computed earlier by another method, as can be seen in Table 20.2.

Chapter 21
Model-Assisted Probability of Detection

21.1 Introduction

To validate the performance of inspection techniques, probability of detection (POD) studies are performed. The preparation of POD samples with real fatigue cracks and the process to acquire a statistically significant number of measurements is often very time consuming and expensive, providing a significant burden for the validation of new inspection techniques. A model-assisted strategy for the design and execution of POD studies has been proposed to help mitigate the validation costs and to improve POD evaluation quality by addressing a wider array of inspection variables [118–121]. Using computer and empirical models to address variables that cannot be easily recreated in experimental samples is a significant opportunity. Two methodologies have been proposed to perform model-assisted POD evaluations: the transfer function approach and the full-model-assisted approach. The transfer function approach transforms an existing POD model for one inspection to address another similar inspection based on a limited number of new experimental measurements studying a specific varying parameter. This approach has been successfully demonstrated to transform POD results for EDM notch measurements in engine components [122] and aircraft structures [123] to parts with real cracking conditions. The full-model-assisted approach uses computer simulations to model the inspection process and determines the POD for the inspection technique through a combination of experimental and simulated data. This chapter will introduce the full-model-assisted POD methodology, discuss the case study evaluation for the inspection of cracks around fastener sites in a multilayer aircraft structure, and lastly discuss special topics.

H.A. Sabbagh et al., *Computational Electromagnetics and Model-Based Inversion*, Scientific Computation, DOI 10.1007/978-1-4419-8429-6_21,

21.2 MAPOD Process

An overview of the MAPOD process is presented in this section based on prior work [121, 124–128]. A block diagram of the model building process for a model-assisted POD evaluation is presented in Fig. 21.1. This process was developed primarily through the efforts of Bruce Thompson and Chuck Annis and can be found in parts in the Appendix H of the MIL-HNBK 1823A [121]. For any POD study, the scope of the POD study must be assessed and all critical factors for the NDT technique, part material, part geometry, and discontinuity characteristics that control signal and noise must be identified. As well, the amount of variability (as distribution functions) and the impact must be assessed through expert knowledge, experimental results, and/or model-based studies.

Once key factors have been determined, an assessment can be made on whether to pursue a model-based POD evaluation based on the cost of a fully experimental-based study and the performance of available NDE simulation tools. Knowledge of the accuracy and speed of NDT simulations is a critical part of this assessment process. Thus, an evaluation of the model quality is often necessary at this stage. For many applications, there will be a mix of simulated and empirical studies that will provide the greatest coverage of the key factors for an inspection technique. For example, certain parameters such as the dimensions and location of a crack in the test part can be more easily controlled and varied through simulated studies. As well, rare events or factor conditions can be readily assessed and incorporated in the MAPOD evaluation. Alternatively, noise data from material specimens and

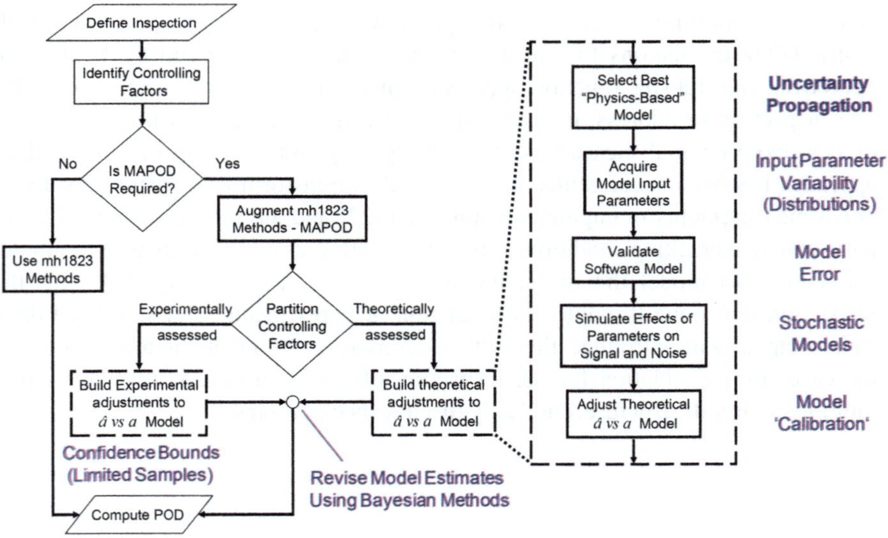

Fig. 21.1 Model-Assisted POD model building process with complete approach to uncertainty propagation in MAPOD from MIL-HNBK 1823A, Appendix H (2009) [121]

measurement system noise can be evaluated quickly through low cost experimental studies. From this perspective, a model-assisted POD evaluation has the potential to not only reduce the cost of POD study through reduced test sample requirements but also to improve the quality of the assessment by including physics-based models that fully address the factors driving the evaluation process.

A critical component of a MAPOD evaluation is the assessment and propagation of variability in parameter conditions and the uncertainty in the POD assessment. Experimental-based POD evaluations must consider the impact of limited samples on the confidence bounds for a POD model fits. Labels were added to Fig. 21.1 to present a complete perspective on uncertainty propagation during the POD study. As part of quantifying the effects of changing the key factors, the amount of variation of each factor must be well understood [124, 125]. Variability in NDE measurements is prevalent due to varying part geometries, material properties, surface conditions, flaw morphology, NDE hardware, and human factors. A typical representation of variability for an input factor or condition is as a probability density function (pdf). In addition, confidence bounds for the pdf may be considered as variation measures of the pdf distribution parameters. Multivariate distributions would thus include model parameters and a variation (covariance) matrix.

Before models can be directly applied in POD evaluations, two steps are needed to ensure their performance. First, model calibration involves model adjustments to variables in a way that mimics the NDE technique procedure. In many cases, gains and thresholds are set based on the desired response to a known calibration standard. Parameters of the model are essentially fit to obtain the best match with a limited set of empirical data acquired according to a calibration procedure. Bayesian calibration approaches have been implemented to facilitate this analysis with limited data [137]. Second, model validation ensures that models are in agreement with well-controlled studies for the appropriate range of conditions expected in practice. Note, uncertainty bounds on the measurements and numerical model error [129] must be tracked and extended to any model-assisted evaluation.

Stochastic numerical models provide a means to efficiently represent random variables in physical systems without excessive computational overhead. New efficient methods such as the probabilistic collocation methods (PCM) will enable the greater use of stochastic studies with existing NDE models [130]. Fundamentally, this POD evaluation including uncertainty bounds becomes a two-level analysis. Parameter variability is associated with the inherent variability or randomness of a factor. Alternatively, uncertainty is associated with imperfect knowledge, often requiring the need for more or better quality data. In practice, parameter variability and uncertainty can be represented by the random variables of a statistical distribution and their associated confidence intervals. Methods such as second order probability analysis can be applied [131]. MAPOD demonstrations using this approach have recently been performed [134, 136].

Lastly, at the final stage of the POD evaluation, experimental and theoretically assessed "$\hat{a}$ vs. a" models must be resolved. Here, an evaluation of the full noise and signal distributions as a function of crack length is performed and the call criteria is applied to evaluation POD and probability of a false call (POFC) or false call rate. Uncertainty (or credibility) bounds on the POD evaluation must be calculated given limited empirical samples, factor variability, uncertainty propagation, and model error. Bayesian methods are ideally suited to incorporate empirical data with NDE models that include prior information/distributions. A typical example of applying Bayesian methods is through the addition of new empirical data to evaluate the posterior distribution, a refinement to the original prior distribution is achieved. Numerical methods such as Markov Chain Monte Carlo (MCMC) methods and Bayes Factors can be applied to perform this evaluation [125, 132]. Note, care must be taken to ensure that all assumptions applied in a Bayesian analysis are valid. Although this process provides significant complexity to address the most complex inspection problems, one must keep in mind the goal to simplify.

21.3 Case Study for MAPOD Evaluation of NDI of Fastener Sites for Sub-surface Fatigue Cracks

The MAPOD case study problem presented here is the detection of cracks under installed countersunk fasteners in an airframe structures (see Fig. 21.2) [124,133,134]. For this problem, the key factors to include in the evaluation were identified from experience with the inspection problem and prior experimental studies. The probe characteristic response, lift-off, and scan resolution are important factors concerning the NDE measurement. The surface condition of the samples, thickness of the layers, type of fastener, fastener fit, and distance between adjacent fasteners and edges are identified as significant factors related to part geometry and material. The dimensions, aspect ratio, location around fastener site, and morphology are

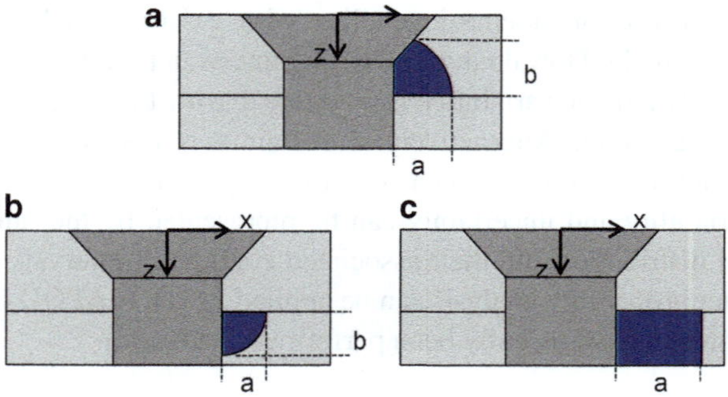

Fig. 21.2 Diagram of a fastener site with (**a**) a first layer corner crack, (**b**) a second layer corner crack, and (**c**) a second layer through crack

important flaw characteristics to be considered. Also, the design of the algorithm for extracting the crack measure from the image data will also play a critical role in the results. Evaluation of the quality of the physics-based models has been the focus of previous work [135]. Because **VIC-3D®** has the capability to handle the generic two-layer structural problem of interest, it was used for this study.

Details on the sample characteristics and eddy-current measurements for the model are as follows. The dimensions for the top and bottom layers measured 3.96 mm (0.156″) and 2.54 mm (0.100″), respectively. Conductivities of 1.87×10^7 S/m for the aluminum layers and 1.79×10^6 S/m for the titanium fasteners were used in the model. The size of the hole was set to a radius of 4.04 mm (0.159″), respectively. The probe frequency was set to 600 Hz and the permeability of 1,000 was used for the ferrite cup core. The coil dimensions had a height of 6.0 mm (0.236″), an inner radius of 3.0 mm (0.118″), and an outer radius of 6.0 mm (0.236″). A corner crack was modeled for the first layer. Two crack conditions in the second layer were simulated: a through crack and a corner crack with the aspect ratio length to width to be 1:1. The transition from through to corner crack was assumed to be at 2.5 mm (0.100″). For the simulations, all cracks were treated as notches of a finite width of 0.25 mm (0.010″), so it was assumed that there is no electrical contact between crack faces. Crack lengths in the experimental samples were available between 0.000″ and 0.169″.

The simulated and experimental data for a given probe and inspection geometry were equated using a calibration procedure developed in prior work [135]. Figure 21.3 presents the calibration results demonstrating good agreement between the model and simulated data for the no-crack fastener site. A Monte Carlo simulation using 5,000 data points was performed using the noise measurement distribution for the fastener site response with no cracks and combined with the corner and/or through crack models to populate the full-model-assisted data sets for POD evaluation. Figure 21.4a and b present the resulting distribution of data points generated by the Monte-Carlo simulation. The POD curves were then fit for the experimental and model-assisted data sets using the conventional maximum likelihood approach for hit–miss data. The confidence bounds were determined using the loglikelihood ratio method. Figure 21.5a and b present a comparison of the POD results for the experimental and full-model-assisted approaches for first and second layer cracks, respectively. For the first and second layer crack inspection, both results are near the confidence bounds of the empirical data and thus near agreement with the experimental solution. Note, these results resolve an error in an assumption of the crack location (to be in second layer but in reality the first) for the experimental data presented in prior results [133, 134]. For first layer cracks, the curve generated with the MAPOD approach is near the upper 95 % confidence bounds of the empirical data, while for second layer cracks, the curve is near the lower bound.

These results are encouraging but indicate that the MAPOD evaluation is not perfect. Potential sources for this remaining difference may include error in the model/experiment fit or subtle details of the probe that are not being modeled precisely. In general, there is good agreement between the two data sets with some error for select experimental data points. However, the transition region for the crack response just exceeding the noise threshold is not in perfect agreement. This difference is the primary source of differences in the two POD calculations. The presence of coherent noise related to surface features (poor paint quality) was observed for select experimental data. In particular, three of the four outliers noted in Fig. 21.5a and b were found to be correlated with the presence of surface related noise. The other outlier was found to be in close proximity to a steel fastener and experienced greater noise as well. It is possible to actually shift the crack measurement either out or in from the optimal location to minimize the effect of the surface noise.

21.4 Bayesian Methods for Estimating Uncertain Parameters in POD/MAPOD Evaluation

To fully address the challenge of performing a quality evaluation with limited experimental samples, a model-based assessment must incorporate the variations of the most significant input factors and appropriately integrate simulated and experimental results. As part of this process, there are several critical components in the evaluation that concern the propagation of varying conditions and uncertainty through the model, or the updating of parameter estimates in the POD assessment. Bayesian analysis is a statistical approach for evaluating a posterior distribution of model parameters through combining information in the prior distribution with new evidence according to Bayes theorem. Bayesian methods can be leveraged here for the evaluation of three key components of the methodology: (1) evaluation of variability for an input factor or condition is as a probability density function (pdf), (2) model calibration [137], and (3) revised measurement model estimates with experimental data. In practice, the posterior distribution can be evaluated, providing a refinement to the original prior distribution through numerical methods such as Markov Chain Monte Carlo (MCMC) simulation [137–139]. In prior work on NDE reliability, Bayesian methods have been generally proposed [140, 141] with several examples for the evaluation of statistical POD models for hit–miss data [142, 143] and ahat-versus-a reliability studies [144]. In particular, Bayesian methods are quite useful to address complex estimation problems and when prior information is available. For example, Li, Meeker and Hovey [144] used Bayesian methods for simultaneously evaluating the noise interference model for POD and the crack distribution.

The estimation of both the calibration parameters of a physics-based model and the distribution of select model factors, including crack aspect ratio and lift-off was recently studied [145]. Code was developed to call a surrogate measurement model representing the numerical results from **VIC-3D®**. The following model estimation problem was studied:

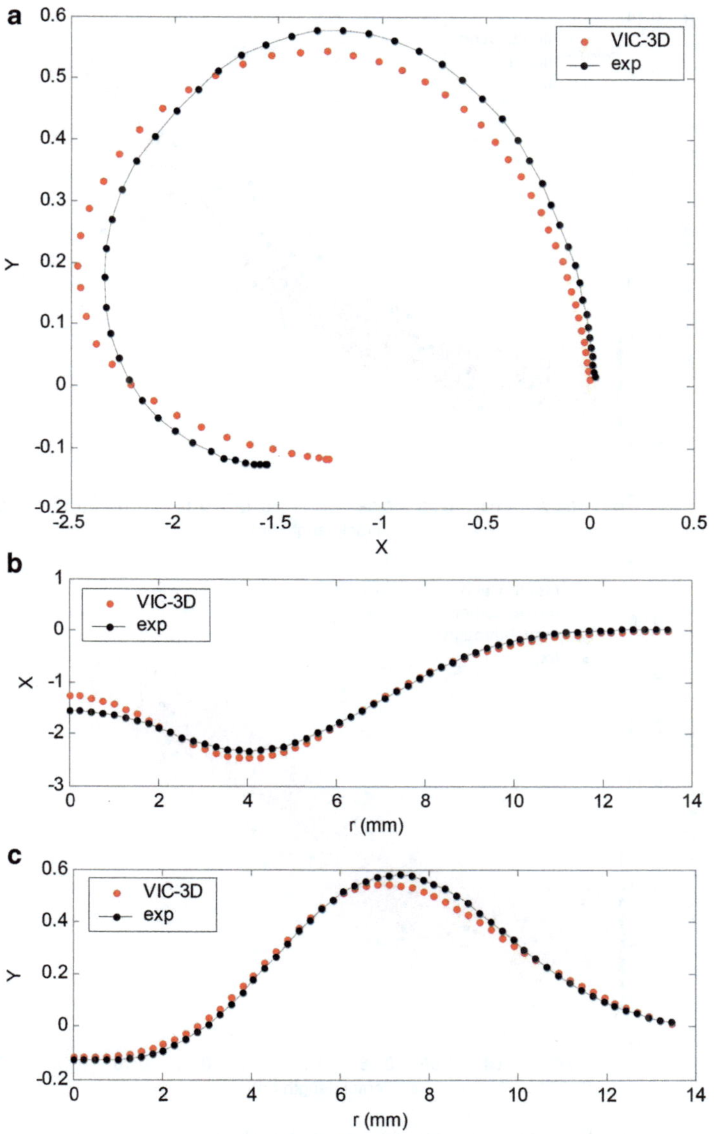

Fig. 21.3 Calibration fit of model with respect to experimental data for titanium fastener site (in Volts). Comparison plots include (**a**) the measurement plane, (**b**) the horizontal component and (**c**) the vertical component responses

$$\hat{a} = \beta_0 + \beta_1 f(a_1, \beta_2) + \varepsilon, \tag{21.1}$$

where $\varepsilon \sim N(0, \sigma_\varepsilon^2)$, $f()$ is a function call for a physics-based model, β_0 and β_1 are model calibration parameters, and β_2 is a random variable associated with crack aspect ratio (b/a). An initial study was performed where the random variables

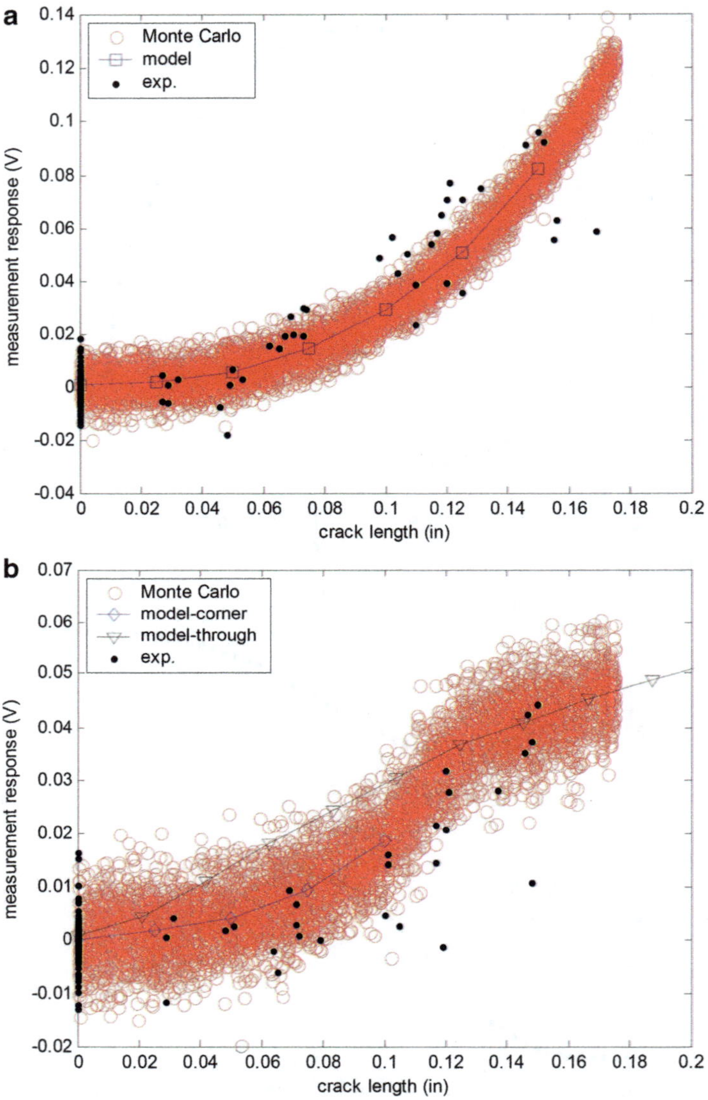

Fig. 21.4 Monte Carlo simulation results for varying length of (**a**) first layer (corner) cracks and (**b**) second layer (through and corner) cracks

β_0, β_1, and β_2 were estimated using a Bayesian (MCMC) approach. A plot of the simulated data is given in Fig. 21.6 for corner flaws located at the faying surface of the first and second layer. Initial results were found to be mixed due to the ill-posedness of estimating β_1 and β_2 simultaneously.

To fully evaluate this model, there is also the need to provide a true estimate of variance for crack aspect ratio random variable. Parametrized eddy-current models

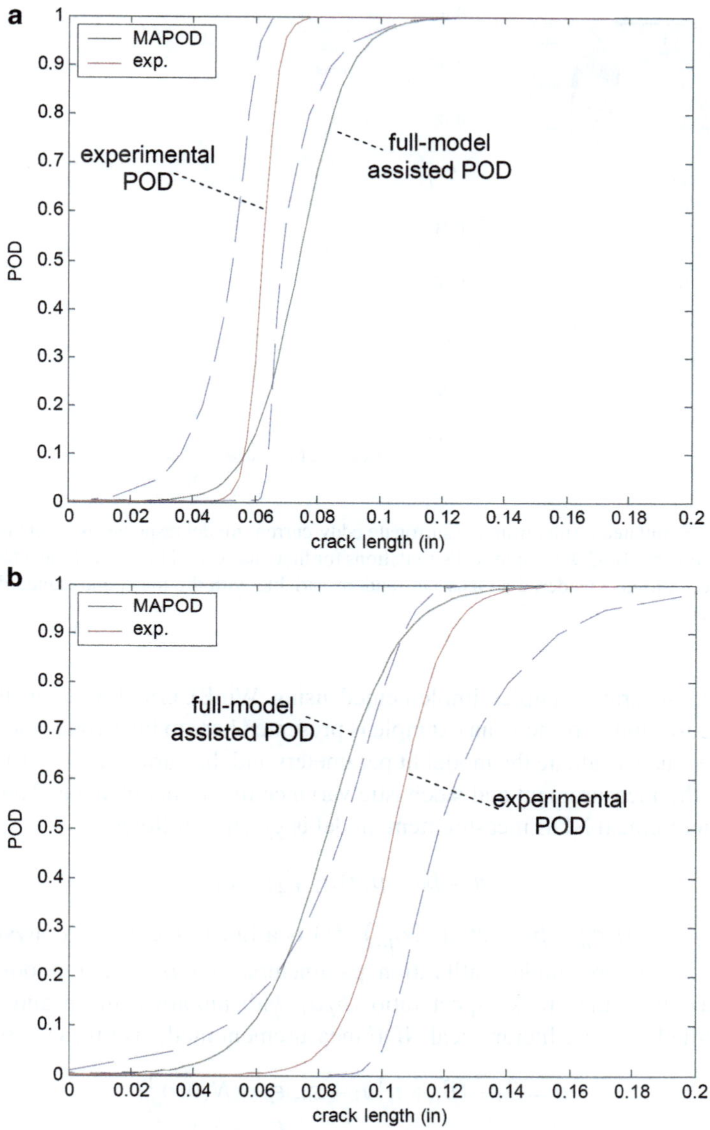

Fig. 21.5 POD evaluation results for experimental and full-model-assisted POD studies for (**a**) first layer corner cracks and (**b**) second layer cracks

that include aspect ratio as a random variable as shown in Fig. 21.6 can reasonably address both the mean response and nonconstant variance trends observed in experimental results. The evaluation of these stochastic model parameters can be achieved through evaluation of hierarchical Bayesian models. Gelman et al introduced hierarchical Bayesian models for these classes of problems [146] and

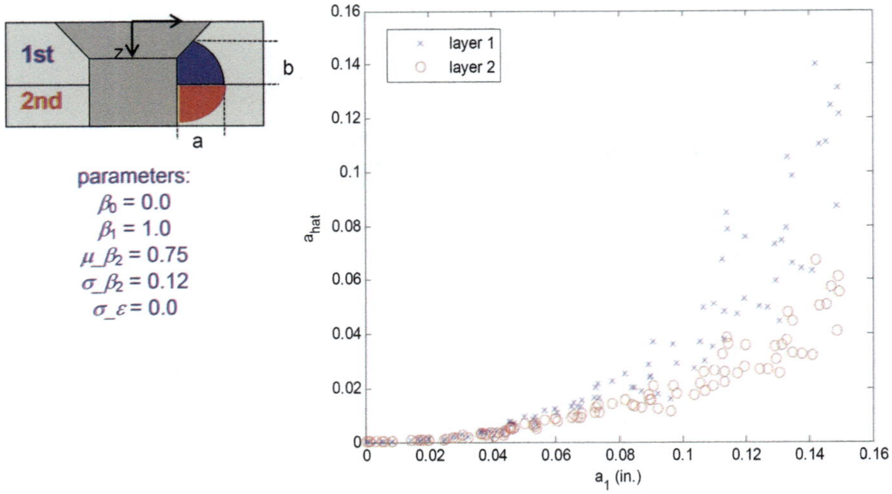

Fig. 21.6 Simulated results from the surrogate eddy-current model response as a function of crack length based on **VIC-3D®** numerical simulations for first and second layer cracks at a fastener site. The aspect ratio is considered a Gaussian random variable with the mean and standard deviation prescribed

presented several examples implemented using WinBUGS. Based on the eddy-current case study problem, an example is presented here to investigate the ability to simultaneously evaluate the model fit parameters and the variance terms for both the noise in the measurement and stochastic variance in the model slope. The physics-based hierarchical NDE measurement model is given as follows:

$$\hat{a} = \beta_0 + \beta_1 f(a_1, \beta_2) + \varepsilon_{\hat{a}}, \tag{21.2}$$

where $\varepsilon_{\hat{a}} \sim N(0, \sigma_{\hat{a}}^2)$, $\beta_2 \sim N(\mu_{\beta_2}, \sigma_{\beta_2}^2)$, $f()$ is a function call for a physics-based model, β_0, β_1 are model calibration parameters, and β_2 is a random variable associated with the crack aspect ratio (b/a) with unknown mean and variance. A simplified test case hierarchical NDE measurement model is given as follows:

$$\begin{aligned} \hat{a} &= \beta_0 + (\beta_1 + \eta)a_1 + \varepsilon_{\hat{a}}, \; \varepsilon_{\hat{a}} \sim N(0, \sigma_{\hat{a}}^2) \\ \eta &= \varepsilon_\eta, \qquad\qquad\qquad\qquad \varepsilon_\eta \sim N(0, \sigma_\eta^2), \end{aligned} \tag{21.3}$$

where β_0 and β_1 are offset and mean slope terms of the model, respectively, η is a random variable associated with varying-slope in the model, and σ_η^2 is the variance in slope parameter. Here, the random variable, η, is used to simulate the increasing variance with increasing flaw size present in the physics-based model shown in Fig. 21.6. The goal of this study is to assess how accurately these four parameters, β_0, β_1, σ_η^2 and, $\sigma_{\hat{a}}^2$, can be simultaneously estimated with respect to known values.

Two example estimation problems are presented in Fig. 21.7. Test case values were selected to closely represent examples in prior experimental and simulated

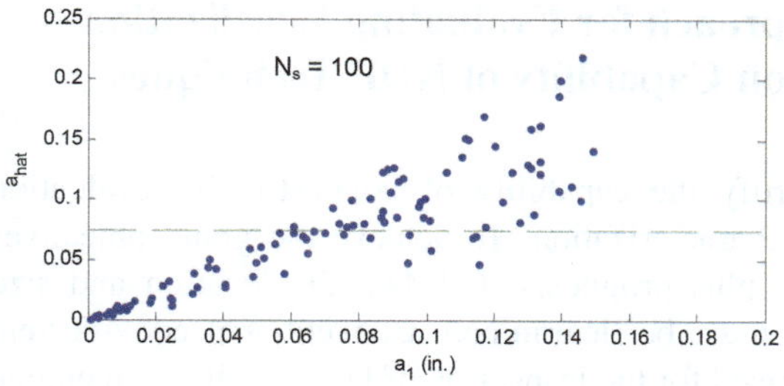

	True values	Bayesian estimates	95% credible bounds
β_0	0.000	-0.00024	(-0.00119,0.00364)
β_1	1.000	1.0312	(0.9848,1.0810)
σ_η	0.300	**0.2580**	(0.2286,0.2923)
σ_ε	0.00100	0.**00139**	(0.00103,0.00210)

a

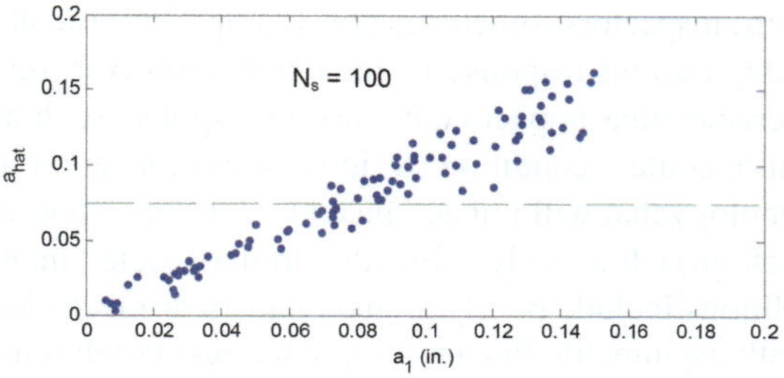

	True values	Bayesian estimates	95% credible bounds
β_0	0.000	0.00193	(0.00007,0.00382)
β_1	1.000	0.9716	(0.9371,1.007)
σ_η	**0.100**	**0.1014**	(0.0795,0.1248)
σ_ε	**0.00500**	**0.00545**	(0.00432,0.00677)

b

Fig. 21.7 Hierarchical model test cases estimating calibration parameters and variation in measurement noise and stochastic model slope. Test cases shown (**a**) for strong model slope variation and (**b**) for varying both slope and measurement noise significant factors

results in Fig. 21.6. The first case (a) investigated the condition where variance as a function of flaw size dominates measurement noise (i.e., the variance independent of flaw size). Using only 100 samples, the estimates for the two variance terms, and, were found to be 0.2580 and 0.00139, respectively, close to the exact values of 0.300 and 0.0010. However, the 95 % credible bounds for both estimates just missed containing the true values. This error may have been due to the specific random sample used with the limited number of points or could be something systematic in the estimation problem. Note, the estimates for the calibration parameters, β_0 and β_1, were found to be in good agreement with the true values for these case studies. The variance terms appear to be the more challenging parameters to estimate in the hierarchical model.

A second case in Fig. 21.7b investigated the condition where variance as a function of flaw size is on a similar order as the measurement noise (variance independent of flaw size). Using only 100 samples, the estimates for the two variance terms, σ_η^2 and $\sigma_{\hat{a}}^2$, were again found to be quite close to the exact values of 0.100 and 0.0050, respectively, and the credible bounds included the true values of the parameters. Both variance parameters are slightly overestimated, while the calibration parameter β_1 was slightly underestimated. All in all, these case studies demonstrate the potential of simultaneously estimating the model calibration parameters, model random variables, and measurement error.

21.5 Model-Assisted Approach for Evaluating Localization and Characterization Capability of NDE Techniques

There is a critical need to certify the capability of nondestructive evaluation techniques to perform damage characterization. To achieve the grand objectives of condition-based maintenance plus prognosis (CBM+), the location and size of damage at any length scale must be determined. Conventional eddy-current methods have been widely employed for the inspection of fastener sites and engine components. For some applications, inspectors sometimes apply simple empirically based formulas to equate the eddy-current response to flaw size. However, real cracks have a variety of characteristics that impact eddy-current response such as varying crack location, aspect ratio, contact condition (residual stress), location in complex parts, and surface morphology that will impact the eddy-current response. Often, the source for fatigue crack growth in early life is due to unexpected manufacturing anomalies. Such conditions include poorly aligned panels or misdrilled holes, leading to fastener sites with asymmetric fits and irregular stress conditions. Thus, more sophisticated inverse methods have the potential to evaluate crack size in the presence of these varying conditions. However, proper performance validation of inverse methods must be demonstrated under real test conditions.

Although standards currently exist for evaluation of measurement systems [147], there are limits to their application for assessing NDE characterization performance. For example, conventional ANOVA-based methods do not address complex underlying relationships well, often wrongly assume parameters are independents neglecting joint probability, and typically do not naturally address both aleatory and epistemic uncertainty. Thus, there is a need to develop a comprehensive procedure for the evaluation of NDE characterization error (CE). This section will briefly discuss an emerging approach for experimental and model-based procedures.

From the perspective of quantifying the reliability of NDE systems, one must evaluate the relationship between the accuracy in estimating the damage or material state estimates $(\hat{a})$ with respect to the actual condition (a). An evaluation of the characterization error (CE), $\hat{e} = \hat{a} - a$, for all critical location and sizing estimates is necessary. Characterization error with confidence bounds is the metric that is proposed for use by condition-based maintenance and prognosis programs to help evaluate remaining life and determine necessary maintenance actions. This problem of evaluating characterization error with confidence bounds as a function of a critical parameter such as flaw size is shown in Fig. 21.8a. This evaluation is generally similar to the current procedure found in MIL-HDBK-1823A for the evaluation of the relationship between an NDE measurement $(\hat{a})$ and a critical flaw size (a) as shown in Fig. 21.8b. The foundation for the experimental-based CE procedure will be MIL-HDBK-1823A.

In practice, NDE and structural health monitoring (SHM) systems may measure multiple values associated with damage location and size. The challenge here is to develop a multidimensional characterization error model with confidence bounds, $\hat{e}_j(a_k)$, where $j = 1 \cdots M$, for locations and/or sizing parameters being estimated,

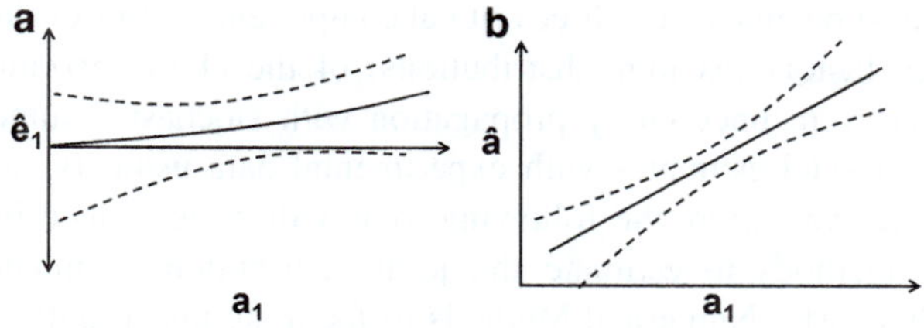

Fig. 21.8 Evaluation of (**a**) characterization error ($\hat{e}_j$) with respect to damage conditions (a_k) is analogous to (**b**) $\hat{a}$-vs-a POD analysis

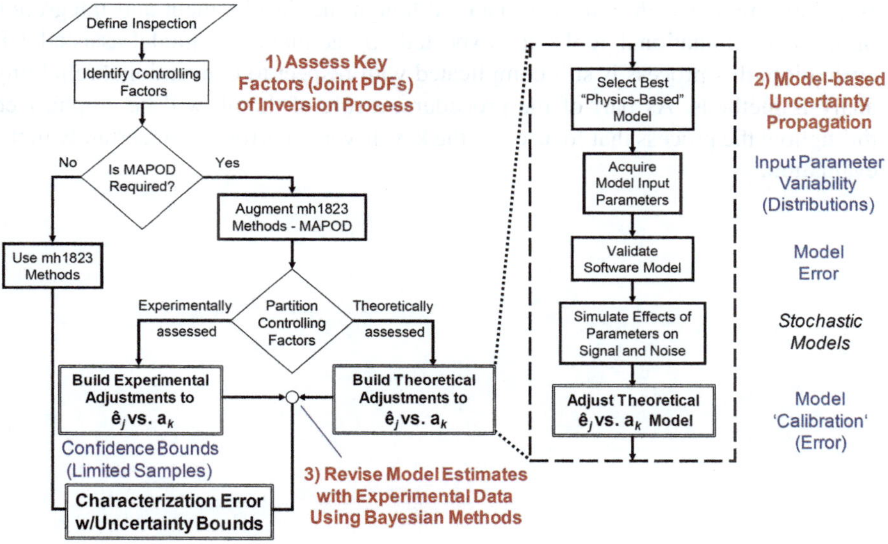

Fig. 21.9 Model-assisted process for NDE characterization error (CE) evaluation

with $k = 1 \cdots N$ key factors, for example the critical flaw size. Gaussian process (GP) models are ideally suited to evaluate this complex multidimensional relationship and provide a useful model to support CBM+. A Gaussian Process model is a principled probabilistic approach to representing data as a collection of random variables with the property that the joint distribution of any finite subset is a Gaussian function. This task will be to develop an empirical-based evaluation procedure following MIL-HDBK-1823A but leveraging Gaussian process models.

Although experimental data are necessary in any measurement system evaluation, due to the large number of factors that often must be addressed and the need for a statistically significant number of test specimens, a model-assisted approach is proposed as an alternative to an empirical-based evaluation of characterization error. This process of model-assisted POD evaluation [121] may be adapted to address a model-assisted approach to characterization error evaluation. A modified process diagram for model-assisted NDE characterization error (CE) evaluation is presented in Fig. 21.9. Throughout this process, proper error evaluation and uncertainty propagation must be managed. Three critical components of this evaluation include: (1) assess key factors (as joint distributions) of the characterization (inversion) process, (2) perform uncertainty propagation with stochastic numerical models, and (3) revise model estimates with experimental data using Bayesian methods. To facilitate the evaluation, the following tools will be leveraged in this process: (a) Bayesian methods to estimate the joint distribution of unknown/uncertain factors, (b) Stochastic Numerical Methods to facilitate uncertainty propagation of input parameter distributions through the model, (c) Gaussian Processes to represent characterization error (CE) as a range of random variables with a multivariate normal distribution, and (d) Bayesian Calibration to revise the GP model using Bayesian method with empirical data. Although the development and integration of these computational tools are expected to facilitate the model-assisted CE evaluation, this process is still complicated with respect to conventional reliability analysis methods. As part of the procedure, a rule-of-thumb will be emphasized throughout the process that focuses on the key drivers of error and uncertainty in the evaluation.

Erratum to:

NLSE: Parameter-Based Inversion Algorithm

Harold A. Sabbagh, R. Kim Murphy, Elias H. Sabbagh, John C. Aldrin, and Jeremy S. Knopp

H.A. Sabbagh et al., *Computational Electromagnetics and Model-Based Inversion*, Scientific Computation, DOI 10.1007/978-1-4419-8429-6_12,

DOI 10.1007/978-1-4419-8429-6_22

The print and online versions of the book contain some errors, and the corrections to these versions are given on the following pages:

H.A. Sabbagh et al., *Computational Electromagnetics and Model-Based Inversion*, Scientific Computation, DOI 10.1007/978-1-4419-8429-6_22,

On page 257, the complete text under Sect. 12.2.1 need to be replaced with the following text:

12.2.1 Overview of the Algorithm: Nonlinear Least-Squares

Let

$$Z = g(p_1, \ldots, p_N, f), \tag{12.1}$$

where $p_1,\ldots,p_N$ are the N parameters of interest, and f is a control parameter at which the impedance, Z, is measured. f can be frequency, scan-position, lift-off, etc. It is, of course, known; it is not one of the parameters to be determined. To be explicit during our initial discussion of the theory, we will call f 'frequency.'

In order to determine $p_1,\ldots,p_N$, we measure Z at M frequencies, $f_1,\ldots,f_M$, where $M > N$:

$$\begin{aligned} Z_1 &= g(p_1, \ldots, p_N, f_1) \\ &\vdots \\ Z_M &= g(p_1, \ldots, p_N, f_M). \end{aligned} \tag{12.2}$$

The right-hand side of (12.2) is computed by applying the volume-integral code to a model of the problem, usually at a discrete number of values of the vector, $\mathbf{p}$, forming a multidimensional interpolation grid.

Because the problem is nonlinear, we use a Gauss–Newton iteration scheme to perform the inversion [97]. First, we decompose (12.2) into its real and imaginary parts, thereby doubling the number of equations (we assume the $p_1,\ldots,p_N$ are real). Then we use the linear approximation to the resistance, R_i, and reactance, X_i, at the ith frequency:

$$\begin{bmatrix} R_1 \\ X_1 \\ \vdots \\ R_M \\ X_M \end{bmatrix} \approx \begin{bmatrix} R_1(p_1^{(q)},\ldots,p_N^{(q)}) \\ X_1(p_1^{(q)},\ldots,p_N^{(q)}) \\ \vdots \\ R_M(p_1^{(q)},\ldots,p_N^{(q)}) \\ X_M(p_1^{(q)},\ldots,p_N^{(q)}) \end{bmatrix} + \begin{bmatrix} \frac{\partial R_1}{\partial p_1} & \cdots & \frac{\partial R_1}{\partial p_N} \\ \frac{\partial X_1}{\partial p_1} & \cdots & \frac{\partial X_1}{\partial p_N} \\ & \vdots & \\ \frac{\partial R_M}{\partial p_1} & \cdots & \frac{\partial R_M}{\partial p_N} \\ \frac{\partial X_M}{\partial p_1} & \cdots & \frac{\partial X_M}{\partial p_N} \end{bmatrix}_{(p_1^{(q)},\ldots,p_N^{(q)})} \begin{bmatrix} p_1 - p_1^{(q)} \\ \vdots \\ p_N - p_N^{(q)} \end{bmatrix}, \tag{12.3}$$

where the superscript (q) denotes the qth iteration, and the partial derivatives are computed numerically by the software. The left side of (12.3) is taken to be the measured values of resistance and reactance. We rewrite (12.3) as

$$0 \approx r + Jp, \tag{12.4}$$

where r is the $2M$-vector of residuals, J is the $2M \times N$ Jacobian matrix of derivatives, and p is the N-dimensional correction vector. Equation (12.4) is solved in a least-squares manner starting with an initial value, $\left(x_1^{(0)}, \ldots, x_N^{(0)}\right)$, for the vector of unknowns, and then continuing by replacing the initial vector with the updated vector $\left(x_1^{(q)}, \ldots, x_N^{(q)}\right)$ that is obtained from (12.3), until convergence occurs.

We are interested in determining a bound for the sensitivity of the residual norm to changes in some linear combination of the parameters. Given an $\varepsilon > 0$ and a unit vector, v, the problem is to determine a sensitivity (upper) bound, σ, such that

$$\|r(x^* + \sigma v)\| \leq (1+\varepsilon)\,\|r(x^*)\|\,. \tag{12.5}$$

We will derive an estimate of σ. Equation (12.5) is equivalent to

$$\|r(x^* + \sigma v)\| - \|r(x^*)\| \leq \varepsilon\,\|r(x^*)\|\,. \tag{12.6}$$

The left-hand side of (12.6) can be approximated to the second order in σ by the second-order Taylor expansion:

$$\|r(x^* + \sigma v)\| - \|r(x^*)\| \approx \sigma v \cdot \nabla\,\|r(x^*)\| + \frac{\sigma^2}{2}\sum_{i,j}\frac{\partial^2\,\|r(x)\|}{\partial x_j \partial x_i}\,|_{x^*} v_i v_j\,, \tag{12.7}$$

where ∇ is the gradient operator in $N-$dimensional space. Even though the gradient vanishes at the minimum point, we will compute it to get the algebra started:

$$\begin{aligned}
\nabla\,\|r(x)\| &= \nabla\left[f_1^2(x) + f_2^2(x) + \cdots + f_{2M}^2(x)\right]^{1/2} \\
&= \frac{1}{\|r(x)\|}\begin{bmatrix} f_1\dfrac{\partial f_1}{\partial x_1} + \cdots + f_{2M}\dfrac{\partial f_{2M}}{\partial x_1} \\ \vdots \\ f_1\dfrac{\partial f_1}{\partial x_N} + \cdots + f_{2M}\dfrac{\partial f_{2M}}{\partial x_N}\end{bmatrix}^T \\
&= \frac{r(x)^T}{\|r(x)\|}\begin{bmatrix} \dfrac{\partial f_1}{\partial x_1} & \cdots & \dfrac{\partial f_1}{\partial x_N} \\ & \vdots & \\ \dfrac{\partial f_{2M}}{\partial x_1} & \cdots & \dfrac{\partial f_{2M}}{\partial x_N}\end{bmatrix} \\
&= e^T(x) \cdot J,
\end{aligned} \tag{12.8}$$

where the superscript T denotes the transpose of a matrix (or vector), and $e(x) = r(x)/\|r(x)\|$ is a unit vector.

The second derivative that we want is the gradient of (12.8):

$$\nabla\nabla\|r(x)\| = -\frac{\nabla\|r(x)\|}{\|r(x)\|^2}\begin{bmatrix} f_1\dfrac{\partial f_1}{\partial x_1}+\cdots+f_{2M}\dfrac{\partial f_{2M}}{\partial x_1} \\ \vdots \\ f_1\dfrac{\partial f_1}{\partial x_N}+\cdots+f_{2M}\dfrac{\partial f_{2M}}{\partial x_N}\end{bmatrix}^T$$

$$+\frac{1}{\|r(x)\|}\nabla\begin{bmatrix} f_1\dfrac{\partial f_1}{\partial x_1}+\cdots+f_{2M}\dfrac{\partial f_{2M}}{\partial x_1} \\ \vdots \\ f_1\dfrac{\partial f_1}{\partial x_N}+\cdots+f_{2M}\dfrac{\partial f_{2M}}{\partial x_N}\end{bmatrix}^T. \quad (12.9)$$

Before going further, we can immediately drop the first term in (12.9) because the gradient of the norm vanishes at the solution x^*. Thus, (12.9) becomes, using index notation,

$$\begin{aligned}\nabla\nabla\|r(x)\| &= \frac{1}{\|r(x)\|}\frac{\partial}{\partial x_j}\left[f_1\frac{\partial f_1}{\partial x_i}+\cdots+f_{2M}\frac{\partial f_{2M}}{\partial x_i}\right], \quad i,j=1,\ldots,N \\ &= \frac{1}{\|r(x)\|}\sum_\alpha\left[\frac{\partial f_\alpha}{\partial x_j}\frac{\partial f_\alpha}{\partial x_i}+f_\alpha\frac{\partial^2 f_\alpha}{\partial x_j\partial x_i}\right], \quad \alpha=1,\ldots,2M. \quad (12.10)\end{aligned}$$

Following [148, p. 523], we discard the second-derivative term in (12.10) by arguing that the residual vector for a good model fit should be small, which would make the second derivative term small. Furthermore, it is likely that the residual vector should have terms that are uncorrelated with each other and with the model, thus tending to cancel the second derivative terms when summed over α. We will call (12.10) the first-order curvature tensor, Γ_{ij}, of the mapping (or deformation) of the parameter space, $\{x_i\}$ into the model-measurement space. If we call the ith column of the Jacobian matrix, c_i, then it follows from (12.10) that

$$\Gamma_{ij}(x^*) = \frac{c_i(x^*)\cdot c_j(x^*)}{\|r(x^*)\|}, \quad (12.11)$$

where we are ignoring the second-derivative term in (12.10).

Substituting this result into (12.7) yields an upper bound for the quadratic term:

$$\begin{aligned}\sigma^2 \sum_{i,j} \frac{\partial^2 \|r(x)\|}{\partial x_j \partial x_i}\Big|_{x^*} v_i v_j &= \frac{\sigma^2}{\|r(x^*)\|} \sum_{\alpha} \left[\sum_{i,j} \frac{\partial f_\alpha}{\partial x_i} v_i \frac{\partial f_\alpha}{\partial x_j} v_j \right]_{x^*} \\ &= \frac{\sigma^2}{\|r(x^*)\|} (J(x^*) \cdot v) \cdot (J(x^*) \cdot v) \\ &= \frac{\sigma^2}{\|r(x^*)\|} \|J(x^*) \cdot v\|^2, \end{aligned} \tag{12.12}$$

and if we equate this to the right-hand side of (12.6), we get the final result

$$\sigma_v = \varepsilon^{1/2} \left(\frac{\|r(x^*)\|}{\|J(x^*) \cdot v\|} \right). \tag{12.13}$$

We will call this the 'first-order' approximation, in the sense that we have truncated the Taylor series expansion with the first nonzero term, and have ignored the second-derivative terms in (12.10). This is the expression that is stated, but not derived, in [97].

Note that if $\|J(x*) \cdot v\|$ is small compared to $\|r(x*)\|$, then σ is large and the residual norm is insensitive to changes in the linear combination of the parameters specified by v. If $v = e_i$, the ith column of the $N \times N$ identity matrix, then (12.13) produces σ_i, the sensitivity bound for the ith parameter. Since σ_i will vary in size with the magnitude of x_i^*, it is better to compare the ratios σ_i/x_i^* for $i = 1,\ldots,N$ before drawing conclusions about the fitness of a solution.

The importance of these results is that we now have metrics for the inversion process: $\Phi = \|r(x*)\|$, the norm of the residual vector at the solution, tells us how good the fit is between the model data and measured data. The smaller this number the better, of course, but the 'smallness' depends upon the experimental setup and the accuracy of the model to fit the experiment. Heuristic judgement based on experience will help in determining the quality of the solution for a given Φ.

The sensitivity coefficient, σ, is more subtle, but just as important. It, too, should be small, but, again, the quality of the 'smallness' will be determined by heuristics based upon the problem. If σ is large in some sense, it suggests that the solution is relatively independent of that parameter, so that we cannot reasonably accept the value assigned to that parameter as being meaningful, as suggested in Fig. 12.1, which shows a system, S, for which the system is sensitive to variable, x_i, at the solution point, x_i^*, and another system, I, for which the system is insensitive to x_i.

An example occurs when one uses a high-frequency excitation, with its attendant small skin depth, to interrogate a deep-seated flaw. The flaw will be relatively invisible to the probe at this frequency, and whatever value is given for its parameters will be highly suspect. When this occurs, we will either choose a new parameter to characterize the flaw, or acquire data at a lower frequency.

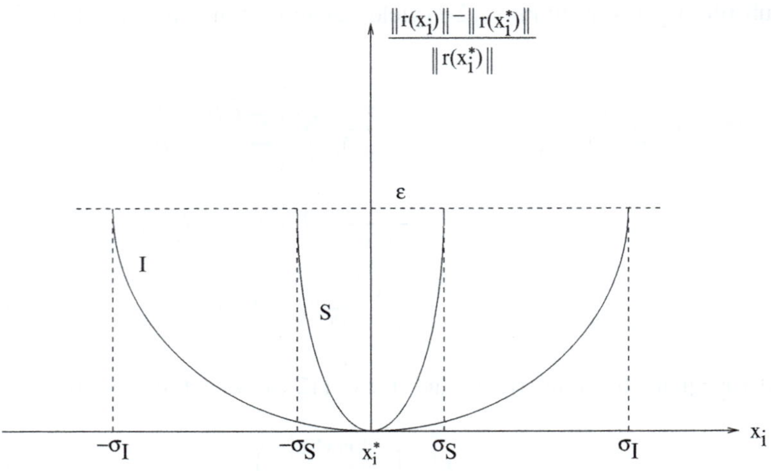

Fig. 12.1 Showing sensitivity parameters for two system responses to x_i. Response S is sensitive to x_i at x_i^*, whereas response I is not

These metrics are not available to us in the current inspection method, in which analog instruments acquire data that are then interpreted by humans using hardware standards. The opportunity to use these metrics is a significant advantage to the model-based inversion paradigm that we propose in this paper.

If the residual norm is relatively insensitive to changes in some linear combination of the parameters, then the Jacobian matrix at the solution is nearly rank-deficient, and it may be useful to determine a set of linearly independent parameters. The covariance matrix $(\mathbf{J}^T\mathbf{J})^{-1}$ can be used for this purpose.

The online version of the original chapter can be found at
http://dx.doi.org/10.1007/978-1-4419-8429-6_12

References

1. Burrows, M.L.: A theory of eddy-current flaw detection. PhD Thesis, University of Michigan, University Microfilms, Inc., Ann Arbor, Michigan (1964)
2. Förster, F., Breitfeld, H.: Theoretische und Experimentelle Grundlagen der zerstörungsfreien Werkstoffprüfung mit Wirbelstromverfahren, (Parts I and II). Z. Metallkd. **43**(5), 163–180 (1952)
3. Förster, F., Breitfeld, H., Stambke, K.: Theoretishe und Experimentelle Grundlagen der zerstörungsfreien Werkstoffprüfung mit Wirbelstromverfahren, (Parts III–VII). Z. Metallkd. **45**(5), 166–199 and 221–226 (1954)
4. Dodd, C.V., Deeds, W.E.: Analytical solutions to eddy-current probe coil problems. J. Appl. Phys. **39**(6), 2829–2838 (1968)
5. Luquire, J.W., Deeds, W.E., Dodd, C.V.: Alternating current distribution between planar conductors. J. Appl. Phys. **41**(10), 3983–3991 (1970)
6. Cheng, C.C., Dodd, C.V., Deeds, W.E.: General analysis of probe coils near stratified conductors. Int. J. Nondestr. Test. **3**, 109–130 (1971)
7. Nestor, C.W., Jr., Dodd, C.V., Deeds, W.E.: Analysis and computer programs for eddy current coils concentric with multiple cylindrical conductors. Report No. ORNL-5220, Oak Ridge National Laboratory, Oak Ridge, TN 37830, July 1979
8. Deeds, W.E., Dodd, C.V., Scott, G.W.: Computer-aided design of multifrequency eddy-current tests for layered conductors with multiple property variations. Report No. ORNL/TM-6858, Oak Ridge National Laboratory, Oak Ridge, TN 37830, October 1979
9. Miller, E.K.: Model-based parameter estimation in electromagnetics: III–applications to EM integral equations. Appl. Comput. Electrom. **10**(3), 9–29 (1995)
10. Murphy, K., Sabbagh, H.A.: A boundary-integral code for electromagnetic nondestructive evaluation. In: Conference Proceedings: 12th Annual Review of Progress in Applied Computational Electromagnetics, Applied Computational Electromagnetics Society, 18–22 March 1996, pp. 171–178
11. Xie, H., Song, J., Yang, M., Nakagawa, N.: A novel boundary integral equation for surface crack model. In: Thompson, D.O., Chimenti, D.E. (eds.) Review of Progress in Quantitative Nondestructive Evaluation, vol. 29, pp. 329–336. American Institute of Physics, Melville (2010)
12. Harrington, R.F.: Field Computation by Moment Methods. The Macmillan Company, New York (1968)
13. Berreman, D.W.: Optics in Stratified and Anisotropic Media: 4 $\times$ 4-Matrix Formulation. J. Opt. Soc. Am. **62**(4), 502–510 (April 1972)
14. Altman, C., Schatzberg, A.: Appl. Phys. B **28**, 327–333 (1982)
15. Altman, C., Schatzberg, A.: Appl. Phys. B **26**, 147–153 (1981)

H.A. Sabbagh et al., *Computational Electromagnetics and Model-Based Inversion*, Scientific Computation, DOI 10.1007/978-1-4419-8429-6,

16. Altman, C., Schatzberg, A., Suchy, K.: IEEE Trans. Antenn. Propag. **AP-32**(11) (November 1984)
17. Schatzberg, A., Altman, C.: J. Plasma Phys. **26**(Part 2), 333–344 (1981)
18. Suchy, K., Altman, C.: J. Plasma Phys. **13**(Part 3), 437–449 (1975)
19. Krowne, C.M.: IEEE Antennas and Propagation Symposium Digest, Boston, MA, 25–29 June 1984, pp. 569–572
20. Krowne, C.M.: IEEE Trans. Microw. Theor. Tech. **MTT-32**(12), 1617–1625 (December 1984)
21. Krowne, C.M.: IEEE Trans. Antenn. Propag. **AP-32**(11), 1224–1230 (November 1984).
22. Roberts, T.M., Sabbagh, H.A., Sabbagh, L.D.: Electromagnetic interactions with an anisotropic slab. IEEE Trans. Magn. **24**(6), 3193–3200 (November 1988)
23. Roberts, T.M., Sabbagh, H.A., Sabbagh, L.D.: Electromagnetic scattering for a class of anisotropic layered media. J. Math. Phys. **29**, 2675–2681 (December 1988)
24. Roberts, T.M.: Explicit eigenmodes for anisotropic media. IEEE Trans. Magn. **26**(6), 3064–3071 (November 1990)
25. Bowler, J.R., Sabbagh, L.D., Sabbagh, H.A.: A theoretical and computational model of eddy-current probes incorporating volume integral and conjugate gradient methods. IEEE Trans. Magn. **25**(3), pp. 2650–2664 (May 1989)
26. Sabbagh, H.A., Bowler, J.R., Sabbagh, L.D.: A model of eddy-current probes with ferrite cores. Nondestr. Test. Eval. **5**(1), 67–79 (1989)
27. Sabbagh, H.A.: A model of eddy-current probes with ferrite cores. IEEE Trans. Magn. **MAG-23**(3), 1888–1904 (May 1987)
28. Sabbagh, H.A., Sabbagh, L.D., Bowler, J.R.: A volume-integral code for eddy-current nondestructive evaluation. Int. J. Comput. Math. Elec. Electron. Eng. **9**(Suppl. A), 67–70 (1990)
29. Bowler, J.R., Sabbagh, L.D., Sabbagh, H.A.: Eddy-current probe impedance due to a surface slot in a conductor. IEEE Trans. Magn. **26**(2), 889–892 (March 1990)
30. Rao, S.M., Wilton, D.R., Glisson, A.W.: Electromagnetic scattering by surfaces of arbitrary shape. IEEE Trans. Antenn. Propag. **AP-30**(3), 409–418 (May 1982)
31. Aubin, J-P.: Approximation of Elliptic Boundary-Value Problems. Wiley-Interscience, New York (1972)
32. Glisson, A.W., Wilton, D.R.: Simple and efficient numerical methods for problems of electromagnetic radiation and scattering from surfaces. IEEE Trans. Antenn. Propag. **AP-29**, 593–603 (1980)
33. Wertgen, W., Jansen, R.H.: Efficient direct and iterative electrodynamic analysis of geometrically complex MIC and MMIC structures. Int. J. Numer. Model. Electron. Network. Dev. Field. **2**(3), 153–186 (September 1989)
34. Yaghjian, A.D.: Electric dyadic Green's functions in the source region. Proc. IEEE. **68**, 248–263 (February 1980)
35. Burke, G.J., Dease, C.G., Didwall, E.M., Lytle, R.J.: Numerical modeling of subsurface communication. UCID-20439 Rev. 1, Lawrence Livermore National Laboratory, August 1985
36. Catedra, M.F., Gago, E., Nuño, L.: A numerical scheme to obtain the RCS of three-dimensional bodies of resonant size using the conjugate gradient method and the fast Fourier transform. IEEE Trans. Antenn. Propag. **37**(5), 528–537 (May 1989)
37. Peter, A., Zwamborn, M., van den Berg, P.M., Mooibroek, J., Koenis, F.T.C.: Computation of three-dimensional electromagnetic-field distributions in a human body using the weak form of the CGFFT method. Appl. Comput. Electrom. **7**(2), 26–42 (Winter 1992)
38. Zwamborn, P., van den Berg, P.M.: The three-dimensional weak form of the conjugate gradient FFT method for solving scattering problems. IEEE Trans. Microw. Theor. Tech. **40**(9), 1757–1766 (September 1992)
39. Sabbagh, H.A.: Splines and their reciprocal-bases in volume-integral equations. IEEE Trans. Magn. **29**(6), 4142–4152 (November 1993)
40. http://www.sabbagh.com
41. Dongarra, J.J., Moler, C.B., Bunch, J.R., Stewart, G.W.: LINPACK Users' Guide. Society for Industrial and Applied Mathematics, Philadelphia (1979)

42. Hestenes, M.: Conjugate Direction Methods in Optimization. Springer, New York (1980)
43. Sarkar, T.P.: Application of the Conjugate Gradient Method in Electromagnetics and Signal Processing. Elsevier, New York (1991)
44. Catedra, M.F., Torres, R.P., Basterrechea, J., Gago, E.: The CG-FFT Method: Application of Signal Processing Techniques to Electromagnetics. Artech House, Boston (1995)
45. Peterson, A., Ray, S., Mittra, R.: Computational Methods for Electromagnetics. IEEE, New York (1998)
46. Chew, W.C., Jin, J.M., Michielsssen, E., Song, J.M.: Fast and Efficient Algorithms in Computational Electromagnetics. Artech House, Boston (2001)
47. Harrington, R.F.: Time-Harmonic Electromagnetic Fields. McGraw-Hill, New York (1961)
48. Collin, R.E.: Field Theory of Guided Waves. McGraw-Hill, New York (1960)
49. Sabbagh, H.A., Sabbagh, E.H., Murphy, R.K.: Recent advances in modeling eddy-current probes. In: Thompson, D.O., Chimenti, D.E. (eds.) Review of Progress in Quantitative Nondestructive Evaluation, vol. 21, pp. 423–429. American Institute of Physics, Melville (2002)
50. Wozencraft, J.M., Jacobs, I.M.: Principles of Communication Engineering. Wiley, New York (1965)
51. Umashankar, K.R., Nimmagadda, S., Taflove, A.: Numerical analysis of electromagnetic scattering by electrically large objects using spatial decomposition technique. IEEE Trans. Antenn. Propag. **40**(8), 867–877 (August 1992)
52. Murphy, K., Sabbagh, H.A., Treece, J.C.: Thickness measurements with eddy-current probes: a simple inversion problem. In: Thompson, D.O., Chimenti, D.E. (eds.) Review of Progress in Quantitative Nondestructive Evaluation, vol. 13, pp. 927–934. Plenum Press, New York (1994)
53. Lawson, C.L., Hanson, R.J.: Solving Least Squares Problems. Prentice-Hall, Inc., Englewood Cliffs (1974)
54. Murphy, K., Sabbagh, H.A., Treece, J.C.: Some inversion problems in nondestructive evaluation. In: Thompson, D.O., Chimenti, D.E. (eds.) Review of Progress in Quantitative Nondestructive Evaluation, vol. 14, pp. 857–861. Plenum Press, New York (1995)
55. Baltzersen, O.: Model-based inversion of plate thickness and liftoff from eddy current probe coil measurements. Mat. Eval. **51**, 72–76 (1993)
56. Förster, F. Libby, H.: Electromagnetic Testing, 2nd edn, pp. 178–179. The American Society or Nondestructive Testing, Columbus (1986)
57. Sabbagh, H.A., Murphy, R.K. Sabbagh, E.H.: Advances in modeling eddy-current NDE of ferromagnetic bodies. In: Thompson, D.O., Chimenti, D.E. (eds.) Review of Progress in Quantitative Nondestructive Evaluation, vol. 22, pp. 383–389. American Institute of Physics, Melville (2003)
58. Rinnooy Kan, A.H.G., Timmer, G.T.: Stochastic global optimization methods. Part I: clustering methods. Math. Program. **39**, 27–56 (1987)
59. Rinnooy Kan, A.H.G., Timmer, G.T.: Stochastic global optimization methods Part II: multi level methods. Math. Program. **39**, 57–78 (1987)
60. Byrd, R.H., Dert, C.L., Rinnooy Kan, A.H.G., Schnabel, R.B.: Concurrent stochastic methods for global optimization. Math. Program. **46**, 1–29 (1990)
61. Nakhkash, M., Huang, Y., Fang, M.T.C.: Application of the multilevel single-linkage method to one-dimensional electromagnetic inverse scattering problem. IEEE Trans. Antenn. Propag. **47**(11), 1658–1668 (November 1999)
62. Golub, G.H., Van Loan, C.F.: Matrix Computations. The Johns Hopkins University Press, Baltimore (1983)
63. Rousseeuw, P., Yohai, V.: Robust regression by means of s-estimators. In: Robust and Nonlinear Time Series Analysis: Proceedings of a Workshop, pp. 256–272, 1984
64. Van Trees, H.L.: Detection, Estimation, and Modulation Theory: Part I. Wiley, New York (1968)
65. Kay, S.M.: Fundamentals of Statistical Signal Processing: Estimation Theory. Prentice-Hall, Inc., Upper Saddle River (1993)

66. Devaney, A.J., Tsihrintzis, G.A.: Maximum likelihood estimation of object location in diffraction tomography. IEEE Trans. Signal Process. **39**(3), 672–682 (March 1991)
67. Herman, G.T., Lent, A., Hurwitz, H.: A storage-efficient algorithm for finding the regularized solution of a large, inconsistent system of equations. J. Inst. Math. Appl. **25**, 361–366 (1980)
68. Censor, Y.: Row-action methods for huge and sparse systems and their applications. SIAM Rev. **23**, 444–446 (October 1981)
69. Herman, G.T., Meyer, L.B.: Algebraic reconstruction techniques can be made computationally efficient. IEEE Trans. Med. Imag. **12**(3), 600–609 (September 1993)
70. Sabbagh, H.A., Sabbagh, L.D., Vernon, S.N.: Verification of an eddy-current flaw inversion algorithm. IEEE Trans. Magn. **22**(6), 1881–1886 (November 1986)
71. Papoulis, A.: A new algorithm in spectral analysis and band-limited extrapolation. IEEE Trans. Circ. Syst. **CAS-22**, 735–742 (1975)
72. Youla, D.C.: Generalized image restoration by the method of alternating orthogonal projections. IEEE Trans. Circ. Syst. **CAS-25**, 694–702 (1978)
73. Youla, D.C., Webb, H.: Image restoration by the method of convex projections: Part 1-Theory. IEEE Trans. Med. Imag. **MI-1**, 81–94 (October 1982)
74. Sezan, M.I., Stark, H.: Image restoration by the method of convex projections: Part 2-applications and numerical results. IEEE Trans. Med. Imag. **MI-1**, 95–101 (October 1982)
75. Oskoui-Fard, P., Stark, H.: Tomographic image reconstruction using the theory of convex projections. IEEE Trans. Med. Imag. **7**(1), 45–58 (March 1988)
76. Bucci, O.M., D'Elia, G., Mazzarella, G., Panariello, G.: Antenna pattern synthesis: a new general approach. Proc. IEEE. **82**(3), 358–371 (March 1994)
77. Oh, S., Marks II, R.J., Atlas, L.E.: Kernel synthesis for generalized time-frequency distributions using the method of alternating projections onto convex sets. IEEE Trans. Signal Process. **42**(7), 1653–1661 (July 1994)
78. Lent, A., Tuy, H.: An iterative method for the extrapolation of band limited functions. J. Math. Anal. Appl. **83**, 554–565 (1981)
79. Kaczmarz, S.: Angenaherte auflosung von systemen linearer gleichungen. Bull. Pol. Acad. Sci. Lett. A, **6–8A**, 355–357 (1937)
80. Tanabe, K.: Projection method for solving a singular system. Numer. Math. **17**, 203–214 (1971)
81. Hounsfield, G.N.: A method of and apparatus for examination of a body by radiation such as x-ray or gamma radiation. Patent Specification 1283915, The Patent Office (1972)
82. Kak, A.C., Slaney, M.: Principles of Computerized Tomographic Imaging. IEEE Press, New York (1988)
83. Goodman, J.W.: Introduction to Fourier Optics. McGraw-Hill, San Francisco (1968)
84. Pratt, W.K.: Digital Image Processing. Wiley, New York (1978).
85. Sabbagh, L.D., Sabbagh, H.A., Klopfenstein, J.S.: Image enhancement via extrapolation techniques: a two dimensional iterative scheme and a direct matrix inversion scheme. In: Thompson, D.O., Chimenti, D.E. (eds.) Review of Progress in Quantitative Nondestructive Evaluation, vol. 5, pp. 473–483. Plenum Press, New York (1986)
86. Sabbagh, L.D., Sabbagh, H.A.: Inversion of eddy current data and the reconstruction of flaws Part 2: inversion of data. In: Thompson, D.O., Chimenti, D.E. (eds.) Review of Progress in Quantitative Nondestructive Evaluation, vol. 6, pp. 619–626. Plenum Press, New York (1987)
87. Sabbagh, L.D., Sabbagh, H.A.: Eddy-current modeling and flaw reconstruction. J. Nondestr. Eval. **7**(1/2), 95–110 (1988)
88. Sabbagh, H.A., Sabbagh, L.D.: An eddy-current model for three-dimensional inversion. IEEE Trans. Magn. **MAG-22**(4), 282–291 (July 1986)
89. Sabbagh, L.D., Sabbagh, H.A.: Eddy current modeling and signal processing in NDE. In: Chen, C.H. (ed.) Signal Processing and Pattern Recognition in Nondestructive Evaluation of Materials: NATO ASI Series, vol. F44, pp. 145–154. Springer, Berlin (1988)
90. Theodoulidis, T.P., Poulakis, N., Bowler, J.R.: Developments in modeling eddy current coil interactions with a right-angled conductive wedge. In: Takahashi, S., Kikuchi, H. (eds.) Electromagnetic Nondestructive Evaluation X, pp 41–48. IOS Press, Amsterdam (2007)

91. Sabbagh, H.A., Sabbagh, L.D., Roberts, T.M.: An eddy-current model and algorithm for three-dimensional nondestructive evaluation of advanced composites. IEEE Trans. Magn. **24**(6), 3201–3212 (November 1988)
92. Murphy, K., Sabbagh, H.A. Treece, J.C.: Verification of a model of eddy-current probes with ferrite cores. In: Thompson, D.O., Chimenti, D.E. (eds) Review of Progress in Quantitative Nondestructive Evaluation, vol. 13, pp. 1089–1093. Plenum Press, New York (1994)
93. Mittleman, D.M., Jacobsen, R.H., Buss, M.C.: T-ray imaging. IEEE J. Sel. Top. Quant. Electron. **2**(3), 679–692 (September 1996)
94. Sze, S.M.: Physics of Semiconductor Devices. Wiley, New York (1969)
95. Sabbagh, H.A., Murphy, R.K., Woo, L.W., Sabbagh, E.H., Krzywosz, K.: Recent advances in modeling eddy-current probe-flaw interactions. In: Thompson, D.O., Chimenti, D.E. (eds.) Review of Progress in Quantitative Nondestructive Evaluation, vol. 15, pp. 331–338. Plenum Press, New York (1996)
96. Burke, S.K., Ditchburn, R.J.: Mutual impedance of planar eddy-current driver-pickup spiral coils. Res. Nondestr. Eval. **19**, 1–19 (2008)
97. Moré, J.J., Garbow, B.S., Hillstrom, K.E.: USER GUIDE FOR MINPACK-1. ANL-80-74, Argonne National Laboratory, August 1980
98. Collin, R.E.: Foundations for Microwave Engineering, Chap. 4. McGraw-Hill Book Company, New York (1966)
99. Sabbagh, H.A., Sabbagh, E.H., Murphy, R.K., Nyenhuis, J.: Assessing thermal barrier coatings by eddy current inversion. Mater. Eval. **59**(11), 1307–1312 (November 2001)
100. Sabbagh, H.A., Sabbagh, E.H., Murphy, R.K., Nyenhuis, J.: Assessing thermal barrier coatings by eddy current inversion. In: Thompson, D.O., Chimenti, D.E. (eds.) Review of Progress in Quantitative Nondestructive Evaluation, vol. 21, pp. 722–727. American Institute of Physics, Melville (2002)
101. Wang, S. Solid-State Electronics. McGraw-Hill Book Company, New York (1966)
102. Knopp, J.S., Aldrin, J.C., Misra, P.: Considerations in the validation and application of models for eddy current inspection of cracks around fastener holes. J. Nondestr. Eval. **25**(3), 123–138 (2006)
103. Carpenter, D.C.: Use of the finite element method in simulation and visualization of electromagnetic nondestructive testing applications. Mater. Eval. **58**(7), 877–881 (2000)
104. Palanisamy, R., Lord, W.: Prediction of eddy current probe signal trajectories. IEEE Trans. Magn. **16**(5), 1083–1085 (1980)
105. Knopp, J.S., Aldrin, J.C., Sabbagh, H.A., Jata, K.V.: Estimation theory metrics in electromagnetic NDE, electromagnetic nondestructive evaluation workshop proceedings, Seoul, Korea, 10–12 June 2008
106. Trefethen, L.N. Bau, D.: Numerical Linear Algebra. SIAM, Philadelphia (1997)
107. Sabbagh, H.A., Murphy, R.K., Sabbagh, E.H. Aldrin, J.C., Knopp, J., Blodgett, M.: Computational electromagnetics and model-based inversion: a modern paradigm for eddy-current nondestructive evaluation. Appl. Comput. Electrom. **24**(6), 533–540 (December 2009)
108. Sabbagh, E.M.: Circuit Analysis. Ronald Press Company, New York (1961)
109. Ramo, S., Whinnery, J.R., Van Duzer, T.: Fields and Waves in Communication Electronics. Wiley, New York (1965)
110. Vernon, S.N.: The universal impedance diagram of the ferrite pot core eddy current transducer. IEEE Trans. Magn. **25**(3), 2639–2645 (May 1989)
111. Bowler, J.R., Sabbagh, L.D., Sabbagh, H.A.: The reduced impedance function for cup-core eddy-current probes. IEEE Trans. Magn. **25**(3), 2646–2649 (May 1989)
112. 'Validation of Direct and Inverse Models of SEACURE Ferritic Tubes With Benchmark Data,' PID069961, Technical Update, October 2010, Electrical Power Research Institute. Prepared by Victor Technologies, LLC
113. Todorov, E., Levesque, S., Ames, N., Krzywosz, K.: Measurement of magnetic properites of ferromagnetic tubes for heat exchangers. Trans. Am. Nucl. Soc. **104**, 297–298 (26–30 June 2011, Hollywood, Florida)

114. Scully, J.R.: Hidden corrosion, what should be measured to improve emerging anticipate and manage strategies. In: 32nd Annual Review of Progress in Quantitative Non-destructive Evaluation, QNDE-Brunswick, Maine (Key Note Lecture for Symposium Kick-off General Session) (2005)
115. Tian, Y., Tamburrino, A., Udpa, S.S., Udpa, L.: Time-of-flight measurements from eddy current tests. In: Thompson, D.O., Chimenti, D.E. (eds.) Review of Progress in Quantitative Nondestructive Evaluation, vol. 22, pp. 593–600. American Institute of Physics, Melville (2003)
116. Liu, Z., Safizadeh, M.-S. Forsyth, D.S., Lepine, B.A.: Data fusion method for the optimal mixing of multi-frequency eddy current signals. In: Thompson, D.O., Chimenti, D.E. (eds.) Review of Progress in Quantitative Nondestructive Evaluation, vol. 22, pp. 577–584. American Institute of Physics, Melville (2003)
117. Liu, X., Deng, Y., Zeng, Z., Udpa, L., Knopp, J.: Model based inversion technique of GMR signals using element-free Galerkin method. In: Conference Proceedings: 24th Annual Review of Progress in Applied Computational Electromagnetics, Applied Computational Electromagnetics Society, pp. 221–226 (March 2008)
118. Gray, J.N., Gray, T.A., Nakagawa, N., Thompson, R.B.: Models for predicting NDE reliability, nondestructive evaluation and quality control, Metals Handbook 17, pp 702–715. ASM International, Ohio, 1989
119. Thompson, R.B.: Using Physical Models of the Testing Process in Determination of Probability of Detection. Mater. Eval. **69**(7), 861–865 (2001)
120. Thompson, R. B.: A unified approach to the model-assisted determination of probability of detection. Mater. Eval. **66**, 667–673 (2008)
121. U.S. Department of Defense, Handbook, Nondestructive Evaluation System Reliability Assessment, MIL-HDBK-1823A, 7 April 2009
122. Smith, K.D., Thompson, R.B., Brasche, L.: Model-Based POD: Successes and Opportunities, 1st Meeting of the MAPOD Working Group, Albuquerque, New Mexico, 23–24 September 2004. Web site: http://www.cnde.iastate.edu/MAPOD/
123. Harding, C., Hugo, G., Bowles, S.: Model-assisted Probability of Detection Validation of Automated Ultrasonic Scanning for Crack Detection at Fastener Holes, Proceedings of the 10th Joint FAA/DoD/NASA Conference on Aging Aircraft, Palm Springs, CA, 16–19 April 2007
124. Aldrin, J.C., Knopp, J.S., Lindgren E.A., Jata, K.V.: Model-assisted probability of detection (MAPOD) evaluation for eddy current inspection of fastener sites. Rev. progr. Quant. Nondestr. Eval. **28**, AIP, 1784–1791 (2009)
125. Aldrin, J.C., Medina, E.A., Lindgren, E.A., Buynak, C., Knopp, J.: Case studies for model-assisted probabilistic reliability assessment for structural health monitoring systems. Rev. progr. Quant. Nondestr. Eval. **30**, AIP, 1589–1596 (2011)
126. Aldrin, J.C., Medina, E.A., Lindgren, E.A., Buynak, C.F., Knopp, J.S.: Protocol for reliability assessment of structural health monitoring systems incorporating model-assisted probability of detection (MAPOD) approach. In: Chang, F.-K. (ed.) Proceedings of the 8th International Workshop on Structural Health Monitoring, Stanford, 13–15 September 2011
127. Aldrin, J.C., Medina, E.A., Santiago, J., Lindgren, E.A., Buynak, C.F., Knopp, J.S.: Demonstration study for reliability assessment of SHM systems incorporating model-assisted probability of detection approach. In: Thompson, D.O., Chimenti, D.E. (eds.) Review of Progress in Quantitative Nondestructive Evaluation, vol. 31, pp. 1543–1550. American Institute of Physics, Melville (2012)
128. Aldrin, J.C., Sabbagh, H.A., Murphy, R.K., Sabbagh, E.H., Knopp, J.S., Lindgren, E.A., Cherry, M.R.: Demonstration of model-assisted probability of detection evaluation methodology for eddy-current nondestructive evaluation. In: Thompson, D.O., Chimenti, D.E. (eds.) Review of Progress in Quantitative Nondestructive Evaluation, vol. 31, pp. 1733–1740. American Institute of Physics, Melville (2012)
129. Oberkampf, W.L., Roy, C.J.: Verification and Validation in Scientific Computing. Cambridge University Press, New York (2010)

130. Knopp, J.S., Aldrin, J.C., Blodgett, M.P.: Efficient propagation of uncertainty in simulations via the probabilistic collocation method. In: Chady, T., Gratkowski, S., Takagi, T., Udpa, S.S. (eds.) Electromagnetic Nondestructive Evaluation (XIV), pp. 141–148. IOS Press, Amsterdam (2011)
131. Frey, H.C.: Quantitative analysis of uncertainty and variability in environmental policy making. Fellowship Program for Environmental Science and Engineering, American Association for the Advancement of Science, Washington, DC (1992)
132. Mahadevan, S., Rebba, R.: Validation of reliability computational models using Bayes networks. Reliab. Eng. Syst. Saf. **87**, 223–232 (2005)
133. Knopp, J.S., Aldrin, J.C., Lindgren, E., Annis, C.: Investigation of a model-assisted approach to probability of detection evaluation. Rev. Prog. Quant. Nondestr. Eval. **26**, 1775–1782 (2007)
134. Aldrin, J.C., Knopp, J.S.: Modeling and simulation for nondestructive testing with applications to aerospace structures. Mater. Eval. **66**(1), 53–59 (2008)
135. Knopp, J.S., Aldrin, J.C., Misra, P.: Considerations in the validation and application of models for eddy current inspection of cracks around fastener holes. J. Nondestr. Eval. **25**(3), 123–138 (2006)
136. Dominguez, N., Feuillard, V., Jenson, F., Willaume, P.: Simulation assisted POD of a phased array ultrasonic inspection in manufacturing. Rev. Progr. Quant. Nondestr. Eval. **31**, AIP, 1765–1772 (2012)
137. Kennedy, M.C., O'Hagan, A.: Bayesian calibration of computer models. J. R. Stat. Soc. B **63**, 425–464 (2001)
138. Gelman, A., Carlin, J.B., Stern, H.S., Rubin, D.B.: Bayesian Data Analysis. CRC, London (2003)
139. Christensen, R., Johnson, W., Branscum, A.: Bayesian Ideas and Data Analysis: An Introduction for Scientists and Statisticians. CRC, Boca Raton (2010)
140. Meeker, W.Q., Escobar, L.A.: Introduction to the use of Bayesian methods for reliability data. In: Statistical Methods for Reliability Data, pp. 343–368. Wiley, New York (1998)
141. Thompson, R.B.: A Bayesian approach to the inversion of NDE and SHM data. Rev. Progr. Quant. Nondestr. Eval. **29**, AIP, 679–686 (2010)
142. Leemans, D.V., Forsyth, D.: Bayesian approaches to using field test data in determining the probability of detection. Mater. Eval. **62**, 855–859 (2004)
143. Knopp, J.S., Zeng, L.: Statistical analysis of hit/miss data. Mater. Eval. **71**(3), 323–329 (2013)
144. Li, M., Meeker, W.Q., Hovey, P.: Joint estimation of NDE inspection capability and flaw-size distribution for in-service aircraft inspections. Res. Nondestr. Eval. **23**, 104–123 (2012)
145. Aldrin, J.C., Knopp, J.S., Sabbagh, H.A.: Bayesian methods in probability of detection estimation and model-assisted probability of detection evaluation. In: Thompson, D.O., Chimenti, D.E. (eds.) Review of Progress in Quantitative Nondestructive Evaluation, vol. 32, pp. 1733–1740. American Institute of Physics, Melville (2013)
146. Gelman, A., Hill, J.: Data Analysis Using Regression and Multilevel/Hierarchical Models. Cambridge University Press, Cambridge (2007)
147. ASTM Standard E2782, 2011: Standard guide for measurement systems analysis (MSA). ASTM International, West Conshohocken (2011). doi:10.1520/E2782-11, www.astm.org
148. Press, W.H., Teukolsky, S.A., Vetterling, W.T., Flannery, B.P.: Numerical Recipes in C, 2nd edn. Cambridge University Press, Cambridge (1992)

Index

H.A. Sabbagh et al., *Computational Electromagnetics and Model-Based Inversion*, Scientific Computation, DOI 10.1007/978-1-4419-8429-6,

9781441984289